RANUNCULALES
MEDICINAL
PLANTS

RANUNCULALES MEDICINAL PLANTS

Biodiversity, Chemodiversity
and Pharmacotherapy

DA-CHENG HAO

ACADEMIC PRESS
An imprint of Elsevier

Academic Press is an imprint of Elsevier
125 London Wall, London EC2Y 5AS, United Kingdom
525 B Street, Suite 1800, San Diego, CA 92101-4495, United States
50 Hampshire Street, 5th Floor, Cambridge, MA 02139, United States
The Boulevard, Langford Lane, Kidlington, Oxford OX5 1GB, United Kingdom

Library of Congress Cataloging-in-Publication Data
A catalog record for this book is available from the Library of Congress

British Library Cataloguing-in-Publication Data
A catalogue record for this book is available from the British Library

ISBN: 978-0-12-814232-5

For information on all Academic Press publications visit our website at
https://www.elsevier.com/books-and-journals

Working together
to grow libraries in
developing countries

www.elsevier.com • www.bookaid.org

Publisher: John Fedor
Acquisition Editor: Glyn Jones
Editorial Project Manager: Naomi Robertson
Production Project Manager: Punithavathy Govindaradjane
Designer: Mark Rogers

Typeset by Thomson Digital

Contents

About the Author

Da-Cheng Hao, Associate Professor/Principle Investigator, School of Environment and Chemical Engineering/Biotechnology Institute, Dalian Jiaotong University, Dalian 116028, China. e-mail: hao@djtu.edu.cn.

Dr. Hao earned his bachelor's degree in medicine, master's degree in science, and PhD degree in biotechnology from Xi'an Jiaotong University, National University of Singapore and Chinese Academy of Sciences respectively. He had postdoctoral training in the Institute of Medicinal Plant Development (IMPLAD), Chinese Academy of Medical Sciences (CAMS), under the supervision of Prof. Pei Gen Xiao and Prof. Shi Lin Chen. He was a visiting scholar of John Innes Centre, UK, for 1 year (2012–13), supported by the Ministry of Education, China.

Studies of Dr. Hao are supported by Liaoning Natural Science Fund (2015020663), Liaoning postgraduate education and teaching reform project (2016), the Scientific Research Foundation for ROCS, Ministry of Education, China (2014), and more. Dr. Hao is the editorial member of the Journal Chinese Herbal Medicines at https://www.journals.elsevier.com/chinese-herbal-medicines/ and Biotechnology Bulletin at http://biotech.caas.cn/.

Dr. Hao has published more than 70 SCI papers and two books, *Medicinal Plants: Chemistry, Biology and Omics* (Elsevier/Woodhead) and *An Introduction of Plant Pharmacophylogeny* (Chemical Industry Press, China). He is the peer reviewer of scores of SCI journals, Elsevier book proposals, and research grant proposals of Poland National Science Centre (NCN) and French National research Agency (ANR).

Preface

The history of medicinal plants is as long as the origin of human beings. For a few millennia, medicinal plants have contributed countless medicinal active ingredients, which are widely used in traditional Chinese medicine (TCM) and worldwide ethnomedicine. The current increasing interests in plant-based medicinal resources promote the upsurge in research and development, and many useful compounds are found from plants of different evolutionary levels, such as steroidal saponins, alkaloids, terpenoids, and glycosides. The expanded studies of chemotaxonomy, molecular phylogeny, and pharmacological activity of medicinal plants are gaining deeper insights. Pharmaphylogeny (pharmacophylogeny) is about the study of medicinal plant genetic relationships (chemical composition), efficacy (pharmacological activity and traditional efficacy), and their correlations, and is a basic tool for research and development of Chinese medicine and phytomedicine resources. In the framework of pharmacophylogenetics, the plant taxa and species genetic relationships are used as the clue in this book, and the phytochemistry, chemotaxonomy, molecular biology, and phylogenetic relationships of representative medicinal tribes and genera, as well as their relevance to therapeutic efficacy, are discoursed systematically.

There are many disciplines related to pharmacophylogeny. In the process of writing, a lot of recent research literature is referenced. In combination with the author's own experimental data, this book attempts to reflect the latest progress in related fields and enrich the connotation of pharmacophylogeny research. The author pays attention to novel concepts and new technology, conforms to the global trend of research and development, and puts forward the concept of pharmacophylogenomics. The author advocates the comprehensive study of molecular phylogenetic inferences and chemotaxonomic results, taking into account the treatment outcome, in order to understand the medicinal plant genetic relationship. Full attention is paid not only to the domestic medicinal species but also to relevant foreign species. The genomics/metabolomics data should be fully used in the inference. The contradiction between the chemotaxonomy and the results of molecular phylogeny is discussed in a proper way. Due to the limitation of space, the book is only an introduction and a compendium. It is intended to be a spur and enlighten thinking and promote the cognition of related researchers on pharmacophylogeny; the development of discipline itself and its application in medical practice are hopefully promoted. In the future, we must study more species of medicinal groups under the guidance of pharmacophylogeny in order to promote the sustainable utilization of TCM resources and find new compounds with potential therapeutic value. The continuous integration of systems biology and various omics concepts and techniques has opened up a broad space for the development of pharmacophylogeny.

Ranunculales is an order of flowering plants and contains the families Ranunculaceae (buttercup family), Berberidaceae, Menispermaceae, Lardizabalaceae, Circaeasteraceae,

Papaveraceae, and Eupteleaceae. Ranunculales belongs to the basal eudicots. It is the most basal clade in this group and is sister to the remaining eudicots. Medicinal plants of this order, for example, poppies and barberries (Fig. 1), and buttercups, provide myriad pharmaceutically active components, which have been commonly used in TCM and worldwide ethnomedicine since the beginning of civilization. Increasing interest in Ranunculales plant-based medicinal resources have led to additional discoveries of many novel compounds, such as alkaloids, saponins, terpenoids, glycosides, and phenylpropanoids, in various species, and to extensive investigations on their chemodiversity, biodiversity, and pharmacotherapy. Based on my studies of plant pharmacophylogeny and my first Elsevier book *Medicinal Plants: Chemistry, Biology and Omics*, this book presents comprehensive commentary on the phytochemistry, chemotaxonomy, molecular biology, and phylogeny of selected Ranunculales families and genera and their relevance to drug efficacy. Exhaustive literature searches are used to characterize the global trend in the flexible technologies being applied and fruitful data therefrom. The interrelationship between Chinese species and between Chinese species

and non-Chinese species is inferred by the molecular phylogeny based on nuclear and chloroplast DNA sequences. The conflict between chemotaxonomy and molecular phylogeny is revealed and discussed within the context of drug discovery and development. It is indispensable to study more Ranunculales species for both the sustainable conservation/utilization of medicinal resources and mining novel chemical entities with potential clinical efficacy. Systems biology and high-resolution omics technologies (genomics, epigenomics, transcriptomics, proteomics, metabolomics, etc.) will accelerate the pharmaceutical research involving bioactive compounds of Ranunculales.

FEATURES OF THIS BOOK

1. Offers the current perception of biodiversity and chemodiversity of Ranunculales medicinal plants.
2. Explains how the conceptual framework of plant pharmacophylogeny benefits the sustainable exploitation of Ranunculales pharmaceutical resources.
3. Describes how Ranunculales medicinal plants work from the chemical level upward.
4. Discusses how polypharmacology of Ranunculales compounds inspire new chemical entity design and development for improved treatment outcome.

The book is written as a reference for graduate and senior undergraduate students, researchers, and professionals in medicinal plant, phytochemistry, pharmacognosy, molecular biology, biotechnology, and agriculture and pharmacy within the academic and industrial sectors. Students and researchers in pharmacology, medicinal chemistry, plant systematics, food and nutrition, clinical medicine, evolution and ecology, as well as professionals in pharmaceutical industries,

FIGURE 1 *Mahonia fortunei* of Berberidaceae, taken in Xi'an Botanical Garden, China.

might also be interested in the plants discussed in this book.

This book is supported by Academic Publication Fund of Dalian Jiaotong University. Friends and colleagues in many parts of the world lent support to this book. We would like to thank all those who have published the findings that we cite in the chapters. Special thanks go to the project editor, Dr. Glyn Jones, from Elsevier and his group for their interest, support, and encouragement.

1

Genomics and Evolution of Medicinal Plants

1.1 INTRODUCTION

There are more than 300,000 species of extant seed plants around the globe (Jiao et al., 2011a,b). About 60% of plants have medicinal use in post-Neolithic human history. Nowadays, people collect plants for medicinal use from not only wild environments but also artificial cultivation, which is an indispensable part of human civilization. There are over 10,000 medicinal plant species in China, accounting for c. 87% of the Chinese materia medica (CMM) (Chen et al., 2010). Medicinal plants are also essential raw materials of many chemical drugs, for example, the blockbuster drugs for antimalarial and anticancer therapies. Currently more than one-third of clinical drugs are from botanical extracts and/or their derivatives. Unfortunately, most medicinal plants have not been domesticated, and currently there is no toolkit

Ranunculales Medicinal Plants. http://dx.doi.org/10.1016/B978-0-12-814232-5.00001-0

FIGURE 1.1   *Taxus cuspidata* **var. nana.** *Source: Taken in Dalian Jiaotong University, China.*

to improve their medicinal attributes for better clinical efficacy. Immoderate harvesting has led to a supply crisis of phytomedicine, exemplifying in taxane-producing *Taxus* plants (Hao et al., 2012a) (Fig. 1.1). On the other hand, successful domestication and improvement are not realistic without deeper insights into the evolutionary pattern of medicinal plant genomes. Artificial selection can be regarded as an accelerated and targeted natural selection. Studies of medicinal plant genome evolution are crucial to not only the ubiquitous mechanisms of plant evolution and phylogeny but also plant-based drug discovery and development, as well as the sustainable utilization of plant pharmaceutical resources. This chapter presents a preliminary examination of the recent developments in medicinal plant genome evolution research and summarizes the benefits, gaps, and prospects of the current research topics.

1.2 EVOLUTION OF GENOME, GENE, AND GENOTYPE

1.2.1 Genome Sequencing

The genomic studies of medicinal plants lag behind those of model plants and important crop plants. The genome sequences encompass essential information of plant origin, evolution, development, physiology, inheritable traits, epigenomic regulation, etc. These elements are the premise and foundation of deciphering genome diversity and chemodiversity (especially various secondary metabolites with potential bioactivities) at the molecular level. The high-throughput sequencing of medicinal plants could not only shed light on the biosynthetic pathways of medicinal compounds, especially secondary metabolites (Boutanaev et al., 2015), and their regulation mechanisms but also play a major role in the molecular breeding of high-yield medicinal cultivars and molecular farming of transgenic medicinal strains.

A few principles should be considered when selecting medicinal plants for whole genome sequencing projects. First, the source plants of well-known and expensive CMMs or important chemical drugs that are in heavy demand have priority, for example, *Panax ginseng* (Chen et al., 2011; Zhao et al., 2015) and *Artemisia annua* (Moses et al., 2015) (Fig. 1.2); second, the representative plants whose pharmaceutical components are relatively unambiguous and that have typical secondary metabolism pathways, for example, *Salvia* medicinal plants (Hao et al., 2015a,b); third, the characteristic plants that are in the large medicinal genus/ family, such as *Glycyrrhiza uralensis* (Chinese liquorice; *Fabaceae*) (Hao et al., 2012b, 2015c)

FIGURE 1.2 *Artemisia annua* of **Asteraceae.** *Source: Taken in Tashilunpo Monastery, Tibet, China.*

and *Lycium chinense* (Chinese boxthorn; *Solanaceae*) (Yao et al., 2011); fourth, the medicinal plants that are potential model plants and have considerable biological data; and last, the medicinal plants whose genetic backgrounds are known, with reasonably small diploid genomes and relatively straightforward genome structures, are preferred.

As there is a lack of comprehensive molecular genetic studies for most medicinal plants, it is vital to have some preliminary genome evaluations before the whole genome sequencing. First, DNA barcoding techniques (Hao et al., 2012c) could be used to authenticate the candidate species; second, karyotypes should be determined by observing metaphase chromosomes; and last, flow cytometry and pulsed field gel electrophoresis (PFGE) (Hao et al., 2011b, 2015b) could be chosen to determine the ploidy level and genome size. For example, flow cytometry was used to determine the genome size of four *Panax* species (Pan et al., 2014), with *Oryza sativa* as the internal standard. *Panax notoginseng* (San Qi in traditional Chinese medicine (TCM)) has the largest genome (2454.38 Mb), followed by *P. pseudoginseng* (2432.72 Mb), *P. vietnamensis* (2018.02 Mb), and *P. stipuleanatus* (1947.06 Mb), but their genomes are smaller than the *Pa. ginseng* genome (~3.2 Gb). A more reliable approach for species without the reference genome is the genome survey via the whole genome shotgun sequencing (Polashock et al., 2014). Such nondeep sequencing (30 × coverage), followed by the bioinformatics analysis, is highly valuable in assessing the genome size, heterozygosity, repeat sequence, guanine/cytosine (GC) content, etc., facilitating the decision-making of the whole genome sequencing approaches. In addition, RAD-Seq (restriction-site associated DNA sequencing; Fig. 1.3) (Rubin et al., 2012) could be chosen to construct a RAD library and perform the low-coverage genome sequencing of reduced representation, which is an effective approach for assessing the heterozygosity of the candidate genome.

The whole genome sequencing platform is chosen based on the budgetary resources and the preliminary evaluation of candidate genomes (Chen et al., 2010). GS FLX or Illumina HiSeq 2500 platforms might be suitable for the small simple genome. However, the majority of the plant genomes belong to the complex genome, which refers to the diploid/polyploidy genome, with >50% repeat sequences and >0.5% heterozygosity. Two or more sequencing platforms could be combined for shotgun and paired-end sequencing, while large insert libraries, for example, BAC (bacterial artificial chromosome) (Hao et al., 2015b), yeast artificial chromosome (Noskov et al., 2011), and Fosmid (Hao et al., 2011b), can be constructed for sequencing, then

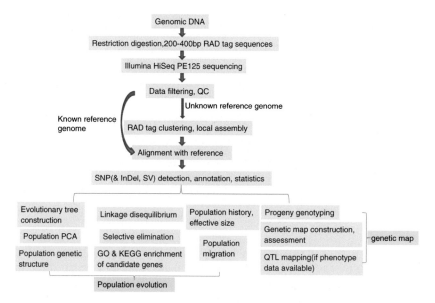

FIGURE 1.3 **Technology roadmap of RAD-Seq and its utility in population evolution and genetic map.** *InDel*, Insertion and deletion; *PCA*, principal component analysis; *PE*, paired end; *QC*, quality control; *QTL*, quantitative trait loci; *SV*, splice variant.

the sophisticated bioinformatics softwares (Cai et al., 2015; Chalhoub et al., 2014; Denoeud et al., 2014; Kim et al., 2014; Qin et al., 2014) can be used for sequence quality control and assembly. For instance, GS FLX and shotgun sequencing can be used for the initial genome assembly to generate 454 contigs, then the paired-end sequencing data from the Illumina HiSeq or SOLiD platform can be used to determine the order and orientation of 454 contigs, thus generating scaffolds. Next, Illumina HiSeq or SOLiD data are used to fill the gaps between some contigs. These steps streamline the genome sequencing pipeline as a whole.

The genetic map and physical map are fundamental tools for the assembly of the complex plant genome and functional genomics research. The genetic linkage map of *Bupleurum chinense* (Bei Chai Hu in TCM) was constructed using 28 ISSR (intersimple sequence repeat) and 44 SSR (microsatellite) markers (Zhan et al., 2010); 29 ISSRs and 170 SRAPs (sequence-related amplified polymorphisms) were mapped to 25 linkage groups of *Siraitia grosvenorii* (Luo Han Guo in TCM) (Liu et al., 2011). These preliminary results are useful in metabolic gene mapping, map-based cloning, and marker-assisted selection of medicinal traits. The high-throughput physical map could be anchored via the BAC-pool sequencing (Cviková et al., 2015), which, along with its integration with high-density genetic maps, could benefit from next generation sequencing (NGS) and high-throughput array platforms (Ariyadasa and Stein, 2012). The development of dense genetic maps of medicinal plants is still challenging, as the parental lines and their progenies with the unambiguous genetic link are not available for most medicinal plants.

1.2.2 Chloroplast Genome Evolution

Chloroplast (cp) is responsible for photosynthesis, and its genome sequences have versatile utility in evolution, adaptation, and robust growth of most medicinal plants. The substitution rate of the cp nucleotide sequence is three to four times faster than that of the mitochondria (mt) sequence (Zhao et al., 2015), implicating more uses of the former in inferring both interspecific and intraspecific evolutionary relationships (Ma et al., 2014; Malé et al., 2012; Qian et al., 2013; Su et al., 2014; Wu et al., 2013; Xu et al., 2012; Zhao et al., 2015).

Pa. ginseng is a "crown" TCM plant and frequently used in health-promoting food and clinical therapy. NGS technology provides insight into the evolution and polymorphism of *Pa. ginseng* cp genome (Zhao et al., 2015). The cp genome length of Chinese *Pa. ginseng* cultivars Damaya (DMY), Ermaya (EMY), and Gaolishen (GLS) was 156,354 bp, while the genome length was 156,355 bp in wild ginseng (YSS), which is smaller than Omani lime (*C. aurantiifolia*; 159,893 bp) (Su et al., 2014) and 12 *Gossypium* cp genomes (159,959–160,433 bp) (Xu et al., 2012) but bigger than *Rhazya stricta* cp genome (154,841 bp) (Park et al., 2014). Gene content, GC content, and gene order in DMY are quite similar to other strains, and nucleotide sequence diversity of inverted repeat region (IR) is lower than that of large single-copy region (LSC) and small single-copy region (SSC). The high-resolution reads were mapped to the genome sequences to investigate the differences of the minor allele, which showed that the cp genome is heterogeneous during domestication; 208 minor allele sites with minor allele frequencies of ≥0.05 were identified. The polymorphism site numbers per kb cp genome of DMY, EMY, GLS, and YSS were 0.74, 0.59, 0.97, and 1.23, respectively. All minor allele sites were in LSC and IR regions, and the four strains showed the same variation types (substitution base or indel) at all identified polymorphism sites. The minor allele sites of the cp genome underwent purifying selection to adapt to the changing environment during domestication. The study of medicinal plant cp genomes with particular focus on minor allele sites would be valuable in probing the dynamics of the cp genomes and authenticating different strains and cultivars.

The genus *Citrus* contains many economically important fruits that are grown worldwide for their high nutritional and medicinal value. Due to frequent hybridizations among species and cultivars, the exact number of natural species and the evolutionary relationships within this genus are blurred. It is essential to compare the *Citrus* cp genomes and to develop suitable genetic markers for both basic research and practical use. A reference-assisted approach was adopted to assemble the complete cp genome of Omani lime (Su et al., 2014), whose organization and gene content are similar to most rosid lineages characterized to date. Compared with the sweet orange (*Ci. sinensis*), three intergenic regions and 94 simple sequence repeats (SSRs) were identified as potentially informative markers for resolving interspecific relationships, which can be harnessed to better understand the origin of domesticated *Citrus* and foster the germplasm conservation. A comparison among 72 species belonging to 10 families of representative rosid lineages also provides new insights into their cp genome evolution.

The monocot family Orchidaceae, evolutionarily more ancient than asterids and rosids, is one of the largest angiosperm families, including many medicinal, horticultural, and ornamental species. Orchid phytometabolites display antinociceptive (Morales-Sánchez et al., 2014), antiangiogenic (Basavarajappa et al., 2014), and antimycobacterial (Ponnuchamy et al., 2014) activities, etc. In South Asia, orchid bulb is used for the treatment of asthma, bronchitis, throat infections, and dermatological infections and also used as a blood purifier (Nagananda and Satishchandra, 2013). Sequencing the complete cp genomes of the medicinal plant *Dendrobium officinale* (Tie Pi Shi Hu in TCM) and the ornamental orchid *Cypripedium macranthos* reveals their gene content and order, as well as potential RNA-editing sites (Luo et al., 2014). The cp genomes of these two species and five known photosynthetic orchids are similar in structure as well as gene order and content, but the organization of the IR/SSC junction and ndh gene is distinct. IRs flanking the SSC region underwent expansion or contraction in different *Orchidaceae* species. Fifteen highly divergent protein-coding genes were identified and are useful in phylogenetic inference of orchids. Cp phylogenomic analysis can be used to resolve the interspecific relationship that cannot be inferred by a few cp markers. Bamboo leaves are used as a component in TCM for the antiinflammatory function (Koide et al., 2011). Medicinal bamboo cupping therapy is applied to reduce fibromyalgia symptoms (Cao et al., 2011). Bamboo extracts exhibit antioxidant effects (Jiao et al., 2011a,b) and are used to treat chronic fever and infectious diseases (Wang et al., 2012). The whole cp genome data sets of 22 temperate bamboos considerably increased resolution along the backbone of tribe *Arundinarieae* (temperate woody bamboo) and afforded solid support for most relationships regardless of the very short internodes and long branches in the tree (Ma et al., 2014). An additional cp phylogenomic study, involving the full cp genome sequences of eight *Olyreae* (herbaceous bamboo) and 10 *Arundinarieae* species, strengthened the soundness of the above study and recovered monophyletic relationship between *Bambuseae* (tropical woody bamboo) and *Olyreae* (Wysocki et al., 2015).

The monocot genus *Fritillaria* (*Liliaceae*) has about 140 species of bulbous perennial plants that embraces taxa of both horticultural and medicinal importance. The bulbs of plants belonging to the *Fritillaria cirrhosa* group have been used as antitussive and expectorant herbs in TCM for thousands of years (Wu et al., 2015). The anticancer activity and cardiovascular effects of *Fritillaria* phytometabolites are well documented (Hao et al., 2015c). *Fritillaria* species have attracted attention also because of their remarkably large genome sizes, with all values recorded to date above 30 Gb (Day et al., 2014). A phylogenetic reconstruction, including the most currently recognized species diversity of the genus, was performed (Day et al., 2014). Three regions of the cp genome were sequenced in 92 species (c. 66% of the genus) and in representatives of nine other genera of *Liliaceae*. Eleven low-copy nuclear genes were screened in selected species, but they had limited utility in phylogenetic reconstruction. Phylogenetic analysis of a combined plastid data set supported the monophyly of the majority of presently identified subgenera. However, the subgenus *Fritillaria*, which is by far the largest subgenera and includes the most important species used in TCM, is found to be polyphyletic. Clade, containing the source plants of Chuan Bei Mu, Hubei Bei Mu, and Anhui Bei Mu, might be treated as a separate subgenus (Hao et al., 2013a). The Japanese endemic subgenus *Japonica*, which contains the species with the largest recorded genome size for any diploid plant, is sister to the largely Middle Eastern and Central Asian subgenus *Rhinopetalum*, which is significantly incongruent with the nuclear internal transcribed spacer (ITS) tree. Convergent or parallel evolution of phenotypic traits may be a common cause of incongruence between morphology-based classifications and the results of molecular phylogeny. While relationships between most major *Fritillaria* lineages can be resolved, these results also highlight the need for data from more independently evolving loci, which is pretty perplexing given the huge nuclear genomes found in these plants.

Medicinal plant diversity, comprised of genetic diversity, medicinal species diversity, ecological system diversity, etc. (Hao et al., 2014a), results from the intricate interactions between medicinal plant and environment, and thus is profoundly influenced by the ecological complex and the relevant versatile ecological processes. The effects of the evolutionary processes have to be taken into full consideration when explaining the link between climatic/ecological factors and medicinal plant diversity, especially where there is strong, uneven differentiation of species. A distinguished example is the "sky islands" of Southwest

China (He and Jiang, 2014), where the extraordinarily rich resources of medicinal plants rose and thrived during the Quaternary Period. To date, many medicinal tribes and genera, for example, *Pedicularis* (Eaton and Ree, 2013) (Fig. 1.4), *Clematis* (Hao et al., 2013b), *Aconitum* (Hao et al., 2013c, 2015d,e), and *Delphinium* (Jabbour and Renner, 2012), are still in the process of rapid radiation and dynamic differentiation. The cp genome sequence can be regarded as the super-barcode of the organelle scale and thus can be used to probe the intraspecific variations (Whittall et al., 2010) and phylogeographic patterns of the same species in the disparate geographic locations (e.g., geoherb or Daodi medicinal materials) (Zhao et al., 2012). The application of the cp genome sequence at the population level may provide clues for the timing and degree of the intraspecific differentiation. Distilling the interpopulation relationship from the cp data set can be considered a more detailed phylogenetic reconstruction.

1.2.3 Mitochondria (mt) Genome Evolution

Some fundamental evolution concepts, such as lateral gene transfer, are bolstered by the inquiry of the origin of mt, while plants are especially useful inelucidating the mechanisms of cytonuclear coevolution. Although the gene order of the mt genome might evolve relatively faster in land plants, the substitution rate of its nucleotide sequence is merely 1/100 that of its animal sequence (Hao et al., 2014a). Therefore, the mt genome sequence is less useful than the cp one in inferring the phylogenetic relationship of medicinal species (Henriquez et al., 2014). Notwithstanding, analyzing genome sequences contributes knowledge about the evolution of the mt genome. Moreover, the terpene synthase has been found in mt (Hsu et al., 2012), highlighting its utility in secondary metabolism.

R. stricta (*Apocynaceae*) is native to arid regions in South Asia and the Middle East and is used extensively in folk medicine. Analyses of the complete cp and mt genomes and a nuclear (nr) transcriptome of *Rhazya* shed light on intercompartmental transfers between genomes and the patterns of evolution among eight asterid mt genomes (Park et al., 2014). The *Rhazya* genome is

FIGURE 1.4 *Pedicularis longiflora* of **Scrophulariaceae.** *Source: Taken in Dingri County, Tibet, China.*

highly conserved with gene content and order identical to the ancestral organization of angiosperms. The 548,608 bp mt genome contains recombination-derived repeats that generate a compound organization; transferred DNA from the cp and nr genomes as well as bidirectional DNA transfers between the mt and the nucleus are also disclosed. The mt genes sdh3 and rps14 have been transferred to the nucleus and have acquired targeting transit peptides. Two copies of rps14 are present in the nucleus; only one has the mt targeting transit peptide and may be functional. Phylogenetic analyses suggest that *Rhazya* has experienced a single transfer of this gene to the nucleus, followed by a duplication event. The phylogenetic distribution of gene losses and the high level of sequence divergence in targeting transit peptides suggest multiple independent transfers of both sdh3 and rps14 across asterids. Comparative analyses of mt genomes of eight asterids indicates a complicated evolutionary history in this thriving eudicot clade, with substantial diversity in genome organization and size, repeat, gene and intron content, and amount of alien DNA from the cp and nr genomes. The genomic data enable a rigorous inspection of the gene transfer events.

1.2.4 Nuclear Genome Evolution

1.2.4.1 Monocots

The whole cp genome data set is not enough to elucidate the phylogenetic relationship of groups undergoing rapid radiation, for example, Zingiberales (Barrett et al., 2014). The cp genome is equivalent to one gene locus, thus it only represents one fulfillment to the coalescent random processes and cannot be used with confidence to reconstruct the evolutionary history of the populations. The most genetic history of any medicinal plant hides in the nr genome.

High-throughput sequencing and the relevant bioinformatics advances have revolutionized contemporary thinking on nuclear genome/transcriptome evolution and provided basic data for further breeding endeavors. *Coix* (*Poaceae*), a closely related genus of *Sorghum* and *Zea*, has 9–11 species with different ploidy levels. The exclusively cultivated *C. lacryma-jobi* (2n = 20) is widely used in East and Southeast Asia as food and traditional medicine. *C. aquatica* has three fertile cytotypes (2n = 10, 20, and 40) and one sterile cytotype (2n = 30), *C. aquatica* HG, which is found in Guangxi of China (Cai et al., 2014). Low coverage genome sequencing (genome survey) showed that around 76% of the *C. lacryma-jobi* genome and 73% of the *C. aquatica* HG genome are repetitive sequences, among which the long terminal repeat (LTR) retrotransposable elements dominate, but the proportions of many repeat sequences vary greatly between the two species, suggesting their evolutionary divergence. A novel 102 bp variant of centromeric satellite repeat CentX and two other satellites are exclusively found in *C. aquatica* HG. The FISH analysis and fine karyotyping showed that *C. lacryma-jobi* is likely a diploidized paleotetraploid species, and *C. aquatica* HG is possibly from a recent hybridization. These *Coix* taxa share more coexisting repeat families and higher sequence similarity with *Sorghum* than with *Zea*, which agrees with the phylogenetic relationship.

The heterozygous genome sequences of the tropical epiphytic orchid *Phalaenopsis equestris* provide insights into the unique crassulacean acid metabolism (CAM) (Cai et al., 2015). The assembled genome contains 29,431 predicted protein-coding genes and is rich in genes that might be involved in self-incompatibility pathways, which ensure the genetic diversity and enhance the fitness and survival. An orchid-specific paleopolyploidy event is disclosed, which preceded the radiation of most orchid clades, and gene duplication might have contributed to the evolution of CAM photosynthesis in *Ph. equestris*. The expanded and diversified families of MADS-box C/D-class, B-class AP3, and AGL6-class genes might contribute to the highly specialized morphology of orchid flowers. LTRs are the most abundant transposable element (Fig. 1.5), followed by long interspersed nuclear elements (LINEs).

1.2.4.2 Basal Eudicots

The *Macleaya cordata* (*Papaveraceae*, Ranunculales) genome covering 378 Mb encodes 22,328 predicted protein-coding genes, with 43.5% being transposable elements (Liu et al., 2017). As a member of basal eudicots, this genome lacks the paleohexaploidy event that occurred in almost all eudicots. From the genomics data, all 16 metabolic genes for sanguinarine and chelerythrine biosynthesis were retrieved, and the biochemical activities of 14 genes were validated. These genomics and metabolic data show the conserved benzylisoquinoline alkaloid (BIA) metabolic pathways in *M. cordata* and provide the knowledge base for future productions of BIAs by crop improvement or microbial pathway reconstruction.

1.2.4.3 Eudicots: Asterids

Whole genome sequencing has been implemented in the representative species of some plant families/genera (Fig. 1.6), for example, *Capsicum annuum* (Kim et al., 2014; Qin et al., 2014), *Coffea canephora* (Denoeud et al., 2014), *Brassica napus* (Chalhoub et al., 2014), and *Ph. equestris* (Cai et al., 2015). The genome sequences of the cultivated pepper Zunla-1 (*Cap. annuum*) and its wild progenitor Chiltepin (*Cap. annuum* var. *glabriusculum*) were compared to provide insights into *Capsicum* domestication and specialization. The pepper genome expanded ~0.3 Mya by a rapid amplification of retrotransposon elements, resulting in a genome containing ~81% repetitive sequences and 34,476 protein-coding genes. Comparison of cultivated and wild pepper genomes with 20 resequencing accessions revealed

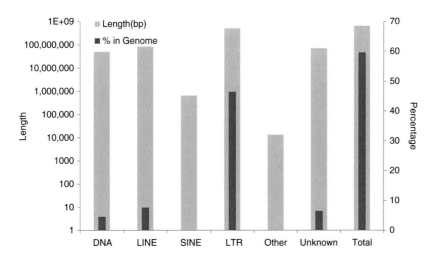

FIGURE 1.5 **Categories of transposable elements predicted in the orchid genome (Cai et al., 2015).** *DNA*, DNA transposon; *LINE*, long interspersed element (retrotransposon); *LTR*, long terminal repeat (retrotransposon); *SINE*, short interspersed nuclear element (retrotransposon).

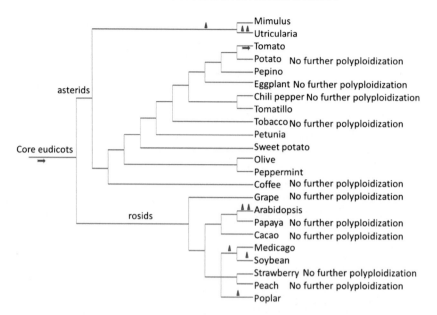

FIGURE 1.6 **Examples of the phylogeny and genome duplication history of core eudicots.** *Arrowheads* indicate hexaploidization; *triangles* indicate tetraploidization. The current evidence does not suggest further polyploidization after speciation in the genomes of potato, eggplant, chili pepper, tobacco, coffee, grape, papaya, cacao, strawberry, and peach. Few genome data are available in pepino, tomatillo, and many other species.

molecular signatures of artificial selection, providing a list of candidate domestication genes (Qin et al., 2014). Dosage compensation effect of tandem duplication genes might contribute to the pungency divergence in pepper (Qin et al., 2014). The *Capsicum* reference genome, along with tomato and potato genomes, provides critical information for the study of the evolution of other *Solanaceae* species, including the well-known *Atropa* medicinal plants.

The highly heterozygous *Salvia miltiorrhiza* (Danshen, *Lamiaceae*) genome was assembled with the help of 395× raw read coverage using Illumina technologies and about 10× raw read coverage using single molecular sequencing technology (Zhang et al., 2015). The final draft genome is approximately 641 Mb, with a contig N50 size of 82.8 kb and a scaffold N50 size of 1.2 Mb. Further analyses predicted 34,598 protein-coding genes and 1644 unique gene families in the Danshen genome, which provides a valuable resource for the investigation of novel bioactive compounds in this traditional Chinese herb.

One of the milestone breakthroughs is the successful sequencing and assembly of the complex heterozygous genome. The heterozygous genome of *Co. canephora* has been deciphered (Denoeud et al., 2014), and it displays a conserved chromosomal gene order among asterid angiosperms. Although it shows no sign of the whole-genome triplication identified in *Solanaceae* species, the genome includes several species-specific gene family expansions, for example, *N*-methyltransferases (NMTs) involved in caffeine biosynthesis, defense-related genes, and alkaloid and flavonoid enzymes involved in secondary metabolite production. Caffeine NMTs expanded through sequential tandem duplications independently and are distinct from those of cacao and tea, suggesting that caffeine in eudicots is of polyphyletic origin and its biosynthesis underwent convergent evolution.

The 3.5-Gb genome of *Pa. ginseng* contains more than 60% repeats and encodes 42,006 predicted genes (Xu et al., 2017). Twenty-two transcriptome data sets and mass spectrometry images of ginseng roots were used to precisely quantify the functional genes. Thirty-one genes were identified to be involved in the ginsenoside biosynthetic mevalonic acid pathway, eight of which were 3-hydroxy-3-methylglutaryl-CoA reductases. A total of 225 UDP-glycosyltransferases (UGTs) were identified, which constitute one of the largest gene families of ginseng. Tandem repeats contributed to the duplication and divergence of UGTs. Molecular modeling of UGTs in the 71st, 74th, and 94th families revealed a regiospecific conserved motif at the *N*-terminus, which captures ginsenoside precursors. The ginseng genome represents a valuable resource for understanding and improving the breeding, cultivation, and synthetic biology of this king of TCM.

Pan. notoginseng experienced a series of genome evolution events that created the unique medicinal properties of this famous medicinal plant (Chen et al., 2017; Zhang et al., 2017). For example, a recent polyploidy event occurred about 26 million years ago, and there were a large number of specific duplications of triterpenoid biosynthesis-related gene families; genes related to triterpenoid saponin biosynthesis formed many gene clusters. Comparative genomics, transcriptomics, and comparative phytochemistry further confirmed that the rapid functional variation and evolution of genomes determines the chemodiversity, which is closely related to the therapeutic efficacy of *Pan. notoginseng*. Most genes associated with the saponin biosynthesis are mainly expressed in flowers and leaves; after synthesis, saponins could be transported and stored in the roots. This discovery subverts the view that the *Pan. notoginseng* saponins are synthesized in the roots.

The highly heterozygous *Erigeron breviscapus* genome was assembled using a combination of PacBio, single-molecular, real-time sequencing and next-generation sequencing methods on the Illumina HiSeq platform (Yang et al., 2017). The final draft genome is around 1.2 Gb, with contig and scaffold N50 sizes of 18.8 kb and 31.5 kb, respectively. Further analyses predicted 37,504 protein-coding genes in the *E. breviscapus* genome and 8172 shared gene families among the *Compositae* species.

1.2.4.4 Gene Family

More than 40 plant genomes have been sequenced, representing a diverse set of taxa of agricultural, energy, medicinal, and ecological importance (Cai et al., 2015; Chalhoub et al., 2014; Denoeud et al., 2014; Kim et al., 2014; Qin et al., 2014). Gene family members are often inferred from DNA sequence homology, but deeper insights into evolutionary processes contributing to gene family dynamics are imperative. In a comparative genomics framework, multiple lines of evidence can be generated by gene synteny, sequence homology, and protein-based hidden Markov modeling (HMM) to extract homologous super-clusters composed of multidomain resistance (R)-proteins of the NB-LRR (nucleotide binding-leucine rich repeat) type, which are involved in plant innate immunity (Hofberger et al., 2014). Twelve eudicot plant genomes were screened to assess the intra- and interspecific diversity of R-proteins, where 2363 NB-LRR genes were found. Half of the R-proteins have tandem duplicates, and 22% of gene copies are left from ancient polyploidy events (ohnologs, whole genome duplication duplicates). The positive Darwinian selection and major differences in molecular evolution rates (K_a/K_s) were detected among tandem (mean = 1.59), ohnolog (1.36), and singleton (1.22) R-gene duplicates. The distribution pattern of all 140 NB-LRR genes present in the model plant *Arabidopsis* is species specific, and four distinct clusters of NB-LRR "gatekeeper" loci sharing syntenic orthologs across all analyzed genomes were identified and could be useful for the gene-edited plant breeding. The near-complete set of multidomain R-protein clusters in a eudicot-wide scale could shed light on evolutionary dynamics underlying diversification of the plant innate immune system. More functional NB-LRR genes could be identified from more sequenced plant species.

The estimated upper limit of extant plants is c. 450,000, indicating the potentially enormous biological space. The multiple and recurrent genome duplications during plant evolution result in the generation of novel biosynthetic pathways of diverse medicinal compounds, which are frequently involved in plant defense and disease resistance and, more importantly, create huge chemical space for drug discovery and development. The duplicated gene copies could explain the diversification processes of the multigene secondary metabolism pathways, such as those involved in the biosynthesis of terpenoids (Boutanaev et al., 2015), benzoxazinoid (Dutartre et al., 2012), steroidal glycoalkaloid (Manrique-Carpintero et al., 2013), and glucosinolates (Hofberger et al., 2013). More than 200,000 secondary metabolites have been found in angiosperms, many of which could stem from the genome duplication-based rapid innovation of complex traits (Hao et al., 2014a).

1.2.4.5 *Single Copy Gene*

Single copy genes are common across angiosperm genomes. Based on 29 sufficiently high-quality sequenced genomes, the large-scale identification and evolutionary characterization of single copy genes among multiple species is possible (Han et al., 2014). A significant negative correlation was found between the number of duplicate blocks and the number of single copy genes. Only 17% of single copy genes are located in organelles, most of which are involved in binding and catalytic activity. Most single copy genes are in nuclear genomes. Single copy genes have a stronger codon bias than nonsingle copy genes in eudicots (Han et al., 2014). The relatively high expression level of single copy genes was partially confirmed by the RNA-Seq (transcriptome sequencing) data. Unlike in most other species, there is a strong negative correlation between N_c (effective number of codons) and GC3 (G + C content at third codon position) of single copy genes in grass genomes. Compared to nonsingle copy genes, single copy genes are of more conservation, as indicated by K_a and K_s values. Selective constraints on alternative splicing are weaker in single copy genes than in low-copy family genes (1–10 paralogs) and stronger than high-copy family genes (>10 paralogs). Using concatenated, shared, single copy genes, a well-resolved phylogenetic tree can be obtained. Addition of intron sequences improved the branch support, but striking incongruences are also obvious. Inclusion of intron sequences might be more appropriate for the phylogenetic reconstruction at lower taxonomic levels. Evolutionary constraints between single copy genes and nonsingle copy genes are distinct and are somewhat species specific, especially between eudicots and monocots.

1.2.5 Transcriptome

The high cost of the whole genome sequencing is still formidable. The accurate sequence assembly is still challenging, especially when the genome is of a high proportion of repeat sequences, high heterozygosity, and nondiploid. RNA-Seq is a powerful tool for the assessment of gene expression and the identification and characterization of molecular markers in nonmodel organisms (Hao et al., 2012b). Unlike genome sequences, the intron sequences are not included in the RNA-Seq data set, and the Unigene (contig) assembly is not disturbed by the repeat sequences and the ploidy level. The global view of the ethnomedicine resources and the accurate delimitation of the novel medicinal taxa cannot be achieved without the molecular phylogeny based on the complete taxon sampling of the relevant tribes/genera. Due to the plummeting cost of RNA-Seq, the dense taxon sampling is now possible in the phylotranscriptomic studies. It is obvious that the large-scale comparative transcriptome studies, including those of medicinal plants, are more feasible than comparative genomics based on whole genome sequencing. As shown in National Centre for Biotechnology Information (NCBI) PubMed, sequence read archive (SRA), and Gene Expression Omnibus (GEO) databases, transcriptomes of hundreds of medicinal plants have been sequenced, for example, *Caryophyllales* (Yang et al., 2015) (Fig. 1.7), *Fabaceae* (Cannon et al., 2015), *Oenothera* (*Onagraceae*) (Hollister et al., 2015), *Rhodiola algida* (*Crassulaceae*) (Zhang et al., 2014), *S. sclarea* (*Lamiaceae*) (Hao et al., 2015a), *Polygonum cuspidatum* (*Polygonaceae*) (Hao et al., 2012d), *Taxus mairei* (*Taxaceae*) (Hao et al., 2011a), etc. The single copy orthologous gene sequences could be extracted from the UniGene data sets of multiple medicinal plants (Hao et al., 2012c), which can be used in the phylogenetic reconstruction and evolutionary analyses (Hao et al., 2012d; Yang et al., 2015). The information uncovered in transcriptome studies could serve in the characterization of important traits related to secondary metabolite formation and in probing the relevant molecular mechanisms (Hao et al., 2011a, 2012c, 2015a; Zhang et al., 2014).

Reconstructing the origin and evolution of land plants and their algal relatives is a vital problem in plant phylogenetics and is essential for understanding how novel adaptive traits, for example, secondary metabolites, arose. Despite advances in molecular systematics, some evolutionary relationships remain poorly resolved. Inferring deep phylogenies with rapid diversification is often tricky, and genome-scale data significantly increase the number of informative characters for analyses. Since the sparse taxon sampling could result in inconsistent results, transcriptome data of 92 streptophyte taxa were generated and analyzed along with 11 published plant genome sequences (Wickett et al., 2014). Phylogenetic reconstructions were conducted using 852 nuclear genes and 1,701,170 aligned sites. Robust support for a sister-group relationship between land plants and one streptophyte green algae, the *Zygnematophyceae*, was obtained. Strong and robust support for a clade comprising liverworts and mosses contradicts the widely accepted view of early land plant evolution. Phylogenetic

FIGURE 1.7 *Stellaria chinensis* of **Caryophyllaceae.** *Source: Taken in Taibai Mountain, Shaanxi, China.*

hypotheses could be tested using phylotranscriptomic approach to give deeper insights and novel arguments into the evolution of fundamental plant traits, including the fascinating chemodiversity.

The transcriptome sequencing also sheds light on other untapped issues of plant evolution. Arbuscular mycorrhizal (AM) are symbiotic systems in nature and have great significance in promoting the growth and stress resistance of medicinal plants (Zeng et al., 2013). AM has multifaceted effects on the active ingredients of TCM plants. The transcriptomes of nine phylogenetically divergent non-AM symbiosis plants were analyzed to reveal the correlation between the loss of AM symbiosis and the loss of many symbiotic genes (Delaux et al., 2014), which was found in four additional plant lineages besides the *Arabidopsis* lineage (Brassicales), implicating the convergent evolution. RNA-Seq was used to outline gene sequence and expression discrepancy between cultivated tomato and five allied wild species (Koenig et al., 2013). Human handling of the genome has profoundly altered the tomato transcriptome via directed admixture and by secondarily choosing nonsynonymous over synonymous substitutions. A hitherto unidentified paleopolyploidy event that arose 20–40 million years ago was uncovered based on the transcriptomes of 11 *Linum* species (Sveinsson et al., 2014), which is specific to a clade enclosing cultivated flax (*L. usitatissimum*) and other mainly blue-flowered species.

1.2.6 Evolution and Population Genetics/Genomics

SSRs play a major role as molecular markers for genome analysis and plant breeding. The microsatellites existing in the complete genome sequences would have a direct role in the genome organization, recombination, gene regulation, quantitative genetic variation, and evolution of genes. Microsatellite markers have been characterized for many medicinal plant families and genera, for example, Acanthaceae family (Kaliswamy et al., 2015), *Artemisia* genus (Karimi et al., 2015), *Camellia* genus (Li et al., 2015), Chinese jujube (Wang et al., 2014), etc. For instance, 11 nuclear SSR loci were used to reveal the relative low genetic diversity of three *Camellia taliensis* (Da Li tea) populations, three *Ci. sinensis* var. *assamica* (Pu Er tea in TCM) populations, and two transitional populations of *Ca. taliensis*. A momentous genetic differentiation was found between *Ci. sinensis* var. *assamica* and *Ca. taliensis* populations. The transitional populations of *Ca. taliensis* stemmed mainly from *Ca. taliensis* and underwent genetic differentiation during domestication. Gene introgression was spotted in the cultivated *Ci. sinensis* var. *assamica* and *Ca. taliensis* of the same tea garden, and genetic material of *Ca. taliensis* seemingly intruded into *Ci. sinensis* var. *assamica*, suggesting that the former was genetically involved in the domestication of the latter. These results are useful for protecting the genetic resources of ancient tea plants. The whole nucleotide sequences—for example, the genomic sequences (Hao et al., 2015b) or the transcriptomic sequences (Hao et al., 2012b, 2015a)—of plant species can be obtained from NCBI databases and screened for the presence of SSRs.

Both ISSR (Hao et al., 2010) and SRAP markers were suitable for discriminating among the studied individuals, and the SRAP markers were more efficient and preferable (Karimi et al., 2015). Multiple regression analysis revealed statistically significant associations between rust resistance and some molecular markers; this can provide clues for identification of the individuals with higher rust resistance. RAPD (randomly amplified polymorphic DNA) and

ISSR markers were used to characterize *Schisandra chinensis* with white fruit (Li et al., 2014). The molecular marker-based study of genetic diversity helps in assessing the studied germplasm, which would be a valuable genetic resource for future breeding. Based on such a study, *in situ* conservation measures or other methods could be recommended to preserve the valuable medicinal plant genetic resources.

Sinopodophyllum hexandrum is an endangered *Berberidaceae* (Ranunculales) medicinal plant, and its genetic diversity must be protected against habitat loss and anthropogenic factors. The Qinling Mountains are an *Si. hexandrum* distribution area, where unique environmental features highly affect the evolution of the species. ISSR analysis of 32 natural populations revealed the genetic diversity and population structure of *Si. hexandrum* in Qinling and provides reference data for evolutionary and conservation studies (Liu et al., 2014). The 32 populations fell into three major groups, where analysis of molecular variance confirmed significant variation among populations. The high genetic differentiation may be attributed to the limited gene flow within the species. The spatial pattern and geographic locations of different populations are not correlated with one another. In light of the low within-population genetic diversity, high differentiation among populations, and the increasing anthropogenic pressure on the species, *in situ* conservation is proposed to preserve *Si. hexandrum* in Qinling, and other populations must be sampled to maintain genetic diversity of the species for the *ex situ* preservation.

SNPs (single nucleotide polymorphisms) are much more abundant than SSRs in most species (Clevenger et al., 2015), including medicinal plants. The mutation rate of SNPs (10^{-9} per locus per generation) is much lower than that of SSRs (10^{-3}–10^{-4}) (Guichoux et al., 2011). Generally there are only two alleles in each SNP site, while more than 10 alleles can be in each SSR. The highly polymorphic SSRs are especially suitable for detecting the hybridization between closely related species and studying the gene flow/introgression (Wee et al., 2015). SSRs are of lower ascertainment bias and are also good for studying the recent population structure. Mining suitable SSR sites via transcriptome sequencing data sets is fast and affordable; for example, 3446 microsatellites are identified from 2718 Unigenes (16.8% of 16,142 assembled sequences) of the *S. sclarea* transcriptome (Fig. 1.8). Trinucleotide (1883) is the predominant microsatellite, followed by dinucleotide (1144) and mononucleotide (315), indicating that many microsatellites are in translated regions of the expressed genes. CCG/CGG is the predominant trinucleotide SSR, followed by AAG/CTT and AGC/GCT. AG/CT is the most common dinucleotide SSR. Of the identified repeats, 601 (19.2%) have sufficient flanking sequence information to allow for PCR primer design. Intriguingly, many SSR motifs are linked with unique sequences encoding enzymes involved in phenylpropanoid/terpenoid metabolism. For instance, SSRs were detected in phenylalanine ammonia-lyase, 4-coumarate-CoA ligase, hydroxyphenylpyruvate dioxygenase, flavonoid 3′-hydroxylase, cinnamyl alcohol dehydrogenase, and lignan glycosyltransferase sequences, which belong to the phenylpropanoid pathway; SSRs were also found in 2-C-methyl-D-erythritol 4-phosphate pathway genes (1-deoxy-D-xylulose 5-phosphate synthase, 1-deoxy-D-xylulose 5-phosphate reductoisomerase, 2-C-methyl-D-erythritol 4-phosphate cytidylyltransferase, 4-hydroxy-3-methylbut-2-enyl diphosphate synthase), mevalonate pathway genes (mevalonate pyrophosphate decarboxylase) and other terpenoid biosynthesis genes (isopentenyl diphosphate isomerase, cytochrome P450 71D18, pinene synthase, squalene synthase, and squalene monooxygenase). These SSRs might be useful in future breeding and ecological studies. One of the

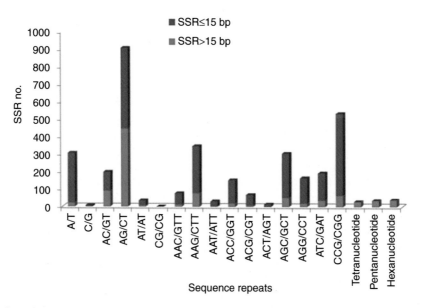

FIGURE 1.8   SSRs predicted from the *Salvia sclarea* transcriptome data set (Hao et al., 2012b). Msatcommander (http://code.google.com/p/msatcommander/) was used to annotate SSRs. BatchPrimer3 (You et al., 2008) was employed to design PCR primers in the flanking regions of the detected SSRs, setting a minimum product size of 100 bp, a minimum primer length of 18 bp, a minimum GC content of 30%, a melting temperature between 50°C and 70°C, and a maximum melting temperature difference between primers of 8°C.

major drawbacks of SSRs is the low universality and poor transferability; that is, usually the species-specific SSR primers have to be developed. The other disadvantage of SSRs is their uncertain mutation model, which is often simplified to be the stepwise mutation model, but the actual mutation pattern might be more complicated.

NGS tool kits could provide grist for the medicinal plant phylogeography mill. Sufficiently abundant SNPs could be identified directly from the genome sequences of the model plants. Most medicinal plants lack the genomic data; therefore, two alternative strategies can be adopted. The faster and cheaper one is mining suitable SNP sites via transcriptome sequencing data sets (Hao et al., 2012b). However, the subsequent PCR primer design might not be successful, as no information about the intergenic sequences and the introns are available from the RNA-Seq data. On the other hand, large amounts of SNPs can be obtained by the simplified genome sequencing, mainly referring to RAD-Seq (Eaton and Ree, 2013) and genotyping-by-sequencing (GBS) (He et al., 2014a); although, their reproducibility and reliability warrant further improvements.

Plants of various evolutionary levels, not only higher plants, are harnessed in TCM and worldwide ethnomedicine. The caterpillar fungus *Ophiocordyceps sinensis* (Dong Chong Xia Cao in TCM) is one of the most valuable medicinal fungi in the world, and host insects of family Hepialidae (Lepidoptera) are a must to complete its life cycle. The genetic diversity and phylogeographic structures of the host insects are characterized using mtCOI (cytochrome oxidase subunit I) sequences (Quan et al., 2014). Abundant haplotype and nucleotide diversity were mainly found in the east edge of the Qinghai-Tibet Plateau (QTP), which is the diversity center or microrefuge of the host insects. The genetic variation of the host insects is negligible among 72.1% of all *O. sinensis* populations. All host insects are monophyletic except for those of four *O. sinensis* populations around Qinghai Lake. Significant phylogeographic structure was revealed for the monophyletic host insects, and the three major phylogenetic groups corresponded to specific geographical areas. The divergence of most host insects might occur at c. 3.7 Ma, shortly before the rapid uplift of the QTP. The geographical distribution and starlike network of the haplotypes implied that most host insects were derived from the relicts of a once-widespread host that subsequently became fragmented. Most host insects underwent recent demographic expansions that began c. 0.118 Ma in the late Pleistocene, suggesting that the genetic diversity and distribution of the present-day insects could be ascribed to effects of the QTP uplift and glacial advance/retreat cycles during the Quaternary ice age. These results provide valuable reference to the conservation and sustainable use of both host insects and *O. sinensis*.

Population genetics can be upgraded to population genomics using the large data set of transcriptomes from multiple species (Hollister et al., 2015). The dearth of extant asexual species might be partially caused by buildup of harmful mutations and intensified elimination risk linked with repressed recombination and segregation in these species, which was tested with a data set of 62 transcriptomes of 29 *Oenothera* species (*Onagraceae*; Hollister et al., 2015). The nonsynonymous polymorphism is more abundant than the synonymous variation within asexual species, implying relaxed purifying selection. Asexual species also displayed more transcripts with premature stop codons. The increased proportion of nonsynonymous mutations was positively associated with divergence time between sexual and asexual species. These results suggest that sex enables selection against deleterious alleles.

1.3 MECHANISMS OF SPECIES EVOLUTION AND DIVERSIFICATION

The incidence of polyploidy in land plant evolution has led to an acceleration of genome variations compared with other crown eukaryotes and is connected with key innovations in plant evolution (Cannon et al., 2015). Increasing genome resources facilitates linking genomic alterations to the origins of novel phytochemical and physiological features of medicinal plants. Ancestral gene contents for key nodes of the plant family tree are inferred (Jiao and Paterson, 2014). The ancestral WGDs (whole genome duplications) concentrating c. 319–192 million years ago expedited the diversification of regulatory genes vital to seed and flower development and were responsible for key innovations followed by the upsurge and ultimate supremacy of seed plants and flowering plants. Widespread polyploidy in angiosperms might be the major factor generating novel genes and expanding some gene families (Hofberger et al., 2013). However, most gene families lose the majority of duplicated copies in an early neutral process, and a few families are actively selected for single-copy status. It is challenging to link genome modifications to speciation, diversification, and the phytochemical and/or physiological innovations that jointly comprise biodiversity and chemodiversity. Ongoing evolutionary genomics investigations may greatly improve the resolution, enabling the identification of specific genes responsible for particular innovations. More concise understanding of plant evolution may enrich the fundamental knowledge of botanical diversity, including medicinally important traits that sustain humanity.

Case studies are important to illustrate the correlation between WGD and the diversification of secondary metabolism pathways. WGD and the tandem duplication facilitated glucosinolate pathway diversification in the mustard family (Brassicaceae) (Hofberger et al., 2013) (Fig. 1.9). In *Arabidopsis thaliana*, at least 52 biosynthetic and regulatory genes are involved in the glucosinolate biosynthesis. *Aethionema arabicum*, basal to other *Brassicales* species, harbors 67glucosinolate biosynthesis genes, most of which have the ortholog in *Ar. thaliana*, displaying the syntenic relationship. In *Ar. thaliana*, 45% of the protein-coding genes have more than one copy, while 95% of *Ar. thaliana* and 97% of *Aethionema* glucosinolate pathway genes possess multiple copies, suggesting the particular diversification of this defense pathway. The sequence alignment and phylogenetic analysis showed that the significant duplications of glucosinolate pathway genes occurred during the last common WGD event. The tandem duplication and the subsequent subfunctionalization and neofunctionalization further increase the genetic diversity and chemodiversity of the glucosinolate secondary metabolites, thus enhancing the phenotypic plasticity and adaptation. More importantly, the chemical space of the diverse secondary metabolites has great potential in drug discovery. Duplicated gene copies also explain the diversification process of terpenoids (Boutanaev et al., 2015), the largest class of plant natural products. Tracing the roots of terpene biosynthesis and diversification in plants reveals that distinct genomic mechanisms of pathway assembly have evolved in eudicots and monocots.

Besides polyploidy, allopatric divergence, climatic oscillation-based divergence, hybridization and introgression, and pollination-mediated isolation are also highlighted as the mechanisms of medicinal species evolution, especially in the hot spot areas of biodiversity, such as QTP (Wen et al., 2014) (Fig. 1.10). Rapid species diversification followed the extensive uplift of QTP and brought about numerous morphologically and phytochemically distinct species. Both morphological and metabolic phenotype innovations are apparently ecologically adaptive, and the underlying molecular mechanisms are still elusive.

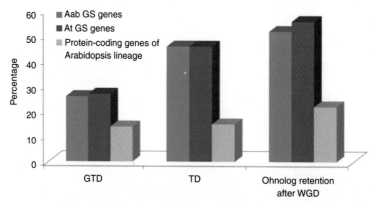

FIGURE 1.9 **Duplicate distribution among *Arabidopsis thaliana* (At) protein-coding genes compared with AtGS (glucosinolate) and *Aethionema arabicum* (Aab) GS loci, according to Hofberger et al., 2013.** Percentage of genes with retained ohnolog (clusters of dose-sensitive genes organized in functional modules), tandem duplicate (TD), and gene transposition duplicate (GTD) are shown. GS metabolic plasticity during lineage evolution arose from a combination of increased ohnolog retention and TD rates.

FIGURE 1.10 **A representative landscape of QTP.** *Source: Taken in July 2017, near Ningjin Kangsha snow mountain, Tibet, China.*

1.4 PHENOTYPE EVOLUTION AND ECOLOGY

Medicinal plants synthesize an arsenal of protective molecules, most of which are secondary metabolites and can be ingested by animals and humans, and then help them antagonize against disadvantageous environmental conditions (Sternberg et al., 2015). The epidemiological (parasite prevalence and virulence) and environmental (medicinal plant toxicity and abundance) conditions that predict the evolution of genetically fixed versus phenotypically plastic forms of animal medication can be identified using the tritrophic interaction between the monarch butterfly, its protozoan parasite, and its food plant *Asclepias* spp. as a test case (Choisy and de Roode, 2014). Analogously, in folk medicine practice people accumulate knowledge about the relative benefits (the antiparasitic/antimicrobial properties of medicinal plants) and costs (side effects of phytomedicine, the costs of searching for medicine) in ethnomedicine practice, which determine whether medication is for therapeutic use or preventive use.

Numerous botanical compounds, as the integral part of plant defense mechanisms, also bind and modify fundamental regulators of animal physiological processes in ways that enhance animal adaptation to the ever-changing environments (Kennedy, 2014). The underlying mechanism might be that animals and fungi, as heterotrophs, are capable of sensing chemical signals produced by plants and responding actively to the biotic/abiotic stress (xenohormesis) (Howitz and Sinclair, 2008). These plant-derived cues offer early warning about fading ecological circumstances, permitting the heterotrophs to get ready for misfortune when conditions are still auspicious. Plant secondary metabolites could activate the evolutionarily conserved cellular stress response and subsequently enhance the cellular adaptation to adversity in both plants themselves and animals that consume them. Xenohormesis could explain TCM pharmacological effects from an evolutionary and ecological perspective (Qi et al., 2013). Medicinal herb, microbial, and human cellular signal transduction pathways have many conserved similarities, enabling beneficial effects of botanical metabolites in humans via a process of "cross-kingdom" signaling (Kennedy, 2014).

Daodi medicinal material (geoherb) is produced in particular geographic regions, where there is defined ecological environment and cultivation pipelines (Zhao et al., 2012). The clinical efficacy of geoherb is superior to that of the same medicinal plant growing in other regions. The special medicinal features of a plant are determined by its genome, while the proper ecological conditions have major effects on the formation of geoherb. For instance, Zhejiang of China is the best production area of the geoherb Bai Shao (*Paeonia lactiflora*), where the paeoniflorin content of *P. lactiflora* roots was positively correlated with soil pH and rhizosphere bacterial diversity (Yuan et al., 2014) but negatively correlated with the organic matter content of the rhizosphere. The rhizosphere soil properties have a close relationship with the geoherbalism of *F. thunbergii* (Zhe Bei Mu in TCM) (Shi et al., 2011) and *Pa. ginseng* (Ying et al., 2012).

The section *Moutan* of the genus *Paeonia* consists of eight species that are distributed in a particular area of China from which various secondary metabolites, including monoterpenoid glucosides, flavonoids, tannins, stilbenes, triterpenoids, steroids, paeonols, and phenols, have been found. The metabolic phenotype evolves in the differentiated niche and in response to the plant-insect and plant-microbe interactions (Yuan et al., 2014), which can be used for the chemotaxonomy of the section *Moutan* (He et al., 2014). Forty-three metabolites were identified from eight species by high-performance liquid chromatography-quadrupole time of flight-mass spectrometry, including 17 monoterpenoid glucosides, 11 galloyl glucoses, five flavonoids, six paeonols, and four phenols. PCA (principal component analysis) and HCA (hierarchical cluster analysis) showed a clear separation between the species based on metabolomic similarities, and four groups were identified, which coincides with conventional classification based on the morphological and geographical distributions. *P. decomposita*, from the geoherb production region of Sichuan, China, was found to be a transition species between two subsections. According to the metabolic fingerprints, *P. ostii* (Feng Dan in TCM) and *P. suffruticosa* (Mu Dan) could be the same species. The metabolic profiles of *P. delavayi* (wild Mu Dan) were highly variable, and no significant difference was found between *P. delavayi* and *P. ludlowii* (yellow Mu Dan), implying that they either have a close evolutionary relationship or underwent the convergent evolution of the specialized metabolism. The combination of metabolomics and multivariate analyses has great potential for guiding chemotaxonomic studies of other medicinal plants (Hao et al., 2012a).

With the surge of NGS technology, it is becoming common to perform the phylogenetic study based on genomic data. However, for most medicinal plants it is not realistic to rely on the whole genome sequencing data. RAD-Seq is easily applied to nonmodel plants for which no reference genome is available (Eaton and Ree, 2013; Fig. 1.3), and it is promising for reconstructing phylogenetic relationships in evolutionarily younger clades in which sufficient numbers of orthologous restriction sites are retained across species (Rubin et al., 2012). Coincidentally, the younger clades are more likely to harbor a wider variety of secondary metabolites, as the chemodiversity often accompanies the rapid radiation and diversification. The evolutionarily young *Pedicularis* section *Cyathophora* is a systematically refractory clade of the broomrape family (Orobanchaceae). The phylogenetic inferences were performed based on the data sets of 40,000 RAD loci (Eaton and Ree, 2013). The maximum likelihood and Bayesian methods generated similar trees that had two major clades: a "rex-thamnophila" clade, comprised of two species and several subspecies with relatively low floral diversity and geographically widespread distributions at lower elevations, and a "superb" clade, comprised of three species with high floral diversity and isolated geographic distributions at higher altitudes. Levels of molecular divergence between subspecies in the rex-thamnophila clade are similar to those between species in the superba clade. The significant introgression among nearly all taxa in the rex-thamnophila clade was identified, while no gene flow was detected between clades or among taxa within the superba clade. The geographic isolation, following the uplift of QTP in the Quaternary Period and the emergence of "sky islands" (He and Jiang, 2014), might be crucial in the advent of species barriers by enabling local adaptation and differentiation without the influence of homogenizing gene flow. *Pedicularis* plants are traditionally used in folk medicine. It is intriguing to study its chemotaxonomy and treat the chemodiversity and biodiversity data in a holistic approach for drug discovery and development.

Understanding which factors determine chemical diversity has the potential to shed light on plant defenses against herbivores and diseases and accelerate drug discovery. Traditionally, *Cinchona* alkaloids were the primary treatment for malaria. The genetic profiles of *Cinchona calisaya* leaf samples were generated from four plastid and ITS regions of 22 *Cin. calisaya* stands in the Yungas region of Bolivia (Maldonado et al., 2017). Climatic and soil parameters were characterized and bark samples were analyzed for content of four major alkaloids to explore the utility of evolutionary history (phylogeny) in determining variation within species under natural conditions. A significant phylogenetic signal was found for the content of quinine and cinchonidine and total alkaloid content. Climatic parameters, primarily driven by changing altitude, predicted 20.2% of the overall alkaloid variation, and geographical separation accounted for a further 9.7%. A clade of high alkaloid–producing trees spanned a narrow range between 1100 and 1350 m. The climate expressed by altitude was not a significant driver when accounting for phylogeny, suggesting that the chemical diversity is primarily driven by phylogeny. Comparisons of the relative effects of both environmental/ecological and genetic variability in determining plant chemical diversity should be performed at the genotypic level if the extensive genotypic variation in plant biochemistry is to be fully understood.

1.5 PHARMACOPHYLOGENY VS. PHARMACOPHYLOGENOMICS

1.5.1 Concept

Diverse new terms are emerging in the genomic era, such as phylogenomics, pharmacophylogenomics, and phylotranscriptomics, which are somewhat overlapping with pharmaphylogeny (pharmacophylogeny/pharmacophylogenetics) (Hao et al., 2014a; Hao and Xiao, 2017). Phylogenomics is the crossing of evolutionary biology and genomics, in which genome data are utilized for evolutionary reconstructions. Pharmacophylogeny, advocated by Pei-gen Xiao since the 1980s (Xiao, 1980; Peng et al., 2006), focuses on the phylogenetic relationship of medicinal plants and aims to foster the sustainable utilization of TCM resources and is thus nurtured by molecular phylogeny, chemotaxonomy, ethnopharmacology, and bioactivity studies (Fig. 1.11). Phylogenomics can be integrated into the pipeline of drug discovery and development and can extend the field of pharmaphylogeny at the omic level, thus the concept of pharmacophylogenomics, initially emphasizing the genomic analysis of the evolutionary history of drug targets (Searls, 2003), could be redefined as an upgraded version of pharmacophylogeny.

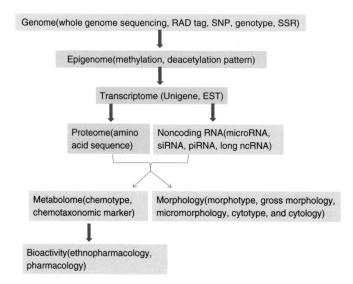

FIGURE 1.11 **Omics data that could be used in the pharmacophylogeny inference.** *EST*, Expressed sequence tag; *RAD*, restriction site associated DNA; *SNP*, single nucleotide polymorphism; *SSR*, simple sequence repeat.

The new conceptual framework of pharmacophylogenomics highlights the comprehensive analysis of the evolutionary history of medicinal organisms (especially the predominant medicinal plants), in particular the congruence and conflict between molecular phylogeny and chemotaxonomy (Day et al., 2014; Hao et al., 2013a–c; He et al., 2014), the orthology and paralogy relationships (Yang et al., 2015; Yang and Smith, 2014), the degree and landscape of evolutionary transformation they have undergone, and the involved evolving metabolic pathways and regulatory networks. More specifically, first, the tree of life of different scales can be constructed based on the genomic information to determine the phylogenomic relationship of medicinal plant groups, for example, the relationship between geoherb (higher content of medicinal compounds and better therapeutic efficacy) (Zhao et al., 2012) and nongeoherb populations; second, the genomic data, in particular those from the RAD-Seq or GBS, can be exploited to estimate the divergence time, reconstruct the geographic distribution, and infer the origin and the spatial distribution pattern of extant medicinal plants/geoherbs (Hao et al., 2014a, 2015d,e) (Fig. 1.3); third, within the context of the temporal tree, the ecological factors, environmental attributes, and evolutionarily innovative traits can be combined to dissect the diversification process and mechanism of medicinal plants; fourth, the origin and structure of the phylogenetic diversity of medicinal plants can be revealed; the diversity of medicinal compounds can be dissected based on biodiversity to promote drug discovery via high-throughput screening (Hao et al., 2015d,e); and last but not least, the dynamic alteration of the medicinal plant diversity can be predicted, and then the appropriate conservation and development strategies can be developed.

1.5.2 Molecular Phylogeny and Therapeutic Utility

During evolution, plants develop tactics of chemical defenses, leading to the evolution of specialized metabolites with diverse potencies. A correlation between phylogeny and biosynthetic pathways could offer a predictive approach enabling more efficient selection of alternative and/or complementary plants for guaranteeing clinical use and novel lead discovery. This relationship has been rigorously tested, and the potential predictive power is subsequently validated (Rønsted et al., 2012). A phylogenetic hypothesis was proposed for the medicinal plant subfamily Amaryllidoideae (Amaryllidaceae) based on parsimony and Bayesian analysis of nuclear, cp, and mt DNA sequences of more than 100 species (Rønsted et al., 2012). It is intriguing to test whether alkaloid diversity and activity in bioassays related to the central nervous system are significantly correlated with molecular phylogeny. Evidence for a significant phylogenetic signal in these traits is found, albeit the effect is not strong. Several genera are nonmonophyletic, highlighting the importance of using phylogeny for understanding character distribution. Lack of congruence between specialized metabolism and molecular phylogeny is not unusual (Day et al., 2014; Hao et al., 2013a–c, 2015c; He et al., 2014), and the prominent factor is convergent evolution.

At least 20,654 phytochemicals from 16,102 plants are associated with 1592 human disease phenotypes. Only 8% of 36,932 phytochemicals are localized in certain parts of the taxonomy (Jensen et al., 2014). For example, the genus *Lens* (*Fabaceae*, Fabales), which includes lentils, and *Citrus* (*Rutaceae*, Sapindales), which includes orange, contain 60 out of 562 compounds and 42 out of 214 compounds, respectively, that are not found anywhere else on the taxonomy. On the other hand, compounds such as β-sitosterol, palmitic acid, and catechin are spread all over the taxonomy. It is possible that the synthesis of small compounds in plants is mainly defined by short-term regulatory rather than long-term evolutionary adaptation to the environment.

Alkaloid diversity and *in vitro* inhibition of acetylcholinesterase and binding to the serotonin reuptake transporter are significantly correlated with phylogeny (Rønsted et al., 2012), illustrating the validity of pharmacophylogeny, which has implications for the use of molecular phylogenies to interpret chemical evolution and biosynthetic pathways to select candidate taxa for lead discovery and to make policies regarding therapeutic use and conservation priorities. The phylogenetic classification was also taken into account in evaluating colchicine and related phenethylisoquinoline alkaloids of the family Colchicaceae (Larsson and Rønsted, 2014). The evolutionary reasoning can be utilized for inferring mechanisms in, for example, drug resistance in cancer and infections, which could exemplify how thinking about evolution influences the plant selection in drug lead discovery and how phylogeny knowledge may be used to evaluate predicted biosynthetic pathways.

The common practice of grouping medicinal plant uses into standardized categories, in terms of systems of the human body, may restrict the relevance of phylogenetic predictions (Ernst et al., 2016), as they only poorly reflect biological responses to the botanical drug. Medicinal plant uses should be interpreted from a perspective of the biological response, revealing different phylogenetic patterns of presumed underlying bioactivity. In the cosmopolitan and pharmaceutically highly relevant genus *Euphorbia* (Fig. 1.12), identifying anti-inflammatory uses highlighted a greater phylogenetic diversity and number of potentially promising species than standardized categories, which allow for a more targeted approach for future phylogeny-guided drug discovery at an early screening stage, possibly resulting in higher discovery rates of novel chemistry with functional bioactivity.

The correlation between the plant molecular phylogeny and therapeutic utility has been suggested (Grace et al., 2015; Leonti et al., 2013; Saslis-Lagoudakis et al., 2012). For instance, bulky, juicy leaves representative of medicinal aloes (Aloeaceae, Liliales) rose during the most recent expansion ~10 million years ago and are powerfully associated with

FIGURE 1.12   *Euphorbia stracheyi* of **Euphorbiaceae.** *Source: Taken in Shangri-La Alpine Botanical Garden, Yunnan, China.*

the molecular phylogeny and correlated to the probability of a species being used for therapy (Grace et al., 2015). A noteworthy, though feeble, phylogenetic hint is apparent in the remedial uses of aloes, signifying that their pharmaceutical properties do not arise stochastically across the clades of the evolutionary tree. The taxonomic clades included in native pharmacopoeias are indeed associated with certain disease groups, and ecology and angiosperm phylogeny, which could be the alternative and/or complementary for chemical kinship and convergence, to a certain extent explain the observed preference of the therapeutic use. For instance, evolutionarily related plants from New Zealand, Nepal, and the Cape of South Africa are used to combat diseases of the same therapeutic spaces (Saslis-Lagoudakis et al., 2012), which powerfully shows the self-determining discovery of the botanical value. A considerably greater fraction of recognized medicinal plants is present in these phylogenetic groups than in haphazard samples, suggesting that screening work be focused on a subgroup of traditionally used plants that are more affluent in medicinal molecules. The phylogenetic/phylogenomic cross-cultural evaluations would invigorate the use of old-fashioned knowledge in bioprospecting. Statistical analysis of the ethnopharmacology data based on Chinese medicinal plants of Magnoliidae (Xiao et al., 1986), Hamamelidae, and Caryophyllidae (Xiao et al., 1989) has been performed to summarize the distribution pattern of ethnomedicine uses across three subclasses. These nearly extinct traditional knowledges, collected nationwide during the TCM resources survey, lay a foundation for further quantitative correlation studies of molecular phylogeny and therapeutic efficacy.

1.5.3 Chinese Medicinal Plants

Chinese medicinal material resource is the foundation of the development of TCM. In the study of sustainable utilization of TCM resources, adopting innovative theories and methods to find new TCM resources is one of the hot spots and is always highlighted (Hao et al., 2015d,e). Pharmacophylogeny interrogates the phylogenetic relationship of medicinal organisms (especially medicinal plants), as well as the intrinsic correlation of morphological taxonomy, molecular phylogeny, chemical constituents, and therapeutic efficacy (ethnopharmacology and pharmacological activity). This new discipline may have the power to change the way we utilize medicinal plant resources and develop plant-based drugs. Phylogenomics can be integrated into the flowchart of drug discovery and development and extends the field of pharmacophylogeny at the omic level. Analogously, phyloproteomics can be used in the proteome-based phylogeny study (Villar et al., 2013); phyloepigenomics could be used to examine the evolutionary relationship at the epigenomic level (Martin et al., 2011); and phylometagenomics is applicable in the exploration of medicinal plant-associated microbiota (Brindefalk et al., 2011).

The theory of TCM's property (Yao Xing in Chinese) is the core part of TCM theory. Meridian (Gui Jing in Chinese) theory is an important part of TCM's property theory. The medicine is selected according to the meridian to which it belongs, which improves the accuracy and pertinence of clinical drug use and is of great significance for guiding the clinical prescription of Chinese medicine. To study the association and distribution of TCMs with different meridian tropism on the phylogenetic tree, and to provide a basis for the interpretation and evaluation of TCM meridian tropism, 2435 herbs and a related 3044 species were screened (Li et al., 2017). Among species of the Viridiplantae, up to 1151 species belong to the liver meridian; among species of the Spermatophyta, up to 1109 species are classified into the liver meridian; among monocots, up to 110 species belong to the lung meridian. The association rules for the same meridian tropism were distributed on the same branch or nearby branch of the tree. For example, *Taxus* is related with the kidney meridian, *Caprifoliaceae* and *Rubia* have a close relationship with the liver meridian, and *Punica* is related to the colorectal meridian. There is a close relationship between Yao Xing and the phylogeny. TCMs with close phylogenetic relationships may have the same meridian tropism, which provides a new index and reference for prediction and evaluation of Yao Xing, selection and compatibility, and clinical application of new TCMs. Marine Chinese medicine (MCM) is one important part of TCM. The exploration of marine organism resources is a good base for the development of MCM, and the evaluation of Yao Xing of the new MCM resource is a key issue of the clinic application of MCM. A total of 613 MCMs and related 1091 marine species were screened (Fu et al., 2015). The MCMs of similar Yao Xing cluster on the phylogenetic tree. The MCMs from the same taxonomic family are more likely to have the same Yao Xing. For example, the marine plantae Chlorophyta, Florideophyceae, and Phaeohpyceaeare related with cold nature (Han Xing in Chinese), while the marine animalia Decapoda, Malacostraca, and Arthropoda are closely related with hot nature (Re Xing). The neutral nature (Ping Xing) was shown in Squamata. These results implied the close relationship between Yao Xing and the phylogeny, which can be used in predicting and evaluating Yao Xing of new MCM.

Many medicinally important tribes and genera, such as *Clematis* (Hao et al., 2013b), *Pulsatilla*, *Anemone*, *Cimicifugeae* (Hao et al., 2013d), *Nigella*, *Delphinieae* (Jabbour and Renner 2012) (Fig. 1.13), *Adonideae*, *Aquilegia*, *Thalictrum* (Zhu and Xiao 1991), and *Coptis*, belong to Ranunculaceae family. Chemical components of this family include several representative groups: BIA, ranunculin, triterpenoid saponin and diterpene alkaloid, etc. Ranunculin and magnoflorine were found to coexist in some genera. Other medicinal compounds also show some intriguing distribution patterns in 5 subfamilies and 10 tribes (Hao et al., 2015d,e; Wang et al., 2009). Compared with other plant families, Ranunculaceae has the most species recorded in China Pharmacopoeia (CP) version 2015. However, many *Ranunculaceae* species, for example, those that are closely related to CP species, as well as those endemic to China,

FIGURE 1.13   *Delphinium yuanum* of **Ranunculaceae.** *Source: Taken in Shangri-La Alpine Botanical Garden, Yunnan, China.*

have not been investigated in depth (Hao et al., 2015d,e), and their phylogenetic relationship and potential in medicinal use remain elusive. As such, it is proposed to select Ranunculaceae to exemplify the utility of pharmacophylogenomics and to elaborate the new concept empirically. It is argued that phylogenetic and evolutionary relationships of medicinally important tribes and genera within Ranunculaceae could be elucidated at the genomic, transcriptomic, and metabolomic levels, from which the intrinsic correlation between medicinal plant genotype and metabolic phenotype, and between genetic diversity and chemodiversity of closely related taxa, could be revealed. This proof-of-concept study would enrich the intension and spread the extension of pharmacophylogeny, promote the development of TCM genomics, and boost the sustainable development of Chinese medicinal plant resources.

1.5.4 Aconitum

Aconitum (Delphinieae, Ranunculaceae) has more than 300 species in the temperate regions of the Northern Hemisphere, over half of which are distributed in China. This genus has two subgenera, *Lycoctonum* and *Aconitum* (Hao et al., 2014a) (Fig. 1.14). The southwest China, particularly Hengduan Mountains, is the most important center of origin and diversity of the genus. Many *Aconitum* species are used as poisonous and medicinal plants. Their anticancer activity, cardioactive effect, analgesic activity, antiinflammatory activity, effect on energy metabolism, and antimicrobial and pesticidal activities—mainly due to the abundant diterpenoid alkaloids—are well-archived (Hao et al., 2013c). The correlation between molecular phylogeny, chemical components, and medicinal uses in *Aconitum* is notable (Hao et al., 2013c; Xiao et al., 2006). Diterpenoid alkaloids belong to four skeletal types: C_{18}, C_{19}, C_{20} and bisditerpenoid alkaloids. The subgenera *Lycoctonum* contain mainly the C_{18} (lappaconine-type and ranaconine-type) and C19 (lycoctonine-type). Roots of *Lycoctonum* plants exhibit a relatively lower toxicity and have been used to combat against rheumatism, pains, irregular menstruation, etc. This subgenus worths a more detailed phytochemical investigation for the new lead discovery and development. The Chinese taxa of section *Aconitum* (predominant in subgenera *Aconitum*) are morphologically divided into 11 series.

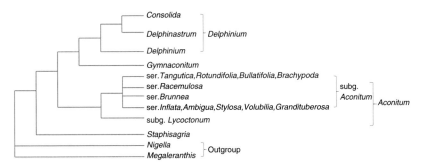

FIGURE 1.14 **Cladogram of the Ranunculaceae tribe Delphinieae, according to Hao et al., 2013c and Wang et al., 2013.** *Gymnaconitum* and *Staphisagria* were regarded as the subgenus of *Aconitum* and *Delphinium* respectively. *Consolida*, usually treated as an independent genus, could belong to the genus *Delphinium*.

The series *Tangutica* and *Rotundifolia* have abundant lactone-type C_{19}-diterpenoid alkaloids (Xiao et al., 2006), which can be considered as the chemical markers of these two series. The toxicity of their roots is much lower than those of the series *Bullatifolia* and *Brachypoda*, and the whole plants are traditionally used in Western China for high fever. The highly toxic aconitine-type diester C_{19} dominate in the series *Stylosa* (Da Wu Tou in TCM). The series *Ambigua* contains mainly the aconitine-type C_{19} with anisoyloxy residues, indicating its close affinity to series *Stylosa*. Several species of the series *Volubilia* have the highly advanced 15-hydroxyl aconitine-type C_{19}, indicating their possible kinship to series *Inflata*, which harbors two of the most widely used TCM/CP aconite species *A. carmichaeli* (Wu Tou in TCM) and *A. kusnezoffii* (Bei Wu Tou in TCM). *A. hemsleyanum* of the series *Volubilia*, as well as many other *Aconitum* herbs, is morphologically polymorphic and displays a substantial interpopulational phytochemical variation. The series *Grandituberosa* is more toxic than series *Inflata*, *Volubilia*, and *Ambigua*.

The morphology-based 11-series classification of section *Aconitum*, subgenus *Aconitum*, is not supported by chemotaxonomy and molecular phylogeny. Molecular phylogeny based on nr and cp DNA sequences divided the nine morphologically similar series into two clusters, which is bolstered by the chemotaxonomic data. Series *Rotundifolia* and *Brachypoda*, as well as *Tangutica* and *Bullatifolia*, are not monophyletic groups and cluster together. The series *Ambigua*, *Stylosa*, *Volubilia*, and *Inflata* are also not monophyletic but are intermingled on the phylogenetic tree (Hao et al., 2013c, 2015c). Series *Grandituberosa* is closer to *Volubilia* than to other series. *A. brunneum* and *A. racemulosum* are distinct in both molecular phylogeny and chemotaxonomy. *Gymnaconitum*, previously regarded as a subgenus of *Aconitum* but distinct phytochemically from *Aconitum*, is between *Aconitum* and the genus *Delphinium* in molecular phylogeny (Fig. 1.7) and now treated as an independent genus (Wang et al., 2013). The high possibility of deriving novel chemical entities from untapped species in traditionally used drug-productive genera/families has been suggested (Zhu et al., 2012). New genomic technologies that discover hidden gene clusters (Boutanaev et al., 2015), pathways, and interspecific crosstalk allow the unearthing of innovative natural products (Zhu et al., 2011). It is critical to assimilate the omic platforms into *Aconitum* studies for both the sustainable utilization of *Aconitum* pharmaceutical resources and finding novel compounds with potential clinical utility and less toxicity.

1.6 CONCLUSION AND PROSPECTS

The trend of integrating genomics and evolution into studies of medicinal plants is perceivable, and therefore, it is time to summarize the current progress in the relevant fields to make full use of evolutionary biology/genomics and revolutionize the roadmap of medicinal plant inquiries. Plants included in the same node of a phylogeny commonly have similar food and medicinal uses, which is called "ethnobotanical convergence" (Garnatje et al., 2017). This phylogenetic approach, together with the "omics" revolution, shows how combining modern technologies with traditional ethnobotanical knowledge could be used to identify potential new applications of plants. This chapter gives a brief analysis of the association and the distinguished features of the multifaceted medicinal plant evolution and genomics studies in the context of the plant-based drug discovery and the sustainable utilization of traditional pharmaceutical resources. A phylogenetic approach along with transcriptomics and other omics has value for understanding the evolution of medicinal plants, and a stronger case for the utility of these methods for future identification of useful genes and/or taxa for medicinal use is warranted.

The welfare of the global human population rests on provisioning services delivered by 12% of the Earth's ~400,000 plant species. People preferentially use large, widespread species rather than small, narrow-ranged species (Cámara-Leret al., 2017), but the latter potentially contain medicinally important compounds. Relying on plant size and availability may prevent the optimal realization of wild-plant services, since ecologically rare but chemically important clades cannot be overlooked. The research paradigms of medicinal plant genome and evolution are evolving, and the use of omics techniques is reshaping the landscape of this dynamic field. Genomics, transcriptomics, proteomics, metabolomics, and other omics platforms generate formidably big data, which cannot be used efficiently in probing plant genome and evolution without the aid of advancing bioinformatics. Medicinal plants evolve new traits to adapt to the changing environments and pave the road for themselves to a better life, while both hypothesis-driven and big data-driven, studies integrate herbal technology, biotechnology, and information technology to pave the road for human to a healthier life.

References

Ariyadasa, R., Stein, N., 2012. Advances in BAC-based physical mapping and map integration strategies in plants. J. Biomed. Biotechnol., 184854.

Barrett, C.F., Specht, C.D., Leebens-Mack, J., et al., 2014. Resolving ancient radiations: can complete plastid gene sets elucidate deep relationships among the tropical gingers (Zingiberales)? Ann. Bot. 113, 119–133.

Basavarajappa, H.D., Lee, B., Fei, X., et al., 2014. Synthesis and mechanistic studies of a novel homoisoflavanone inhibitor of endothelial cell growth. PLoS One 9 (4), e95694.

Boutanaev, A.M., Moses, T., Zi, J., 2015. Investigation of terpene diversification across multiple sequenced plant genomes. Proc. Natl. Acad. Sci. USA 112 (1), E81–E88.

Brindefalk, B., Ettema, T.J., Viklund, J., et al., 2011. A phylometagenomic exploration of oceanic alphaproteobacteria reveals mitochondrial relatives unrelated to the SAR11 clade. PLoS One 6, e24457.

Cai, Z., Liu, H., He, Q., et al., 2014. Differential genome evolution and speciation of *Coixlacryma-jobi* L. and *Coix aquatica* Roxb. hybrid guangxi revealed by repetitive sequence analysis and fine karyotyping. BMC Genomics (15), 1025.

Cai, J., Liu, X., Vanneste, K., et al., 2015. The genome sequence of the orchid *Phalaenopsis equestris*. Nat. Genet. 47 (1), 65–72.

Cámara-Leret, R., Faurby, S., Macía, M.J., et al., 2017. Fundamental species traits explain provisioning services of tropical American palms. Nat. Plants 3, 16220.

Cannon, S.B., McKain, M.R., Harkess, A., et al., 2015. Multiple polyploidy events in the early radiation of nodulating and nonnodulating legumes. Mol. Biol. Evol. 32 (1), 193–210.

Cao, H., Hu, H., Colagiuri, B., et al., 2011. Medicinal cupping therapy in 30 patients with fibromyalgia: a case series observation. Forsch. Komplementmed. 18 (3), 122–126.

Chalhoub, B., Denoeud, F., Liu, S., et al., 2014. Early allopolyploid evolution in the post-Neolithic *Brassica napus* oilseed genome. Science 345 (6199), 950–953.

Chen, S.L., Sun, Y., Xu, J., et al., 2010. Strategies of the study on Herb Genome Program. Acta Pharm. Sin. 45 (7), 807–812.

Chen, S.L., Luo, H., Li, Y., et al., 2011. 454 EST analysis detects genes putatively involved in ginsenoside biosynthesis in *Panax ginseng*. Plant Cell. Rep. 30 (9), 1593–1601.

Chen, W., Kui, L., Zhang, G., et al., 2017. Whole-genome sequencing and analysis of the Chinese herbal plant *Panax notoginseng*. Mol. Plant. 10 (6), 899–902.

Choisy, M., de Roode, J.C., 2014. The ecology and evolution of animal medication: genetically fixed response versus phenotypic plasticity. Am. Nat. 184(S1), S31–S46.

Clevenger, J., Chavarro, C., Pearl, S.A., et al., 2015. Single nucleotide polymorphism identification in polyploids: a review, example, and recommendations. Mol. Plant. 8 (6), 831–846.

Cviková, K., Cattonaro, F., Alaux, M., et al., 2015. High-throughput physical map anchoring via BAC-pool sequencing. BMC Plant Biol. 15 (1), 99.

Day, P.D., Berger, M., Hill, L., et al., 2014. Evolutionary relationships in the medicinally important genus *Fritillaria* L. (Liliaceae). Mol. Phylogenet. Evol. 80, 11–19.

Delaux, P.M., Varala, K., Edger, P.P., et al., 2014. Comparative phylogenomics uncovers the impact of symbiotic associations on host genome evolution. PLoS Genet. 10 (7), e1004487.

Denoeud, F., Carretero-Paulet, L., Dereeper, A., et al., 2014. The coffee genome provides insight into the convergent evolution of caffeine biosynthesis. Science 345 (6201), 1181–1184.

Dutartre, L., Hilliou, F., Feyereisen, R., 2012. Phylogenomics of the benzoxazinoid biosynthetic pathway of Poaceae: gene duplications and origin of the Bx cluster. BMC Evol. Biol. 12, 64.

Eaton, D.A., Ree, R.H., 2013. Inferring phylogeny and introgression using RADseq data: an example from flowering plants (Pedicularis: Orobanchaceae). Syst. Biol. 62 (5), 689–706.

Ernst, M., Saslis-Lagoudakis, C.H., Grace, O.M., et al., 2016. Evolutionary prediction of medicinal properties in the genus *Euphorbia* L. Sci. Rep. 6, 30531.

Fu, X.J., Wang, Z.G., Wang, C.Y., et al., 2015. The distribution and association relationships of marine Chinese medicine with different nature in the phylogenetic tree of marine organisms. World Sci. Technol./Mode Trad. Chin. Med. Mate Med. 17, 2189–2196.

Garnatje, T., Peñuelas, J., Vallès, J., 2017. Ethnobotany, phylogeny, and 'omics' for human health and food security. Trends Plant. Sci. 22 (3), 187–191.

Grace, O.M., Buerki, S., Symonds, M.R., et al., 2015. Evolutionary history and leaf succulence as explanations for medicinal use in aloes and the global popularity of aloe vera. BMC Evol. Biol. 15, 29.

Guichoux, E., Lagache, L., Wagner, S., et al., 2011. Current trends in microsatellite genotyping. Mol. Ecol. Resour. 11 (4), 591–611.

Han, F.M., Peng, Y., Xu, L., et al., 2014. Identification, characterization, and utilization of single copy genes in 29 angiosperm genomes. BMC Genomics 15, 504.

Hao, D.C., Xiao, P.G., 2017. An Introduction of Plant Pharmacophylogeny. Chemical Industry Press, Beijing.

Hao, D.C., Chen, S.L., Xiao, P.G., et al., 2010. Authentication of medicinal plants by DNA-based markers and genomics. Chin. Herb. Med. 2 (4), 250–261.

Hao D C, Ge, G., Xiao, P.G., et al., 2011a. The first insight into the tissue specific taxus transcriptome via Illumina second generation sequencing. PLoS One 6 (6), e21220.

Hao, D.C., Yang, L., Xiao, P.G., 2011b. The first insight into the Taxus genome via fosmid library construction and end sequencing. Mol. Genet. Genomics 285 (3), 197–205.

Hao, D.C., Xiao, P.G., Ge, G.B., et al., 2012a. Biological, chemical, and omics research of *Taxus* medicinal resources. Drug Dev. Res. 73, 477–486.

Hao, D.C., Chen, S.L., Xiao, P.G., et al., 2012b. Application of high-throughput sequencing in medicinal plant transcriptome studies. Drug Dev. Res. 73, 487–498.

Hao, D.C., Ma, P., Mu, J., et al., 2012c. De novo characterization of the root transcriptome of a traditional Chinese medicinal plant *Polygonum cuspidatum*. Sci. China Life Sci. 55 (5), 452–466.

Hao, D.C., Xiao, P.G., Peng, Y., et al., 2012d. Evaluation of the chloroplast barcoding markers by mean and smallest interspecific distances. Pak. J. Bot. 44 (4), 1271–1274.

Hao, D.C., Gu, X.J., Xiao, P.G., et al., 2013a. Phytochemical and biological research of *Fritillaria* medicine resources. Chin. J. Nat. Med. 11 (4), 330–344.

Hao, D.C., Gu, X.J., Xiao, P.G., 2013b. Chemical and biological research of *Clematis* medicinal resources. Chin. Sci. Bull. 58, 1120–1129.

Hao, D.C., Gu, X.J., Xiao, P.G., 2013c. Recent advances in the chemical and biological studies of *Aconitum* pharmaceutical resources. J. Chin. Pharm. Sci. 22 (3), 209–221.

Hao, D.C., Gu, X.J., Xiao, P.G., 2013d. Recent advance in chemical and biological studies on *Cimicifugeae* pharmaceutical resources. Chin. Herb. Med. 5 (2), 81–95.

Hao, D.C., Xiao, P.G., Liu, M., et al., 2014a. Pharmaphylogeny vs. pharmacophylogenomics: molecular phylogeny, evolution and drug discovery. Yao Xue Xue Bao 49 (10), 1387–1394.

Hao, D.C., Chen, S.L., Osbourn, A., et al., 2015a. Temporal transcriptome changes induced by methyl jasmonate in *Salvia sclarea*. Gene 558 (1), 41–53.

Hao, D.C., Vautrin, S., Song, C., et al., 2015b. The first insight into the *Salvia* (Lamiaceae) genome via BAC library construction and high-throughput sequencing of target BAC clones. Pak. J. Bot. 47 (4), 1347–1357.

Hao, D.C., Gu, X.J., Xiao, P.G., 2015c. Medicinal Plants: Chemistry, Biology and Omics, first ed. Elsevier-Woodhead, Oxford, ISBN 9780081000854.

Hao, D.C., Xiao, P.G., Liu, L.W., et al., 2015d. Essentials of pharmacophylogeny: knowledge pedigree, epistemology and paradigm shift. China J. Chin. Mat. Med. 40 (13), 1–8.

Hao, D.C., Xiao, P.G., Ma, H.Y., et al., 2015e. Mining chemodiversity from biodiversity: pharmacophylogeny of medicinal plants of the *Ranunculaceae*. Chin. J. Nat. Med. 13 (7), 507–520.

He, K., Jiang, X.L., 2014. Sky islands of southwest China I. An overview of phylogeographic patterns. Chin. Sci. Bull. 59, 585–597.

He, C.N., Peng, B., Dan, Y., et al., 2014. Chemical taxonomy of tree peony species from China based on root cortex metabolic fingerprinting. Phytochemistry 107, 69–79.

He, J., Zhao, X., Laroche, A., et al., 2014a. Genotyping-by-sequencing (GBS), an ultimate marker-assisted selection (MAS) tool to accelerate plant breeding. Front. Plant Sci. 5, 484.

Henriquez, C.L., Arias, T., Pires, J.C., et al., 2014. Phylogenomics of the plant family Araceae. Mol. Phylogenet Evol. 75, 91–102.

Hofberger, J.A., Lyons, E., Edger, P.P., et al., 2013. Whole genome and tandem duplicate retention facilitated glucosinolate pathway diversification in the mustard family. Genome Biol. Evol. 5, 2155–2173.

Hofberger, J.A., Zhou, B., Tang, H., et al., 2014. A novel approach for multi-domain and multi-gene family identification provides insights into evolutionary dynamics of disease resistance genes in core eudicot plants. BMC Genomics 15, 966.

Hollister, J.D., Greiner, S., Wang, W., et al., 2015. Recurrent loss of sex is associated with accumulation of deleterious mutations in *Oenothera*. Mol. Biol. Evol. 32 (4), 896–905.

Howitz, K.T., Sinclair, D.A., 2008. Xenohormesis: sensing the chemical cues of other species. Cell 133 (3), 387–391.

Hsu, C.Y., Huang, P.L., Chen, C.M., et al., 2012. Tangy scent in *Toona sinensis* (Meliaceae) leaflets: isolation, functional characterization, and regulation of TsTPS1 and TsTPS2, two key terpene synthase genes in the biosynthesis of the scent compound. Curr. Pharm. Biotechnol. 13 (15), 2721–2732.

Jabbour, F., Renner, S.S., 2012. A phylogeny of *Delphinieae* (Ranunculaceae) shows that *Aconitum* is nested within *Delphinium* and that Late Miocene transitions to long life cycles in the Himalayas and Southwest China coincide with bursts in diversification. Mol. Phylogenet. Evol. 62 (3), 928–942.

Jensen, K., Panagiotou, G., Kouskoumvekaki, I., 2014. Integrated text mining and chemoinformatics analysis associates diet to health benefit at molecular level. PLoS Comput. Biol. 10 (1), e1003432.

Jiao, Y., Paterson, A.H., 2014. Polyploidy-associated genome modifications during land plant evolution. Philos. Trans. R. Soc. Lond. B. Biol. Sci. 369 (1648)doi: 10.1098/rstb.2013.0355.

Jiao, J., Lü, G., Liu, X., et al., 2011a. Reduction of blood lead levels in lead-exposed mice by dietary supplements and natural antioxidants. J. Sci. Food Agric. 91 (3), 485–491.

Jiao, Y., Wickett, N.J., Ayyampalayam, S., et al., 2011b. Ancestral polyploidy in seed plants and angiosperms. Nature 473 (7345), 97–100.

Kaliswamy, P., Vellingiri, S., Nathan, B., et al., 2015. Microsatellite analysis in the genome of *Acanthaceae*: an *in silico* approach. Pharmacogn. Mag. 11 (41), 152–156.

Karimi A, Hadian, J., Farzaneh, M., et al., 2015. Evaluation of genetic variability, rust resistance and marker-detection in cultivated *Artemisia dracunculus* from Iran. Gene 554 (2), 224–232.

Kennedy, D.O., 2014. Polyphenols and the human brain: plant "secondary metabolite" ecologic roles and endogenous signaling functions drive benefits. Adv. Nutr. 5 (5), 515–533.

Kim, S., Park, M., Yeom, S.I., et al., 2014. Genome sequence of the hot pepper provides insights into the evolution of pungency in *Capsicum* species. Nat. Genet. 46 (3), 270–278.

Koenig, D., Jiménez-Gómez, J.M., Kimura, S., et al., 2013. Comparative transcriptomics reveals patterns of selection in domesticated and wild tomato. Proc. Natl. Acad. Sci. USA 110 (28), E2655–E2662.

Koide, C.L., Collier, A.C., Berry, M.J., et al., 2011. The effect of bamboo extract on hepatic biotransforming enzymes—findings from anobese-diabetic mouse model. J. Ethnopharmacol. 133 (1), 37–45.

Larsson, S., Rønsted, N., 2014. Reviewing Colchicaceae alkaloids—perspectives of evolution on medicinal chemistry. Curr. Top. Med. Chem. 214 (2), 274–289.

Leonti, M., Cabras, S., Castellanos, M.E., et al., 2013. Bioprospecting: evolutionary implications from a post-olmec pharmacopoeia and the relevance of widespread taxa. J. Ethnopharmacol. 147 (1), 92–107.

Li, X.K., Wang, B., Zheng, Y.C., et al., 2014. Molecular characters of *Schisandra chinensis* with white fruit by RAPD and ISSR makers. Zhong Yao Cai. 37 (4), 568–572.

Li, M.M., Kasun, M., Yan, L., et al., 2015. Genetic involvement of *Camellia taliensis* in the domestication of *C. sinensis* var. *assamica* (Assimica Tea) revealed by nuclear microsatellite markers. Plant Diversity Resour. 37 (1), 29–37.

Li, J.Y., Fu, X.J., Li, X.B., et al., 2017. Association relationships of traditional Chinese medicine with different meridian tropism in phylogenetic tree. Chin. J. Expe. Trad. Med. Form 23 (12), 194–200.

Liu, L., Ma, X., Wei, J., et al., 2011. The first genetic linkage map of Luohanguo (*Siraitia grosvenorii*) based on ISSR and SRAP markers. Genome 54 (1), 19–25.

Liu W, Yin, D., Liu, J., et al., 2014. Genetic diversity and structure of *Sinopodophyllum hexandrum* (Royle) Ying in the Qinling Mountains, China. PLoS One 9 (10), e110500.

Liu, X., Liu, Y., Huang, P., et al., 2017. The genome of medicinal plant *Macleaya cordata* provides new insights into benzylisoquinoline alkaloids metabolism. Mol. Plant 10 (7), 975–989.

Luo, J., Hou, B.W., Niu, Z.T., et al., 2014. Comparative chloroplast genomes of photosynthetic orchids: insights into evolution of the Orchidaceae and development of molecular markers for phylogenetic applications. PLoS One 9 (6), e99016.

Ma, P.F., Zhang, Y.X., Zeng, C.X., et al., 2014. Chloroplast phylogenomic analyses resolve deep-level relationships of an intractable bamboo tribe Arundinarieae (poaceae). Syst. Biol. 63 (6), 933–950.

Maldonado, C., Barnes, C.J., Cornett, C., et al., 2017. Phylogeny Predicts the Quantity of Antimalarial Alkaloids within the Iconic Yellow Cinchona Bark (Rubiaceae: *Cinchona calisaya*). Front Plant Sci. 8, 391.

Malé, P.J., Bardon, L., Besnard, G., et al., 2012. Phylogenomics and a posteriori data partitioning resolve the Cretaceous angiosperm radiation Malpighiales. Proc. Natl. Acad. Sci. USA 109, 17519–17524.

Manrique-Carpintero, N.C., Tokuhisa, J.G., Ginzberg, I., et al., 2013. Sequence diversity in coding regions of candidate genes in the glycoalkaloid biosynthetic pathway of wild potato species. G3 (Bethesda) 3, 1467–1479.

Martin, D.I., Singer, M., Dhahbi, J., et al., 2011. Phyloepigenomic comparison of great apes reveals a correlation between somatic and germline methylation states. Genome Res. 21, 2049–2057.

Morales-Sánchez, V., Rivero-Cruz, I., Laguna-Hernández, G., et al., 2014. Chemical composition, potential toxicity, and quality control procedures of the crude drug of *Cyrtopodium macrobulbon*. J. Ethnopharmacol. 154 (3), 790–797.

Moses, T., Pollier, J., Shen, Q., et al., 2015. OSC2 and CYP716A14v2 catalyze the biosynthesis of triterpenoids for the cuticle of aerial organs of *Artemisia annua*. Plant Cell. 27 (1), 286–301.

Nagananda, G.S., Satishchandra, N., 2013. Antimicrobial activity of cold and hot successive pseudobulb extracts of *Flickingerianodosa* (Dalz) Seidenf. Pak. J. Biol. Sci. 16 (20), 1189–1193.

Noskov, V.N., Chuang, R.Y., Gibson, D.G., et al., 2011. Isolation of circular yeast artificial chromosomes for synthetic biology and functional genomics studies. Nat. Protoc. 6 (1), 89–96.

Pan, Y.Z., Zhang, Y.C., Gong, X., et al., 2014. Estimation of genome size of four *Panax* species by flow cytometry. Plant Diversity Resour. 36 (2), 233–236.

Park, S., Ruhlman, T.A., Sabir, J.S., et al., 2014. Complete sequences of organelle genomes from the medicinal plant *Rhazya stricta* (Apocynaceae) and contrasting patterns of mitochondrial genome evolution acrossasterids. BMC Genomics 15, 405.

Peng, Y., Chen, S.B., Liu, Y., et al., 2006. A pharmacophylogenetic study of the Berberidaceae (s.l). Acta Phytotaxo. Sin. 44 (3), 241–257.

Polashock, J., Zelzion, E., Fajardo, D., et al., 2014. The American cranberry: first insights into the whole genome of a species adapted to bog habitat. BMC Plant Biol. 14, 165.

Ponnuchamy, S., Kanchithalaivan, S., Ranjith Kumar, R., et al., 2014. Antimycobacterial evaluation of novel hybrid arylidene thiazolidine-2,4-diones. Bioorg. Med. Chem. Lett. 24 (4), 1089–1093.

Qi, H.Y., Li, L., Yu, J., 2013. Xenohormesis: understanding biological effects of traditional Chinese medicine from an evolutionary and ecological perspective. Zhongguo Zhong Yao Za Zhi 38 (19), 3388–3394.

Qian, J., Song, J., Gao, H., et al., 2013. The complete chloroplast genome sequence of the medicinal plant *Salvia miltiorrhiza*. PLoS One 8 (2), e57607.

Qin, C., Yu, C., Shen, Y., et al., 2014. Whole-genome sequencing of cultivated and wild peppers provides insights into Capsicum domestication and specialization. Proc. Natl. Acad. Sci. USA 111 (14), 5135–5140.

Quan, Q.M., Chen, L.L., Wang, X., et al., 2014. Genetic diversity and distribution patterns of host insects of caterpillar fungus *Ophiocordyceps sinensis* in the Qinghai-Tibet Plateau. PLoS One 9 (3), e92293.

Rønsted, N., Symonds, M.R., Birkholm, T., et al., 2012. Can phylogeny predict chemical diversity and potential medicinal activity of plants? A case study of Amaryllidaceae. BMC Evol. Biol. 12, 182.

Rubin, B.E., Ree, R.H., Moreau, C.S., 2012. Inferring phylogenies from RAD sequence data. PLoS One 7 (4), e33394.

Saslis-Lagoudakis, C.H., Savolainen, V., Williamson, E.M., et al., 2012. Phylogenies reveal predictive power of traditional medicine in bioprospecting. Proc. Natl. Acad. Sci. USA 109 (39), 15835–15840.

Searls, D.B., 2003. Pharmacophylogenomics: genes, evolution and drug targets. Nat. Rev. Drug Discov. 2 (8), 613–623.

Shi, J.Y., Yuan, X.F., Lin, H.R., et al., 2011. Differences in soil properties and bacterial communities between the rhizosphere and bulk soil and among different production areas of the medicinal plant *Fritillaria thunbergii*. Int. J. Mol. Sci. 12 (6), 3770–3785.

Sternberg, E.D., de Roode, J.C., Hunter, M.D., 2015. Trans-generational parasite protection associated with paternal diet. J. Anim. Ecol. 84 (1), 310–321.

Su, H.J., Hogenhout, S.A., Al-Sadi, A.M., et al., 2014. Complete chloroplast genome sequence of Omani lime (*Citrus aurantiifolia*) and comparative analysis within the rosids. PLoS One 9 (11), e113049.

Sveinsson, S., McDill, J., Wong, G.K., et al., 2014. Phylogenetic pinpointing of a paleopolyploidy event within the flax genus (*Linum*) using transcriptomics. Ann. Bot. 113 (5), 753–761.

Villar, M., Popara, M., Mangold, A.J., et al., 2013. Comparative proteomics for the characterization of the most relevant Amblyomma tick species as vectors of zoonotic pathogens worldwide. J. Proteomics 105, 204–216.

Wang, W., Lu, A.M., Ren, Y., et al., 2009. Phylogeny and classification of Ranunculales evidence from four molecular loci and morphological data. Persp. Plant. Ecol. Evol. Syst. 11, 81–110.

Wang, J., Yue, Y.D., Tang, F., et al., 2012. Screening and analysis of the potential bioactive components in rabbit plasma after oral administration of hot-water extracts from leaves of *Bambusa textilis* McClure. Molecules 17 (8), 8872–8885.

Wang, W., Liu, Y., Yu, S.X., et al., 2013. *Gymnaconitum*, a new genus of Ranunculaceae endemic to the Qinghai-Tibetan Plateau. Taxon 62, 713–722.

Wang, S., Liu, Y., Ma, L., et al., 2014. Isolation and characterization of microsatellite markers and analysis of genetic diversity in Chinese jujube (*Ziziphus jujuba* Mill). PLoS One 9 (6), e99842.

Wee, A.K., Takayama, K., Chua, J.L., et al., 2015. Genetic differentiation and phylogeography of partially sympatric species complex *Rhizophoramucronata* Lam. and *R. stylosa* Griff. using SSR markers. BMC Evol. Biol. 15, 57.

Wen, J., Zhang, J.Q., Nie, Z.L., et al., 2014. Evolutionary diversifications of plants on the Qinghai-Tibetan Plateau. Front. Genet. 5, 4.

Whittall, J.B., Syring, J., Parks, M., et al., 2010. Finding a (pine) needle in a haystack: chloroplast genome sequence divergence in rare and widespread pines. Mol. Ecol. 19(S1), 100–114.

Wickett, N.J., Mirarab, S., Nguyen, N., et al., 2014. Phylotranscriptomic analysis of the origin and early diversification of land plants. Proc. Natl. Acad. Sci. USA 111 (45), E4859–E4868.

Wu, C.S., Chaw, S.M., Huang, Y.Y., 2013. Chloroplast phylogenomics indicates that *Ginkgo biloba* is sister to cycads. Genome Biol. Evol. 5, 243–254.

Wu, K., Mo, C., Xiao, H., et al., 2015. Imperialine and verticinone from bulbs of *Fritillaria wabuensis* inhibit pro-inflammatory mediators in LPS-stimulated RAW 264.7 macrophages. Planta Med. 81 (10), 821–829.

Wysocki, W.P., Clark, L.G., Attigala, L., et al., 2015. Evolution of the bamboos (Bambusoideae; Poaceae): a full plastomephylogenomic analysis. BMC Evol. Biol. 15, 50.

Xiao, P.G., 1980. A preliminary study of the correlation between phylogeny, chemical constituents and pharmaceutical aspects in the taxa of Chinese Ranunculaceae. Acta Phytotaxo Sin 18 (2), 142–153.

Xiao, P.G., Wang, L.W., Lv, S.J., 1986. Statistical analysis of the ethnopharmacologic data based on Chinese medicinal plants by electronic computer I Magnoliidae. Chin. J. Integ. Trad. West Med. 6 (4), 253–256.

Xiao, P.G., Wang, L.W., Qiu, G.S., et al., 1989. Statistical analysis of the ethnopharmacologic data based on Chinese medicinal plants by electronic computer II *Hamamelidae* and *Caryophyllidae*. Chin. J. Integ. Trad. West. Med. 9 (7), 429–432.

Xiao, P.G., Wang, F.P., Gao, F., et al., 2006. A pharmacophylogenetic study of *Aconitum* L (Ranunculaceae) from China. Acta Phytotaxon Sin 44 (1), 1–46.

Xu, Q., Xiong, G., Li, P., et al., 2012. Analysis of complete nucleotide sequences of 12 Gossypium chloroplast genomes: origin and evolution of allotetraploids. PLoS One 7 (8), e37128.

Xu, J., Chu, Y., Liao, B., et al., 2017. *Panax ginseng* genome examination for ginsenoside biosynthesis. Gigascience 6 (11), 1–15.

Yang, Y., Smith, S.A., 2014. Orthology inference in nonmodel organisms using transcriptomes and low-coverage genomes: improving accuracy and matrix occupancy for phylogenomics. Mol. Biol. Evol. 31 (11), 3081–3092.

Yang, Y., Moore, M.J., Brockington, S.F., et al., 2015. Dissecting molecular evolution in the highly diverse plant clade Caryophyllales using transcriptome sequencing. Mol. Biol. Evol. 32 (8), 2001–2014.

Yang, J., Zhang, G., Zhang, J., et al., 2017. Hybrid de novo genome assembly of the Chinese herbal fleabane *Erigeron breviscapus*. Gigascience 6 (6), 1–7.

Yao, X., Peng, Y., Xu, L.J., et al., 2011. Phytochemical and biological studies of *Lycium* medicinal plants. Chem. Biodivers. 8 (6), 976–1010.

Ying, Y.X., Ding, W.L., Li, Y., 2012. Characterization of soil bacterial communities in rhizospheric and nonrhizospheric soil of *Panax ginseng*. Biochem. Genet. 50 (11–12), 848–859.

You, F.M., Huo, N., Gu, Y.Q., et al., 2008. BatchPrimer3: a high throughput web application for PCR and sequencing primer design. BMC Bioinformatics 9 (253).

Yuan, X.F., Peng, S.M., Wang, B.L., et al., 2014. Effects of growth years of *Paeonia lactiflora* on bacterial community in rhizosphere soil and paeoniflorin content. Zhongguo Zhong Yao Za Zhi 39 (15), 2886–2892.

Zeng, Y., Guo, L.P., Chen, B.D., et al., 2013. Arbuscular mycorrhizal symbiosis for sustainable cultivation of Chinese medicinal plants: a promising research direction. Am. J. Chin. Med. 41 (6), 1199–1221.

Zhan, Q.Q., Sui, C., Wei, J.H., et al., 2010. Construction of genetic linkage map of *Bupleurum chinense* DC. using ISSR and SSR markers. Yao Xue Xue Bao 45 (4), 517–523.

Zhang, F., Gao, Q., Khan, G., et al., 2014. Comparative transcriptome analysis of aboveground and underground tissues of *Rhodiolaalgida*, an important ethno-medicinal herb endemic to the Qinghai-Tibetan Plateau. Gene 553 (2), 90–97.

Zhang, G., Tian, Y., Zhang, J., 2015. Hybrid de novo genome assembly of the Chinese herbal plant danshen (*Salvia miltiorrhiza* Bunge). Gigascience 4, 62.

Zhang, D., Li, W., Xia, E.H., et al., 2017. The medicinal herb *Panax notoginseng* genome provides insights into ginsenoside biosynthesis and genome evolution. Mol. Plant. 10 (6), 903–907.

Zhao, Z.Z., Guo, P., Brand, E., 2012. The formation of daodi medicinal materials. J. Ethnopharmacol. 140 (3), 476–481.

Zhao, Y., Yin, J., Guo, H., et al., 2015. The complete chloroplast genome provides insight into the evolution and polymorphism of *Panax ginseng*. Front. Plant. Sci. 5, 696.

Zhu, M., Xiao, P.G., 1991. Chemosystematic studies on *Thalictrum* L. in China. Acta Phytotaxon Sin 29 (4), 358–369.

Zhu, F., Qin, C., Tao, L., et al., 2011. Clustered patterns of species origins of nature-derived drugs and clues for future bioprospecting. Proc. Natl. Acad. Sci. USA 108 (31), 12943–12948.

Zhu, F., Ma, X.H., Qin, C., et al., 2012. Drug discovery prospect from untapped species: indications from approved natural product drugs. PLoS One 7 (7), e39782.

CHAPTER

2

Mining Chemodiversity From Biodiversity: Pharmacophylogeny of Ranunculaceae Medicinal Plants

Ranunculales Medicinal Plants. http://dx.doi.org/10.1016/B978-0-12-814232-5.00002-2
Copyright © 2018 Elsevier Ltd. All rights reserved.

2.1 INTRODUCTION

Ranunculaceae, which is mostly herbs, some of which are small shrubs or woody vines, have about 60 genera and 2200 species. Plants of this family are distributed worldwide, mainly in the temperate region of the Northern Hemisphere. 42 genera and around 720 species are distributed throughout China, most of which are in the southwest mountainous region (Wu et al., 2003). The morphological feature of this family is the primitive character such as poly-carpellary. Plants of this family contain a variety of chemical components. At least 30 genera and about 220 species have medicinal use in China, among which Rhizoma coptidis, monks-hood, rhizoma cimicifugae, and caulis clematidis armandii have a long history in traditional Chinese medicine. Ranunculaceae plants are complicated in chemical composition, most of which have taxonomic implication. There are numerous reports about plant systematics, phytochemistry, chemotaxonomy, and pharmacology of this family (Xiao, 1980). With the development of science and technology, a large number of new chemical components and bioactivities, as well as therapeutic uses, have been found, presenting new evidence in che-motaxonomy. A comprehensive review of relevant research from recent years is presented here, and the correlation between phylogeny, chemical ingredients, and pharmaceutical aspects (i.e., pharmaphylogeny) of Ranunculaceae is discussed.

2.2 SYSTEMATICS OF RANUNCULACEAE

Tamura (1993) recognized five subfamilies, mainly based on chromosome and floral characteristics (Hydrastidoideae, Thalictroideae, Isopyroideae, Ranunculoideae, and Hel-leboroideae) (Tamura, 1993). Takhtajan excluded *Hydrastis* and *Glaucidium* from the family Ranunculaceae (Takhtajan, 1997). Peng et al. agree with the taxonomy proposed by Tamura and divide Ranunculaceae into six subfamilies (Peng et al., 2006a): Helleboroideae, Ranuncu-loideae, Cimicifugoideae, Isopyroideae, Thalictroideae, and Coptidoideae. Based on the com-prehensive analysis of phylogeny, chemotaxonomy, ethnopharmacology, and bioactivities, it seems plausible to treat Cimicifugoideae as a separate subfamily.

The cladogram (Fig. 2.1) proposed by the APG II system, which is based on molecular phy-logeny (Angiosperm Phylogeny and Group, 2003), consists of five subfamilies, the basal Glau-cidiaceae and Hydrastidoideae, Coptoideae, and the evolutionarily young Thalictroideae and Ranunculoideae. Wang et al. present an updated classification based on four molecular loci and 65 morphological characters (Wang et al., 2009), including cytology data and four chemotaxo-nomic markers. This classification agrees with the APG II cladogram and recognizes 10 tribes within the subfamily Ranunculoideae (Fig. 2.1). The subfamily Thalictroideae includes genera such as *Isopyrum*, *Dichocarpum*, and *Aquilegia*, which belong to the previously proposed subfam-ily Isopyroideae (Peng et al., 2006b). Pharmacophylogeny is discussed based on this system.

2.3 THE CHEMICAL COMPOSITION OF RANUNCULOIDEAE PLANTS

2.3.1 Adonideae

The tribe is divided into the genus *Adonis*, *Calathodes*, *Trollius*, and *Megaleranthis*. *Adonis* is basal to other genera. Magnoflorine (Fig. 2.2A) and cardiac glycoside (e.g., Fig. 2.2B

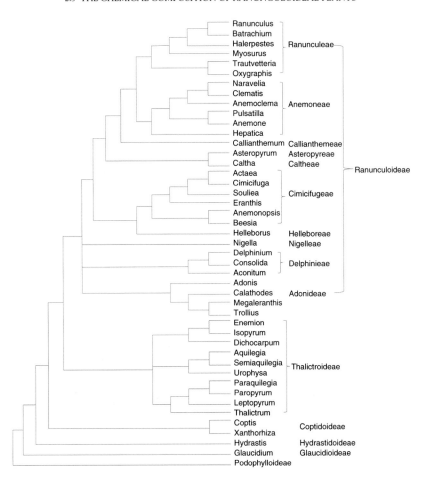

FIGURE 2.1 Ranunculaceae cladogram according to the APG system.

and C) are abundant in *Adonis* (Kubo et al., 2012a). Cardenolides adonioside A, adonioside B, tupichinolide, oleandrine, cryptostigmin II, cymarin Ns, and amurensiosides L-P were isolated from roots of *Adonis amurensis* (Kubo et al., 2015). Flavonoids and lactones are also found in *Adonis* (Dai et al., 2010). Linolenic acid (45.83%) and oleic acid (47.54%) were the most abundant fatty acids in leaves and stems of *Adonis wolgensis*, respectively (Mohadjerani et al., 2014).

Flavonoids are abundant in *Trollius*. Much more flavone C-glycosides were found than O-glycosides. 2″-O-feruloylisoswertiajaponin and (2E)-2-methyl-1-O-vaniloyl-4-β- D-glucopyranoside-2-butene, and an indole alkaloid and other flavonoids, were isolated from flowers of *Trollius chinensis* (Jie-Shi et al., 2017). In human colon adenocarcinoma cell line (Caco-2) monolayers and everted gut sacs, the dominant position of flavonoids was replaced by phenolic acids after absorption (Guo et al., 2017a). Besides flavonoids, which are usually considered the dominant, phenolic acids and alkaloids are also responsible for the therapeutic efficacy of *Trollius* flowers.

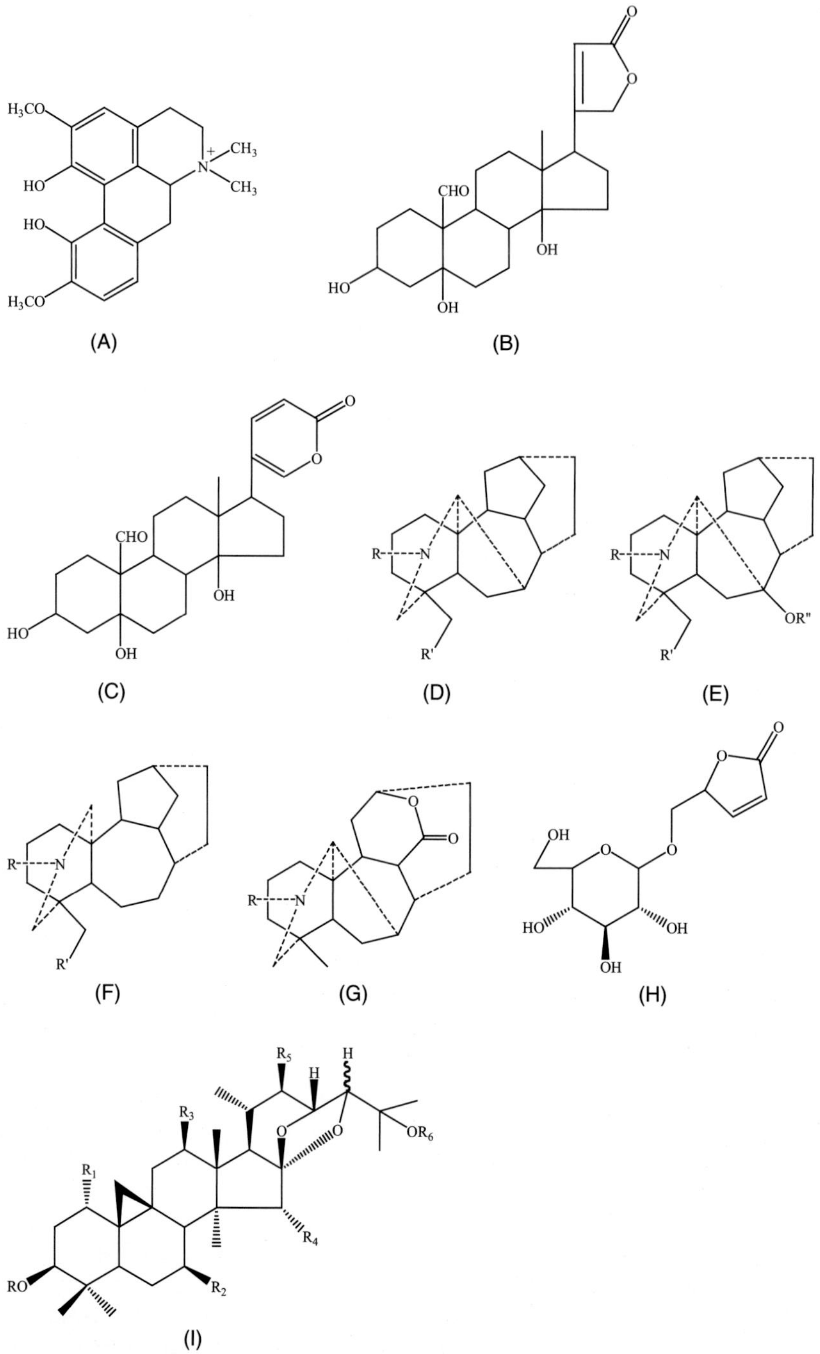

FIGURE 2.2   **Representative examples of medicinal compounds found in Ranunculaceae.** (A) magnoflorine, (B) strophanthidin, (C) hellebrigenin, (D) aconitine (III), (E) lycoctonine (IV), (F) 7, 17-seco type (V) (G) lactones (VI), (H) ranunculin, (I) cimigenol type, (J) hydroshengmanol type, (K) shengmanol type, (L) oleanolic type, (M) olean-3β, 28-diol type, (N) hederagenin type, (O) hederagenin-11,13-dien type, (P) aporphine, (Q) protopine, (R) pavine, (S) phenanthrene, (T) bisbenzylisoquinoline, and (U) cyanogenic glycoside.

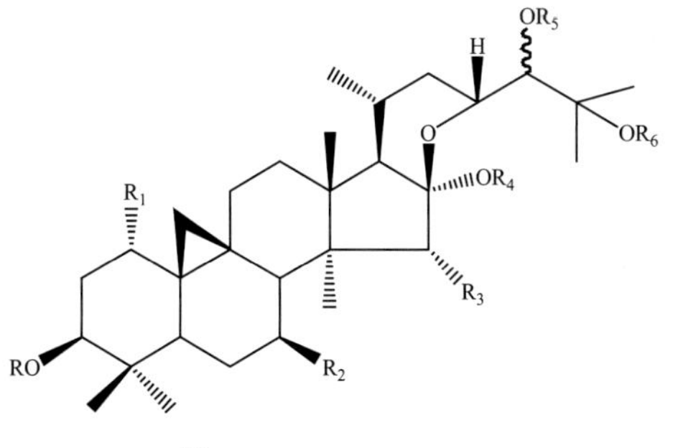

(J)

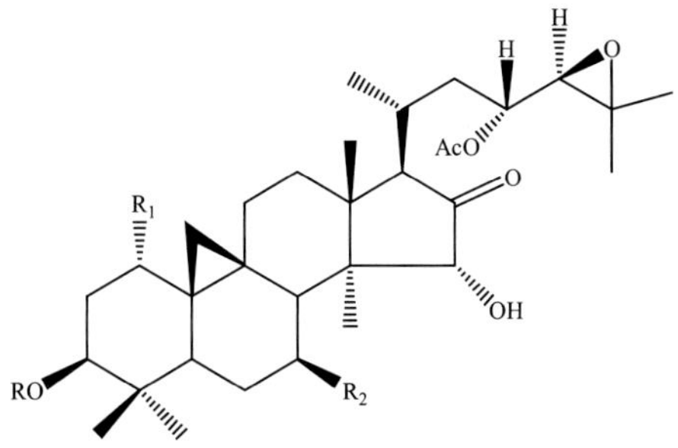

(K)

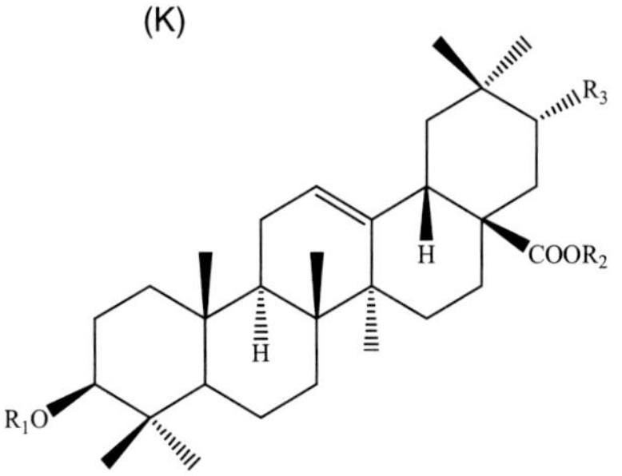

(L)

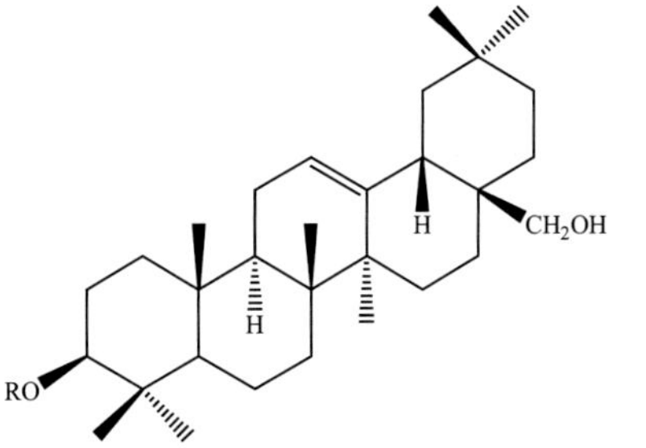

(M)

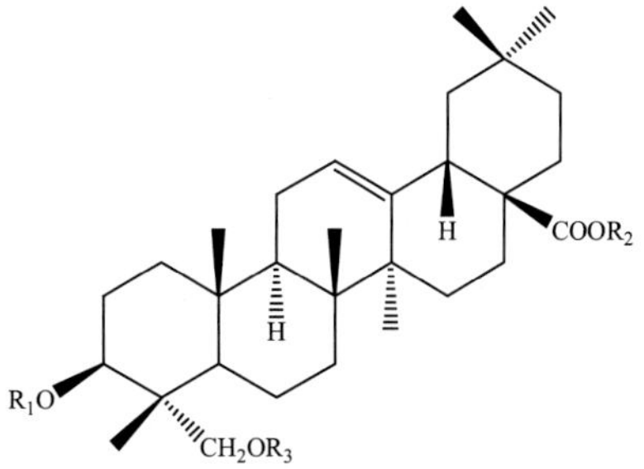

(N)

FIGURE 2.2   (*Cont.*)

The main type of alkaloid is pyrrolizidine alkaloid, which includes senecionine, squalidine, and trolline. The known organic acids are globeflowery acid, palmitic acid, veratric acid, and proglobeflowery acid. The representative volatile organic compounds are sesquiterpenoids,

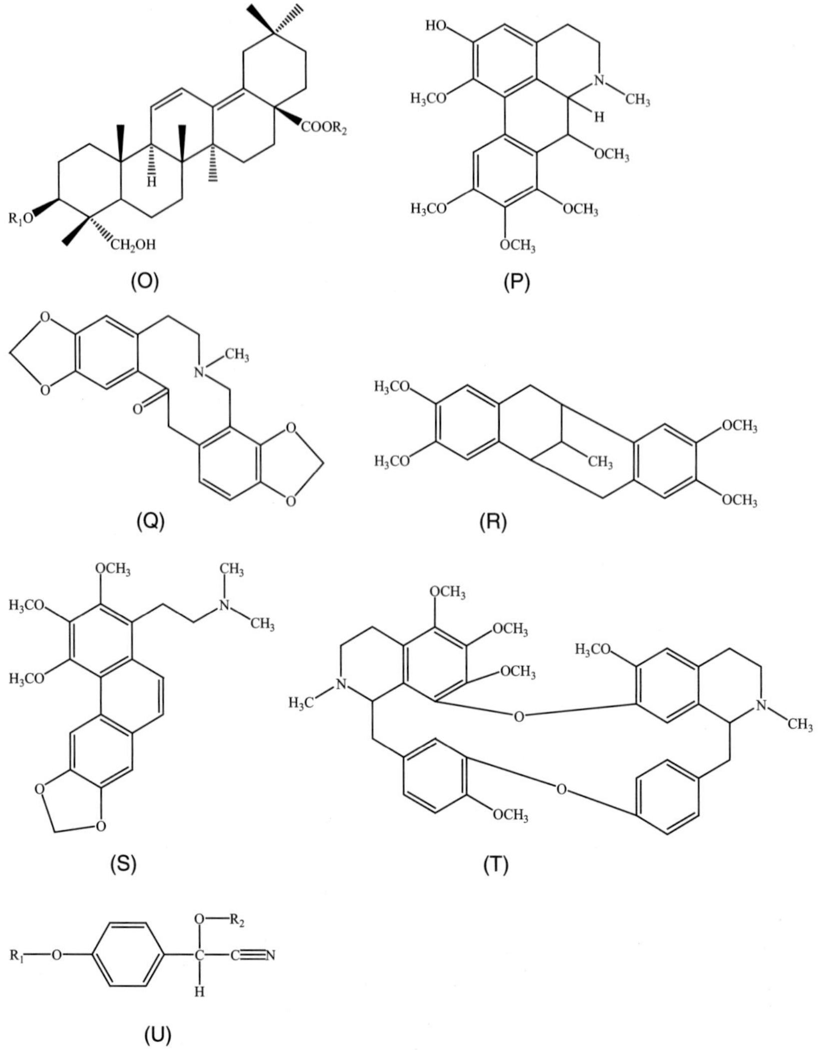

(O)

(P)

(Q)

(R)

(S)

(T)

(U)

FIGURE 2.2   *(Cont.)*

fatty acid derivatives, and nitrogen-containing compounds (Jürgens and Dötterl, 2004). *Trollius* essential oil mainly contains compounds from the group of benzenoids, nitrogen-containing compounds, monoterpenoids and sesquiterpenoids, irregular terpenes, and macrocyclic epoxide (Witkowska-Banaszczak, 2015). Labdane-type diterpenoid (Zou et al., 2006) and phenylethanoid glycosides (Wu et al., 2013) are found in *Trollius*.

2.3.2 Delphinieae

In this tribe, *Delphinium* (Fig. 2.3) is more closely related to *Consolida* than to *Aconitum*. This tribe is rich in diterpenoid alkaloids and flavonoids (Lin et al., 2014). More C19 and C20 diterpenoid alkaloids are found than C18 ones (Hao et al., 2013a). The lycoctonine-type

FIGURE 2.3   *Delphinium*, taken in Gongbo'gyamda County, Tibet, China.

C19-diterpenoid alkaloid, caerudelphinine A, was isolated from *Delphinium caeruleum* (Lin et al., 2017). C19-diterpenoid alkaloids were isolated from *Delphinium tianshanicum* (Zhang et al., 2017c). Tianshanitine A is the first natural diterpenoid alkaloid containing a benzoyl group at C1 position. Tianshanitine B is a rare natural diterpenoid alkaloid bearing an OH group at C16 position. C18-diterpenoid alkaloids, for example, anthriscifoltines C-G, were isolated from *Delphinium anthriscifolium* var. *majus* (Shan et al., 2017). More C18-, C19-, and C20-diterpenoid alkaloids were identified from *D. anthriscifolium*, *D. majus*, and *D. laxicymosum* var. *pilostachyum* respectively (Chen et al., 2014; Shan et al., 2015; Zhao et al., 2015), illustrating great chemodiversity of these compounds and great potential for biodiversity. The oils from *Delphinium* seeds consist chiefly of octadecadienoic (linoleic), hexadecanoic (palmitic), and octadecenoic (oleic) acids (Kokoska et al., 2012).

In *Consolida*, lycoctonine type (C19 type IV) and denudatine and hetisine type (C20 type VIII and X) are known diterpenoid alkaloids (Hao et al., 2013a). Octadecenoic and hexadecanoic acids are the main essential oil constituents (Kokoska et al., 2012).

Diterpenoid alkaloids, polysaccharides, flavonoids, and phenolic acids are abundant in *Aconitum* (Hao et al., 2013a) (Fig. 2.4). More than 450 alkaloids have been identified in various species (Khan et al., 2017). About 60 species and 15 varieties of *Aconitum* are distributed in the Tibetan region of China, and no less than 339 compounds are found in them (Yang et al., 2016). The aconitine type (C19 type III) is the predominant diterpenoid alkaloid (Fig. 2.2D). C19-diterpenoid alkaloids, 14-benzoylliljestrandisine, and 14-anisoylliljestrandisine were isolated from roots of *Aconitum tsaii* (Li et al., 2017b). C19-diterpenoid alkaloids, 14-anisoyllasianine, 14-anisoyl-N-deethylaconine, N-deethylaljesaconitine A, and N-deethylnevadensine, together with C20-diterpenoid alkaloids, were isolated from rhizomes of *Aconitum japonicum subsp. subcuneatum* (Yamashita et al., 2017). The denudatine-type C20-diterpenoid alkaloid sinchianine, together with diterpenoid alkaloids 12-acetyl-12-epi-napelline, 12-epi-napelline, neoline, talatisamine, 14-O-acetylsenbusine A, benzoylaconine, songorine, and aconitine, were isolated from the whole herb of *Aconitum sinchiangense* (Samanbay et al., 2017). Bisditerpenoid alkaloids, for example, navicularines A–C, were isolated from the ground parts of *Aconitum naviculare* (He et al., 2017). Flavonols, for example, kaempferol and quercetin derivatives, are prevalent in *Aconitum*.

FIGURE 2.4 *Aconitum*, taken in Gongbo'gyamda County, Tibet, China.

Ortho-phthalic acid esters were found in lipophilic extract from the cell culture of *Aconitum baicalense* (Semenov et al., 2016). Please refer to Hao et al. (2015) for more details on *Aconitum*.

2.3.3 Nigelleae

Nigella contains triterpenoid saponin, magnoflorine, and other benzylisoquinoline alkaloids. Norditerpenoid alkaloids and pyrroloquinoline alkaloids are also found (Chen et al., 2014a). Thymoquinone, fatty acids, and monoterpene hydrocarbons are the main components of *Nigella* essential oil (Kokoska et al., 2012; Singh et al., 2014).

2.3.4 Helleboreae

Helleborus has magnoflorine and other benzylisoquinoline alkaloids, tetracyclic and pentacyclic triterpenoid saponins, ranunculin (Fig. 2.2H), and cardiac glycosides. Lactones, such as protoanemonin and bufadienolides (Alshatwi, 2014), are abundant. Flavonoids, β-ecdysone, furostanol, and spirostanol steroidal saponins are also found (Duckstein and Stintzing, 2014; Watanabe et al., 2003). Minor spirostanol glycosides were isolated from *Helleborus thibetanus* (Zhang et al., 2017d). The acetylated polyhydroxy hellebosaponins A and D were abundant in *Helleborus niger* roots (Duckstein et al., 2014).

Cardiac steroids (Hellebrin), cysteine-rich proteins (Hellethionins), and several steroidal saponins were identified in *Helleborus* of Southeast Europe (Franz et al., 2017), and the nonproteinogenic pipecolic acid was abundant in the root and rootstock of *Helleborus purpurascens*. Carbohydrate, glycoside, saponins, flavonoid, phytosterols, tannins, and phenolic compounds are present in *H. niger* root (Kumar and Lalitha, 2017), which is used in Ayurvedic and Unani medicine.

2.3.5 Cimicifugeae

There are two main clades in this tribe. *Anemonopsis* and *Beesia* are basal to four other genera, among which *Cimicifuga* (Fig. 2.5) is closer to *Actaea* than to *Souliea* and *Eranthis*. *Cimicifuga* and *Actaea* contain cycloartane triterpenoid glycosides, other tetracyclic triterpenoid saponins, and phenylpropanoids (Hao et al., 2013b). Chemical constituents of *Cimicifuga* include triterpenoid glycosides, phenylpropanoids, nitrogenous compounds, chromones, flavonoids, and 4α-methyl steroid (Guo et al., 2017b). To date, investigation of seven *Cimicifuga* species led to the identification of more than 457 compounds. More than 140 cycloartane triterpene glycosides have been isolated from *Cimicifuga* (Su et al., 2016).

Souliea contains soulieosides A–E, M, and Q (cycloartane type) (Wu et al., 2016, 2017a); prim-O-glucosylcimifugin; and alkaloid soulieotine. The cyclolanostane triterpenoid glycoside soulieoside O was isolated from *Souliea vaginata* (Wu et al., 2017b).

Eranthis contains magnoflorine and triterpenoid saponin (Watanabe et al., 2003a). *Beesia* contains cycloartane saponin beesiosides I-V and A-P. Cimigenol type (type A) is the predominant cimicifugeae triterpenoid saponin (Fig. 2.2I). Phenolic compounds and chromones are found in this tribe. Please refer to Hao et al. (2015) for more details on Cimicifugeae.

2.3.6 Caltheae

Caltha contains oleanane-type saponin polypetalosides A-I, triterpenoid lactones caltholide, epicaltholide, and protoanemonin. In *Caltha* sesquiterpenoids, fatty acid derivatives (e.g., protoanemonin) and monoterpenoids dominate the flower scent (Jürgens and Dötterl, 2004).

FIGURE 2.5   *Cimicifuga foetida*, taken in Shangri-La Alpine Botanical Garden, Yunnan, China.

2.3.7 Asteropyreae

The water-soluble quarternary ammonium-type alkaloids magnoflorine, palmatine, columbamine, and jatrorrhizine are found in *Asteropyrum* (Ma et al., 1992). Berberine and berberrabine are also present (Xu, 2000).

2.3.8 Callianthemeae

Ranunculin is found in this tribe (Peng et al., 2006a). Flavonol glycoside and other phenolic compounds are present in *Callianthemum* (Wang et al., 2012).

2.3.9 Anemoneae

In this tribe, *Hepatica* is the basalmost, followed by *Pulsatilla* and *Anemone* (Fig. 2.6), and *Clematis* (Fig. 2.7) is closer to *Naravelia* than to *anemoclema*. Identified *Anemone* compounds include triterpenoids, steroids, lactones, fats and oils, saccharide, and alkaloids (Sun et al., 2011). Oleanolic acid triterpene saponin is abundant in *Anemone*. *Anemone* contains ranunculin, anemonin, and protoanemonin, which are characteristic constituents of *Pulsatilla* and illustrate the close relationship between these two genera. *Anemone* also contains coumarin and flavonoids. Benzenoids, fatty acid derivatives (e.g., pentadecane and nonanal), and sesquiterpenoids are the dominant volatile compounds in *Anemone* (Jürgens and Dötterl, 2004).

Anemonin, okinalin, and okinalein are abundant in *Pulsatilla*. This genus has lupane and oleanane-type pentacyclic triterpene saponins, lignans, and daucosterol; (+)-pinoresinol, 13-peltatin, peltatin, and triterpenoic acid (e.g., anemonic acid) are found. Fatty acid derivatives (e.g., protoanemonin and pentadecane) and monoterpenoids are the dominant volatile compounds in *Pulsatilla* (Jürgens and Dötterl, 2004).

Clematis contains saponin, coumarin, flavonoid, anthocyanin, and alkaloid (Hao et al., 2013c). Lignans, steroids, macrocyclic compounds, phenolic glycosides, and volatile oils are also found (Chawla et al., 2012; Qiu et al., 2017). The common aglycones belong to pentacyclic triterpene (e.g., Fig. 2.2L–O), such as oleanolic acid, hederagenin, and epihederagenin.

FIGURE 2.6 *Anemone vitifolia*, taken in Ninong Canyon, Deqin County, Yunnan, China.

FIGURE 2.7   *Clematis brevicaudata*, taken in Ninong Canyon, Deqin County, Yunnan, China.

Oleanane-type triterpenoid saponins, clematangoticosides A-H, were isolated from the whole plants of *Clematis tangutica* (Zhao et al., 2016). Clematangoticosides D-G are unusual 23, 28-bidesmosidic glycosides.

The representative coumarins are scopoletin, umbelliferone, and clemastanin A and B. Aporphinoid alkaloids and terpenoid alkaloids, such as magnoflorine, corytuberine, elatine, and aconitine, are found. Ranunculin, anemonin, and protoanemonin are present in this genus. Please refer to Hao et al. (2015) for more details on *Clematis*.

Anemoclema contains saponin anemoclemosides A and B (Li et al., 1995). *Naravelia* contains simple benzamides (Jaroszewski et al., 2005). Please refer to Chapter 8 for more details on *Anemone*.

2.3.10 Ranunculeae

In this tribe, *Oxygraphis* and *Trautvetteria* are basal to four other genera, while *Ranunculus* is closer to *Batrachium* than to *Halerpestes*. Lactones, including protoanemonin, anemonin, ranunculin, isoranunculin, and ternatolide, are widely distributed in *Ranunculus*. The aglycone of triterpene saponin is hederagenin or oleanolic acid. Alkaloid is abundant in *Ranunculus*, most of which are berberine type and aporphinoid alkaloids. The dominant volatile compound in *Ranunculus* is protoanemonin (Jürgens and Dötterl, 2004).

Please refer to Chapter 9 for more details on *Ranunculus*. Refer to Chapter 4 for details on biopharmaceutics and pharmacokinetics of Ranunculaceae.

2.4 THE CHEMICAL COMPOSITION OF THALICTROIDEAE, COPTIDOIDEAE, HYDRASTIDOIDEAE, AND GLAUCIDIOIDEAE

2.4.1 *Thalictrum* Clade

In this group, *Thalictrum* is basal to three other genera. *Paraquilegia* contains magnoflorine and bisbenzylisoquinoline alkaloid, such as fangchinoline and dimethyl tubocurarine. This genus is also a cyanide-containing compound. Flavone C-glycosides and triterpene saponins are present (Xu et al., 2010, 2011).

Thalictrum has alkaloid, saponin, and flavonoids. More than 300 alkaloids are found in this genus (e.g., Fig. 2.2K–T), many of which are bisbenzylisoquinoline alkaloid and aporphinoid alkaloid. Atisine- and hetidine-type (C20 type VII and IX) diterpenoid alkaloids are found in this genus (Hao et al., 2013a). The genus also contains tetracyclic and pentacyclic triterpenoid saponins.

The isoquinoline alkaloid leptopyrine was isolated from the aerial parts of *Leptopyrum fumarioides*, which is of Mongolian origin (Doncheva et al., 2015). Protopine and thalifoline were also isolated. Please refer to Chapter 7 for more details on *Thalictrum*.

2.4.2 *Aquilegia* Clade

In this group, *Aquilegia* (Fig. 2.8) is closer to *Semiaquilegia* than to *Urophysa*. *Aquilegia* contains flavonoid, phenolic acid, triterpenoid saponin (cycloartane glycoside), and alkaloid (magnoflorine and aporphinoid alkaloid) (Bylka et al., 2004; Chen et al., 2002; Nishida et al., 2003). Labdane diterpene glycosides and megastigmane glycoside are also found (Yoshimitsu et al., 2008). The dominant volatile compounds of *Aquilegia* are fatty acid derivatives (e.g., octanal, protoanemonin and nonanal) and benzenoids (Jürgens and Dötterl, 2004). Organic acids, fatty acids, amino acids, esters, sugars, and unknown compounds were found in flowers of *A. canadensis* and *A. pubescens* (Noutsos et al., 2015). The mean abundances of 25 metabolites were significantly different between two species.

Lactone, magnoflorine, phenolic glycoside, cyanogenic glycoside (Fig. 2.2U), diterpenoid, and benzoic acid derivatives are found in *Semiaquilegia* (Lee et al., 2012; Niu et al., 2006; Su et al., 2004, 2005).

2.4.3 *Isopyrum* Clade

In this clade, *Isopyrum* is closer to *Enemion* than to *Dichocarpum*. *Isopyrum* is rich in bisbenzylisoquinoline alkaloids (Xiao, 1980), for example, berbamine, tetrandrine, penduline, and other benzylisoquinoline alkaloids, such as magnoflorine, corytuberine, and coptisine. *Isopyrum* contains flavonoids.

FIGURE 2.8   *Aquilegia*, taken in Meili snow mountain, Deqin County, Yunnan, China.

2.4.4 Coptidoideae

Coptis contains protoberberine-type alkaloids (Xiao, 1980), such as berberine, palmatine, coptisine, and jatrorrhizine, which are quaternary ammonium-type alkaloids. Magnoflorine is an aporphinoid alkaloid.

2.4.5 Hydrastidoideae

The major alkaloids in *Hydrastis* are berberine, palmatine, hydrastine, hydrastinine, and canadine (Chen et al., 2013). The following were found in the leaves of *Hydrastis canadensis* (Leyte-Lugo et al., 2017): 3,4-dimethoxy-2-(methoxycarbonyl)benzoic acid; 3,5,3'-trihydroxy-7,4'-dimethoxy-6,8-C-dimethyl-flavone; (±)-chilenine; (2R)-5,4'-dihydroxy-6-C-methyl-7- methoxy-flavanone; 5,4'-dihydroxy-6,8-di-C-methyl-7-methoxy-flavanone; noroxyhydrastinine; oxyhydrastinine; and 4',5'-dimethoxy-4-methyl-3'-oxo-(1,2,5,6-tetra-hydro-4H-1,3-dioxolo-[4',5':4,5]-benzo[1,2-e]-1,2-oxazocin)-2-spiro-1'-phtalan.

2.4.6 Glaucidioideae

Glaupalol-related coumarins are present in *Glaucidium palmatum* (Morita et al., 2004).

2.5 ETHNOPHARMACOLOGY AND BIOACTIVITY

Different taxonomic groups have characteristic bioactivities and ethnopharmacological uses dependending on their different chemical profiles. During the past decades, we have collected more than 30,000 cards that record the ethnopharmacological use of more than 5000 Chinese medicinal plant species. Traditional remedy index $(\text{TRI}) = C_1^2/C_2 \times 100$ (Xiao et al., 1986); C_1 is the number of cards on which a specific ethnopharmacological use is recorded in China for a single genus, and C_2 is the number of species that has the same ethnopharmacological use in the same genus. $\text{TRI} \geq 300$ is considered significant. The higher the TRI, the more credible it is that the genus has the specific ethnopharmacological use. Distribution density of ethnopharmacological use $(\beta) = SP_1/SP_2 \times 100$; $SP_1 = C_2$, and $SP_2 =$ number of species in China in this genus. The anticancer mechanisms of Ranunculaceae are detailed in Chapter 6.

2.5.1 Adonideae

Adonis is used in heat-clearing and damp-drying (TRI 500, β 14), cardiac insufficiency (TRI 533, β 42), arrhythmia (TRI 514, β 28), dysentery (TRI 320, β 14), ulcer disease and sores (TRI 500, β 14) (Table 2.2), conjunctivitis (TRI 320, β 14), and vomiting (TRI 320, β 14). In anesthetized sheep, bradycardia and ECG alterations induced by the *Adonis aestivalis* extract could justify the traditional use of *A. aestivalis* in treating cardiovascular insufficiency (Maham and Sarrafzadeh-Rezaei, 2014). Cardenolide glycoside, strophanthidin glycoside, pregnane glycosides, cymarin, and cymarol exhibit anticancer activity (Kubo et al., 2012b; Kuroda et al., 2010; You et al., 2003) (Table 2.3). Cymarilic acid shows antiangiogenic activity (You et al., 2003).

Astaxanthin, a carotenoid of *A. amurensis*, has antioxidant capacity and other health functions (Zhang et al., 2015). Hydro-methanolic extract of *A. wolgensis*, containing a high amount of total phenolics (9.20 ± 0.011 mg GAE per dry matter), is a potent antioxidant and antibacterial (Mohadjerani et al., 2014). Methanol extract of *Adonis coerulea* shows acaricidal activity against *Psoroptes cuniculi* (Shang et al., 2017).

Trollius has antibacterial effects and is used in tonsillitis (TRI 514, β 42), otitis media (TRI 514, β 42), red eye with swelling and pain (TRI 450, β 57), and scrofula (TRI 400, β 57). Three species of *Trollius* are traditionally used to treat upper respiratory tract infections, pharyngitis, tonsillitis, bronchitis, cold with fever, acute tympanitis, aphthae, mouth sore, gum hemorrhage and pain, acute lymphangitis, and acute periostitis (Witkowska-Banaszczak, 2015). Flavonoids and phenolics have major antiinflammatory and antioxidant components (Sun et al., 2013; Wang et al., 2014). Antiinflammatory effects of orientin-2″-O-galactopyranoside on lipopolysaccharide-stimulated microglia were revealed (Zhou et al., 2014). Flavone glycosides have antiviral (Cai et al., 2006), anticomplement (Liu et al., 2013) and DNA-binding activities (Song et al., 2013). Effects of orientin and vitexin of *T. chinensis* on the growth and apoptosis of esophageal cancer EC-109 cells were revealed (An et al., 2015). The antitumor effects of orientin were stronger than that of vitexin of the same concentration. Total flavonoids of *T. chinensis* had growth inhibition and apoptotic effects on human breast cancer MCF-7 cells (Wang et al., 2016b). A C-glycosyl-flavone is a defense molecule against developing larvae (Ibanez et al., 2009).

Phenolic compounds, for example, trolliusol A, from the flowers of *T. chinensis* exhibited antimicrobial activity (Li et al., 2014). Organic acid from *Trollius* showed antiviral activities (Li et al., 2002).

The whole plant of *Calathodes* is used in rheumatic disease and numbness (Xiao, 1980).

2.5.2 Delphinieae

2.5.2.1 *Delphinium and Consolida*

Delphinium is used for wind-dispelling and damp-drying (TRI 750, β 20), interior-warming and cold-dissipating (TRI 756, β 17), pesticides (TRI 1056, β 27), arthritis (TRI 1031, β 20), arthralgia and myalgia (TRI 327, β 13), anemofrigid cold (TRI 500, β 13), diarrhea (TRI 427, β 24), internal lesions caused by overexertion (TRI 327, β 13), toothaches (TRI 514, β 3), stroke and paralysis (TRI 377, β 6), ulcer disease and sores (TRI 528, β 20), and snake bites (TRI 400, β 10). The plant of *Delphinium* has broad-spectrum bioactivities, such as antiinflammatory, immunosuppressive, analgesic, antitumor, cardiotonic, antihypertensive, and vasodilative. Diterpenoid alkaloids shows anticancer activity (Lin et al., 2014). C19-diterpenoid alkaloids exhibit cardiac activity (Jian et al., 2012). Isoxylitone, an anticonvulsant, modulates voltage-gated sodium channel inactivation and prevents kindling-induced seizures (Ashraf et al., 2013). *Delphinium* extracts and alkaloids show antiinflammatory effects against arthritis (Nesterova et al., 2009). Alkaloids show antiparasitic and insecticidal activities (Reina et al., 2007). Flavonoids from *Delphinium* show leishmanicidal and trypanocidal activities (Marín et al., 2011; Ramírez-Macías et al., 2012). *Delphinium* extract shows protective effects against Parkinson's disease via antioxidant activity (Ahmad et al., 2006) and a significant effect against morphine-induced tolerance and dependence in mice (Zafar et al., 2002).

Delphinium denudatum is a rich source of diterpenoid alkaloids and is widely used for the treatment of various neurological disorders, such as epilepsy, sciatica, and Alzheimer's disease. Isotalatazidine hydrate of D. denudatum is a potent dual cholinesterase inhibitor and can be used as a target drug in Alzheimer's disease (Ahmad et al., 2017c). The traditionally used D. staphisagria seeds promote hair growth through induction of angiogenesis (Koparal and Bostancıoğlu, 2016).

The whole plant of *Consolida* is used as an analgesic and against rheumatic disease and numbness (Xiao, 1980). The seed and the whole plant are used against parasites and insects.

2.5.2.2 *Aconitum*

Aconitum is used in wind-dispelling and damp-drying (TRI 5582, β 64), blood-activating and stasis-removing (TRI 999, β 30), interior-warming and cold-dissipating (TRI 313, β 4), traumatic injury (TRI 4489, β 59), arthritis (TRI 6488, β 66), neuropathic pain (TRI 433, β 19), stroke and paralysis (TRI 1230, β 16), cold and pain of stomach (TRI 400, β 4), gastroenteritis (TRI 582, β 11), menstrual disorder (TRI 320, β 4), and ulcer disease and sores (TRI 2701, β 52). The anticancer activity, cardioactive effect, analgesic activity, antiinflammatory activity, effect on energy metabolism, antimicrobial, and pesticidal activities are well-archived (Hao et al., 2013a). Hypaconitine inhibits TGF-β1-induced epithelial-mesenchymal transition and suppresses adhesion, migration, and invasion of lung cancer A549 cells (Feng et al., 2017). C19-diterpenoid alkaloid sinchiangensine A and its analogue, isolated from the root of *A. sinchiangense*, had significant antitumor activities against HL-60, A-549, SMCC-7721, MCF-7, and SW480 cells (Liang et al., 2017), with IC_{50} comparable to cisplatin, and significant antibacterial activities against *Staphylococcus aureus* ATCC-25923. *Aconitum coreanum* polysaccharide and its sulphated derivative affected the migration of human breast cancer MDA-MB-435s cells (Zhang et al., 2017b).Songorine, a C20 diterpenoid alkaloid and 12-keto analog of napelline isolated from *Aconitum soongaricum*, and its derivatives have many pharmacological effects, including antiarrhythmic, anticardiac-fibrillation, excitation of synaptic transmission, anxiolytic, antinociceptive, antiinflammatory, antiarthritic, and regenerative effects in a skin excision wound animal model (Khan et al., 2017). Chasmanthinine of *Aconitum franchetii* var. *villosulum* showed highly potent antifeedant activity against *Spodoptera exigua* (Zhang et al., 2017a).

Antioxidant and anticholinesterase potentials of diterpenoid alkaloids from *Aconitum heterophyllum* were revealed (Ahmad et al., 2017a). The lycoctonine-type swatinine-C (1); three norditerpenoid alkaloids, hohenackerine (2), aconorine (5), and lappaconitine (6); and two benzene derivatives, methyl 2-acetamidobenzoate (3) and methyl 4-[2-(methoxycarbonyl) anilino]-4-oxobutanoate (4), were isolated from roots of *Aconitum laeve* (Ahmad et al., 2017b). Compounds 1 and 2 showed competitive inhibition against AChE and BChE. Compounds 5 and 6 showed noncompetitive inhibition against AChE. Compounds 3 and 4 had weak inhibition against AChE and BChE. Aconitine inhibited the progression of systemic lupus erythematosus and ameliorated the pathologic lesion (Li et al., 2017c). Antirheumatic effects of *Aconitum leucostomum* on human fibroblast-like synoviocyte rheumatoid arthritis cells were revealed (Yang et al., 2017).

2.5.3 Nigelleae

In India, the seeds are used as a carminative and stimulant to ease bowel and indigestion problems and are given to treat intestinal worms and nerve defects and to reduce flatulence and induce sweating. Dried pods are sniffed to restore a lost sense of smell and are also used to repel insects. The seed of *Nigella* is used in diuresis (Xiao, 1980).

Nigella has antiinflammatory, analgesic, heat dissipation, antibacterial, and antitumor effects, as well as antitussive and expectorant effects (Xiao, 1980). Thymoquinone attenuates lipid peroxidation and shows neuroprotective effects (Sedaghat et al., 2014). Thymoquinone also inhibits topoisomerase IIα activity (Ashley and Osheroff, 2014), induces apoptosis, and displays anticancer activity. Essential oil and oleoresins showed in vitro antioxidant and antimicrobial activities (Singh et al., 2014). *Nigella* is a source of bacterial urease inhibitor (Biglar et al., 2014). *Nigella* seeds have hypolipidemic effects in menopausal women (Ibrahim et al., 2014). The seed oil improves semen quality in infertile men (Kolahdooz et al., 2014). Increased 5-HT levels following repeated administration of *Nigella* oil produces antidepressant effects in rats (Perveen et al., 2013). *Nigella* extract prevents scopolamine -induced spatial memory deficits and decreases the acetylcholinesterase (ACE) activity as well as oxidative stress of brain tissues in rats (Hosseini et al., 2015). High ACE-inhibitory activity of hydrolyzed protein fractions was found in globulin fractions of *Nigella damascena* (Alu'datt et al., 2017). High antioxidant activities of hydrolyzed protein fractions were found in glutelin-2 for *Nigella damascene* and *Nigella arvensis*. Seed extract showed immunomodulatory effects on human peripheral blood mononuclear cells (Alshatwi, 2014).

2.5.4 Helleboreae

Helleborus is used in traumatic injury (TRI 500, β 100), urinary tract infection (TRI 500, β 100), ulcer disease and sores (TRI 500, β 100), and internal lesions caused by overexertion (TRI 320, β 100). Bufadienolides exhibits potent cytotoxic activities against cancer cells (Cheng et al., 2014) as well as antibacterial activity (Puglisi et al., 2009). Polyphenolic extracts exhibit inhibitory activity against urease and low inhibition against α-chymotrypsin and could be used in ulcer treatment (Paun et al., 2014). *Helleborus* fraction modulates HMGB1 cytokine and attenuates septic shock in mice (Apetrei et al., 2011). *Helleborus* also shows antioxidant and antiproliferative activities (Cakar et al., 2011), as well as antiinflammatory and antinociceptive activities (Erdemoglu et al., 2003). MCS-18, a macrocyclic carbon suboxide $(C_3O_2)_n$ derivative, exerts immunosuppressive (Seifarth et al., 2011), immunomodulatory (Littmann et al., 2008), and analgesic activities (Neacsu et al., 2010).

2.5.5 Cimicifugeae

Cimicifuga is used for wind-heat dispersing (TRI 1350, β 57), heat-clearing and detoxification (TRI 1370, β 71), benefiting Qi and raising Yang (TRI 1350, β 57), swelling and pain in throat (TRI 776, β 85), variola and exanthema (TRI 400, β 71), archoptosis (TRI 1112, β 85), and uterine prolapse (TRI 704, β 57). *Actaea* is used in wind-heat dispersing (TRI 300, β 50), heat-clearing, and detoxification (TRI 300, β 50). *Souliea* is used in heat-clearing and detoxification (TRI 320, β 100) and ulcer disease and sores (TRI 500, β 100). *Beesia* is used in wind-dispelling, damp-drying (TRI 400, β 100), and arthritis (TRI 400, β 100).

Chromones from the tubers of *Eranthis cilicica* have antioxidant activity (Kuroda et al., 2009). *Eranthis hyemalis* lectin is a cytotoxic effector in the life cycle of *Caenorhabditis elegans* (McConnell et al., 2015). The whole plant of *Eranthis* is used in diuresis and for urinary stones (Xiao, 1980).

Cimicifugeae has antiinflammatory, antipyretic, analgesic, antivirus, detoxification, and estrogenlike activities. The anticancer activity, effects on menopausal symptoms and the cardiovascular system, osteoprotective effects, and immunosuppressive activity are well archived (Hao et al., 2013b). The cycloartane triterpenoids show anticancer, immunomodulatory, and hepatoprotective activities, as well as effects on the cardiovascular system (Tian et al., 2006). 9,19-Cycloartenol glycoside G3 of *Cimicifuga simplex* regulated immune responses by modulating Th17/Treg ratio (Su et al., 2017). Induction of mast cell degranulation by triterpenoid saponins of *Cimicifuga* rhizome was revealed (Choi et al., 2016). Cycloartane glycosides, isolated from roots of *Cimicifuga dahurica*, had inhibition potential against soluble epoxide hydrolase (Thao et al., 2017a). Actein ameliorated hepatic steatosis and fibrosis in high fat diet–induced non-alcoholic fatty liver disease (NAFLD) by regulation of insulin and leptin resistance (Chen and Liu, 2017). Deoxyactein protected pancreatic β-cells against methylglyoxal-induced oxidative cell damage by upregulating mitochondrial biogenesis (Suh et al., 2017). Cimicifugamide from *Cimicifuga* rhizomes functions as a nonselective β-AR agonist for cardiac and sudorific effects (Wang et al., 2017). Two dimeric prenylindole alkaloids with a unique indole-benzoindolequinone skeleton, cimicifoetones A and B, had antiproliferative activity on seven tumor cell lines (Zhou et al., 2017a,b). Cimicifoetone B induces cell apoptosis via death receptor–mediated extrinsic and mitochondrial-mediated intrinsic pathways. Isoferulic acid acts against glycation-induced changes in structural and functional attributes of human high-density lipoprotein (Jairajpuri and Jairajpuri, 2016).

Cimicifuga foetida extract is safe and effective for the treatment of menopausal symptoms in postmenopausal women (Gao et al., 2017). Soluble epoxide hydrolase inhibitors were also found from indolinone alkaloids and phenolic derivatives of *C. dahurica* (Thao et al., 2017b). The piscidic acid derivatives of *C. dahurica* have marked neuroprotective effects at certain concentrations (Lv et al., 2017). Polyphenolic compounds of *C. dahurica* have antioxidant potential and neuroprotective effects (Qin et al., 2016).

2.5.6 Caltheae

Caltha is used in arthritis (TRI 400, β 100). Polysaccharide fraction shows antiarthritis and immunomodulatory activities (Suszko and Obmi ska-Mrukowicz, 2013). *C. palustris* polysaccharides modulate macrophage function and exert beneficial effects on the clinical course of collagen-induced arthritis in mice (Suszko and Obmi ska-Mrukowicz, 2017), which was comparable to that of methotrexate treatment for certain parameters. Anthelmintic, antimicrobial, antioxidant, and cytotoxic activities of *C. palustris* var. *alba* of Kashmir were revealed (Mubashir et al., 2014).

2.5.7 Asteropyreae

Asteropyrum is used by minority ethnic groups of Southwest China in heat-clearing and detoxification, damp-drying and diuresis, dysentery, jaundice, ulcer disease and sores, and traumatic injury (Xu, 2000).

2.5.8 Callianthemeae

Callianthemum has detoxifying and antiinflammatory activity and is used in various diseases, including children's pneumonia and drug-fire eyesight (Wang et al., 2012).

2.5.9 Anemoneae

2.5.9.1 *Anemone*

Anemone is used for heat-clearing and detoxification (TRI 424, β 30), wind-dispersing and damp-eliminating (TRI 476, β 35), warming and orifice-opening (TRI 700, β 15), pesticides (TRI 400, β 30), dysentery (TRI 1051, β 46), malaria (TRI 356, β 30), tinea (TRI 445, β 46), ulcer disease and sores (TRI 1932, β 84), arthritis (TRI 896, β 76), traumatic injury (TRI 930, β 53), pharyngolaryngitis (TRI 327, β 7), parasitic disease (TRI 424, β 30), and hepatitis (TRI 445, β 7). A broad spectrum of pharmacological activity includes antitumor, antimicrobial, antiinflammatory, sedative, and analgesic activities, as well as anticonvulsant and antihistamine effects (Sun et al., 2011). Triterpenoid saponins have anticancer, antibacterial, and antiarthritis properties. Anhuienoside C of Anemone flaccida ameliorated collagen-induced arthritis through inhibition of MAPK and NF-κB signaling pathways (Liu et al., 2017b).

2.5.9.2 *Pulsatilla*

Pulsatilla is used for heat-clearing and blood-cooling (TRI 1125, β 50), heat-clearing and detoxification (TRI 845, β 50), dysentery (TRI 1570, β 100), hemorrhoids (TRI 405, β 50), and nasal hemorrhage (TRI 337, β 33). *Pulsatilla* has antitumor, antibacterial, antiinflammatory, immune-enhancing, and antitrichomonal effects (Xiao, 1980). Anemoside A3 of *Pulsatilla chinensis* ameliorates experimental autoimmune encephalomyelitis by modulating T helper 17 cell response (Ip et al., 2017). Anemoside B4 of *P. chinensis* upregulates IL-2 expression in PRRSV-induced endothelial cells (Hu et al., 2016). *Pulsatilla* saponin A induces apoptosis and inhibits tumor growth of human colon cancer cells with/without 5-FU (Xu et al., 2017). Pulchinenoside inhibited the fibroblast-like synoviocyte apoptosis in adjuvant arthritis rats (Miao et al., 2015).

Modified *Pulsatilla* powder is used in the treatment of enterotoxigenic *Escherichia coli* O101-induced diarrhea in mice (Yu et al., 2017). *Pulsatilla* decoction inhibits vulvovaginal *Candida albicans* proliferation and reduces inflammatory cytokine levels in vulvovaginal candidiasis mice (Xia et al., 2016). Modified *Pulsatilla* decoction attenuated oxazolone-induced colitis in mice through suppression of inflammation and epithelial barrier disruption (Wang et al., 2016c).

2.5.9.3 *Clematis*

Clematis is used for wind-dispelling and damp-eliminating (TRI 3788, β 52), blood-activating and stasis-dispelling (TRI 1067, β 33), damp-filtering and diuresis (TRI 1824, β 27), heat-clearing and damp-drying (TRI 578, β 20), swell-reducing and detoxification (TRI 678, β 29), arthritis (TRI 4848, β 58), traumatic injury (TRI 1778, β 37), urinary tract infection (TRI 1972, β 45), bones stuck in the throat (TRI 1536, β 12), agalactia (TRI 1869, β 18), abnormal menstruation (TRI 1393, β 33), nephritis and edema (TRI 845, β 25), dysentery (TRI 632, β 18), ulcer disease and sores (TRI 1000, β 37), stomatitis (TRI 612, β 14), toothache caused by wind-fire, (TRI 511, β 14) and snake bites (TRI 352, β 18). *Clematis* has anticancer, antibacterial, antiinflammatory, analgesic, sedative, and plant hormone–like effects (Hao et al., 2013c) and is used in myocardial ischemia and chronic cholecystitis. Diuretic, antiarthritis, hepatoprotective, hypotensive, and HIV-1 protease inhibitor activities are also revealed (Chawla et al., 2012). The synovial succinate accumulation and HIF-1α induction might be therapeutic targets of clematichinenoside AR, a triterpene saponin isolated from the root of *Clematis manshurica* for the prevention of fibrosis in arthritis (Li et al., 2016). Apigenin-7-O-β-D-(-6″-p-coumaroyl)-glucopyranoside of *C. tangutica* attenuates myocardial ischemia/reperfusion injury via activating PKCε signaling (Zhu et al., 2017). Boehmenan, a lignan of *Clematis armandii*, induces apoptosis in lung cancer cells through modulation of EGF-dependent pathways (Pan et al., 2016b). Ikshusterol 3-O-glucoside of *Clematis gouriana* had a potent snake-venom neutralizing capacity and might be useful for the treatment for snakebite (Chinnasamy et al., 2017). The antioxidant potential of phenolic- and flavonoid-rich fractions of *Clematis orientalis* and *Clematis ispahanica* were revealed (Karimi et al., 2017). Antifungal, molluscicidal, and larvicidal activities of anemonin and *C. flammula* extract against mollusc *Galba truncatula*, intermediate host of *Fasciola hepatica*, were revealed (Saidi et al., 2017). Noteworthy antimicrobial activity of *Clematis brachiata* justifies its traditional use in oral infections (Akhalwaya et al., 2018). Lignans of *Clematis mandshurica*, for example, clemomanshurinane C and D, inhibit lipoxygenase (Fu et al., 2017). (7R,8S)-9-acetyl-dehydrodiconiferyl alcohol, a lignan isolated from *C. armandii* stem, inhibit inflammation and migration in lipopolysaccharide (LPS)-stimulated macrophages (Pan et al., 2016a). (7R,8S)-dehydrodiconiferyl alcohol, isolated from dried stems of *C. armandii*, suppresses LPS-induced inflammatory responses in BV2 microglia by inhibiting MAPK signaling (Liu et al., 2016).

The whole plant of *Naravelia* is used in Qi-moving and pain relief (Xiao 1980). The root and stem of *Hepatica* are used in traumatic injury, internal lesion caused by overexertion, arthralgia, and myalgia (Xiao, 1980).

2.5.10 Ranunculeae

Ranunculus (Fig. 2.9) is used for parasiticide (TRI 785, β 21), swell-reducing and detoxification (TRI 326, β 50), malaria (TRI 3291, β 57), scrofula (TRI 1525, β 64), arthritis (TRI 568, β 42), asthma (TRI 785, β 21), jaundice (TRI 589, β 28), and ulcer disease and sores (TRI 944, β 50). *Ranunculus* has anticancer, antiinflammatory, antioxidant, analgesic, antimicrobial, antiparasitic, and cardiovascular effects. *Halerpestes* is used for wind-dispelling and damp-eliminating (TRI 400, β 100), arthritis (TRI 400, β 100), and edema (TRI 400, β 100). The whole plant of *Oxygraphis* is used in wind-dispelling and cold-expelling, orifice-opening, and collateral-dredging (Xiao, 1980).

2.5.11 Thalictroideae

2.5.11.1 *Thalictrum Clade*

Thalictrum (Fig. 2.10) is used for heat-clearing and detoxification (TRI 1731, β 54), heat-clearing and damp-drying (TRI 682, β 33), heat-clearing and blood-cooling (TRI 300, β 6), dysentery (TRI 2807, β 69), gastroenteritis (TRI 1340, β 48), red eye with swelling and pain (TRI

FIGURE 2.9   *Ranunculus*, taken in Nyingchi, Tibet, China.

FIGURE 2.10   *Thalictrum*, taken in Meili snow mountain, Deqin County, Yunnan, China.

2625, β 63), hepatitis (TRI 306, β 12), diarrhea (TRI 327, β 12), jaundice (TRI 804, β 27), ulcer disease and sores (TRI 1225, β 54), stomatitis (TRI 776, β 30), abdominal pain (TRI 583, β 33), vomiting (TRI 465, β 30), infantile convulsion (TRI 492, β 12) and child's indigestion (TRI 320, β 6). *Thalictrum* has anticancer, antibacterial, antihypertensive, antiarrhythmia, spasmolytic, analgesic, and sedative effects (Peng et al., 2006a,b). Saponin has immunomodulatory activity.

Leptopyrum is used in gastrointestinal diseases. *L. fumarioides* protects the DNA damage induced by catechol (Boldbaatar et al., 2014), most likely acting as potent antioxidants. *L. fumarioides* extract regulates choleresis in white rats with toxic hepatitis (Nikolaev et al., 2012).

Paraquilegia is used in blood-activating and stasis-resolving, as well as for traumatic injury. In Tibetan medicine, it is used for uterine hemorrhaging and when giving birth to a stillborn (Peng et al., 2006b).

2.5.11.2 Aquilegia Clade

Semiaquilegia is used for heat-clearing and detoxification (TRI 1164, β 100), damp-filtering and diuresis (TRI 476, β 50), ulcer disease and sores (TRI 1473, β 100), scrofula (TRI 1473, β 100), traumatic injury (TRI 550, β 100), snake bites (TRI 805, β 50), and urinary stones (TRI 305, β 50). *Semiaquilegia* has antitumor and antiinflammatory activities (Duan et al., 2013; Lee et al., 2012). *Semiaquilegia adoxoides* n-butanol extract has protective effects against hydrogen peroxide–induced oxidative stress in human lens epithelial cells (Liang et al., 2016).

Aquilegia is used for menstrual disorders (TRI 600, β 60) and uterine hemorrhages (TRI 400, β 40). *Aquilegia* flavonoids have antioxidant, antimicrobial, and hepatoprotective effects (Hassan et al., 2010). The underground parts of *Aquilegia fragrans* are traditionally used for the treatment of wounds and various inflammatory diseases like bovine mastitis (Mushtaq et al., 2016). *Aquilegia* is also used for epilepsy and as a hypnotic. Saponins show immunosuppressive activity (Nishida et al., 2003a). Alkaloids have cytotoxic activity (Chen et al., 2002). The root of *Urophysa* is used for blood-activating and stasis-removing (Xiao, 1980).

2.5.11.3 Isopyrum Clade

Isopyrum has antibacterial activity and is used for reducing swelling and resolving mass, as well as detoxification. *Dichocarpum* is used for swell-reducing and detoxification (TRI 300, β 20), ulcer disease and sores (TRI 300, β 20), and child's dyspepsia (TRI 300, β 20).

2.5.12 Coptidoideae

Coptis is used for heat-clearing and detoxification (TRI 847, β 60), heat-eliminating and fire-purging (TRI 588, β 60), heat-clearing and damp-drying (TRI 588, β 20), antimicrobial and antiinflammatory (TRI 300, β 20), dysentery (TRI 1032, β 100), gastroenteritis (TRI 712, β 80), red eye with swelling and pain (TRI 758, β 100), stomatitis (TRI 526, β 100), and ulcer disease and sores (TRI 889, β 100). *Coptis* has antibacterial, antiviral, antipyretic, anticancer, immunomodulatory, spasmolytic, antidiarrheal, anti–gastric ulcer, hypoglycemic, antiinflammatory, antihypertensive, antiplatelet aggregation, antiarteriosclerosis, and antiarrhythmia activities (Peng et al., 2006a).

Berberine (BBR) ameliorates diabetic neuropathy via PKC pathway and TRPV1 modulation (Zan et al., 2017). BBR attenuates depressive-like behaviors by suppressing neuroinflammation in stressed mice (Liu et al., 2017a,b). 8-Hydroxy-7,8-dihydrocoptisine exerts antiinflammatory activity in mice through inhibition of NF-κB and MAPK signaling pathways (Chen et al., 2017). Palmatine inhibits *Helicobacter pylori* and urease (Zhou et al., 2017a). Columbamine of Rhizoma coptidis displays the antihypercholesterolemic effect in high fat and high cholesterol (HFHC)-diet-induced hamsters through HNF-4α/FTF-mediated CYP7A1 activation (Wang et al., 2016a). Isoquinoline alkaloids of *Coptis japonica* stimulates the myoblast differentiation via p38 MAPK and Akt signaling pathways (Lee et al., 2017). *Coptis chinensis* polysaccharide shows the inhibiting effects on type II diabetic mice (Cui et al., 2016). *C. chinensis* inhibits glioma cells by reducing HDAC3 and downregulating phosphorylation of STAT3 (Li et al., 2017a). *C. chinensis* has neuroprotective effects in 1-methyl-4-phenylpyridinium (MPP) and 1-methyl-4-phenyl-1,2,3,6-tetrahydropyridine (MPTP)-induced Parkinson's disease models (Friedemann et al., 2016).

Alkaloids of *Xanthorhiza simplicissima* have antimicrobial properties (Okunade et al., 1994).

2.5.13 Hydrastidoideae and Glaucidioideae

Hydrastis has anticatarrhal, antiinflammatory, and antiseptic activities and is used as an astringent, bitter tonic, laxative, antidiabetic, and muscular stimulant. The astringent effect is on mucous membranes of the upper respiratory tract, the gastrointestinal tract, the bladder, the rectum, and the skin. *Hydrastis* stimulates the appetite and bile secretion and aids digestion.

3,3′-Dihydroxy-5,7,4′-trimethoxy-6,8-C-dimethylflavone of *Hydrastis canadensis* (goldenseal), when tested in combination with BBR, lowers the IC_{50} of BBR against *S. aureus* from 132.2 ± 1.1–91.5 ± 1.1 µM (Britton et al., 2017). 3,5,3′-Trihydroxy-7,4′-dimethoxy-6,8-C-dimethyl-flavone of goldenseal shows bacterial efflux pump inhibitory activity against *S. aureus* (Leyte-Lugo et al., 2017).

(-)-β-Hydrastine suppresses the proliferation and invasion of human lung adenocarcinoma cells by inhibiting p21-activated kinase 4 activity (Guo et al., 2016). Ultra-highly diluted extracts of *H. canadensis* and *Marsdenia condurango* induce epigenetic modifications and alter gene expression profiles in HeLa cells in vitro (Saha et al., 2015).

Glaupalol-related coumarins from *Glaucidium* show antimitotic activity (Morita et al., 2004).

2.6 DISCUSSION

2.6.1 The Relationship Between Ranunculaceae Chemical Composition and Systematics

There are about 60 genera and 2200 species in Ranunculaceae, leading to complex chemical composition. The most prominent alkaloid compounds are benzylisoquinoline alkaloid, bisbenzylisoquinoline alkaloid, apophinoid alkaloid (e.g., magnoflorine), and protoberberine alkaloid. Triterpenoid saponin, lactone (e.g., ranunculin), and cyanogenic glycoside, are also commonly found in some genera (Table 2.1); flavonoids are present in most genera. Other compounds are distributed selectively, reflecting the phylogenetic relationship between taxonomic groups.

It was believed that ranunculin and magnoflorine appear alternately (Xiao, 1980); however, they are now found to coexist in *Clematis*, *Caltha*, and *Helleborus*. Both α- and β-magnoflorine were isolated from the aerial parts of *Clematis parviloba* (Chen et al., 2009). Magnoflorine was also isolated from *Clematis recta* and *Helleborus viridis* (Slavík et al., 1987). *Caltha* species contain the irritant glycoside ranunculin (Knight, 2007). These results suggest that it is reasonable to put these genera into the same subfamily, Ranunculoideae.

The chemical profiles of *Cimicifuga*, *Actaea*, *Souliea*, and *Beesia* are similar. Triterpenoid saponins, especially the cycloartane-type ones, are abundant in these genera. Magnoflorine and ranunculin are not found in them; this is distinct from other tribes of Ranunculoideae plants, justifying treatment of a separate tribe, Cimicifugeae.

No or very few benzylisoquinoline alkaloid is found in Ranunculeae, Anemoneae, and Callianthemeae (Table 2.1). Ranunculeae and Anemoneae are rich in saponins, especially the pentacyclic triterpenoid ones, embodying the close relationship between these tribes.

Cyanogenic glycoside is the outstanding feature of the subfamily Thalictroideae. Thalictroideae and Coptidoideae are rich in alkaloids. Benzylisoquinoline alkaloid and bisbenzylisoquinoline alkaloid are abundant in Thalictroideae, while protoberberine alkaloid is copious in Coptidoideae. *Thalictrum*, *Paraquilegia* and *Aquilegia* contain triterpenoid saponin (e.g., cycloartane saponin), indicating the close relationship between Thalictroideae and Ranunculoideae. The sterile stamens of *Aquilegia* and *Semiaquilegia* are morphologically very similar, and these two genera have very similar flavonoid C-glycosides, suggesting their close phylogenetic relationship.

TABLE 2.1 Distribution of Ranunculaceae Compounds Detected by TLC

Taxon	Alkaloid			Ranunculin	Triterpene		Cardiac glycoside	Cyanogenic glycoside
	Diterpene alkaloid	Magnoflorine	Other isoquinoline alkaloids		Tetracyclic	Pentacyclic		
Ranunculoideae								
Ranunculeae								
Ranunculus		−	−	++		+		+
Batrachium		−	−	+				
Anemoneae								
Clematis		−	−	++		++		
Pulsatilla		−	−	+		+		
Anemone		−	−	++		++		
Hepatica		−	−	+				
Callianthemeae		−	−	−				
Asteropyreae	+	+	++	−				
Caltheae	+	+	+	−				
Cimicifugeae								
Actaea		−	−	−	+			
Cimicifuga		−	−	−	+			
Eranthis		−	−	−				
Helleboreae		−	+	+			+	
Nigelleae	+	+	−	−				
Delphinieae								
Delphinium	++	+		−				
Consolida	+	+		−				
Aconitum	++	+	+	+[a]				
Adonideae								

(*Continued*)

TABLE 2.1 Distribution of Ranunculaceae Compounds Detected by TLC (cont.)

Taxon	Alkaloid			Ranunculin	Triterpene		Cardiac glycoside	Cyanogenic glycoside
	Diterpene alkaloid	Magnoflorine	Other isoquinoline alkaloids		Tetracyclic	Pentacyclic		
Adonis	+	+		–			++	
Trollius	+	+		–				
Thalictroideae								
Isopyrum		+	+	–				+
Aquilegia		+	+	–				+
Semiaquilegia		+		–				+
Leptopyrum		+		–				+
Thalictrum		+	++	–				+
Coptidoideae		+	++	–				

+, Detected by thin layer chromatography (TLC); ++, more abundant than "+"; –, not detected by TLC.
a Found in the root of A. scaposum var. vaginatum from Sichuan, China.

Benzylisoquinoline alkaloids are of the monophyletic origin (Liscombe et al., 2005; Zhu and Xiao, 1991). Ranunculin is only abundant in the tribes Ranunculeae, Anemoneae, and Helleboreae (Table 2.1) (Xiao, 1980). Diterpene alkaloids are the characteristic of the tribe Delphinieae (Xiao, 1980), although a few types and a little amount of diterpene alkaloids have been detected in *Nigella* and *Thalictrum* (Chen et al., 2014; Hao et al., 2013a). Ferulic + sinapic acid is abundant in *Trollius* and *Adonis* and is present in *Thalictrum* in small amounts (Jensen, 1995). These chemotaxonomic markers, along with molecular markers, morphology, and cytology data, have shown their utility in the comprehensive analysis of the phylogeny of Ranunculales (Wang et al., 2009). Similarly, volatile compounds, such as fatty acid derivatives, benzenoids, phenylpropanoids, nitrogen-containing compounds, and terpenes, are promising chemotaxonomic markers (Jürgens and Dötterl, 2004). In the future, based on the accumulated metabolomic data of the respective genus, triterpenoid saponin, cardiac glycoside, and cyanogenic glycoside could be useful in the phylogenetic analysis of Ranunculaceae.

2.6.2 The Relationship Between Chemical Composition and Therapeutic Effect of Ranunculaceae

Many Ranunculaceae plants are traditionally used for heat-clearing and detoxification (Table 2.2), ulcer disease and sores, antimicrobial, and antiinflammation, but the effective ingredients may vary. For example, ranunculin and protoanemonin lactone, triterpenoid saponin, benzylisoquinoline alkaloid, and protoberberine have antibacterial activity.

Cimicifugeae is used in wind-dispersing and detoxification, as well as promoting eruption. The use of Cimicifugeae in benefiting Qi and raising Yang, and treating archoptosis and uterine prolapsed, is unique among Ranunculaceae plants and is distinct from the ethnopharmacological use of the closely related tribe Helleboreae. The estrogen-like effects of *Cimicifuga* (*Actaea*) *racemosa* are exceptional, as no plants of other tribes/subfamilies show such activities.

Many saponins have anticancer/cytotoxic activity (Table 2.3). Most saponin-containing plants, for example, those of Ranunculeae, Anemoneae, Cimicifugeae, Nigelleae, and Thalictroideae, are used for swell-reducing and pain-relieving and ulcer disease and sores. These plants are extremely valuable for researching druglike anticancer compounds.

The tribe Delphinieae is rich in unique diterpenoid alkaloid, which has broad spectrum pharmacological effects (Table 2.3), not limited to analgesics and anti-arthritis. Small amounts of diterpenoid alkaloid are found in *Thalictrum*, *Nigella*, and *Clematis*, which might contribute to the versatile therapeutic use of these genera, but their exact modes of action await further study.

Thalictroideae and Coptidoideae are rich in benzylisoquinoline alkaloids that make them valuable for antibacterial, antiviral, and anticancer compounds. In addition, Thalictroideae with bisbenzylisoquinoline is used in heat-clearing and blood-cooling. They might be promising lead compounds for use against hypertension and other cardiovascular diseases.

The common-folk perception of Ranunculaceae plants might therefore require reconsideration, as they may possess more general properties and/or novel bioactivities. In the broader context, ethnopharmacologic indications for all herbal remedies should be revisited in light of the explosion in understanding of modes of action of small molecule effectors for which Ranunculaceae compounds are only representative examples.

TABLE 2.2 Top 11 Ethnopharmacological Uses of Ranunculaceae Plants

Taxon	Alkaloid			Ranunculin	Triterpene		Cardiac glycoside	Cyanogenic glycoside
	Diterpene alkaloid	Magnoflorine	Other isoquino-line alkaloids		Tetracyclic	Pentacyclic		
Ranunculoideae								
Ranunculeae								
Ranunculus		+			+	+	+	
Batrachium								
Halerpestes				+	+			
Oxygraphis				+				
Anemoneae								
Clematis		+	+	+	+	+	+	+
Naravelia								
Pulsatilla	+						+	
Anemone	+	+	+	+	+	+	+	+
Hepatica			+					
Callianthemeae	+	+						
Asteropyreae	+	+						
Caltheae				+	+			
Cimicifugeae								
Actaea	+	+						
Cimicifuga	+		+					
Souliea	+							
Eranthis								
Beesia	+		+	+	+			

Helleboreae						+		+		+	
Nigelleae	+	+							+		
Delphinieae											
Delphinium	+	+			+	+	+	+	+	+	
Consolida		+					+				
Aconitum	+	+		+	+	+	+	+		+	
Adonideae											
Adonis			+							+	
Trollius						+			+		
Calathodes							+				
Thalictroideae											
Isopyrum				+					+		
Dichocarpum				+						+	
Aquilegia						+		+		+	
Semiaquilegia				+				+		+	
Urophysa						+					
Paraquilegia						+					
Leptopyrum											
Thalictrum	+		+					+	+	+	
Coptidoideae			+						+	+	
Frequency	6	6	6	7	8	8	10	11	12	13	13

TABLE 2.3 Top Nine Bioactivities of Ranunculaceae Plants

Taxon	Alkaloid			Ranunculin	Triterpene		Cardiac glycoside	Cyanogenic glycoside
	Diterpene alkaloid	Magnoflorine	Other isoquinoline alkaloids		Tetracyclic	Pentacyclic		
Ranunculoideae								
Ranunculeae								
Ranunculus	+	+	+	+	+		+	
Anemoneae								
Clematis	+	+	+	+	+			+
Pulsatilla	+	+	+				+	
Anemone	+	+	+	+				
Callianthemeae								
Caltheae								
Cimicifugeae								
Actaea		+	+					
Cimicifuga				+	+			+
Anemonopsis	+	+	+	+				
Beesia	+		+					
Helleboreae	+	+	+	+		+		
Nigelleae	+	+	+	+		+		
Delphinieae								
Delphinium	+	+	+	+	+	+	+	
Aconitum	+	+		+			+	

Adonideae									
Adonis	+								
Trollius		+			+		+		
Thalictroideae									
Isopyrum		+							
Aquilegia	+	+			+			+	+
Semiaquilegia	+		+						
Thalictrum	+	+		+		+			
Coptidoideae	+	+	+	+	+	+			
Hydrastidoideae			+						+
Glaucidioideae	+								
Frequency	17	15	14	11	11	6	5	5	4

2.7 CONCLUSION

Biodiversity and its compositional chemical diversity have served as one of the richest sources of bioprospecting, resulting in the discovery of some of the most important clinical drugs. Ranunculaceae plants have a huge reservoir of chemical constituents, which are not distributed randomly and thus have taxonomic implication. Alkaloids, saponin, and ranunculin are the main chemical features and could be useful in chemotaxonomy. The new chemical profile data, based on the more sensitive analytical technology, show that the distribution of ranunculin and magnoflorine are not mutually exclusive. This book agrees with Wang et al.'s system (2009) that divides Ranunculaceae into 5 subfamilies and the subfamily Ranunculeae into 10 tribes. More species could be discovered in biodiversity hot spots, and there is a lack of chemical data in many Ranunculaceae genera, for example, *Dichocarpum, Paraquilegia, Urophysa, Hepatica, Naravelia, Oxygraphis*, and *Halerpestes*, implying that the biological and chemical space for exploration is still wide open. Pharmacophylogeny, as a fascinating approach of mining chemodiversity from biodiversity, could be further developed and used in the future to accelerate the pace of drug discovery.

References

Ahmad, M., Yousuf, S., Khan, M.B., et al., 2006. Protective effects of ethanolic extract of *Delphinium denudatum* in a rat model of Parkinson's disease. Hum. Exp. Toxicol. 25 (7), 361–368.

Ahmad, H., Ahmad, S., Shah, S.A.A., et al., 2017a. Antioxidant and anticholinesterase potential of diterpenoid alkaloids from *Aconitum heterophyllum*. Bioorg. Med. Chem. 25 (13), 3368–3376.

Ahmad, H., Ahmad, S., Shah, S.A.A., et al., 2017b. Selective dual cholinesterase inhibitors from *Aconitum laeve*. J. Asian Nat. Prod. Res.doi: 10.1080/10286020.2017.1319820.

Ahmad, H., Ahmad, S., Khan, E., et al., 2017c. Isolation, crystal structure determination and cholinesterase inhibitory potential of isotalatizidine hydrate from *Delphinium denudatum*. Pharm. Biol. 55 (1), 680–686.

Akhalwaya, S., van Vuuren, S., Patel, M., 2018. An in vitro investigation of indigenous South African medicinal plants used to treat oral infections. J. Ethnopharmacol. 210, 359–371.

Alshatwi, A.A., 2014. Bioactivity-guided identification to delineate the immunomodulatory effects of methanolic extract of *Nigella sativa* seed on human peripheral blood mononuclear cells. Chin. J. Integr. Med.doi: 10.1007/s11655-013-1534-3.

Alu'datt, M.H., Rababah, T., Alhamad, M.N., et al., 2017. Molecular characterization and bio-functional property determination using SDS-PAGE and RP-HPLC of protein fractions from two Nigella species. Food Chem. 230, 125–134.

An, F., Wang, S., Tian, Q., et al., 2015. Effects of orientin and vitexin from *Trollius chinensis* on the growth and apoptosis of esophageal cancer EC-109 cells. Oncol. Lett. 10 (4), 2627–2633.

Angiosperm Phylogeny Group, 2003. An update of the Angiosperm Phylogeny Group classification for the orders and families of flowering plants: APG II. Bot. J. Linn. Soc. 141 (4), 399–436.

Apetrei, N.S., C lug ru, A., Kerek, F., et al., 2011. A highly purified vegetal fraction able to modulate HMGB1 and to attenuate septic shock in mice. Roum. Arch. Microbiol. Immunol. 70 (3), 114–123.

Ashley, R.E., Osheroff, N., 2014. Natural products as topoisomerase II poisons: effects of thymoquinone on DNA cleavage mediated by human topoisomerase IIα. Chem. Res. Toxicol. 27 (5), 787–793.

Ashraf, M.N., Gavrilovici, C., Shah, S.U., et al., 2013. A novel anticonvulsant modulates voltage-gated sodium channel inactivation and prevents kindling-induced seizures. J. Neurochem. 126 (5), 651–661.

Biglar, M., Sufi, H., Bagherzadeh, K., et al., 2014. Screening of 20 commonly used Iranian traditional medicinal plants against urease [J]. Iran. J. Pharm. Res. 13 (Suppl), 195–198.

Boldbaatar, D., El-Seedi, H.R., Findakly, M., et al., 2014. Antigenotoxic and antioxidant effects of the Mongolian medicinal plant *Leptopyrum fumarioides* (L.): an in vitro study. J. Ethnopharmacol. 155 (1), 599–606.

Britton, E.R., Kellogg, J.J., Kvalheim, O.M., et al., 2017. Biochemometrics to identify synergists and additives from botanical medicines: a case study with *hydrastis canadensis* (goldenseal). J. Nat. Prod. 7, b00654. doi: 10.1021/acs. jnatprod.

Bylka, W., Szaufer-Hajdrych, M., Matławska, I., et al., 2004. Antimicrobial activity of isocytisoside and extracts of *Aquilegia vulgaris* L. Lett. Appl. Microbiol. 39 (1), 93–97.

Cai, S.Q., Wang, R., Yang, X., et al., 2006. Antiviral flavonoid-type C-glycosides from the flowers of *Trollius chinensis*. Chem. Biodivers. 3 (3), 343–348.

Cakar, J., Paric′, A., Vidic, D., et al., 2011. Antioxidant and antiproliferative activities of *Helleborus odorus* Waldst. & Kit, *H. multifidus* Vis. and *H. hercegovinus* Martinis. Nat. Prod. Res. 25 (20), 1969–1974.

Chawla, R., Kumar, S., Sharma, A., 2012. The genus *Clematis* (Ranunculaceae): chemical and pharmacological perspectives. J. Ethnopharmacol. 143 (1), 116–150.

Chen, H.J., Liu, J., 2017. Actein ameliorates hepatic steatosis and fibrosis in high fat diet-induced NAFLD by regulation of insulin and leptin resistant. Biomed. Pharmacother. 97, 1386–1396.

Chen, S.B., Gao, G.Y., Li, Y.S., et al., 2002. Cytotoxic constituents from *Aquilegia ecalcarata*. Planta Med. 68 (6), 554–556.

Chen, J.H., Du, Z.Z., Shen, Y.M., et al., 2009. Aporphine alkaloids from *Clematis parviloba* and their antifungal activity. Arch. Pharm. Res. 32 (1), 3–5.

Chen, S., Wan, L., Couch, L., et al., 2013. Mechanism study of goldenseal-associated DNA damage. Toxicol. Lett. 221 (1), 64–72.

Chen, D.L., Tang, P., Chen, Q.H., et al., 2014. New C20-diterpenoid alkaloids from *Delphinium laxicymosum* var. *pilostachyum*. Nat. Prod. Commun. 9 (5), 623–625.

Chen, Q.B., Xin, X.L., Yang, Y., et al., 2014a. Highly conjugated norditerpenoid and pyrroloquinoline alkaloids with potent PTP1B inhibitory activity from *Nigella glandulifera*. J. Nat. Prod. 77 (4), 807–812.

Chen, H.B., Luo, C.D., Liang, J.L., et al., 2017. Anti-inflammatory activity of coptisine free base in mice through inhibition of NF-κB and MAPK signaling pathways. Eur. J. Pharmacol. 811, 222–231.

Cheng, W., Tan, Y.F., Tian, H.Y., et al., 2014. Two new bufadienolides from the rhizomes of *Helleborusthibetanus* with inhibitory activities against prostate cancer cells. Nat. Prod. Res. 28 (12), 901–908.

Chinnasamy, S., Nagarajan, S., Sivaraman, T., et al., 2017. Computational and in vitro insights on snake venom phospholipase A2 inhibitor of phytocompound ikshusterol 3-O-glucoside of *Clematis gouriana* Roxb. ex DC. J. Biomol. Struct. Dyn., 1409653. doi: 10.1080/07391102. 2017.

Choi, J.Y., Jeon, S.J., Son, K.H., et al., 2016. Induction of mast cell degranulation by triterpenoidal saponins obtained from *Cimicifugae rhizoma*. Immunopharmacol. Immunotoxicol. 38 (5), 311–318.

Cui, L., Liu, M., Chang, X., et al., 2016. The inhibiting effect of the *Coptis chinensis* polysaccharide on the type II diabetic mice. Biomed. Pharmacother. 81, 111–119.

Dai, Y., Zhang, B.B., Xu, Y., et al., 2010. Chemical constituents of *Adonis coerulea* Maxim. Nat. Prod. Res. Dev. 22 (4), 594–596.

Doncheva, T., Solongo, A., Kostova, N., et al., 2015. Leptopyrine, new alkaloid from *Leptopyrum fumarioides* L. (Ranunculaceae). Nat. Prod. Res. 29 (9), 853–856.

Duan, S.P., Jin, C.L., Hao, J., et al., 2013. A study on the inhibitory effect of Radix Semiaquilegiae extract on human hepatoma HepG-2 and SMMC-7721 cells. Afr. J. Tradit. Complement. Altern. Med. 10 (5), 336–340.

Duckstein, S.M., Stintzing, F.C., 2014. Comprehensive study of the phenolics and saponins from *Helleborusniger* L. Leaves and stems by liquid chromatography/tandem mass spectrometry. Chem. Biodivers. 11 (2), 276–298.

Duckstein, S.M., Lorenz, P., Conrad, J., et al., 2014. Tandem mass spectrometric characterization of acetylated polyhydroxy hellebosaponins, the principal steroid saponins in *Helleborus niger* L. roots. Rapid Commun. Mass Spectrom. 28 (16), 1801–1812.

Erdemoglu, N., Küpeli, E., Yes̗ilada, E., 2003. Anti-inflammatory and antinociceptive activity assessment of plants used as remedy in Turkish folk medicine. J. Ethnopharmacol. 89 (1), 123–129.

Feng, H.T., Zhao, W.W., Lu, J.J., et al., 2017. Hypaconitine inhibits TGF-β1-induced epithelial-mesenchymal transition and suppresses adhesion, migration, and invasion of lung cancer A549 cells. Chin. J. Nat. Med. 15 (6), 427–435.

Franz, M.H., Birzoi, R., Maftei, C.V., et al., 2017. Studies on the constituents of Helleborus purpurascens: analysis and biological activity of the aqueous and organic extracts. Amino Acidsdoi: 10.1007/s00726-017-2502-6.

Friedemann, T., Ying, Y., Wang, W., et al., 2016. Neuroprotective effect of *Coptis chinensis* in MPP and MPTP-induced Parkinson's disease models. Am. J. Chin. Med. 44 (5), 907–925.

Fu, Q., Zhou, C., Ma, Y., et al., 2017. Lipoxygenase-inhibiting lignans from *Clematis mandshurica*. J. Asian. Nat. Prod. Res. 19 (9), 884–889.

Gao, L., Zheng, T., Xue, W., et al., 2017. Efficacy and safety evaluation of *Cimicifuga foetida* extract in menopausal women. Climacteric, 1406913. doi: 10.1080/13697137. 2017.

Guo, B., Li, X., Song, S., et al., 2016. (-)-β-hydrastine suppresses the proliferation and invasion of human lung adenocarcinoma cells by inhibiting PAK4 kinase activity. Oncol. Rep. 35 (4), 2246–2256.

Guo, L., Qiao, S., Hu, J., et al., 2017a. Investigation of the effective components of the flowers of *Trollius chinensis* from the perspectives of intestinal bacterial transformation and intestinal absorption. Pharm. Biol. 55 (1), 1747–1758.

Guo, Y., Yin, T., Wang, X., et al., 2017b. Traditional uses, phytochemistry, pharmacology and toxicology of the genus *Cimicifuga*: a review. J. Ethnopharmacol. 209, 264–282.

Hao, D.C., Gu, X.J., Xiao, P.G., et al., 2013a. Recent advances in the chemical and biological studies of *Aconitum* pharmaceutical resources. J. Chin. Pharm. Sci. 22 (3), 209–221.

Hao, D.C., Gu, X.J., Xiao, P.G., et al., 2013b. Recent advance in chemical and biological studies on *Cimicifugeae* pharmaceutical resources. Chin. Herb. Med. 5 (2), 81–95.

Hao, D.C., Gu, X.J., Xiao, P.G., Peng, Y., 2013c. Chemical and biological research of *Clematis* medicinal resources. Chin. Sci. Bull. 58 (10), 1120–1129.

Hao, D.C., Gu, X., Xiao, P.G., 2015. Medicinal Plants: Chemistry, Biology and Omics, first ed. Elsevier-Woodhead, Oxford, ISBN 9780081000854.

Hassan, A.M., Mohamed, S.R., El-Nekeety, A.A., et al., 2010. *Aquilegia vulgaris* L. extract counteracts oxidative stress and cytotoxicity of fumonisin in rats. Toxicon 56 (1), 8–18.

He, J.B., Luan, J., Lv, X.M., et al., 2017. Navicularines A-C: new diterpenoid alkaloids from *Aconitum naviculare* and their cytotoxic activities. Fitoterapia 120, 142–145.

Hosseini, M., Mohammadpour, T., Karami, R., et al., 2015. Effects of the hydro-alcoholic extract of *Nigella sativa* on scopolamine-induced spatial memory impairment in rats and its possible mechanism. Chin. J. Integr. Med. 21 (6), 438–444.

Hu, Y., Mao, A., Tan, Y., et al., 2016. Role of 5 Saponins in secretion of cytokines by PRRSV-induced endothelial cells. Drug. Res. 66 (7), 357–362.

Ibanez, S., Gallet, C., Dommanget, F., Després, L., 2009. Plant chemical defence: a partner control mechanism stabilizing plant--seed-eating pollinator mutualisms. BMC Evol. Biol. 9, 261.

Ibrahim, R.M., Hamdan, N.S., Mahmud, R., et al., 2014. A randomised controlled trial on hypolipidemic effects of *Nigella sativa* seeds powder in menopausal women. J. Transl. Med. 12 (1), 82.

Ip, F.C.F., Ng, Y.P., Or, T.C.T., et al., 2017. Anemoside A3 ameliorates experimental autoimmune encephalomyelitis by modulating T helper 17 cell response. PLoS One 12 (7), e0182069.

Jairajpuri, D.S., Jairajpuri, Z.S., 2016. Isoferulic acid action against glycation-induced changes in structural and functional attributes of human high-density lipoprotein. Biochemistry 81 (3), 289–295.

Jaroszewski, J.W., Staerk, D., Holm-Móller, S.B., et al., 2005. *Naravelia zeylanica* occurrence of primary benzamides in flowering plants. Nat. Prod. Res. 19 (3), 291–294.

Jensen, U., 1995. Secondary compounds of the Ranunculiflorae. Plant Syst. Evol. 9 (Suppl), 85–97.

Jian, X.X., Tang, P., Liu, X.X., et al., 2012. Structure-cardiac activity relationship of C19-diterpenoid alkaloids. Nat. Prod. Commun. 7 (6), 713–720.

Jie-Shi, Y., Wei-Sang, L., Yan, R., et al., 2017. Two new compounds from *Trollius chinensis* Bunge. J. Nat. Med. 71 (1), 281–285.

Jung, J.W., Baek, N.I., Hwang-Bo, J., et al., 2015. Two new cyotoxic cardenolides from the whole plants of *Adonis multiflora* Nishikawa & Koki Ito. Molecules 20 (11), 20823–20831.

Jürgens, A., Dötterl, S., 2004. Chemical composition of anther volatiles in Ranunculaceae: genera-specific profiles in *Anemone, Aquilegia, Caltha, Pulsatilla, Ranunculus*, and *Trollius* species. Am. J. Bot. 91 (12), 1969–1980.

Karimi, E., Ghorbani Nohooji, M., Habibi, M., et al., 2017. Antioxidant potential assessment of phenolic and flavonoid rich fractions of *Clematis orientalis* and *Clematis ispahanica* (Ranunculaceae). Nat. Prod. Res., 1359171.

Khan, H., Nabavi, S.M., Sureda, A., et al., 2017. Therapeutic potential of songorine, a diterpenoid alkaloid of the genus *Aconitum*. Eur. J. Med. Chem.doi: 10.1016/j. ejmech. 2017. 10.065.

Knight, A., 2007. A Guide to Poisonous House and Garden Plants. CRC Press.

Kokoska, L., Urbanova, K., Kloucek, P., et al., 2012. Essential oils in the ranunculaceae family: chemical composition of hydrodistilled oils from *Consolidaregalis, Delphinium elatum, Nigella hispanica*, and *N. nigellastrum* seeds. Chem. Biodivers. 9 (1), 151–161.

Kolahdooz, M., Nasri, S., Modarres, S.Z., et al., 2014. Effects of *Nigella sativa* L. seed oil on abnormal semen quality in infertile men: a randomized, double-blind, placebo-controlled clinical trial. Phytomedicine 21 (6), 901–905.

Koparal, A.T., Bostancıoğlu, R.B., 2016. Promotion of hair growth by traditionally used *Delphinium staphisagria* seeds through induction of angiogenesis. Iran. J. Pharm. Res. 15 (2), 551–560.

Kubo, S., Kuroda, M., Matsuo, Y., et al., 2012a. New cardenolides from the seeds of *Adonis aestivalis*. Chem. Pharm. Bull. 60 (10), 1275–1282.

Kubo, S., Kuroda, M., Matsuo, Y., et al., 2012b. New cardenolides from the seeds of *Adonis aestivalis*. Chem. Pharm. Bull. 60 (10), 1275–1282.

Kubo, S., Kuroda, M., Yokosuka, A., et al., 2015. Amurensiosides L-P, five new cardenolide glycosides from the roots of *Adonis amurensis*. Nat. Prod. Commun. 10 (1), 27–32.

Kumar, V.K., Lalitha, K.G., 2017. Pharmacognostical and phytochemical studies of *Helleborus niger* L. root. Anc. Sci. Life 36 (3), 151–158.

Kuroda, M., Uchida, S., Watanabe, K., et al., 2009. Chromones from the tubers of *Eranthis cilicica* and their antioxidant activity. Phytochemistry 70 (2), 288–293.

Kuroda, M., Kubo, S., Uchida, S., et al., 2010. Amurensiosides A-K11 new pregnane glycosides from the roots of *Adonis amurensis* [J]. Steroids 75 (1), 83–94.

Lee, C.L., Hwang, T.L., Peng, C.Y., et al., 2012. Anti-neutrophilic inflammatory secondary metabolites from the traditional Chinese medicine, Tiankuizi. Nat. Prod. Commun. 7 (12), 1623–1626.

Lee, H., Tuong, L.T., Jeong, J.H., et al., 2017. Isoquinoline alkaloids from *Coptis japonica* stimulate the myoblast differentiation via p38 MAP-kinase and Akt signaling pathway. Bioorg. Med. Chem. Lett. 27 (6), 1401–1404.

Leyte-Lugo, M., Britton, E.R., Foil, D.H., et al., 2017. Secondary metabolites from the leaves of the medicinal plant goldenseal (Hydrastis canadensis). Phytochem. Lett. 20, 54–60.

Li, X.C., Yang, C.R., Liu, Y.Q., et al., 1995. Triterpenoid glycosides from *Anemoclema glaucifolium*. Phytochemistry 39 (5), 1175–1179.

Li, Y.L., Ma, S.C., Yang, Y.T., et al., 2002. Antiviral activities of flavonoids and organic acid from *Trollius chinensis* Bunge. J. Ethnopharmacol. 79 (3), 365–368.

Li, D.Y., Wei, J.X., Hua, H.M., et al., 2014. Antimicrobial constituents from the flowers of *Trollius chinensis*. J. Asian Nat. Prod. Res. 16 (10), 1018–1023.

Li, Y., Zheng, J.Y., Liu, J.Q., et al., 2016. Succinate/NLRP3 inflammasome induces synovial fibroblast activation: therapeutical effects of clematichinenoside AR on arthritis. Front. Immunol. 7, 532.

Li, J., Ni, L., Li, B., et al., 2017a. *Coptis chinensis* affects the function of glioma cells through the down-regulation of phosphorylation of STAT3 by reducing HDAC3. BMC Complement. Altern. Med. 17 (1), 524.

Li, G.Q., Zhang, L.M., Zhao, D.K., et al., 2017b. Two new C19-diterpenoid alkaloids from *Aconitum tsaii*. J. Asian Nat. Prod. Res. 19 (5), 457–461.

Li, X., Gu, L., Yang, L., et al., 2017c. Aconitine: a potential novel treatment for systemic lupus erythematosus. J. Pharmacol. Sci. 133 (3), 115–121.

Liang, B., Wei, W., Wang, J., et al., 2016. Protective effects of *Semiaquilegia adoxoides* n-butanol extract against hydrogen peroxide-induced oxidative stress in human lens epithelial cells. Pharm. Biol. 54 (9), 1656–1663.

Liang, X., Chen, L., Song, L., et al., 2017. Diterpenoid alkaloids from the root of *Aconitum sinchiangense* W T. Wang with their antitumor and antibacterial activities. Nat. Prod. Res. 31 (17), 2016–2023.

Lin, C.Z., Zhao, Z.X., Xie, S.M., et al., 2014. Diterpenoid alkaloids and flavonoids from *Delphinium trichophorum*. Phytochemistry 97, 88–95.

Lin, C.Z., Liu, Z.J., Bairi, Z.D., et al., 2017. A new diterpenoid alkaloid isolated from *Delphinium caeruleum*. Chin. J. Nat. Med. 15 (1), 45–48.

Liscombe, D.K., MacLeod, B.P., Loukanina, N., et al., 2005. Evidence for the monophyletic evolution of benzylisoquinoline alkaloid biosynthesis in angiosperms. Phytochemistry 66 (20), 2501–2520.

Littmann, L., Rössner, S., Kerek, F., et al., 2008. Modulation of murine bone marrow-derived dendritic cells and B-cells by MCS-18 a natural product isolated from *Helleborus purpurascens*. Immunobiology 213 (9-10), 871–878.

Liu, J.Y., Li, S.Y., Feng, J.Y., et al., 2013. Flavone C-glycosides from the flowers of *Trollius chinensis* and their anticomplementary activity. J. Asian Nat. Prod. Res. 15 (4), 325–331.

Liu, S.Y., Xu, P., Luo, X.L., et al., 2016. (7R,8S)-dehydrodiconiferyl alcohol suppresses lipopolysaccharide-induced inflammatory responses in BV2 microglia by inhibiting MAPK signaling. Neurochem. Res. 41 (7), 1570–1577.

Liu, Y.M., Niu, L., Wang, L.L., et al., 2017a. Berberine attenuates depressive-like behaviors by suppressing neuroinflammation in stressed mice. Brain Res. Bull. 134, 220–227.

Liu, Q., Xiao, X.H., Hu, L.B., et al., 2017b. Anhuienoside C ameliorates collagen-induced arthritis through inhibition of MAPK and NF-κB signaling pathways. Front. Pharmacol. 8, 299.

Lv, C., Yang, F., Qin, R., et al., 2017. Bioactivity-guided isolation of chemical constituents against H2O2-induced neu-rotoxicity on PC12 from *Cimicifuga dahurica* (Turcz) Maxim. Bioorg. Med. Chem. Lett. 27 (15), 3305–3309.

Ma, Y.C., Luo, M., Peng, J., Lin, X.H., 1992. Quantitative comparison of the active components of *Asteropyrum*. Chin. J. Chin. Mat. Med. 17 (11), 679.

Maham, M., Sarrafzadeh-Rezaei, F., 2014. Cardiovascular effects of *Adonis aestivalis* in anesthetized sheep. Vet. Res. Forum. 5 (3), 193–199.

Marín, C., Ramírez-Macías, I., López-Céspedes, A., et al., 2011. In vitro in vivo trypanocidal activity of flavonoids from *Delphinium staphisagria* against Chagas disease. Nat. J.Prod. 74 (4), 744–750.

McConnell, M.T., Lisgarten, D.R., Byrne, L.J., et al., 2015. Winter aconite (*Eranthis hyemalis*) lectin as a cytotoxic effec-tor in the lifecycle of *Caenorhabditis elegans*. Peer J. 3, e1206.

Miao, C., Zhou, G., Qin, M., et al., 2015. Pulchinenoside inhibits the fibroblast-like synoviocytes apoptosis in adju-vant arthritis rats. Zhong Nan Da Xue Xue Bao Yi Xue Ban 40 (2), 144–149.

Mohadjerani, M., Tavakoli, R., Hosseinzadeh, R., 2014. Fatty acid composition, antioxidant and antibacterial activi-ties of *Adonis wolgensis* L. extract. Avicenna J. Phytomed. 4 (1), 24–30.

Morita, H., Dota, T., Kobayashi, J., 2004. Antimitotic activity of glaupalol-related coumarins from *Glaucidiumpalma-tum*. Bioorg. Med. Chem. Lett. 14 (14), 3665–3668.

Mubashir, S., Dar, M.Y., Lone, B.A., et al., 2014. Anthelmintic, antimicrobial, antioxidant and cytotoxic activity of *Caltha palustris* var. *alba* Kashmir, India. Chin. J. Nat. Med. 12 (8), 567–572.

Mushtaq, S., Aga, M.A., Qazi, P.H., et al., 2016. Isolation, characterization and HPLC quantification of compounds from *Aquilegia fragrans* Benth: their in vitro antibacterial activities against bovine mastitis pathogens. J. Ethno-pharmacol. 178, 9–12.

Neacsu, C., Ciobanu, C., Barbu, I., et al., 2010. Substance MCS-18 isolated from *Helleborus purpurascens* is a potent antagonist of the capsaicin receptor, TRPV1, in rat cultured sensory neurons. Physiol. Res. 59 (2), 289–298.

Nesterova, Y.V., Povetieva, T.N., Nagornyak, Y.G., et al., 2009. Correction of adjuvant arthritis with delphinium ex-tracts and alkaloids. Bull. Exp. Biol. Med. 147 (6), 711–714.

Nikolaev, S.M., Zandanov, A.O., Sambueva, Z.G., et al., 2012. Effect of Leptopyrum fumarioides (Ranunculaceae) extract on choleresis in white rats with toxic hepatitis. Eksp. Klin. Gastroenterol. 4, 21–24.

Nishida, M., Yoshimitsu, H., Okawa, M., et al., 2003. Four new cycloartane glycosides from *Aquilegia vulgaris* and their immunosuppressive activities in mouse allogeneic mixed lymphocyte reaction. Chem. Pharm. Bull. 51 (6), 683–687.

Nishida, M., Yoshimitsu, H., Okawa, M., et al., 2003a. Four new cycloartane glycosides from *Aquilegia vulgaris*. Chem. Pharm. Bull. 51 (8), 956–959.

Niu, F., Chang, H.T., Jiang, Y., et al., 2006. New diterpenoids from *Semiaquilegia adoxoides*. J. Asian Nat. Prod. Res. 8 (1–2), 87–91.

Noutsos, C., Perera, A.M., Nikolau, B.J., et al., 2015. Metabolomic profiling of the nectars of *Aquilegia pubescens* and *A. canadensis*. PLoS One 10 (5), e0124501.

Okunade, A.L., Hufford, C.D., Richardson, M.D., et al., 1994. Antimicrobial properties of alkaloids from *Xanthorhiza simplicissima*. J. Pharm. Sci. 83 (3), 404–406.

Pan, L.L., Zhang, Q.Y., Luo, X.L., et al., 2016a. (7R,8S)-9-acetyl-dehydrodiconiferyl alcohol inhibits inflammation and migration in lipopolysaccharide-stimulated macrophages. Phytomedicine 23 (5), 541–549.

Pan, L.L., Wang, X.L., Zhang, Q.Y., et al., 2016b. Boehmenan, a lignan from the Chinese medicinal plant *Clematis ar-mandii*, induces apoptosis in lung cancer cells through modulation of EGF-dependent pathways. Phytomedicine 23 (5), 468–476.

Paun, G., Litescu, S.C., Neagu, E., et al., 2014. Evaluation of Geranium spp., *Helleborus* spp. and *Hyssopus* spp. Poly-phenolic extracts inhibitory activity against urease and α-chymotrypsin. J. Enzyme. Inhib. Med. Chem. 29 (1), 28–34.

Peng, Y., Chen, S.B., Chen, S.L., et al., 2006a. Preliminary pharmaphylogenetic study on Ranunculaceae. Chin. J. Chin. Mat. Med. 31 (13), 1124–1128.

Peng, Y., Chen, S.B., Liu, Y., et al., 2006b. Pharmaphylogenetic study on Isopyroideae (Ranunculaceae). Chin. J. Chin. Mat. Med. 31 (14), 1210–1214.

Perveen, T., Haider, S., Zuberi, N.A., et al., 2013. Increased 5-HT levels following repeated administration of *Nigella sativa* L. (black seed) oil produce antidepressant effects in rats. Sci. Pharm. 82 (1), 161–170.

Puglisi, S., Speciale, A., Acquaviva, R., et al., 2009. Antibacterial activity of *Helleborus bocconei* Ten. subsp. siculus root extracts. J. Ethnopharmacol. 125 (1), 175–177.

Qin, R., Zhao, Y., Zhao, Y., et al., 2016. Polyphenolic compounds with antioxidant potential and neuro-protective effect from *Cimicifuga dahurica* (Turcz) Maxim. Fitoterapia 115, 52–56.

Qiu, L., Yuan, H.M., Liang, J.M., et al., 2017. Clemochinenosides C and D, two new macrocyclic glucosides from *Clematis chinensis*. J. Asian Nat. Prod. Res., 1387780. doi: 10.1080/10286020. 2017.

Ramírez-Macías, I., Marín, C., Díaz, J.G., et al., 2012. Leishmanicidal activity of nine novel flavonoids from *Delphinium staphisagria*. Sci. World J. 2012, 203646.

Reina, M., Mancha, R., Gonzalez-Coloma, A., et al., 2007. Diterpenoid alkaloids from *Delphinium gracile*. Nat. Prod. Res. 21 (12), 1048–1055.

Saha, S.K., Roy, S., Khuda-Bukhsh, A.R., 2015. Ultra-highly diluted plant extracts of *Hydrastis canadensis* and *Marsdenia condurango* induce epigenetic modifications and alter gene expression profiles in HeLa cells in vitro. J. Integr. Med. 13 (6), 400–411.

Saidi, R., Khanous, L., Khadim Allah, S., et al., 2017. Antifungal, molluscicidal and larvicidal assessment of anemonin and *Clematis flammula* L. extracts against mollusc *Galba truncatula*, intermediate host of *Fasciola hepatica* in Tunisia. Asian Pac. J. Trop. Med. 10 (10), 967–973.

Samanbay, A., Zhao, B., Aisa, H.A., 2017. A new denudatine type C20-diterpenoid alkaloid from *Aconitum sinchiangense* W T. Wang. Nat. Prod. Res.doi: 10.1080/14786419.1410814.

Sedaghat, R., Roghani, M., Khalili, M., 2014. Neuroprotective effect of thymoquinone, the nigella sativa bioactive compound, in 6-hydroxydopamine-induced hemi-parkinsonian rat model. Iran. J. Pharm. Res. 13 (1), 227–234.

Seifarth, C., Littmann, L., Resheq, Y., et al., 2011. MCS-18, a novel natural plant product prevents autoimmune diabetes. Immunol. Lett. 139 (1–2), 58–67.

Semenov, A.A., Enikeev, A.G., Snetkova, L.V., et al., 2016. Ortho-phthalic acid esters in lipophilic extract from the cell culture of *Aconitum baicalense* Turcz ex Rapaics 1907. Dokl. Biochem. Biophys. 471 (1), 421–422.

Shan, L., Zhang, J., Chen, L., et al., 2015. Two new C18-diterpenoid alkaloids from *Delphinium anthriscifolium*. Nat. Prod. Commun. 10 (12), 2067–2068.

Shan, L.H., Zhang, J.F., Gao, F., et al., 2017. C18-diterpenoid alkaloids from *Delphinium anthriscifolium* var. *majus*. J. Asian Nat. Prod. Res.doi: 10.1080/10286020.1335309.

Shang, X., Guo, X., Yang, F., et al., 2017. The toxicity and the acaricidal mechanism against *Psoroptes cuniculi* of the methanol extract of *Adonis coerulea* Maxim. Vet. Parasitol. 240, 17–23.

Singh, S., Das, S.S., Singh, G., et al., 2014. Composition vitro antioxidant antimicrobial activities of essential oil oleoresins obtained from black cumin seeds (*Nigella sativa* L.). Biomed. Res. Int., 918209.

Slavík, J., Bochor̆áková, J., Slavíková, L., 1987. Occurrence of magnoflorine and corytuberine in some wild or cultivated plants of Czechoslovakia. Collect. Czech. Chem. Commun. 52, 804–812.

Song, Z., Wang, H., Ren, B., et al., 2013. On-line study of flavonoids of *Trollius chinensis* bunge binding to DNA with ethidium bromide using a novel combination of chromatographic, mass spectrometric and fluorescence techniques. J. Chromatogr. A 1282, 102–112.

Su, Y., Zhang, Z., Guo, C., 2004. A new nitroethylphenolic glycoside from *Semiaquilegia adoxoides*. Fitoterapia 75 (3–4), 420–422.

Su, Y.F., Zhang, Z.X., Guo, C.Y., et al., 2005. A nobel cyanogenic glycoside from *Semiaquilegia adoxoides*. J. Asian Nat. Prod. Res. 7 (2), 171–174.

Su, Y., Chi, W.C., Wu, L., et al., 2016. Photochemistry and pharmacology of 9 19-cyclolanostane glycosides isolated from genus *Cimicifuga*. Chin. J. Nat. Med. 14 (10), 721–731.

Su, Y., Wu, L., Mu, G., et al., 2017. 9 19-Cycloartenol glycoside G3 from *Cimicifuga simplex* regulates immune responses by modulating Th17/Treg ratio. Bioorg. Med. Chem. 25 (17), 4917–4923.

Suh, K.S., Choi, E.M., Jung, W.W., et al., 2017. Deoxyactein protects pancreatic β-cells against methylglyoxal-induced oxidative cell damage by the upregulation of mitochondrial biogenesis. Int. J. Mol. Med. 40 (2), 539–548.

Sun, Y.X., Liu, J.C., Liu, D.Y., 2011. Phytochemicals and bioactivities of *Anemone raddeana* Regel: a review. Pharmazie 66 (11), 813–821.

Sun, Y., Yuan, H., Hao, L., et al., 2013. Enrichment and antioxidant properties of flavone C-glycosides from trollflowers using macroporous resin. Food Chem. 141 (1), 533–541.

Suszko, A., Obmi ska-Mrukowicz, B., 2013. Influence of polysaccharide fractions isolated from *Calthapalustris* L. on the cellular immune response in collagen-induced arthritis (CIA) in mice A comparison with methotrexate. J. Ethnopharmacol. 145 (1), 109–117.

Suszko, A., Obmi ska-Mrukowicz, B., 2017. Effects of polysaccharide fractions isolated from *Caltha palustris* L. on the activity of phagocytic cells & humoral immune response in mice with collagen-induced arthritis: a comparison with methotrexate. Indian J. Med. Res. 145 (2), 229–236.

Takhtajan, A., 1997. Diversity and Classification of Flowering Plants. Columbia University Press, New York.

Tamura, M., 1993. Ranunculaceae. In: Kubitzki, K., Rohwer, J.G., Bittrich, V. (Eds.), The Families and Genera of Vascular Plants II. Springer, Berlin, pp. 563–583.

Thao, N.P., Luyen, B.T.T., Lee, J.S., et al., 2017a. Inhibition potential of cycloartane-type glycosides from the roots of *Cimicifuga dahurica* against soluble epoxide hydrolase. J. Nat. Prod. 80 (6), 1867–1875.

Thao, N.P., Luyen, B.T., Lee, J.S., et al., 2017b. Soluble epoxide hydrolase inhibitors of indolinone alkaloids and phenolic derivatives from *Cimicifuga dahurica* (Turcz) Maxim. Bioorg. Med. Chem. Lett. 27 (8), 1874–1879.

Tian, Z., Xiao, P.G., Wen, J., et al., 2006. Review of bioactivities of natural cycloartane triterpenoids. Chin. J. Chin. Mat. Med. 31 (8), 625–629.

Wang, W., Lu, A.M., Ren, Y., et al., 2009. Phylogeny and classification of ranunculales evidence from four molecular loci and morphological data. Persp. Plant Ecol. Evol. Syst. 11, 81–110.

Wang, D.M., Pu, W.J., Wang, Y.H., et al., 2012. A new isorhamnetin glycoside and other phenolic compounds from *Callianthemum taipaicum*. Molecules 17 (4), 4595–4603.

Wang, R., Wu, X., Liu, L., et al., 2014. Activity directed investigation on anti-inflammatory fractions and compounds from flowers of *Trollius chinensis*. Pak. J. Pharm. Sci. 27 (2), 285–288.

Wang, Y., Han, Y., Chai, F., et al., 2016a. The antihypercholesterolemic effect of columbamine from Rhizoma coptidis in HFHC-diet induced hamsters through HNF-4α/FTF-mediated CYP7A1 activation. Fitoterapia 115, 111–121.

Wang, S., Tian, Q., An, F., et al., 2016b. Growth inhibition and apoptotic effects of total flavonoids from *Trollius chinensis* on human breast cancer MCF-7 cells. Oncol. Lett. 12 (3), 1705–1710.

Wang, X., Fan, F., Cao, Q., 2016c. Modified Pulsatilla decoction attenuates oxazolone-induced colitis in mice through suppression of inflammation and epithelial barrier disruption. Mol. Med. Rep. 14 (2), 1173–1179.

Wang, Z., Wang, Q., Zhang, M., et al., 2017. Cimicifugamide from Cimicifuga rhizomes functions as a nonselective β-AR agonist for cardiac and sudorific effects. Biomed. Pharmacother. 90, 122–130.

Watanabe, K., Mimaki, Y., Sakagami, H., et al., 2003. Bufadienolide and spirostanol glycosides from the rhizomes of *Helleborus orientalis*. J. Nat. Prod. 66 (2), 236–241.

Watanabe, K., Mimaki, Y., Sakuma, C., et al., 2003a. Eranthisaponins A and B, two new bisdesmosidic triterpene saponins from the tubers of *Eranthis cilicica*. J. Nat. Prod. 66 (6), 879–882.

Witkowska-Banaszczak, E., 2015. The genus *Trollius*-review of pharmacological and chemical research. Phytother. Res. 29 (4), 475–500.

Wu, Z.Y., Lu, A.M., Tang, Y.C., et al., 2003. The Families and Genera of Angiosperms in China, A Comprehensive Analysis. Science Press, Beijing.

Wu, L.Z., Zhang, X.P., Xu, X.D., et al., 2013. Characterization of aromatic glycosides in the extracts of *Trollius* species by ultra high-performance liquid chromatography coupled with electrospray ionization quadrupole time-of-flight tandem mass spectrometry. J. Pharm. Biomed. Anal. 75, 55–63.

Wu, H.F., Zhang, G., Wu, M.C., et al., 2016. A new cycloartane triterpene glycoside from *Souliea vaginata*. Nat. Prod. Res. 30 (20), 2316–2322.

Wu, H.F., Liu, X., Zhu, Y.D., et al., 2017a. A new cycloartane triterpenoid glycoside from *Souliea vaginata*. Nat. Prod. Res. 31 (21), 2484–2490.

Wu, H.F., Li, P.F., Zhu, Y.D., et al., 2017b. Soulieoside O, a new cyclolanostane triterpenoid glycoside from *Souliea vaginata*. J. Asian Nat. Prod. Res. 19 (12), 1177–1182.

Xia, D., Zhang, M., Shi, G., et al., 2016. Pulsatilla decoction inhibits vulvovaginal *Candida albicans* proliferation and reduces inflammatory cytokine levels in vulvovaginal candidiasis mice. Xi Bao Yu Fen Zi Mian Yi Xue Za Zhi 32 (2), 153–157.

Xiao, P.G., 1980. A preliminary study of the correlation between phylogeny, chemical constituents and pharmaceutical aspects in the taxa of Chinese *Ranunculaceae*. Acta Phytotax. Sin. 18 (2), 142–153..

Xiao, P.G., Wang, L.W., Lv, S.J., et al., 1986. Statistical analysis of the ethnopharmacologic data based on Chinese medicinal plants by electronic computer I *Magnoliidae*. Chin. J. IntegratedTrad. Western Med. 6 (4), 253–256.

Xu, H.L., 2000. Studies on alkaloids of *Asteropyrum cavaleriei* (Lévl. etVant) Drumm. Et Hutch. Chin. J. Chin. Mat. Med. 25 (8), 486–488.

Xu, K., Zhang, P., Liao, X., et al., 2010. Two new triterpene saponins from the aerial parts of *Paraquilegia microphylla*. Fitoterapia 81 (6), 581–585.

Xu, K.J., Xu, X.M., Deng, W.L., et al., 2011. Three new flavone C-glycosides from the aerial parts of *Paraquilegia microphylla*. J. Asian Nat. Prod. Res. 13 (5), 409–416.

Xu, L., Cheng, G., Lu, Y., et al., 2017. An active molecule from *Pulsatilla chinensis*, *Pulsatilla saponin* A, induces apoptosis and inhibits tumor growth of human colon cancer cells without or with 5-FU. Oncol. Lett. 13 (5), 3799–3802.

Yamashita, H., Takeda, K., Haraguchi, M., et al., 2017. Four new diterpenoid alkaloids from *Aconitum japonicum* subsp. subcuneatum. J. Nat. Med.doi: 10.1007/s11418-017-1139-9.

Yang, L.H., Lin, L.M., Wang, Z.M., et al., 2016. Advance on chemical compounds of Tibetan medicinal plants of *Aconitum* genus. Chin. J. Chin. Mat. Med. 41 (3), 362–376.

Yang, J., Zhao, F., Nie, J., 2017. Anti-rheumatic effects of *Aconitum leucostomum* Worosch. on human fibroblast-like synoviocyte rheumatoid arthritis cells. Exp. Ther. Med. 14 (1), 453–460.

Yoshimitsu, H., Nishida, M., Nohara, T., 2008. Two labdanediterpene and megastigmane glycosides from *Aquilegia hybrid*. Chem. Pharm. Bull. 56 (7), 1009–1012.

You, Y.J., Kim, Y., Nam, N.H., et al., 2003. Inhibitory effect of *Adonis amurensis* components on tube-like formation of human umbilical venous cells. Phytother. Res. 17 (5), 568–570.

Yu, J., Zhang, Y., Song, X., et al., 2017. Effect of modified Pulsatilla powder on enterotoxigenic *Escherichia coli* O101-induced diarrhea in mice. Evid. Based Complement. Alternat. Med. 2017, 3687486.

Zafar, S., Ahmad, M.A., Siddiqui, T.A., 2002. Effect of roots aqueous extract of *Delphinium denudatum* on morphine-induced tolerance in mice. Fitoterapia 73 (7–8), 553–556.

Zan, Y., Kuai, C.X., Qiu, Z.X., et al., 2017. Berberine ameliorates diabetic neuropathy: TRPV1 modulation by PKC pathway. Am. J. Chin. Med.doi: 10.1142/S0192415X17500926.

Zhang, L.H., Peng, Y.J., Xu, X.D., et al., 2015. Determination of other related carotenoids substances in astaxanthin crystals extracted from *Adonis amurensis*. J. Oleo Sci. 64 (7), 751–759.

Zhang, J.F., Chen, L., Huang, S., et al., 2017a. Diterpenoid alkaloids from two Aconitum species with antifeedant activity against *Spodoptera exigua*. J. Nat. Prod. 7, b00380. doi: 10.1021/acs. jnatprod.

Zhang, Y., Wu, W., Kang, L., et al., 2017b. Effect of *Aconitum coreanum* polysaccharide and its sulphated derivative on the migration of human breast cancer MDA-MB-435s cell. Int. J. Biol. Macromol. 103, 477–483.

Zhang, J.F., Shan, L.H., Gao, F., et al., 2017c. Five new C19 -diterpenoid alkaloids from *Delphinium tianshanicum* WT. Wang. Chem. Biodivers.doi: 10.1002/cbdv.201600297.

Zhang, H., Su, Y.F., Yang, F.Y., et al., 2017d. New minor spirostanol glycosides from *Helleborus thibetanus*. Nat. Prod. Res. 31 (8), 925–931.

Zhao, Q., Gou, X.J., Liu, W., et al., 2015. Majusine D: a new C19-diterpenoid alkaloid from *Delphinium majus*. Nat. Prod. Commun. 10 (12), 2069–2070.

Zhao, M., Da-Wa, Z.M., Guo, D.L., 2016. Cytotoxic triterpenoid saponins from *Clematis tangutica*. Phytochemistry 130, 228–237.

Zhou, X., Gan, P., Hao, L., et al., 2014. Antiinflammatory effects of orientin-2-O-galactopyranoside on lipopolysaccharide-stimulated microglia. Biol. Pharm. Bull. 37 (8), 1282–1294.

Zhou, J.T., Li, C.L., Tan, L.H., et al., 2017a. Inhibition of *Helicobacter pylori* and its associated urease by palmatine: i.nvestigation on the potential mechanism. PLoS One 12 (1), e0168944.

Zhou, C.X., Yu, Y.E., Sheng, R., et al., 2017b. Cimicifoetones A and B, dimeric prenylindole alkaloids as black pigments of *Cimicifuga foetida*. Chem. Asian J. 12 (12), 1277–1281.

Zhu, M., Xiao, P.G., 1991. Distribution of benzylisoquinolines in *magnoliidae* and other taxa. Acta Phytotax. Sin. 29, 142–155.

Zhu, Y., Di, S., Hu, W., et al., 2017. A new flavonoid glycoside (APG) isolated from *Clematis tangutica* attenuates myocardial ischemia/reperfusion injury via activating PKCε signaling. Biochim. Biophys. Acta 1863 (3), 701–711.

Zou, J.H., Yang, J.S., Zhou, L., et al., 2006. A new labdane type diterpenoid from *Trollius ledebouri*. Nat. Prod. Res. 20 (12), 1031–1035.

Mining Chemodiversity From Biodiversity: Pharmacophylogeny of Ranunculales Medicinal Plants (Except Ranunculaceae)

Ranunculales Medicinal Plants. http://dx.doi.org/10.1016/B978-0-12-814232-5.00003-4

3.1 INTRODUCTION

Ranunculales is an order of angiosperms containing seven families, Ranunculaceae (at least 50 genera, >2000 species), Berberidaceae (17 genera, 650 species), Menispermaceae (65 genera, >350 species), Lardizabalaceae (9 genera, 50 species), Circaeasteraceae (2 species), Papaveraceae (38 genera, >700 species), and Eupteleaceae (2 species) (Fig. 3.1A and B). Ranunculales belongs to the basal eudicots, in which it is the most basal clade. Widely known members include poppies, barberries, and buttercups.

The chemodiversity of Ranunculaceae, the flagship family of Ranunculales, and their correlation with biodiversity and pharmacotherapy are summarized in Chapter 2. Besides Ranunculaceae, plants of other families also provide humankind with its needs in foods, flavors, fragrances, and not the least, medicines. Ranunculales plants form the basis of sophisticated traditional medicine systems, for example, traditional Chinese medicine (TCM). These systems of medicine give rise to some important Ranunculales-based drugs still in use today. Ethnobotany and ethnopharmacognosy lend support to the current search for new molecules of different sources and classes. The flora of the tropics, Southwest China, and Himalayas is fantastic and contains the most diverse Ranunculales taxa. The biodiversity of these hot-spot regions plays a significant role in providing new leads, although the sovereignty and property rights should be addressed along with the Convention for Biological Diversity.

This chapter highlights the topics mentioned previously, provides an overview of the classes of molecules present in Ranunculales plants, and gives examples of the types of molecules and secondary metabolites that have led to the development of pharmacologically active extracts. Ranunculales plant products can be used to develop functional foods. The validation of Ranunculales extracts is indispensible, and safety, efficacy, and quality of phytomedications should always be emphasized.

3.2 SYSTEMATICS AND EVOLUTION OF RANUNCULALES

Around 75% of all angiosperm species belong to the eudicot clade; this is strongly supported by molecular data and united morphologically by the single synapomorphy-triaperturate pollen. Ranunculales is a major clade of eudicots, which are sister to all other eudicots (Soltis and Soltis, 2004), for example, Saxifragales, Caryophyllales, rosids, and asterids. The majority of gene duplications were placed after the divergence of the Ranunculales and core eudicots, indicating that the gamma, a major polyploidy event, appears to be restricted to core eudicots (Jiao et al., 2012). The duplication events were intensely concentrated around 117 million years ago. Despite rapid progress in resolving angiosperm relationships, the branching order among basal eudicots remains a problem, and fossils should be integrated with extant taxa into a comprehensive tree of angiosperm phylogeny. Many phylogenetic studies of Ranunculales or angiosperms suggest that Eupteleaceae, an East Asian family of two species of trees with perianthless flowers, are sister to all remaining Ranunculales, with Papaveraceae branching next (Fig. 3.1B) (Hilu et al., 2003; Qiu et al., 2006; Soltis et al., 2011;

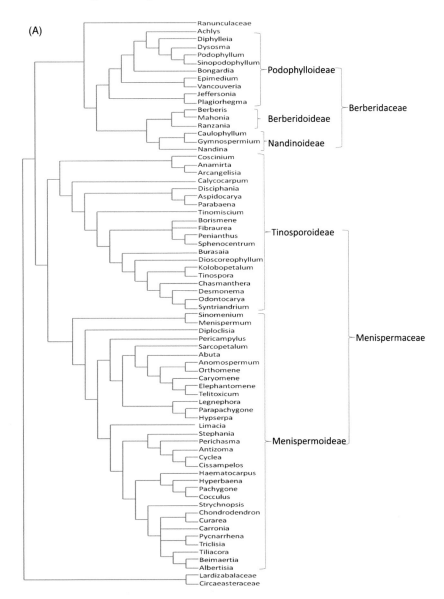

FIGURE 3.1 **Phylogenetic relationships of families and genera within Ranunculales, according to the APG III system (Angiosperm Phylogeny Group, 2009; Sauquet et al., 2015; Wang et al., 2009; Wefferling et al., 2013).** (A) Berberidaceae and Menispermaceae and (B) Lardizabalaceae and Papaveraceae.

(B)

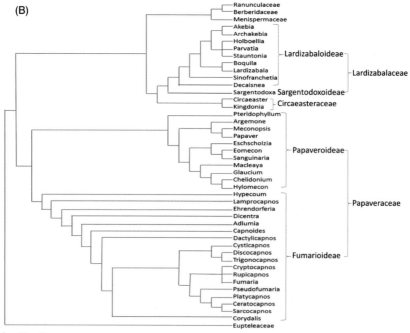

FIGURE 3.1   *(Cont.)*

Sun et al., 2016a; Wang et al., 2009). However, other studies suggest that Papaveraceae are sister to all remaining Ranunculales, with Eupteleaceae branching next (Bell et al., 2010; Hoot et al., 1999; Ren et al., 2007; Sauquet et al., 2015; Soltis et al., 2000). Branch support for these relationships was very low in all studies.

The use of several fossil constraints increased the probability of approaching the true ages of basal eudicots (Anderson et al., 2005), suggesting a rapid diversification during the late Early Cretaceous. All the lineages of basal eudicots emerged during the latest part of the Early Cretaceous. The age of Ranunculales was around 120 my, older than Proteales (119 my), Sabiales (118 my), Buxales (117 my), and Trochodendrales (116 my). Ranunculales, Trochodendrales, Buxales, Proteales, and Sabiaceae are early-diverging eudicots. In recent years the complete plastome sequences of many early-diverging eudicot taxa, for example, *Epimediumsagittatum* (Berberidaceae), *Eupteleapleiosperma* (Eupteleaceae), *Akebia trifoliata* (Lardizabalaceae), *Stephania japonica* (Menispermaceae), and *Papaver somniferum* (Papaveraceae), are determined to elucidate their evolution of plastome structure and the phylogenetic correlation. All of the newly sequenced plastomes share the same 79 protein-coding genes, 4 rRNA genes, and 30 tRNA genes, except for that of *Epimedium*, in which infA is pseudogenized and clpP is highly divergent and possibly a pseudogene (Sun et al., 2016a). The boundaries of the plastid inverted repeat (IR) vary significantly across early-diverging eudicots; IRs ranged from 24.3–36.4 kb in length and contained from 18 to 33 genes. Based on gene content, the IR was classified into six types, with shifts among types characterized by high levels of homoplasy. Maximum likelihood phylogenetic analysis of a 79-gene, 97-taxon data set that included all available early-diverging eudicots and representative sampling of remaining angiosperm diversity largely agreed with previous estimates of early-diverging eudicot relationships (Fig. 3.2), but resolved Trochodendrales rather than Buxales as sister to Gunneridae. Proteales was sister to Sabiaceae with the highest support (bootstrap > 90%).

Menispermaceae consists of two major subfamilies: Tinosporoideae and Menispermoideae (Fig. 3.1A) (Wefferling et al., 2013). Within Tinosporoideae, tribe Coscineae is basal. Within Menispermoideae, tribe Menispermeae is basal. Tinosporoideae consists mainly of taxa with apical style scars, bilateral curvature, subhemispherical condyles, and foliaceous cotyledons with divaricate or imbricate orientation. Menispermoideae consists almost entirely of taxa with basal or subbasal-style scars, dorsoventral curvature, bilaterally and/or dorsoventrally compressed condyles, and subterete or fleshy cotyledons oriented dorsoventrally or laterally. In terms of molecular phylogeny, divergence times, and morphology, the three *Tinospora* clades are recognized as three different genera, including *Tinospora sensu stricto*, a new genus (*Paratinospora*) for *Tinospora dentata* and *Tinospora sagittata*, and *Hyalosepalum* resurrected (Wang et al., 2017a).

Fumarioideae (20 genera, 593 species) is a clade of Papaveraceae (Ranunculales) characterized by flowers that are either disymmetric (i.e., two perpendicular planes of bilateral symmetry) or zygomorphic (i.e., one plane of bilateral symmetry). In contrast, the subfamily Papaveroideae (23 genera, 230 species) has actinomorphic flowers (i.e., more than two planes of symmetry). Six plastid markers and one nuclear marker were used to infer the phylogenetic relationship of *Pteridophyllum*, 18 genera and 73 species of Fumarioideae, 11 genera and 11 species of Papaveroideae, and a wide selection of outgroup taxa (Sauquet et al., 2015). *Pteridophyllum* is not nested in Fumarioideae. Fumarioideae are monophyletic, and *Hypecoum*

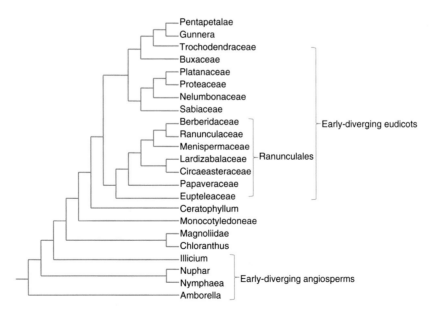

FIGURE 3.2 **Evolutionary relationship of early-diverging eudicots inferred from the plastome sequences.** *According to Sun et al. (2016a).*

(18 species) is the sister group of the remaining genera. Relationships within the core Fumarioideae are well resolved and supported (Fig. 3.1B). *Dactylicapnos* and all zygomorphic genera form a well-supported clade nested among disymmetric taxa. Disymmetry of the corolla is a synapomorphy of Fumarioideae and is strongly correlated with changes in the androecium and differentiation of middle and innertepal shape. Zygomorphy subsequently evolved from disymmetry either once or twice (*Capnoides*, other zygomorphic Fumarioideae) and might be correlated with the loss of one nectar spur.

3.3 THE CHEMICAL COMPOSITION OF BERBERIDACEAE PLANTS

3.3.1 Nandinoideae Subfamily

Nandinaceae is rich in various benzylisoquinoline alkaloids (BIAs), for example, berberine, palmatine (PAL), jatrorrhizine, coptisine, magnoflorine, domesticine, nandinine, and protopine. Steroidal alkaloid, for example, nandsterine, is also found in the fruit (Peng et al., 2014) (Fig. 3.3). Phenolic 1-benzyl-N-methyltetrahydroisoquinolines are present in cell cultures of *Corydalis, Macleaya,* and *Nandina* (Iwasa et al., 2009). The alkaloid higenamine exists in *Aconitum* (Hao et al., 2015), *Tinospora crispa, Nandina domestica* (Zhang et al., 2017a), *Gnetum parvifolium, Asarum heterotropoides,* and *Nelumbo nucifera.*

The existence of the cyanogenic compound (Forrester, 2017) nandinin, biflavonoidamentoflavone, and benzaldehyde-4-O-glucoside in this family indicates its relatively distant relation with the other three families. The cycloartane-type triterpenoid is found in the fruits of *Nandina* (Kodai et al., 2010).

22 compounds represent 82.79% of the *N. domestica* fruit essential oil (Bi et al., 2016). The major compounds are 3-hexen-1-ol (12.9%); linalool (12.3%); 2-methoxy-4-vinylphenol (9.9%); oleic acid (8.0%); furfural (5.8%); and 2,6-di-tert-butyl-4-methylphenol (5.7%). 79 compounds represent 87.06% of total floral oil (Bajpai et al., 2008). The oil contained mainly 1-indolizino carbazole (19.65%); 2-pentanone (16.4%); mono phenol (12.1%); aziridine (9.01%); methylcarbinol (4.6%); ethanone (3.3%); furfural (2.96%); 3,5-dimethylpyrazole (1.29%); and 2(5H)-furanone (1.32%).

Magnoflorine, taspine, and boldine, belonging to aporphine (AP) alkaloids, are found in *Caulophyllum* (Xia et al., 2014b). Typical quinolizidine alkaloids are also present in this genus. Alkaloids with piperidine-acetophenone conjugates, rare in the plant kingdom, are only found in *Caulophyllum* and *Boehmeria.* Fluorenone alkaloids and one dihydroazafluoranthene alkaloid are isolated from the roots of *Caulophyllum robustum* (Wang et al., 2011).

Seven classes of triterpene saponins, including malonyl-triterpene saponins, are isolated from *C. robustum* (Xia et al., 2014a, 2016). Twelve kinds of aglycones are discovered, for example, oleanolic acid, hederagenin, echinocystic acid, caulophyllogenin, and erythrodiol. These saponins generally bear one (monodesmosidic) or two (bidesmosidic) carbohydrate chains that are directly attached to the hydroxyl groups in position C-3 for monodesmosidic saponins and to positions C-3 and C-28 in the bidesmosidic saponins (Xia et al., 2014b). Oligosaccharide moieties are: Rha → (4)Glc → (6)Glc → (4)Rha → (4)Glc → (6)Glc, Glc → (4)Glc → (6) Glc → (4)Rha → (4)Glc → (6)Glc, Rha → Glc → Glc (Glc) → (2,3)Ara, Glc → Glc (Glc) → (2,3) Ara, and Glc (Glc) → (2,3)Ara (Xia et al., 2015).

3.3.2 Berberidoideae

Berberis (Fig. 3.4) and *Mahonia* (Fig. 3.5) contain mainly BIAs, for example, berberine; PAL; jatrorrhizine; columbamine [Protoberberine (PBB) type]; magnoflorine (AP type), particularly a higher content of bisbenzylisoquinoline (BBI) alkaloid berbamine (Jia et al., 2017); isotetrandrine (He and Mu, 2015); and oxyacanthine. Phenolics, flavonoids, and tannins are abundant

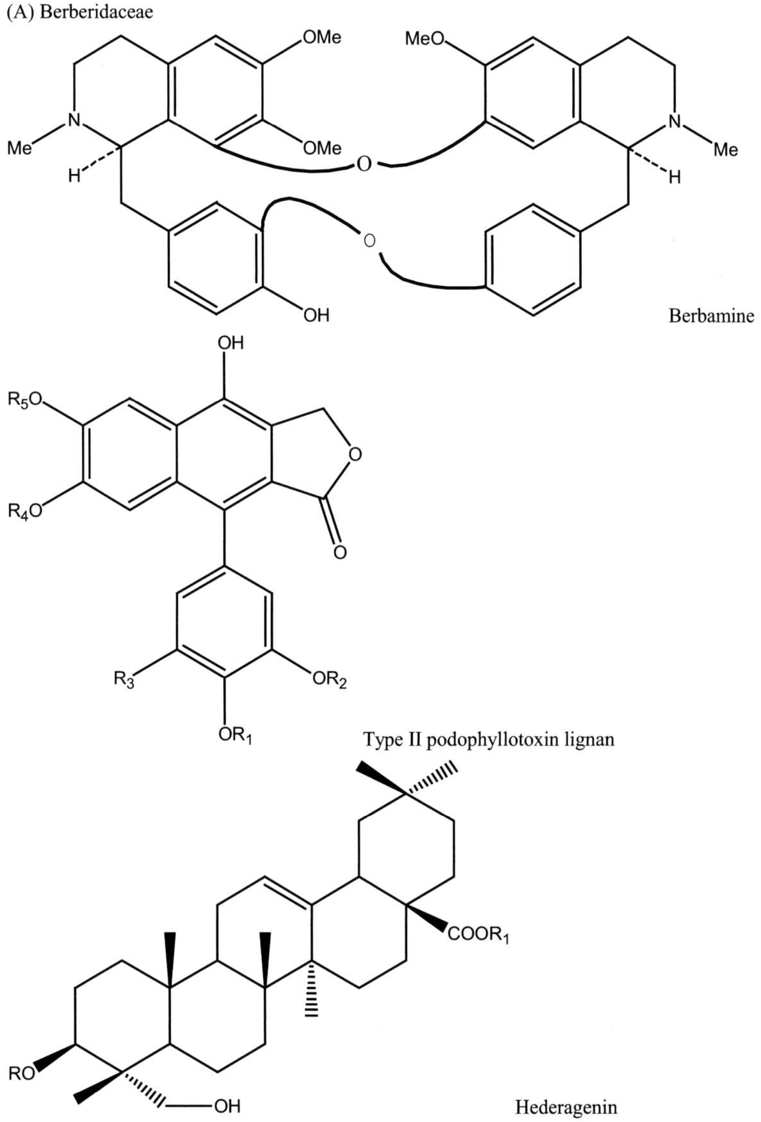

FIGURE 3.3 **Representative medicinal compounds of Ranunculales:** (A) Berberidaceae; (B) Menispermaceae; (C) Lardizabalaceae and Circaeasteraceae; and (D) Papaveraceae and Eupteleaceae.

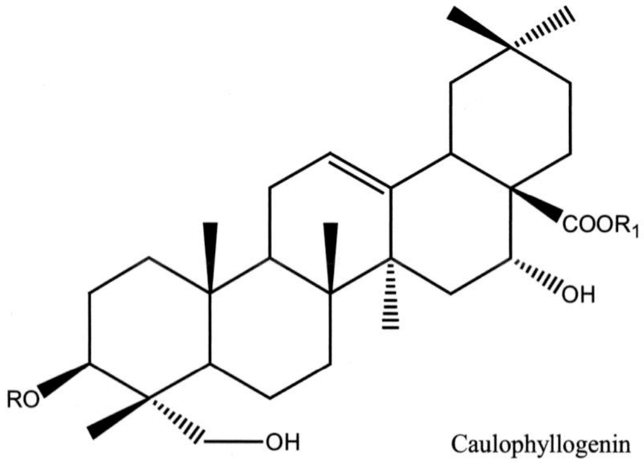

Caulophyllogenin

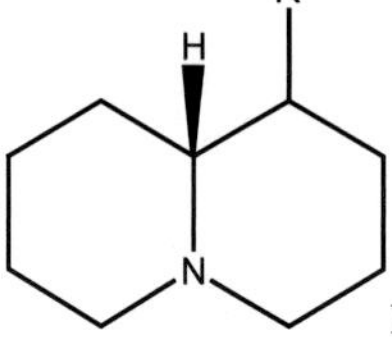

Lupinine-type quinolizidine alkaloid

(B) Menispermaceae

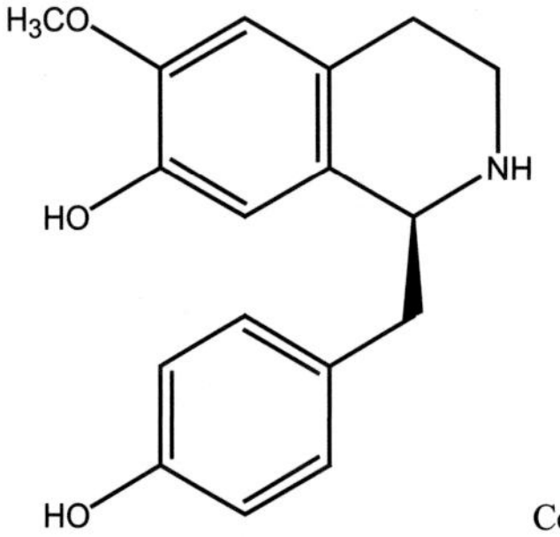

Coclaurine

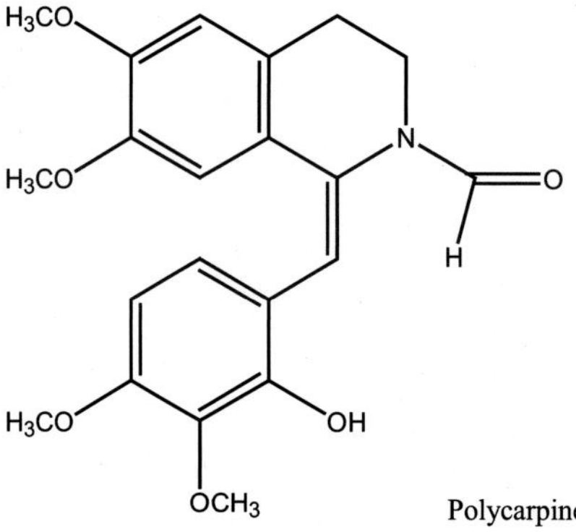

Polycarpine

FIGURE 3.3   *(Cont.)*

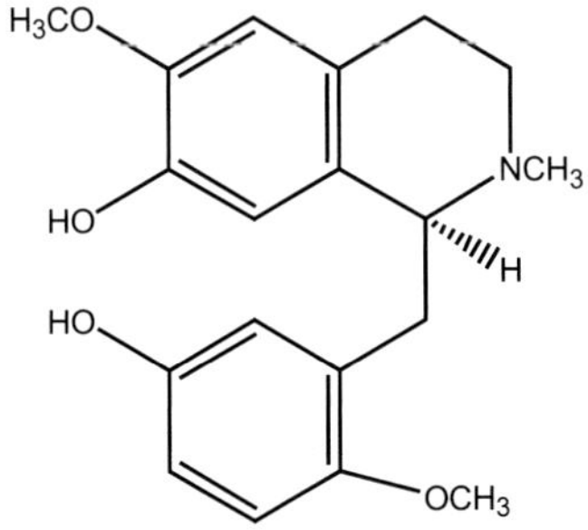

Dehassiline

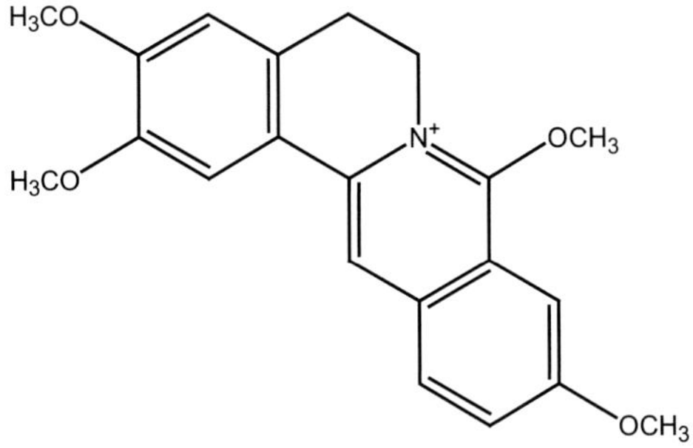

Palmatine

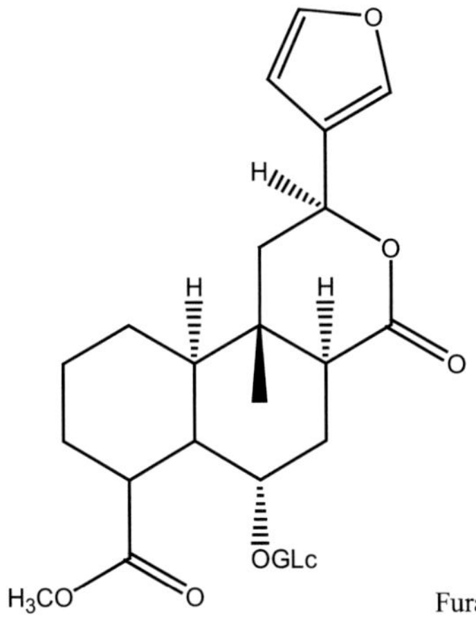

Furanoid diterpene glycoside

FIGURE 3.3 (*Cont.*)

in *Berberis* (Belwal et al., 2017). Berberine is the principle component for many medicinal plants, e.g., *Coptis chinensis*, *Phellodendron chinense* (Huang Bai in TCM), and *Mahonia bealei*.

3.3.3 Podophylloideae

The tribe Podophylleae, consisting of *Podophyllum* (including *Sinopodophyllum* and *Dysosma*) and *Diphylleia*, is rich in various podophyllotoxin lignans (Zhao et al., 2014). Deoxypodophyllotoxin (DPPT) is extracted from *Podophyllum peltatum*, *Podophyllum pleianthum*, *Podophyllum emodi* (*Podophyllum hexandrum*), and *Diphylleia grayi* (Hu et al., 2016a).

(C) Lardizabalaceae and Circaeasteraceae

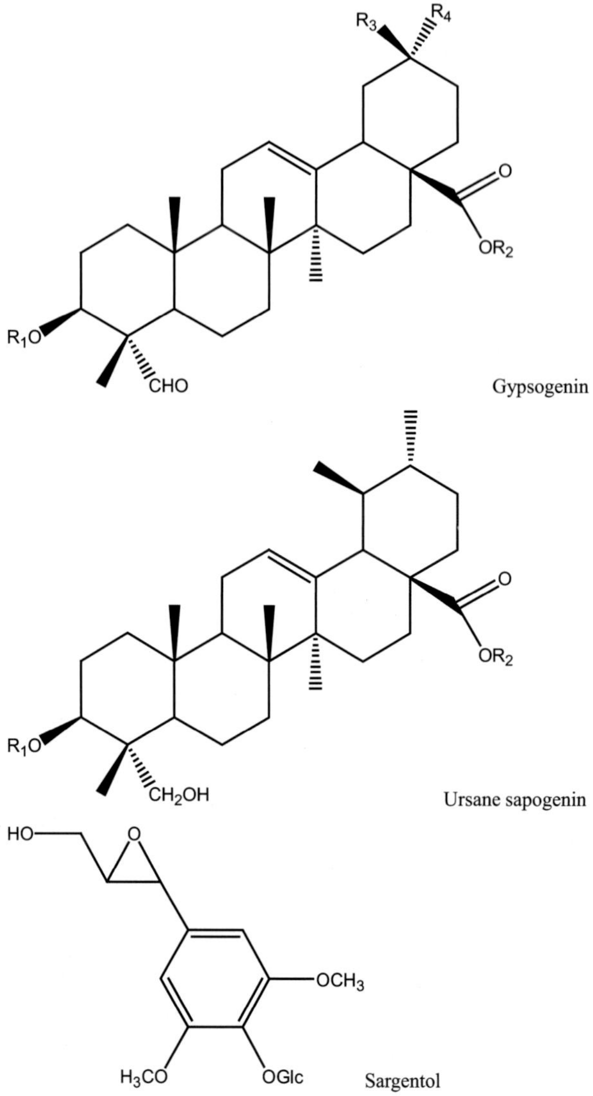

R_3
R_4
O
OR_2
R_1O
CHO
Gypsogenin
O
OR_2
R_1O
CH_2OH
Ursane sapogenin
HO
O
OCH_3
H_3CO
OGlc
Sargentol

FIGURE 3.3 *(Cont.)*

Prenylated flavonoids, flavonoid glycoside, and labdane diterpenes are isolated from the fruits of *Sinopodophyllum hexandrum* (Sun et al., 2016b; Wang et al., 2017b) (Fig. 3.6). Prenylated biflavonoids are isolated from the roots and rhizomes of *Sinopodophyllum* (Sun et al., 2015). Quercetin and kaempferol are more abundant in the samples from Shangri-La, Yunnan, and Nyingchi, Tibet, when compared with those of Jingyuan, Ningxia, and Yongdeng, Gansu, where the production of podophyllotoxin and lignans are favorable (Liu et al., 2015a).

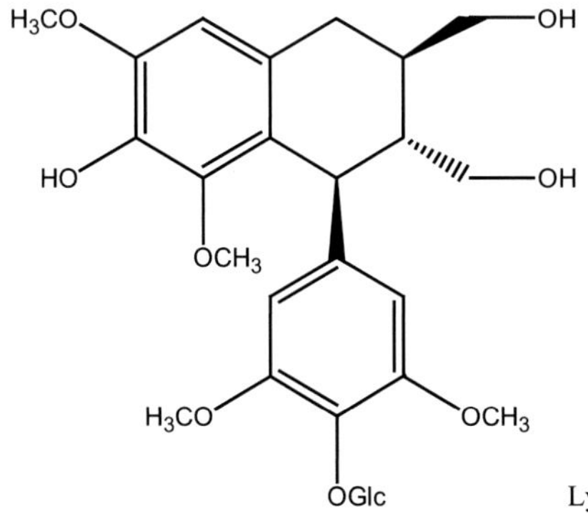

Lyoniresin-4'-yl-β-glucopyranoside

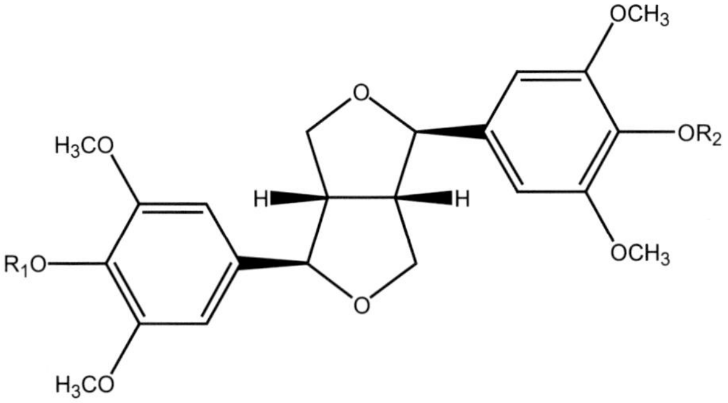

R1=R2=Glc

Acanthoside D

(D) Papaveraceae and Eupteleaceae

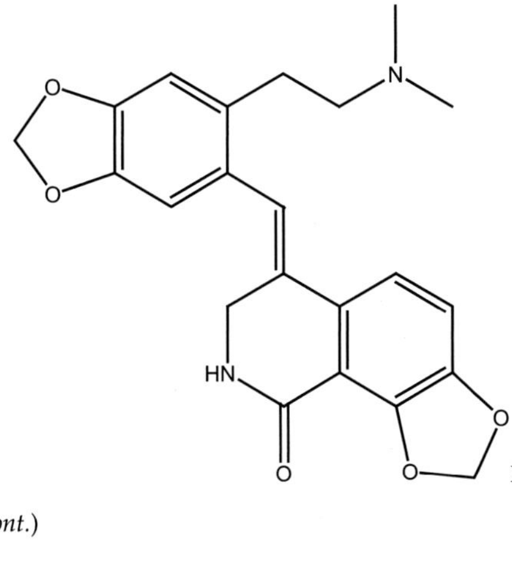

Fumaramine

FIGURE 3.3 (*Cont.*)

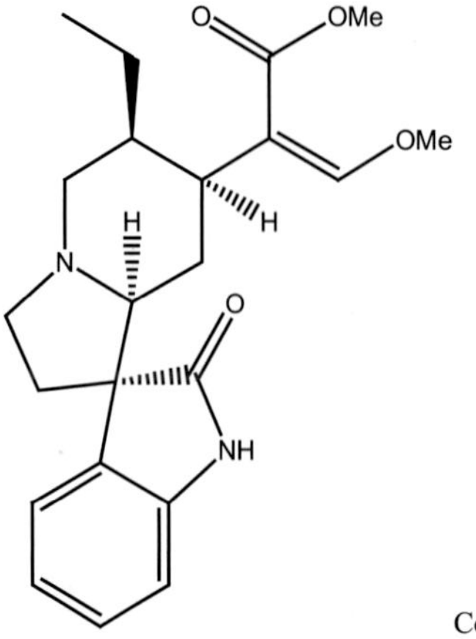

Corynoxine

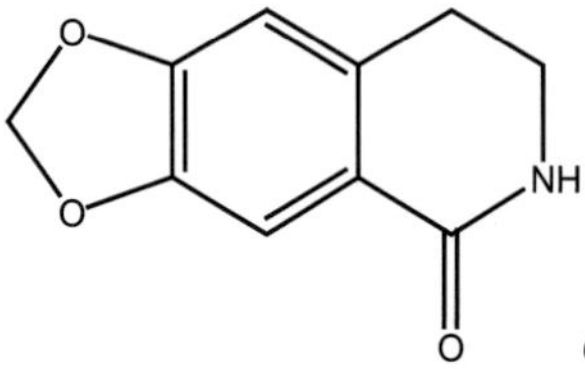

6,7-Methylenedioxy-1(2H)-isoquinolinone Noroxyhydrastinine

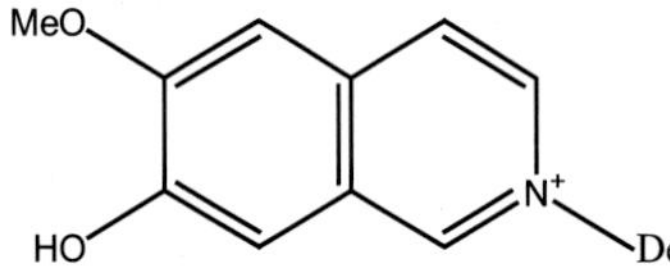

Dehydrocorypalline

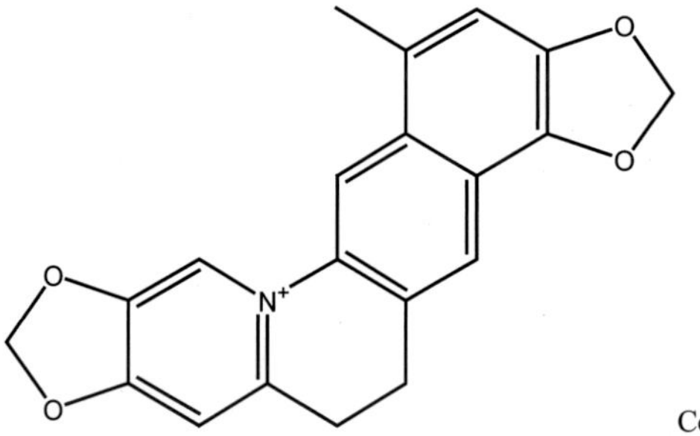

Corysamine

FIGURE 3.3 (*Cont.*)

Both *Epimedium* and *Vancouveria* contain predominately bioactive icariin flavonoids, the characteristic chemical constituents of this group. Neolignans, for example, 8-O-4' neolignan, are isolated from *Epimedium pseudowushanese* (Ti et al., 2017).

Gymnospermium, Leontice, Caulophyllum, and *Bongardia* contain mainly β-amyrin triterpenoids and quinolizidine alkaloids.

FIGURE 3.4   *Berberis*, taken in Yarlung Zangbo Grand Canyon, Tibet, China.

FIGURE 3.5   *Mahonia fortunei*, taken in Xi'an Botanical Garden, Shaanxi, China.

FIGURE 3.6   *Sinopodophyllum hexandrum*, taken in Potatso National Park, Shangri-La, Yunnan, China.

3.4 THE CHEMICAL COMPOSITION OF MENISPERMACEAE

Menispermaceae is a medium-sized family of 70 genera and 420 extant species, mostly climbing plants. It has various medicinal properties, which are used in TCM and the Ayurvedic system of medicine. Menispermaceae plants are rich in alkaloids, especially BBI types (Thavamani et al., 2013). BBI alkaloids (Fig. 3.7 and Table 3.1), morphine alkaloids, AP alkaloids, syringaresinol, and aristolochic acid I could be marker compounds of this family (Sim et al., 2013).

3.4.1 Stephania

Alkaloids of this genus fall into six major types, that is, hasubanan, AP, proaporphine (PAP), PBB, BBI, and morphinandienone.

Stephania tetrandra and other related species of Menispermaceae are the major source of the BBI alkaloid tetrandrine (Bhagya and Chandrashekar, 2016). Magnoflorine is found in *S. tetrandra* (Sim et al., 2013). An aporphine glycoside angkorwatine and eight alkaloids—oblongine, stepharine, asimilobine-β-d-glucopyranoside, isocorydine, tetrahydropalmatine (THP), jatrorrhizine, PAL, and roemerine (ROE)—are isolated from the tuber of *Stephania cambodica* (Dary et al., 2017).

Tetrahydroprotoberberine-, quaternary protoberberine-, aporphine-, PAP-, benzylisoquinoline-, or BBI-type alkaloids were identified in *Stephania hainanensis* (He et al., 2016).

3.4.2 Cyclea

Cissampentine A, an enantiomer of cissampentin, and cycleatjehenine-type BBI alkaloids, cissampentine B–D, are isolated from the roots of *Cyclea tonkinensis* (Wang et al., 2015a). Curine-type BBI alkaloids are found in the roots of *Cyclea wattii* (Wang et al., 2010a). Racemosidine A, isolated from the roots of *Corydalis racemosa*, is a BBI alkaloid that has diphenyl ether bridges at C-11/C-7′ and C-8/C-12′ and a benzyl-phenyl ether bridge at C-7/C-11′ (Wang et al., 2010b). Phytosterols and alkaloids are major phytoconstituents in petroleum ether extract of the *Cyclea peltata* leaf (Hullatti et al., 2011). The ethanolic extract of the *C. peltata* leaf showed the presence of alkaloids, flavonoids, tannins, diterpenes, and saponins.

The roots of *Cissampelos pareira* have a high concentration of alkaloids, especially the presence of high concentration of berberine, which is present in very low concentration in *S. japonica* and absent in roots of *C. peltata* (Hullatti and Sharada, 2010). The roots of *C. peltata* contain a high concentration of saponins, compared with a low concentration in *C. pareira* and absence in roots of *S. japonica*.

3.4.3 *Sinomenium* and *Menispermum*

Alkaloids laudanosoline-1-O-xylopyranose, 6-O-methyl-laudanosoline-1-O-glucopyranoside, menisperine, sinomenine, laurifoline, magnoflorine, and norsinoacutin were identified in *Sinomenium acutum* (Qing Feng Teng in TCM) (Sun et al., 2016c). Pyrrolo[2,1-a]isoquinoline and pyrrole alkaloids are isolated from *S. acutum* (Lv et al., 2017).

Thirteen isoquinoline alkaloids, for example, salutaridine, dauricumine, cheilanthifoline, dauriporphine, 8-demethoxycephatonine, and 7(R)-7,8-dihydrosinomenine, are isolated from the rhizome of *S. acutum* (Lee et al., 2016).

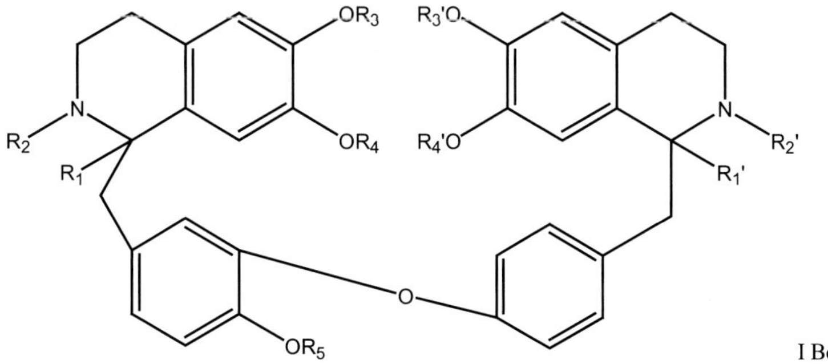

I Berbamunine

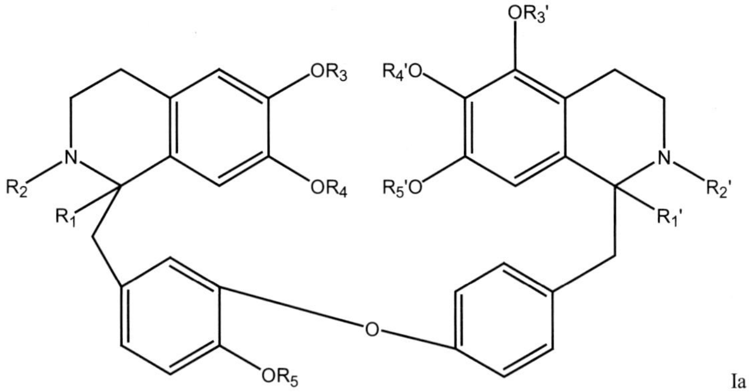

Ia Thaliracebine

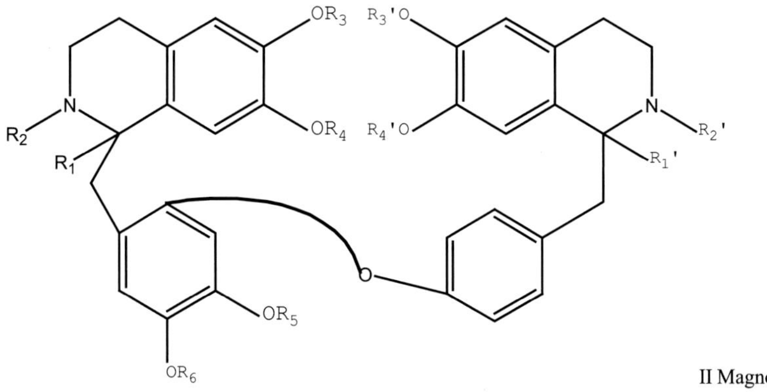

II Magnolamine

FIGURE 3.7 Classification of plant BBI alkaloids. (I) Berbamunine, (Ia) Thaliracebine, (II) Magnolamine, (III) Methothalistyline, (IV) Dirosine, (V) Isoliensine, (VI) Aromoline, (VII) Thaligosine, (VIII) Berbamine, (IX) Hernandezine, (X) Isotenuipine, (XI) Belarine, (XII) Thafoetidine, (XIII) Thalfine, (XIV) Dryadine, (XV) Panurensine, XVI Nemuarine, (XVII) Thalibrunimine, (XVIII) Dinklacorine, (XIX) Tiliamosine, (XX) Cycleanine, (XXI) Curine, (XXII) Cissampareine, (XXIII) Cocsoline, (XXIV) Cocsulinine, (XXV) Repanduline, and (XXVI) Insulanoline.

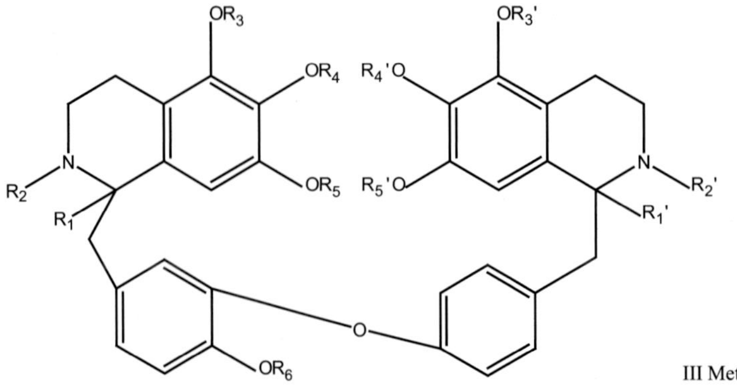

III Methothalistyline

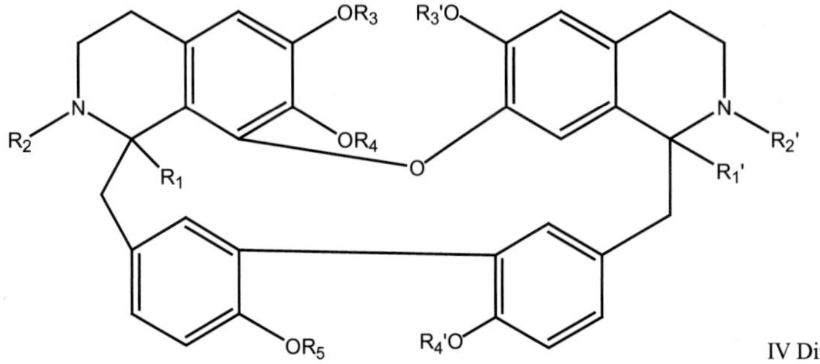

IV Dirosine

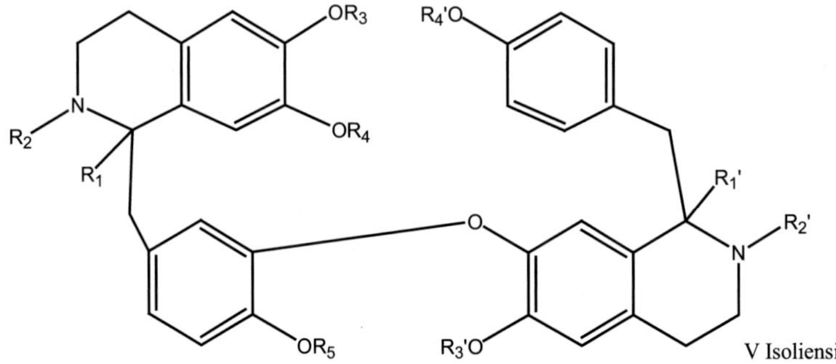

V Isoliensine

FIGURE 3.7 (*Cont.*)

Dauricine is the major bioactive component isolated from the roots of *Menispermum dauricum* (Dong et al., 2014). Daurisoline is a bis-benzylisoquinoline alkaloid isolated from the rhizomes of *M. dauricum* (Liu et al., 2012).

3.4.4 *Tinospora*

Alkaloids, terpenoids, steroids, volatile oils, amino acids, inorganic elements, and other compounds are found in this genus. Crispene A, B, C, and D; clerodane-type furanoid diterpenes; and a furanoid diterpene glucoside borapetoside E are isolated from the stem of *T. crispa* (Hossen et al., 2016). Clerodane-type diterpenoids are also isolated from tuberous

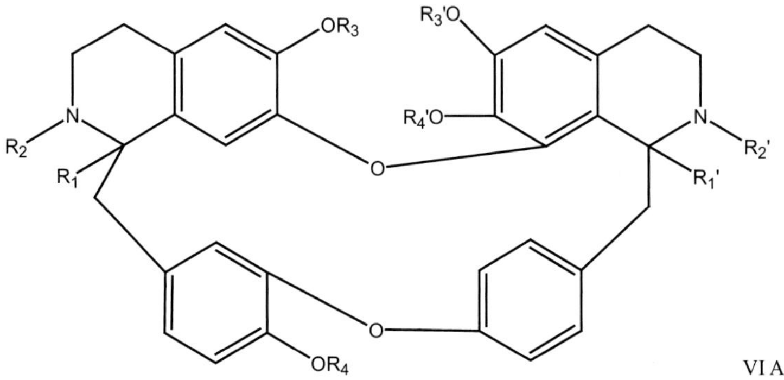

VI Aromoline

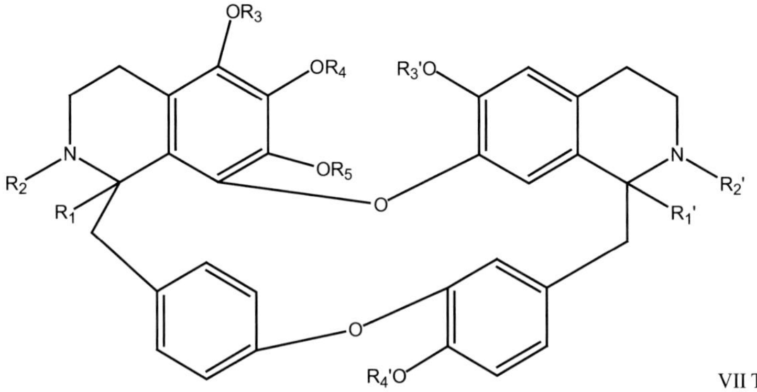

VII Thaligosine

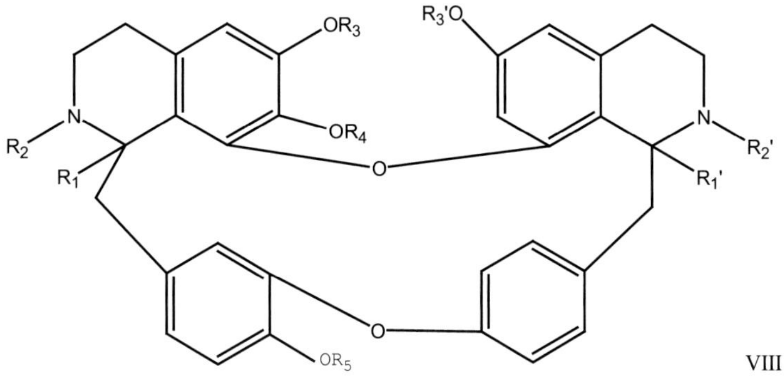

VIII Berbamine

FIGURE 3.7 *(Cont.)*

roots of *T. sagittata* (Zhang et al., 2016a). Clerodane diterpenoid glucosides are isolated from an ethyl acetate (EtOAc) extract of the stems of *Tinospora sinensis* (Jiang et al., 2017).

3.4.5 Cocculus

Wattisine A, O-methylcocsoline, (+) cocsoline, (+) cocsuline, magnoflorine, sino-coculine, isosinococuline, (−) coclaurine, daucosterol, β-sitosterol, and 1-oleioyl-3-(9Z, 12Z-arachoyl)

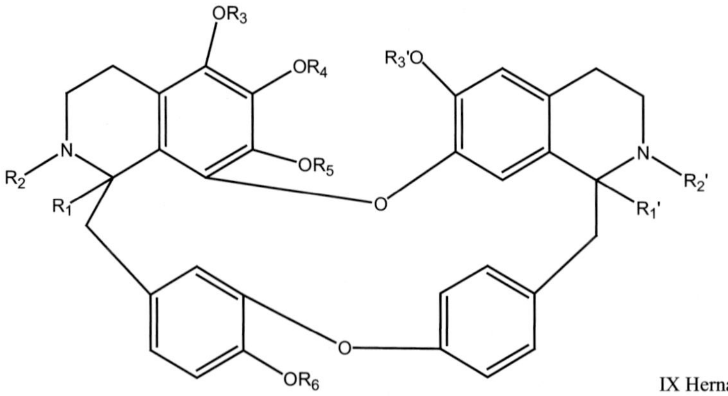

IX Hernandezine

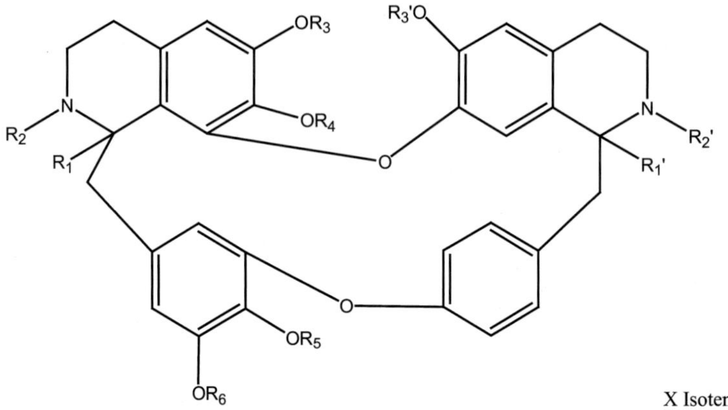

X Isotenuipine

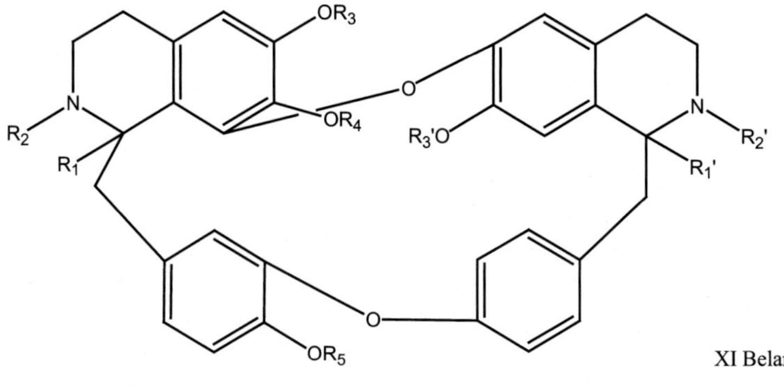

XI Belarine

FIGURE 3.7   (Cont.)

glycerol are isolated from the root of *Cocculus orbiculatus* var. *mollis* (Liao et al., 2014). Syringaresinol is found in *Corydalis trilobus* (Sim et al., 2013).

3.4.6 Other Genera

Alkaloids reticuline, asimilobine, acutumine, dihydroxyprotoberberine, and stepholidine are isolated from the vine stems of *Diploclisia affinis* (Wang et al., 2008). Ecdysteroid, (2-Nitro

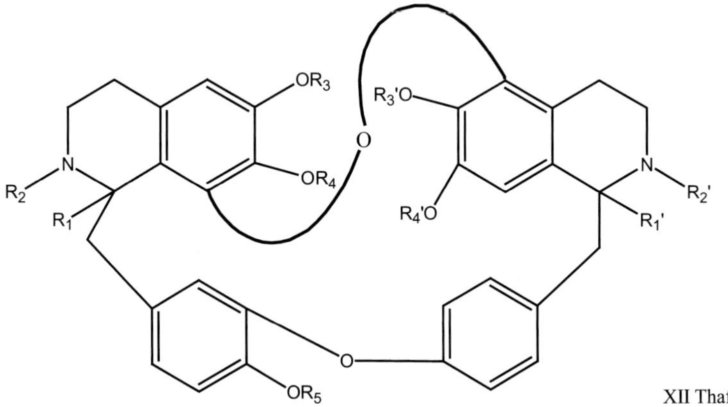

XII Thafoetidine

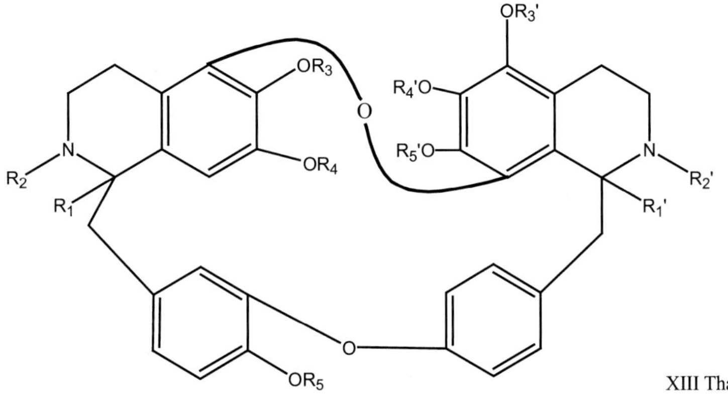

XIII Thalfine

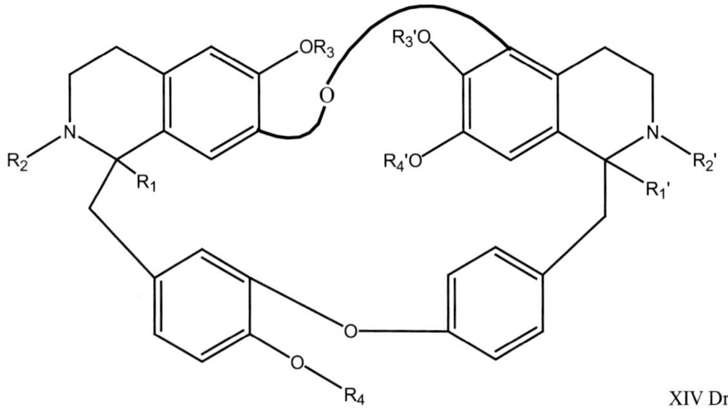

XIV Dryadine

FIGURE 3.7 (*Cont.*)

ethyl)phenyl and cyanophenyl glycosides, and steroidal and triterpenoid saponins are found in the fruits of *Diploclisia glaucescens* (Jayasinghe et al., 2003a, 2005, 2007). Ecdysteroid and oleanane glycosides are isolated from the leaves of *D. glaucescens* (Jayasinghe et al., 2002a,b).

A β-carboline alkaloid sacleuximine A, together with PAL, isotetrandrine, trans-N-feruloyl-tyramine, trans-N-caffeoyltyramine, yangambin, syringaresinol, sesamin, (+)epi-quercitol,

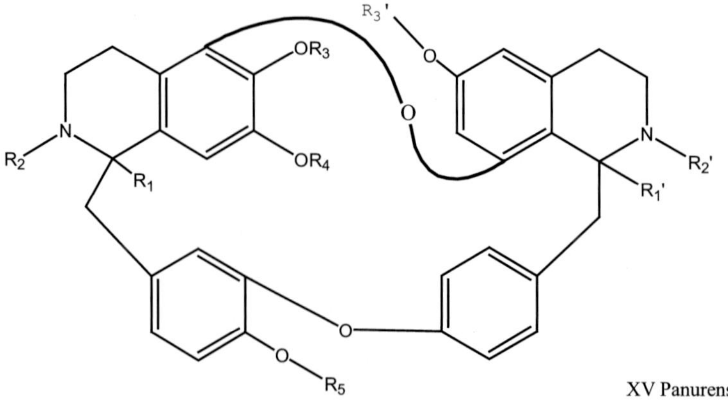

XV Panurensine

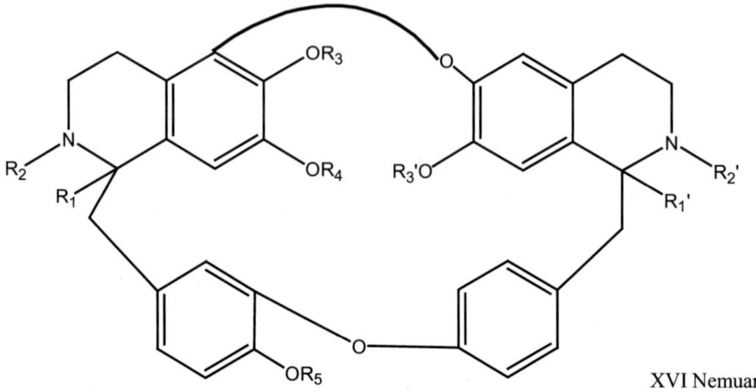

XVI Nemuarine

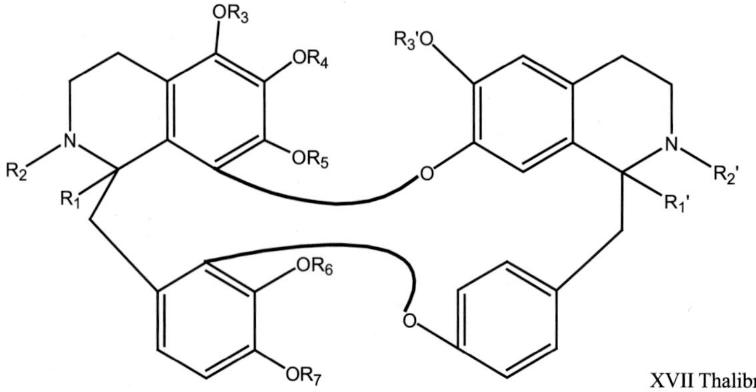

XVII Thalibrunimine

FIGURE 3.7 (*Cont.*)

4-hydroxybenzaldehyde, β-sitosterol, quercetin3-O-rutinoside, and myricetin 3-O-β-glucose (1 → 6) α-rhamnoside, are isolated from methanol extract of *Triclisia sacleuxii* aerial parts (Samita et al., 2017).

PBB isoquinoline alkaloids are isolated from *Arcangelisia gusanlung* and *Arcangelisia flava* (Subeki et al., 2005; Yu et al., 2014). Fibaruretin B (2β,3α-dihy-droxy-2,3,7,8α-tetra-hydro-

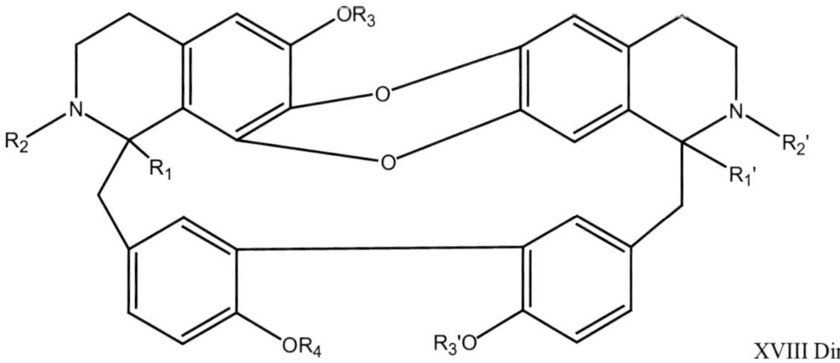

XVIII Dinklacorine

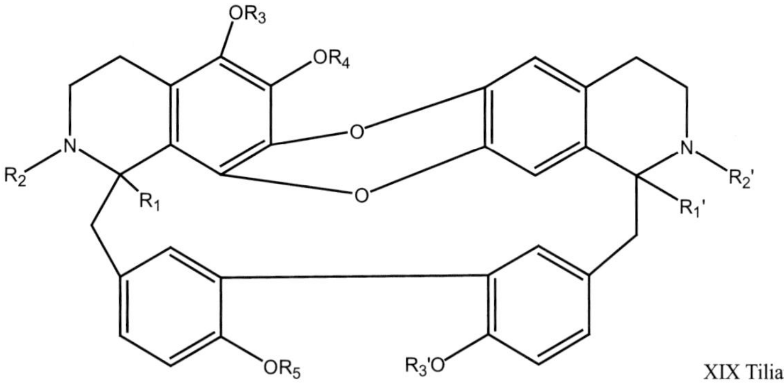

XIX Tiliamosine

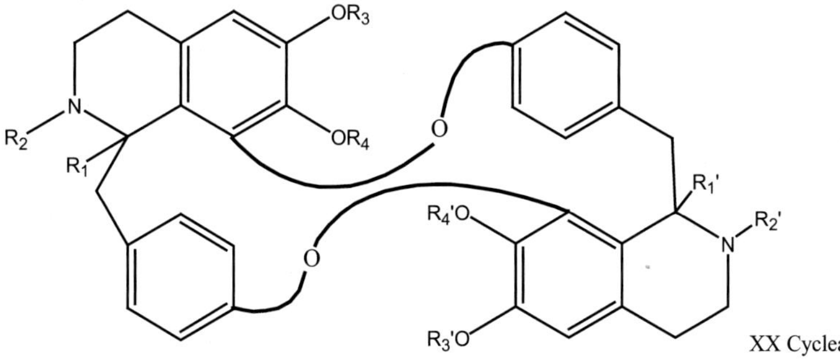

XX Cycleanine

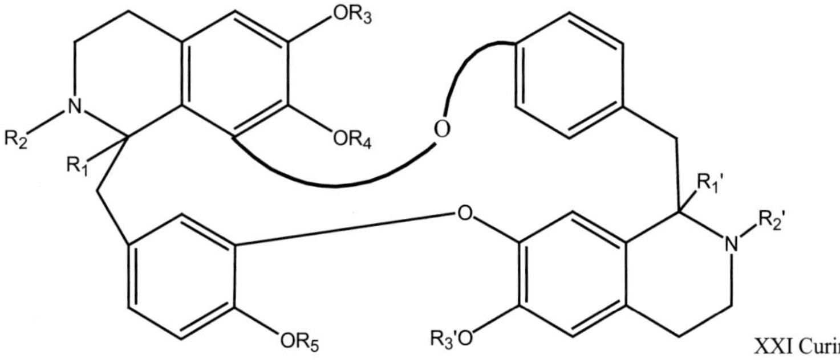

XXI Curine

FIGURE 3.7   (*Cont.*)

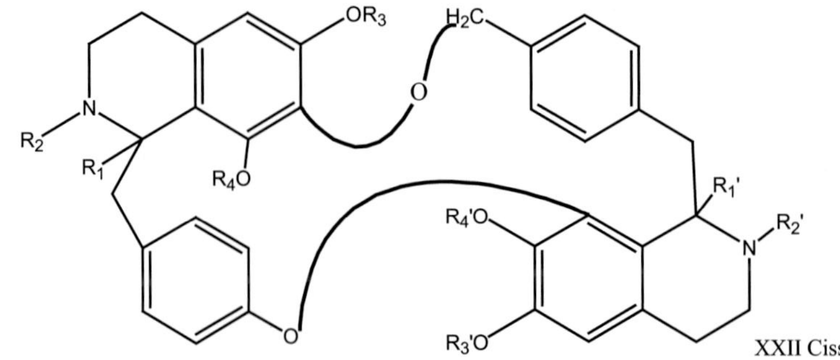

XXII Cissampareine

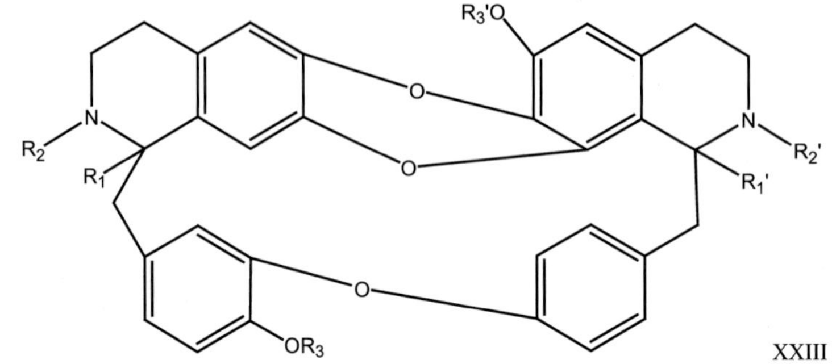

XXIII Cocsoline

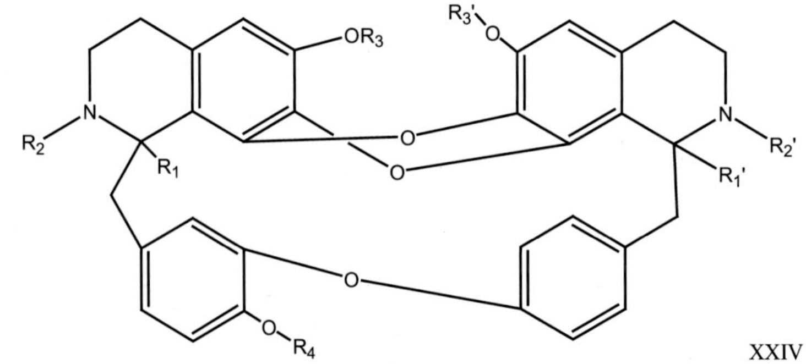

XXIV Cocsulinine

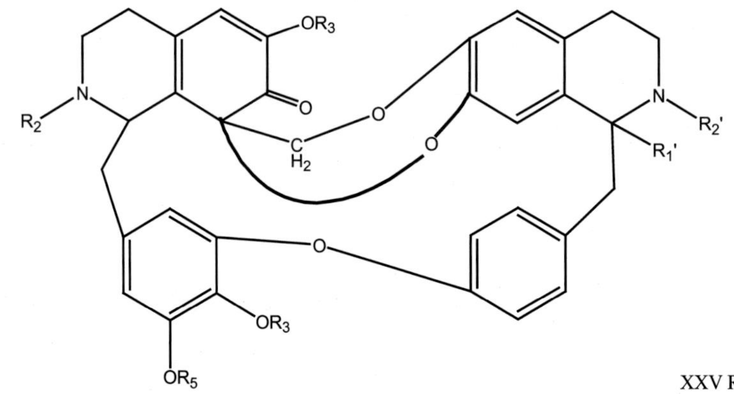

XXV Repanduline

FIGURE 3.7 *(Cont.)*

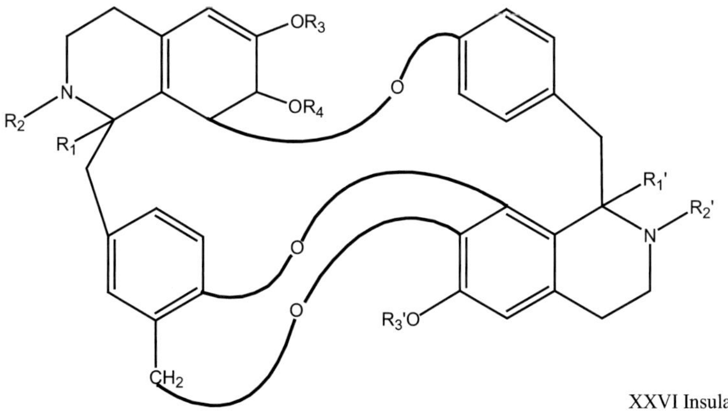

XXVI Insulanoline

FIGURE 3.7   *(Cont.)*

penianthic acid lactone) is isolated from the roots of *A. flava* (Fun et al., 2011). Furanoditerpenes are found in *A. flava* (Kawakami et al., 1987).

PAL, fibrecisine, berberines, tetrahydroberberines, and aporphine derivatives are found in *Fibraurea recisa* (Qiao et al., 2015; Rao et al., 2009). Furanoditerpenoids, for example, epi-8-hydroxycolumbin and fibaruretin B, C, E, and F, are isolated from the stems of *Fibraurea tinctoria* (Su et al., 2008).

3.5 THE CHEMICAL COMPOSITION OF LARDIZABALACEAE AND CIRCAEASTERACEAE PLANTS

3.5.1 *Sargentodoxa*

Lignans (Han et al., 1986), phenolics (e.g., catechin) (Tian et al., 2005), and two triterpene saponins, rosamultin and kajiichigoside F1 (Rücker et al., 1991), are isolated from *Sargentodoxa cuneata* (Da XueTeng in TCM). 3,4-Dihydroxyphenylethyl alcohol glycoside, salidroside, chlorogenic acid, and liriodendrin in the stem of *S. cuneata* are quantified simultaneously by high performance liquid chromatography (HPLC) coupled with evaporative light-scattering detection (Li et al., 2016a). Determining the total phenols, total saponins, and marker constituents salidroside, chlorogenic acid, and 3, 4-dihydroxy-phenylethyl-β-D-glucopyranoside in *Sargentodoxae Caulis* is required by the quality standard of the crude material recorded in the Chinese *Pharmacopoeia* (2015 Edition) (Li et al., 2015a).

3.5.2 Akebia

Diterpene glycoside is isolated from the stems of *Akebia quinata* (An et al., 2016). Hederagenin and saponins are isolated from *A. quinata* fruit (Kim et al., 2017b; Ling et al., 2016). Akebia saponin D (ASD) is present in both *Akebia* and Dipsacaceae (Gong et al., 2016). Pentacyclic triterpene saponins, for example, akebiaoside K and akebiaoside N, are isolated from

TABLE 3.1 Occurrence of Natural BBI Alkaloids in Menispermaceae

Genus	I	IV	V	VI	VII	VIII	IX	XIV	XV	XVIII	XIX	XX	XXI	XXII	XXIII	XXIV	XXVI
Abuta	8			4		4			2			1	1				
Albertisia	2			9		1									11		
Anisocycla				9		4						6			14		
Anomospermum													1				
Arcangelisia	1			1		1											
Caryomene				2		2											
Chondodendron												8	10				
Cissampelos			1									4	11	9			2
Cocculus		1		5		8									35	9	
Curarea				2,		5,											
Cyclea			3	5		20						11	13	5			6
Epinetrum												9					
Limacia	2			3		2											
Limaciopsis						6						1					
Menispermum	9																
Pachygone	3,			4		7				1	3	2	1		13		1
Pleogyne				1								1	1				
Pycnarrhena				8		23						6					
Sciadotenia	1			1													
Spirospermum						1											
Stephania	1			50	3	31						12	1		5		1
Strychnopsis						1											
Synclisia												3			2		
Tiliacora		7				3				19	7						
Triclisia				3	1	5									7	2	
Thalictrum Ranunculaceae							3	1									

Numbers of compounds are shown. I, berbamunine type (Fig. 3.7); IV, dirosine; V, isoliensine; VI, aromoline; VII, thaligosine; VIII, berbamine; IX, hernandezine; XIV, dryadine; XV, panurensine; XVIII, dinklacorine; XIX, tiliamosine; XX, cycleanine; XXI, curine; XXII, cissampareine; XXIII, cocsoline; XXIV, cocsulinine; XXVI, insulanoline.

the leaves of *A. trifoliata* (Xu et al., 2016). Oleanane-type triterpenoids, nortriterpene saponins, and phenolic glycosides are isolated from the pericarps of *A. trifoliata* (Matsuzaki et al., 2014; Wang et al., 2015b).

Caffeoylquinic acid derivatives, for example, cryptochlorogenic acid methyl ester, neo-chlorogenic acid methylester, and chlorogenic acid methyl ester, are isolated from stems of *A. trifoliata* (Wang et al., 2014a).

3.5.3 Holboellia

30-Noroleanane triterpenoid saponins, for example, akebonoic acid 28-O-β-d-glucopyranosyl-(1″→6′)-β-d-glucopyranosyl ester, holboelliside A, and holboelliside B, are isolated from the stems of *Holboellia coriacea* (Ding et al., 2016). Triterpenoid saponins, fargosides A, B, C, D, and E, are isolated from the roots of *Holboellia fargesii* (Fu et al., 2001).

3.5.4 Stauntonia

Triterpenoids (e.g., nor-oleanane triterpenoid saponins) (Liu et al., 2016a,b), flavonoids, lignanoids, and phenylethanoid glycosides were identified in *Stauntonia brachyanthera* (Liu et al., 2015b), *Stauntonia chinensis* (Hu et al., 2014; Wang et al., 2010c), and *Stauntonia obovatifoliola* (Lu et al., 2014a,b). Their structural types are much similar to those of *A. quinata*, that is, either penta-saccharidic or hexa-saccharidic bidesmoside triterpenoid glycosides (Gao et al., 2009). Calceolarioside B is abundant in both *Akebia* and *Stauntonia*. *S. brachyanthera* could be a succedaneum of *Akebia caulis*, which experienced a supply crisis in recent years.

Triterpenoid glycosides (Wang et al., 1993), phenolic acids, flavonoids, organic acids, and calceolariosides are isolated from *Stauntonia hexaphylla* (Hwang et al., 2017). Polysaccharide are found in *S. chinensis* (Yang et al., 2014).

3.5.5 Circaeasteraceae

The medicinal compounds of *Circaeaster agrestis* and *Kingdonia uniflora* have not been reported.

3.6 THE CHEMICAL COMPOSITION OF PAPAVERACEAE AND EUPTELEACEAE PLANTS

3.6.1 Fumarioideae: *Corydalis* and *Dicentra*

Isoquinoline alkaloids are the major bioactive ingredients of *Corydalis* (Fig. 3.8). Eight types of alkaloids are found in Chinese *Corydalis*: protopine; PBB; phthalide; benzo-phenanthridine; aporphine; spirobenzylisoquinoline; benzylisoquinoline; and others. Alkaloids were isolated from *Rhizoma Corydalis Decumbentis* (Xia Tian Wu in TCM) (Mao et al., 2017). Tetrahydroprotoberberines (THPBs) represent a series of compounds extracted from the Chinese herb *Corydalis ambigua* and various species of *Stephania* (Menispermaceae) (Mo et al., 2007). Levo-tetrahydropalmatine (l-THP) is found in both *Stephania* and *Corydalis* (Wang and Mantsch, 2012).

FIGURE 3.8   *Corydalis*, taken in Gama Valley, Tingri County, Tibet, China.

Volatile oils, steroids, and flavonoids are found in *Corydalis*. The content of flavonoids and polyphenol is greater than that of alkaloids (Shang et al., 2014).

3.6.2 Fumarioideae: *Fumaria, Hypecoum,* and *Pteridophyllum*

Alkaloids of the benzylisoquinoline type were identified in *Corydalis cava* or *Fumaria vaillantii* (Meyer and Imming, 2011). *Fumaria kralikii, Fumaria rostellata,* and *Fumaria thuretii* contain more than 50% spirobenzylisoquinoline alkaloids of the crude alkaloid mixtures (Doncheva et al., 2016). *F. rostellata* and *Fumaria schleicherii* contain more than 40% protopine alkaloids and relatively more phthaldeisoquinoline alkaloids (11%–19%). Phthaldeisoquinoline alkaloids were not detected in *F. kralikii, F. rostellata,* and *F. thuretii.* The principle alkaloids of five Bulgaria species are protopine, cryptopine, sinactine, parfumine, fumariline, fumarophycine, and fumaritine (Vrancheva et al., 2016).

N-feruloyltyramine, oxohydrastinine, hydroprotopine, leptopidine, and hypecocarpine are isolated from *Hypecoum leptocarpum* (Zhang et al., 2015a). Protopine and PBB-type protopine, allocryptopine, (–)-N-methylcanadine, and (–)-N-methylstylopine are isolated from *Hypecoum lactiflorum* (Philipov et al., 2009).

3.6.3 Papaveroideae: *Papaver, Meconopsis,* and Chelidonieae

These are summarized in Hao et al. (2015) (Fig. 3.9).

3.6.4 Papaveroideae: *Glaucium, Argemone,* and *Eschscholtzia*

The yellow-horned poppy (*Glaucium flavum*) contains the aporphine alkaloid glaucine, which displays C8-C6′ coupling and four O-methyl groups at C6, C7, C3′, and C4′ as numbered on the 1-benzylisoquinoline scaffold (Chang et al., 2015). Protopine is the major root alkaloid of *G. flavum* (Bournine et al., 2013a). Crabbine and other alkaloids were identified from the aerial parts of *Glaucium paucilobum* (Shafiee and Morteza-Semnani, 1998).

Benzo[c]phenanthridine alkaloids, that is, sanguinarine, chelirubine, macarpine, chelerythrine, dihydrosanguinarine, dihydrochelirubine, dihydromacarpine, and dihydrochelerythrine, are identified in *Eschscholtzia californica* cell culture (Son et al., 2014). Six structural

FIGURE 3.9 *Meconopsis*, **taken in Gama Valley, Tingri County, Tibet, China.**

types (pavinane, protopine, benzylisoquinoline, benzophenanthridine, aporphine, and PBB) are identified in three *Argemone* and four *Eschscholtzia* species (Cahlíková et al., 2012). *E. californica, Chelidonium majus*, and *Macleaya cordata* cultures produced similar amounts of sanguinarine (Bilka et al., 2012).

Berberine, protopine, chelerithrine, sanguinarine, coptisine, PAL, magnoflorine, and galanthamine are identified in *Argemone mexicana* roots (Kukula-Koch and Mroczek, 2015). Berberine and sanguinarine are equally distributed in roots and aerial tissues of developing plantlets of *A. mexicana*, while in juvenile plants, sanguinarine was only detected in roots (Xool-Tamayo et al., 2017). Terpenoids, flavonoids, and phenolics are found in *A. mexicana* whole plant extracts (Ghosh et al., 2015).

3.6.5 Eupteleaceae

Nortriterpene and triterpene oligoglycosides are isolated from the fresh leaves of *Euptelea polyandra* (Murakami et al., 2001). Refer to Chapter 5 for details of biopharmaceutics and pharmacokinetics of Ranunculales. Refer to Chapter 2 for differences and correlations between Ranunculaceae and other Ranunculales families.

3.7 ETHNOPHARMACOLOGY AND BIOACTIVITY

Different taxonomic groups have characteristic bioactivities and ethnopharmacological uses, depending on their different chemical profiles. In past decades, our group has collected more than 30,000 cards that record the ethnopharmacological use of more than 5000 Chinese medicinal plant species. The traditional remedy index (TRI) = $C1^2/C_2 \times 100$ (Xiao et al., 1986); C_1 is the number of cards on which a specific ethnopharmacological use is recorded in China for a single genus, and C_2 is the number of species that has the same ethnopharmacological use in the same genus. TRI ≥ 300 is considered significant. The higher the TRI, the more credible it is that the genus has the specific ethnopharmacological use. Distribution density of ethnopharmacological use (β) = $SP_1/SP_2 \times 100$, $SP_1 = C_2$, SP_2 = number of species in China in this genus.

3.7.1 Berberidaceae

The genus *Berberis* (Fig. 3.10) is used for heat-clearing and detoxification (TRI 3114, β 59) (Table 3.2), heat-clearing and fire-purging (TRI 643, β 30), heat-clearing and damp-drying (TRI 514, β 19), dysentery (TRI 3430, β 56), gastroenteritis (TRI 3512, β 65) (Table 3.2), redness swelling and pain of the eye (TRI 4483, β 73), liver disease (TRI 880, β 32), jaundice (465, 21), traumatic injury (536, 26), ulcer and carbuncles (2604, 54), and scald (331, 30). *Berberis vulgaris*, commonly known as "Aghriss" in Moroccan pharmacopoeia, is used to cure liver disorders and other diseases. Its aqueous extracts have significant hepatoprotective effects against lead-induced oxidative stress and liver dysfunction (Laamech et al., 2017). *Berberis integerrima* ingredients suppress T-cell response and shift immune responses toward Th2 (Fateh et al., 2015). Galangin and berberine have synergistic anticancer effects through apoptosis induction and proliferation inhibition in esophageal carcinoma cells (Ren et al., 2016) (Table 3.3).

Berbamine exerts antiinflammatory effects via inhibition of NF-κB and mitogen activated protein kinase (MAPK) signaling pathways (Jia et al., 2017). Breaking the resistance of *Escherichia coli* is involved in the antimicrobial activity of *Berberis lycium* (Malik et al., 2017). Berberine inhibits enterovirus 71 replication by downregulating the MAPK/extracellular signal-regulated kinase/extracellular signal-regulated kinase (MEK/ERK) signaling pathway and autophagy (Wang et al., 2017c). Berberine ameliorates intrahippocampal kainate-induced status epilepticus and consequent epileptogenic process in the rat (Sedaghat et al., 2017). *Berberis aristata* effectively manages peritonitis induced by carbapenem-resistant *E. coli* in a rat model (Thakur et al., 2017). Berberine prevents the oxidized low density lipoprotein (oxLDL) and TNFα-induced oxLDL receptor 1 (LOX1) expression and oxidative stress, key events that lead to nicotinamide adenine dinucleotide phosphate-oxidase (NOX), MAPK/Erk1/2, and NF-κB activation linked to endothelial dysfunction (Caliceti et al., 2017).

The genus *Caulophyllum* is used for promoting blood circulation to remove stasis (909, 100), dispelling wind and dampness (582, 100), and arthritis (736, 100). The roots and rhizomes of *Caulophyllum thalictroides* (blue cohosh) are used traditionally by Native Americans for inducing childbirth, easing the pain of labor, rectifying delayed or irregular menstruation, and alleviating heavy bleeding and pain during menstruation (Xia et al., 2014b). The antibacterial activity, antiinflammatory (Lee et al., 2012) and analgesic effects, antioxidant effects, anti-acetylcholinesterase activity, effects on atherosclerosis and myocardial ischemia (Si et al., 2010), anticancer activity (Zhang et al., 2011), cytochrome P450 (CYP) and topoisomerase inhibitory activity, and effects on wound healing are well documented.

The genus *Diphylleia* is used for ulcers and carbuncles (500, 100) and arthritis (500, 100). DPPT induces apoptosis of human prostate cancer cells (Hu et al., 2016a).

FIGURE 3.10   *Berberis wilsonae,* taken in Ninong Canyon, Deqin County, Yunnan, China.

TABLE 3.2 Top Ten Ethnopharmacological Uses of Ranunculales Plants

Genus	Arthritis	Heat-clearing and detoxification	Ulcers and carbuncles	Traumatic injury	Dispelling wind and dampness	Urinary tract infections	Dysentery	Redness/swelling and pain of the eye	Blood-activating and stasis-removing	Snake bites
Berberis		1	1	1			1	1		
Caulophyllum	1				1				1	
Diphylleia	1		1							
Dysosma		1	1	1					1	1
Epimedium	1				1					
Plagiorhegma							1	1		
Mahonia	1	1	1	1			1	1		
Nandina		1								
Sinopodophyllum	1			1						
Arcangelisia	1	1			1		1			
Cocculus	1	1	1			1				1
Cyclea	1	1	1			1				1
Diploclisia	1	1				1				1
Fibraurea	1	1	1				1	1		
Menispermum		1								
Sinomenium	1				1					
Stephania	1				1					
Sargentodoxa				1	1				1	
Akebia						1			1	
Holboellia						1				
Stauntonia				1						
Frequency	12	10	7	6	6	5	5	4	4	4

The genus *Dysosma* is used for blood-activating and stasis-removing (1694, 87), dispersing swelling and resolving toxins (574, 87), heat-clearing and detoxification (417, 50), ulcers and carbuncles (2669, 100), snake bites (2144, 87), mumps (903, 75), and traumatic injury (312, 50). Podophyllotoxins and flavonoids have cytotoxic activities (Yang et al., 2016a). DPPT has antineoplastic effects on glioblastoma U-87 MG and SF126 cells (Guerram et al., 2015). Podophyllotoxone inhibits human prostate cancer cells (Li et al., 2015b). Kaempferol of *Dysosma versipellis* inhibits angiogenesis through vascular endothelial growth factor (VEGF) and fibroblast growth factor (FGF) pathways (Liang et al., 2015). Flavonol dimers from callus cultures of *D. versipellis* display in vitro neuraminidase inhibitory activities (Chen et al., 2015). The coexisting flavonoids alleviate the podophyllotoxin toxicity (Li et al., 2013).

The genus *Epimedium* is used for strengthening yang (1988, 70), dispelling wind and dampness (1838, 70), impotence (2597, 100), arthritis (1062, 72), neurasthenia (662, 80), and sterility (323, 60). The total flavonoid fraction of the *Epimedium koreanum* extract has neuroprotective effects on dopaminergic neurons (Wu et al., 2017a). Icariin has neuroprotective properties in 1-methyl-4-phenyl-1,2,3,6-tetrahydropyridine (MPTP)-induced mouse models of Parkinson's disease (Chen et al., 2017a). The protective mechanisms of icariin against oxygen-glucose, deprivation-induced injury may be related to downregulating the expression of HIF-1α, HSP-60, and HSP-70 (Mo et al., 2017). *Epimedium* improves neuroplasticity and accelerates functional recovery of the diseased brain (Cho et al., 2017). *Epimedium grandiflorum* inhibits the growth in a model for the Luminal A molecular subtype of breast cancer (Telang et al., 2017). *Epimedium* improves the bone metabolic disorder through inducing osteogenic differentiation from bone marrow-derived mesenchymal stem cells (Kim et al., 2017a).

Total flavonoids of *Epimedium* reduce ageing-related oxidative DNA damage in testes of rats via p53-dependent pathway (Zhao et al., 2017). Icariin inhibits AMP-activated protein kinase (AMPK)-dependent autophagy and adipogenesis in adipocytes in vitro and in a model of Graves' orbitopathy in vivo (Li et al., 2017a). Icariin promotes mouse hair follicle growth by increasing insulin-like growth factor 1 expression in dermal papillary cells (Su et al., 2017). Icariin ameliorates IgA nephropathy by inhibiting nuclear factor κb/Nlrp3 pathway (Zhang et al., 2016b). Icariin has protective effects against the homocysteine-induced neurotoxicity in the primary embryonic cultures of rat cortical neurons (Li et al., 2016b). Icariside II, a broad-spectrum, anticancer agent, reverses β-amyloid-induced cognitive impairment by reducing inflammation and apoptosis in rats (Deng et al., 2017).

The genus *Plagiorhegma* (*Jeffersonia*) is used for dysentery (400, 100) and redness, swelling, and pain of the eye (400, 100). The PBB alkaloid jatrorrhizine, the lignane glucosides dehydrodiconiferyl-alcohol-4-β- D-glucoside, and its isomer dehydrodiconiferyl-alcohol-γ-β- D-glucoside, isolated from *Plagiorhegma dubium* cell culture, are antiinflammatory (Arens et al., 1985).

The genus *Mahonia* is used in heat-clearing and detoxification (2178, 61), dysentery (3095, 100), gastroenteritis (2579, 94), redness and swelling and pain of the eye (3553, 100), ulcers and carbuncles (1829, 94), tuberculosis (2625, 89), hepatitis (741, 83), traumatic injury (1325, 89), arthritis (896, 83), and common colds (962, 89). 1R, 1'S-isotetrandrine, an alkaloid found in Mahonia, possesses antiinflammatory, antioxidant, antibacterial, and antiviral properties (Wang et al., 2016a). Isotetrandrine reduces astrocyte cytotoxicity in neuromyelitis optica (NMO) by blocking the binding of NMO-IgG to aquaporin 4 (Sun et al., 2016c). PAL from *M. bealei* attenuates gut tumorigenesis in ApcMin/+ mice via inhibition of inflammatory

TABLE 3.3 Top Nine Pharmacological Activities of Ranunculales Plants

Genus	Anticancer/cytotoxic	Antimicrobial	Antiinflammatory	Antioxidant	Analgesic/sedative	Antidiabetic	Neuroprotective	Hepatoprotective	Antiulcer
Berberis	1	1	1	1				1	
Caulophyllum	1	1	1	1	1				
Diphylleia	1								
Dysosma	1	1							
Epimedium	1		1	1			1		
Plagiorhegma			1						
Mahonia	1	1	1	1			1		1
Nandina	1	1	1	1					
Sinopodophyllum	1								
Arcangelisia	1	1	1	1					
Tinospora	1	1	1	1	1	1	1	1	1
Cocculus	1	1	1		1	1			
Cyclea	1	1		1		1	1	1	1
Diploclisia	1								
Fibraurea	1	1							
Menispermum	1		1				1		
Sinomenium	1	1	1	1	1	1	1		
Stephania	1		1				1		
Sargentodoxa	1	1	1	1					

(Continued)

TABLE 3.3 Top Nine Pharmacological Activities of Ranunculales Plants (*cont.*)

Genus	Anticancer/cytotoxic	Antimicrobial	Antiinflammatory	Antioxidant	Analgesic/sedative	Antidiabetic	Neuroprotective	Hepatoprotective	Antiulcer
Akebia	1	1	1			1			
Stauntonia	1	1	1		1	1		1	
Corydalis	1	1	1	1	1	1		1	1
Fumaria	1	1	1				1		1
Hypecoum	1			1				1	
Glaucium	1	1	1		1	1			
Argemone	1	1			1				
Euptelea	1	1							1
Frequency	22	16	16	11	8	8	7	6	6

cytokines (Ma et al., 2016). The dichloromethane fraction from *M. bealei* leaves exerts an anti-inflammatory effect both in vitro and in vivo (Hu et al., 2016b). Alkaloids of *M. bealei* have anti-H^+/K^+-ATPase and antigastrin effects on pyloric ligation-induced gastric ulcer in rats (Zhang et al., 2014a).

The genus *Nandina* is used in heat-clearing and detoxification (405, 100), cough suppression (605, 100), bronchitis (405, 100), and pertussis (845, 100). Caffeoyl glucosides of *N. domestica* inhibit lipopolysaccharide (LPS)-induced endothelial inflammatory responses (Kulkarni et al., 2015). The essential oil of *Nandina* exhibits significant antioxidant activities (Bi et al., 2016). Essential oil and various organic extracts of *Nandina* have antifungal potential against skin-infectious fungal pathogens (Bajpai et al., 2009). Extract from *N. domestica* inhibits LPS-induced cyclooxygenase-2 expression in human pulmonary epithelial A549 cells (Ueki et al., 2012). Higenamine and nantenine induced the biphasic tracheal relaxation (Ueki et al., 2011). The *Nandina* extract inhibits histamine- and serotonin-induced contraction in isolated guinea pig trachea (Tsukiyama et al., 2007).

The genus *Sinopodophyllum* is used for cough suppression (357, 100), traumatic injury (514, 100), stomachache (514, 100), arthritis (357, 100), and menoxenia (357, 100). Flavonoids of *Sinopodophyllum* have anticancer activities against human breast cancer cells (Wang et al., 2017b). *S. hexundrum* promotes K562 cell apoptosis possibly by inhibiting BCR/ABL-STAT5 survival signal pathways and activating the mitochondrion-associated apoptotic pathways (Zhou et al., 2016).

3.7.2 Menispermaceae

The genus *Arcangelisia* is used for heat-clearing and detoxification (320, 100), dysentery (500, 100), gastroenteritis (320, 100), and malaria (500, 100). N-trans-feruloyltyramine of *A. gusanlung* is an active phenylpropanoid compound. It possesses antioxidant, antimicrobial, antimelanogenesis, immunomodulative, and anticancer activities (Jiang et al., 2015). Gusanlungionosides A-D, the megastigmane glycoside from the stems of *A. gusanlung*, are potential tyrosinase inhibitors (Yu et al., 2011). Berberine, extracted from *A. flava*, inhibited *Plasmodium* telomerase activity in a dose-dependent manner (Sriwilaijareon et al., 2002). *A. flava* and *F. tinctoria* have antiplasmodial activity (Nguyen-Pouplin et al., 2007). PBB alkaloids and 20-hydroxyecdysone of *A. flava* have antibabesial activity against *Babesia gibsoni* in culture (Subeki et al., 2005).

Tinospora is useful in countering various disorders and is an antioxidant, antihyperglycemic, and antihyperlipidemic, as well as a hepatoprotective (Lee et al., 2017a), cardiovascular protective (Priya et al., 2017), neuroprotective (Yu et al., 2017), osteoprotective (Abiramasundari et al., 2017), radioprotective, antianxiety, adaptogenic, analgesic, antiinflammatory (Jacob et al., 2017), antipyretic, thrombolytic, antidiarrheal, antiulcer, antimicrobial, and anticancer agent (Bajpai et al., 2017; Dhama et al., 2017;). Epoxy clerodane diterpene inhibits MCF-7 human breast cancer cell growth by regulating the expression of the functional apoptotic genes Cdkn2A, Rb1, mdm2, and p53 (Subash-Babu et al., 2017). Reactive oxygen species (ROS) mediates proapoptotic effects of *Tinospora cordifolia* on breast cancer cells (Ansari et al., 2017).

The regimen containing the dried stem powder of *T. cordifolia* reduces anxietylike behavior in middle-age female rats and associated neuroinflammation by ameliorating key inflammatory cytokines and modulated stress response (Singh et al., 2017). *T. cordifolia* shows immunomodulatory effects in resisting the viral multiplication and immunosuppression inflicted by chicken infectious anaemia virus (CIAV) in chicks (Latheef et al., 2017). *T. cordifolia* extract attenuated the quorum sensing-regulated behavior, and the active constituents might be 2,3,4-triacetyloxybutyl acetate, methyl 16-methyl heptadecanoate, 2-(5-ethenyl-5-methyloxolan-2-yl)propan-2-ol, methyl hexadecanoate, and 2-methoxy-4-vinyl phenol (Gala et al., 2016).

The genus *Cocculus* is used in dispelling wind and dampness (1521, 100), heat-clearing and detoxification (605, 100), arthritis (1184, 100), urinary tract infection (492, 50), ulcers and carbuncles (720, 100), and snake bites (637, 100). Haiderine, shaheenine, and coclaurine have good interaction with Aurora kinase (Thavamani et al., 2016); coclaurine, lirioresinol, and haiderine possess good binding with c-Kit; coclaurine, haiderine, and hisutine have good interaction with NF-kB, suggesting their anticancer potentials. *Cocculus hirsutus* and *C. pareira* have in vitro cytotoxic activity against HeLa cell lines (Thavamani et al., 2013). BBI alkaloids of *Cocculus pendulus* show inhibitory activities against acetyl- and butyrylcholinesterases (Atta-ur-Rahman et al., 2009).

C. hirsutus is used in Indian folk medicine for rheumatism, eczema, diabetics, inflammation, and neuralgia. It also shows anticancer activity against Dalton's lymphoma ascite cells in mice (Thavamani et al., 2014). *C. hirsutus* leaf extract exhibits nephroprotective activity in 5/6 nephrectomized rat models (Gadapuram et al., 2013). The ethanolic extract of *C. hirsutus* leaves has acute and chronic diuretic effects in normal rats (Badole et al., 2009). The acetone extract of *C. hirsutus* shows adulticidal activity against malarial vector, *Anopheles subpictus* (Elango et al., 2011). *C. hirsutus* shows repellent activity against *Culex tritaeniorhynchus* and Japanese encephalitis (Elango et al., 2010).

The genus *Cyclea* is used in heat-clearing and detoxification (847, 83), urinary tract infection (313, 50), ulcers and carbuncles (408, 50), snake bites (400, 67), and sore throats (476, 83). The roots of *C. pareira* var. *hirsuta* are used in the treatment of various diseases like stomach pain, fever, and skin disease in Ayurveda and is commonly known as Patha. Two other species, *C. peltata* and *S. japonica*, of the same family are being used as the source of Patha in various parts of India (Vijayan et al., 2014). (−)-Curine, a BBI alkaloid isolated from the roots of *C. wattii*, induces cell cycle arrest and cell death in hepatocellular carcinoma cells in a p53-independent manner (Gong et al., 2017).

C. peltata displays a protective effect on cisplatin-induced nephrotoxicity and oxidative damage (Vijayan et al., 2007). *C. peltata* inhibits the stone formation induced by ethylene glycol treatment in rats (Christina et al., 2002). The alkaloid extract of *C. peltata* roots ameliorates the APAP/CCl$_4$-induced liver toxicity in Wistar rats and shows in vitro, free-radical scavenging property (Shine et al., 2014). Aqueous extract of *C. peltata* at 40 and 60 mg/kg dose significantly decreases both the fasting and postprandial blood glucose of type-2 diabetic rats (Kirana and Srinivasan, 2010). *C. peltata* shows gastric antisecretory and antiulcer activities in rats (Shine et al., 2009). *C. peltata* is potentially a good source of antibacterial agents (Raja et al., 2011).

The genus *Diploclisia* is used for urinary tract infections (320, 100), arthritis (320, 100), and snake bites (500, 100). *D. glaucescens* shows strong antifeedant activity (Jayasinghe et al., 2003b).

The genus *Fibraurea* is used in heat-clearing and damp-drying (500, 100), heat-clearing and detoxification (320, 100), dysentery (417, 100), sore throats (417, 100), redness/swelling and pain of the eye (417, 100), and ulcers and carbuncles (500, 100). The aporphine alkaloid

ROE from the fresh rattan stem of *F. recisa* has antifungal activity (Ma et al., 2015). The methanol-water fraction of *F. tinctoria* shows higher antiproliferative activities than its methanolic extract (Manosroi et al., 2015). Fibrarecisin, a triterpenoid from *F. recisa*, has antitumor activity (Fu et al., 2007). PAL and jatrorrhizine of *Fibraurea* shows inhibitory effects against CYP3A4 (Su et al., 2007).

The genus *Menispermum* is used in heat-clearing and detoxification (1108, 100) and sore throats (931, 100). Oxoisoaporphine of *M. dauricum* is a potent telomerase inhibitor (Wei et al., 2016). Phenolic alkaloids from *M. dauricum* inhibit BxPC-3 pancreatic cancer cells by blocking the Hedgehog signaling pathway (Zhou et al., 2015). They also inhibit gastric cancer in vivo (Zhang et al., 2014b). *M. dauricum* induces apoptosis in human cervical carcinoma Hela cells (Wang et al., 2014b). Two acidic polysaccharides from the rhizoma of *M. dauricum* have antiovarian cancer potential (Lin et al., 2013). Dauricine inhibits the proliferation of urinary tract tumor cells (Wang et al., 2012).

The rhizome extracts of *M. dauricum* show the intestinal antiinflammatory effects on trinitrobenzene sulfonic acid–induced ulcerative colitis in mice (Su et al., 2016). Sinomenine, norsinoacutin, N-norsinoacutin-β-D-glucopyranoside, 6-O-methyl-laudanosoline-13-O-glucopyranoside, magnoflorine, laurifloline, and dauricinoline, isolated from the dried root of *M. dauricum*, are potential NF-κB inhibitors (Sun et al., 2014). Phenolic alkaloids from *M. dauricum* rhizome protect against brain ischemia injury via regulation of GLT-1, EAAC1, and ROS generation (Zhao et al., 2012a). Dauricine has antiarrhythmic effects and can prolong the action potential duration, which has been attributed to its ability to modulate Ca^{2+} and several K^+ channels (Zhao et al., 2012b).

The genus *Sinomenium* is used in dispelling wind and dampness (514, 100) and arthritis (514, 100). Sinomenine protects PC12 neuronal cells against H_2O_2-induced cytotoxicity and oxidative stress via an ROS-dependent upregulation of endogenous antioxidant systems (Fan et al., 2017). Sinomenine treatment may suppress the upregulation and activation of the P2X3 receptor and relieve the hyperalgesia potentiated by the activation of P38MAPK in diabetic rats (Rao et al., 2017). Laudanosoline-1-O-xylopyranose, 6-O-methyl-laudanosoline-1-O-glucopyranoside, menisperine, sinomenine, laurifoline, magnoflorine, and norsinoacutin are potential NF-κB inhibitors (Sun et al., 2016d). Sinomenine, magnoflorine, and laurifoline are potential β2-adrenergic receptor agonists. Salutaridine, dauricumine, cheilanthifoline, and dauriporphine have significant inhibitions on the receptor activator of NF-κB ligand-induced differentiation of mouse bone marrow–derived macrophages into multinucleated osteoclasts (Lee et al., 2016).

Sinomenine attenuates angiotensin II-induced autophagy via inhibition of P47-Phox translocation to the membrane and influences ROS generation in podocytes (Wang et al., 2016b). Sinomenine, a COX-2 inhibitor, induces cell cycle arrest and inhibits growth of human colon carcinoma cells in vitro and in vivo (Yang et al., 2016b). Vascular normalization induced by sinomenine hydrochloride results in suppressed mammary tumor growth and metastasis (Zhang et al., 2015b).

Sinomenine exerts antiinflammation effects via α7 nicotinic acetylcholine receptors in macrophages stimulated by lipopolysaccharide (Yi et al., 2015). Sinomenine potentiates P815 cell degranulation via upregulation of Ca^{2+} mobilization through the Lyn/PLCγ/IP3R pathway (Wang et al., 2016c). Sinomenine potentiates degranulation of RBL-2H3 basophils via upregulation of phospholipase A2 phosphorylation by Annexin A1 cleavage and ERK phosphorylation without influencing on calcium mobilization (Huang et al., 2015). Intra-articular delivery of sinomenium ameliorates osteoarthritis by effectively regulating autophagy (Chen et al., 2016a). Sinomenine and magnoflorine, major constituents of *Sinomenium* rhizome, show potent protective effects against membrane damage induced by lysophosphatidylcholine in rat erythrocytes (Sakumoto et al., 2015).

The genus *Stephania* is used in heat-clearing and detoxification (4102, 94) and dispelling wind and dampness (1945, 44). The BBI alkaloid tetrandrine inhibits the proliferation of human osteosarcoma cells by upregulating the phosphatase and tensin homolog (PTEN) pathway (Tian et al., 2017). Tetrandrine inhibits glioma stemlike cells by repressing β-catenin expression (Zhang et al., 2017b). Tetrandrine, an agonist of aryl hydrocarbon receptors, reciprocally modulates the activities of STAT3 and STAT5 to suppress Th17 cell differentiation (Yuan et al., 2017). Tetrandrine exhibits antihypertensive and sleep-enhancing effects in spontaneously hypertensive rats (Huang et al., 2016). Tetrandrine induces lipid accumulation through blockades of autophagy in a hepatic stellate cell line (Miyamae et al., 2016).

The biscoclaurine alkaloid cepharanthine induces autophagy, apoptosis, and cell cycle arrest in breast cancer cells (Gao et al., 2017). Fangchinoline has antimetastatic activity in human gastric cancer AGS cells (Chen et al., 2017b). (–)-Stepholidine has dopamine receptor D1 agonistic and dopamine receptor D2 antagonistic activities (Li et al., 2016c).

3.7.3 Lardizabalaceae and Circaeasteraceae

The genus *Sargentodoxa* is used for promoting blood circulation to remove stasis (2181, 100), dispelling wind and dampness (1706, 100), traumatic injury (632, 100), arthritis (1290, 100), menoxenia (632, 100), appendicitis (323, 100), and ascariasis (545, 100). Liriodendrin of *S. cuneata* has a protective role in sepsis-induced acute lung injury (Yang et al., 2016c). Phenolics and phenolic glycosides from *S. cuneata* have antimicrobial and cytotoxic activities (Zeng et al., 2015). The aqueous extracts of *S. cuneata* inhibit Coxsackie virus B3 and B5, polio virus I, Echo virus 9, and Echo virus 29 (Guo et al., 2006).

The genus *Akebia* is used for promoting blood circulation to remove stasis (455, 100), heat-clearing and damp-drying (368, 100), arthritis (655, 100), and urinary tract infection (455, 100). Hederagenin, isolated from *A. quinata* fruit, alleviates the proinflammatory and apoptotic response to alcohol in rats (Kim et al., 2017b). *A. quinata* extract exerts antiobesity and hypolipidemic effects in high-fat diet–fed mice and 3T3-L1 adipocytes (Sung et al., 2015). Akebiasaponin D decreases hepatic steatosis through autophagy modulation (Gong et al., 2016). Protein tyrosine phosphatase 1B inhibitors from the stems of *A. quinata* might be chemopreventative agents for breast cancer (An et al., 2016). *A. trifoliata* seed extract inhibits the proliferation of human hepatocellular carcinoma cells via inducing endoplasmic reticulum stress (Lu et al., 2014c). Fermented *A. quinata* extracts alleviate alcoholic hangover and reduce plasma ethanol concentrations (Jung et al., 2016). *A. quinata* fruit extracts show ameliorating effects on skin aging induced by advanced glycation end products (Shin et al., 2015). Oleanane-type triterpenoids from pericarps of *A. trifoliata* have antibacterial activity (Wang et al., 2015b).

The genus *Holboellia* is used for arthritis (800, 100), menoxenia (313, 100), breast-milk stoppage (450, 100), urinary tract infection (613, 100), urinary stone (313, 100), and beriberi (613, 100).

The genus *Stauntonia* is used for arthritis (613, 100) and traumatic injury (514, 75). *S. hexaphylla* leaf constituents inhibit rat lens aldose reductase and the formation of advanced glycation end products (Hwang et al., 2017). Triterpenoid saponins from *S. chinensis* (Ye Mu Gua in TCM) ameliorate insulin resistance via the AMPK and IR/IRS-1/PI3K/Akt pathways in insulin-resistant HepG2 cells (Hu et al., 2014). *S. chinensis* polysaccharide shows the protective effect on CCl_4-induced acute liver injuries in mice (Yang et al., 2014). Nor-oleanane triterpenoid saponins from *S. brachyanthera* inhibit UDP-glucuronosyltransferases (Liu et al., 2016a) and have antigout activity (Liu et al., 2016b). *S. hexaphylla* leaf methanol extract inhibits osteoclastogenesis and bone resorption activity via proteasome-mediated degradation of c-Fos protein and suppression of NFATc1 expression (Cheon et al., 2015). The extracts of *S. chinensis* have antinociceptive and antiinflammatory activities (Ying et al., 2014; Gao et al., 2009). Triterpenoids from *S. obovatifoliola* subsp. *intermedia* inhibit HIV-1 protease (Wei et al., 2008).

3.7.4 Papaveraceae

Around 428 species of genus *Corydalis* are distributed worldwide, 298 of which are in China, and 10 groups and 219 species are endemic in China. *Corydalis* is widely employed in folk medicines in China and adjacent countries, especially in traditional Tibetan medicines, for the treatment of fever, hepatitis, edema, gastritis, cholecystitis, hypertension, and other diseases (Shang et al., 2014).

Corydalis has pharmacological effects in cardiovascular diseases and the central nervous system and shows antibacterial, analgesic (Kang et al., 2016), antiinflammatory (Liu et al., 2017), antioxidant and hepatoprotective effects (Liang et al., 2016; Wu et al., 2017b), etc. *Corydalis yanhusuo* shows inhibitory effects on rabbit platelet aggregation induced by adenosine diphosphate (ADP), thrombin, or arachidonic acid (Li et al., 2017b). Dehydrocorydaline, an alkaloid isolated from *Corydalis turtschaninovii* tuber, has anticoronary artery disease, antiinflammatory, apoptotic, antiallergic, anti-acetylcholinesterase, antitumor, and antimetastatic effects (Lee et al., 2017b). Lignanamides from the aerial parts of *Corydalis saxicola* have antitumor activity (Zhang et al., 2016c). 13-Methyl-palmatrubine, isolated from *C. yanhusuo*, induces apoptosis and cell cycle arrest in A549 cells in vitro and in vivo (Chen et al., 2016b).

A furochromoneheterocarpin and its source *Corydalis heterocarpa* could be a functional food with potential proapoptotic activity against cancer cells (Kong et al., 2016). *Corydalis edulis* promotes insulin secretion via the activation of protein kinase Cs in mice and pancreatic β cells (Zheng et al., 2017). Levo-tetrahydroberberrubine produces anxiolytic-like effects in mice through the 5-HT1A receptor (Mi et al., 2017). l-THP reduces nicotine self-administration and reinstatement in rats (Faison et al., 2016). It also inhibits the acquisition of ketamine-induced conditioned place preference by regulating the expression of ERK and cAMP-response element binding protein (CREB) phosphorylation in rats, and thus could be useful in the treatment of ketamine addiction (Du et al., 2017).

Allocryptopine (ALL) of *Corydalis decumbens* is antiarrhythmic (Fu et al., 2016). Dehydrocorydaline promotes myogenic differentiation via p38 MAPK activation (Yoo et al., 2016). *Corydalis crispa* showed anthelmintic properties (Wangchuk et al., 2016). Cavidine of *Corydalis impatiens* has antiulcerogenic effect against ethanol-induced acute gastric ulcer in mice (Li et al., 2016d).

The total alkaloid extract of *Fumaria capreolata* has antiinflammatory effects in the DNBS model of mice colitis and intestinal epithelial CMT93 cells (Bribi et al., 2016). Alkaloid extracts of five Bulgaria species significantly inhibited acetylcholinesterase activity (Vrancheva et al., 2016). Alkaloids parfumidine and sinactine of *Fumaria officinalis* exhibits potent prolyl oligopeptidase inhibition activities (Chlebek et al., 2016). *Fumaria indica* shows antisecretory and gastroprotective activities in rats (Chandra et al., 2015; Shakya et al., 2016). *Fumaria parviflora* causes bronchodilation via dual blockade of muscarinic receptors and Ca^{2+} influx (Khan and Gilani, 2015).

H. leptocarpum is a traditional Tibetan drug (Chen and Fang, 1985). Leptopidine of *H. leptocarpum* could suppress growth and induce cytotoxicity in breast cancer cells via inhibiting fatty acid synthase expression (Zhang et al., 2015a). *Hypecoum erectum* extract has the hepatoprotective effect on D-galactosamine-induced damage of rat liver (Nikolaev et al., 2014). *H. erectum* extract inhibits lipid peroxidation, increases the activity of the host endogenous antioxidant system, normalizes hepatocyte energy provision, and limits liver degeneration (Toropova et al., 2014).

The Persian endemic species *Glaucium vitellinum* has antibacterial and antifungal activities (Mehrara et al., 2015). In Algeria, *G. flavum* is used to treat warts (Bournine et al., 2013b). Bocconoline of *G. flavum* root has anticancer activity against breast cancer cells (Bournine et al., 2013a). *Glaucium* is used in Iranian herbal medicine as laxative, hypnotic, narcotic, and antidiabetic agents and also in the treatment of dermatitis. The methanolic extract and total alkaloids of *G. paucilobum* possess analgesic activity (Morteza-Semnani et al., 2006). The topical preparation of *Glaucium grandiflorum* has antiinflammatory and analgesic activity (Morteza-Semnani et al., 2004). The extracts of *Corydalis solida* subsp. *Solida* and *Glaucium corniculatum* inhibit acetylcholinesterase (Orhan et al., 2004).

A. mexicana has antibacterial, anticancer, sedative, and probably antianxiety properties (Arcos-Martínez et al., 2016). *A. mexicana* silver nanoparticles are suitable for bioformulation against mosquitoes and microbes (Kamalakannan et al., 2016). Alkaloids isolated from *A. mexicana* are cytotoxic to the SW480 human colon cancer cell line (Singh et al., 2016). Argemonine and berberine, isolated from the EtOAc fraction of *Argemone gracilenta*, displays high antiproliferative activity on multiple cancer cell lines (Leyva-Peralta et al., 2015). *A. mexicana* leaves have in vitro antiurolithiasis potentials (Chilivery et al., 2016). It is used in the treatment of epileptic disorders in Indian traditional systems of medicine (Asuntha et al., 2015).

3.7.5 Eupteleaceae

Nortriterpene oligoglycosides isolated from the fresh leaves of *E. polyandra* have gastroprotective activity (Yoshikawa et al., 2000). Triterpenes from *E. polyandra* inhibit Epstein-Barr virus activation (Konoshima et al., 1987).

3.8 DISCUSSION AND CONCLUSION

BIAs, used as the taxonomic marker, consist of more than 2500 diverse structures, mainly generated by the order Ranunculales and the eumagnoliids. BIAs also occur in the Rutaceae, Lauraceae, Cornaceae, and Nelumbonaceae and sporadically throughout the order Piperales. Several BIAs function in the defense of plants against herbivores and pathogens, thus BIA biosynthesis might play an important role in the reproductive fitness of certain plants. Biochemical and molecular phylogenetic approaches were used to investigate the evolution of BIA biosynthesis in basal angiosperms (Liscombe et al., 2005). The occurrence of (S)-norcoclaurine synthase (NCS; EC 4.2.1.78) activity in 90 diverse plant species was compared to the distribution of BIAs superimposed on a molecular phylogeny, supporting the monophyletic origin of BIA biosynthesis prior to the emergence of the eudicots. Phylogenetic analyses of NCS, berberine bridge enzyme, and several O-methyltransferases suggest an underlying molecular fingerprint for BIA biosynthesis in angiosperms not known to accumulate such alkaloids. The limited occurrence of BIAs outside the Ranunculales and eumagnoliids suggests the requirement for a highly specialized but evolutionarily unstable cellular platform to accommodate or reactivate the pathway in divergent taxa.

Cycloidea-like genes belong to the TCP family of transcriptional regulators and control different aspects of shoot development in various angiosperm lineages. CYC-like genes underwent duplications in Fumariaceae and Papaveraceae (Kölsch and Gleissberg, 2006). Phylogenetic analyses of CYC-like genes support the phylogenetic inference of Ranunculales based on molecular markers (Hilu et al., 2008) and chemotaxonomy. The observed duplications originated from a single CYC gene present in all Ranunculales.

The pharmacophylogenetic studies support the circumscription of the family proposed by Wu Z-Y et al. (Peng et al., 2006), and the Berberidaceae can be treated as four independent families: Nandinaceae, Berberidaceae (s.s.), Podophyllaceae, and Leonticaceae. The monotypic family Nandinaceae is rich in various BIAs, for example, berberine, PAL, jatrorrhizine, coptisine, magnoflorine, domesticine, nandinine, and protopine. The existence of the cyanogenic compound nandinin, biflavonoidamentoflavone, and benzaldehyde-4-O-glucoside in this family indicates its relatively distant relation with the other three families. *Nandina indica* is ethnopharmacologically used for clearing heat and counteracting toxins or as an antitussive. The Berberidaceae (s.s.) consists of *Berberis* and *Mahonia*, containing mainly BIAs, for example, berberine, PAL, jatrorrhizine, columbamine, and magnoflorine, particularly a higher content of berbamine and oxyacanthine. Ethnopharmacologically the plants of this family are mainly used as medicines for clearing heat and counteracting toxins. Plants in both Berberis and Mahonia have long been used as the main sources of the drugs berberine and berbamine (Xiao et al., 2016). The Podophyllaceae can be divided into two tribes. The tribe Podophylleae, consisting of *Podophyllum* (including *Sinopodophyllum* and *Dysosma*) and *Diphylleia*, is rich in various podophyllotoxin lignans, and the plants in this tribe are used as the most important source for producing anticancer drugs (i.e., podophyllotoxin's derivatives). Ethnopharmacologically the plants are mainly used for activating blood, revolving stasis, relieving swelling, removing toxins, and clearing heat. The tribe Epimedieae, consisting of *Epimedium, Vancouveria, Achlys, Jeffersonia,* and *Ranzania*, has diversified chemical constituents. Both *Epimedium* and *Vancouveria* contain predominately bioactive icariin flavonoids, the characteristic chemical constituents of this group. Ethnopharmacologically the plants in Epimediumare are used as a male sexual tonic and as medicines for dispelling wind and removing dampness. The phytochemistry of *Achlys, Jeffersonia,* and *Ranzania* has not been thoroughly investigated. *Jeffersonia dubia* is used for dysentery and inflammatory eye pain by the Korean minority nationality of Northeast China. The Leonticaceae, including *Gymnospermium, Leontice, Caulophyllum,* and *Bongardia*, contain mainly β-amyrin triterpenoids and quinolizidine alkaloids and are used as medicines for activating blood, revolving stasis, dispelling wind, and removing dampness.

BIAs are a diverse class of plant-specialized metabolites sharing a common biosynthetic origin beginning with tyrosine. Many BIAs have potent pharmacological activities, and plants accumulating them have long histories of use in traditional medicine. The major BIA pathways, elucidated by both conventional biochemical methods and the uproaring high-throughput genome/transcriptome sequencing technologies (Hagel et al., 2015a), are shared in many Ranunculales plants, and more useful BIAs, besides morphine, sanguinarine, and berberine, await to be mined within the context of biodiversity. Moreover, these nonmodel species are a rich source of catalyst diversity valuable to plant biochemists and synthetic biology endeavors. The fast-evolving metabolome analysis platforms facilitate mining of both primary and secondary metabolites from Ranunculales taxa (Hagel et al., 2015b). Far-reaching metabolite profiling based on multiple analytical platforms enabled a more inclusive picture of overall metabolism occurring in selected species. Traditionally important medicinal species and their close relatives should be subjected to a metabolomics approach. Coupled with genomics data, these metabolomics resources are key for mining BIA, saponin, terpenoid, flavonoid, and others.

References

Abiramasundari, G., Gowda, C.M., Pampapathi, G., et al., 2017. Ethnomedicine based evaluation of osteoprotective properties of *Tinospora cordifolia* on in vitro and in vivo model systems. Biomed. Pharmacother. 87, 342–354.

An, J.P., Ha, T.K., Kim, J., 2016. Protein tyrosine phosphatase 1B inhibitors from the stems of *Akebia quinata*. Molecules 21 (8), E1091. doi: 10.3390/molecules21081091.

Anderson, C.L., Bremer, K., Friis, E.M., 2005. Dating phylogenetically basal eudicots using rbcL sequences and multiple fossil reference points. Am. J. Bot. 92 (10), 1737–1748.

Angiosperm Phylogeny Group, 2009. An update of the Angiosperm Phylogeny Group classification for the orders and families of flowering plants: APG III. Bot. J. Linn. Soc. 161 (2), 105–121.

Ansari, J.A., Rastogi, N., Ahmad, M.K., et al., 2017. ROS mediated pro-apoptotic effects of *Tinospora cordifolia* on breast cancer cells. Front. Biosci. 9, 89–100.

Arcos-Martínez, A.I., Muñoz-Muñiz, O.D., Domínguez-Ortiz MÁ., et al., 2016. Anxiolytic-like effect of ethanolic extract of *Argemone mexicana* and its alkaloids in Wistar rats. Avicenna J. Phytomed. 6 (4), 476–488.

Arens, H., Fischer, H., Leyck, S., et al., 1985. Antiinflammatory compounds from *Plagiorhegma dubium* cell culture. Planta Med. 51 (1), 52–56.

Asuntha, G., Raju, Y.P., Sundaresan, C.R., et al., 2015. Effect of *Argemone mexicana* (L.) against lithium-pilocarpine induced status epilepticus and oxidative stress in Wistar rats. Indian J. Exp. Biol. 53 (1), 31–35.

Atta-ur-Rahman, Atia-tul-Wahab, Zia Sultani, S., et al., 2009. Bisbenzylisoquinoline alkaloids from *Cocculus pendulus*. Nat. Prod. Res. 23 (14), 1265–1273.

Badole, S.L., Bodhankar, S.L., Patel, N.M., et al., 2009. Acute and chronic diuretic effect of ethanolic extract of leaves of *Cocculus hirsutus* (L.) Diles in normal rats. J. Pharm. Pharmacol. 61 (3), 387–393.

Bajpai, V., Kumar, S., Singh, A., et al., 2017. Chemometric based identification and validation of specific chemical markers for geographical, seasonal and gender variations in *Tinospora cordifolia* stem using HPLC-ESI-QTOF-MS analysis. Phytochem. Anal. 28 (4), 277–288.

Bajpai, V.K., Rahman, A., Kang, S.C., 2008. Chemical composition and inhibitory parameters of essential oil and extracts of *Nandina domestica* Thunb.to control food-borne pathogenic and spoilage bacteria. Int. J. Food Microbiol. 125 (2), 117–122.

Bajpai, V.K., Yoon, J.I., Kang, S.C., 2009. Antifungal potential of essential oil and various organic extracts of *Nandina domestica* Thunb. against skin infectious fungal pathogens. Appl. Microbiol. Biotechnol. 83 (6), 1127–1133.

Bell, C.D., Soltis, D.E., Soltis, P.S., 2010. The age and diversification of the angiosperms re-revisited. Am. J. Bot. 97, 1296–1303.

Belwal, T., Giri, L., Bhatt, I.D., et al., 2017. An improved method for extraction of nutraceutically important polyphenolics from *Berberis jaeschkeana* C. K. Schneid. fruits. Food Chem. 230, 657–666.

Bhagya, N., Chandrashekar, K.R., 2016. Tetrandrine--A molecule of wide bioactivity. Phytochemistry 125, 5–13.

Bi, S.F., Zhu, G.Q., Wu, J., et al., 2016. Chemical composition and antioxidant activities of the essential oil from *Nandina domestica* fruits. Nat. Prod. Res. 30 (3), 362–365.

Bilka, F., Balážová, A., Bilková, A., et al., 2012. Comparison of sanguinarine production in suspension cultures of the *Papaveraceae* plants. Ceska Slov. Farm. 61 (6), 267–270.

Bournine, L., Bensalem, S., Peixoto, P., et al., 2013a. Revealing the anti-tumoral effect of Algerian *Glaucium flavum* roots against human cancer cells. Phytomedicine 20 (13), 1211–1218.

Bournine, L., Bensalem, S., Wauters, J.N., et al., 2013b. Identification and quantification of the main active anticancer alkaloids from the root of *Glaucium flavum*. Int. J. Mol. Sci. 14 (12), 23533–23544.

Bribi, N., Algieri, F., Rodriguez-Nogales, A., et al., 2016. Intestinal anti-inflammatory effects of total alkaloid extract from *Fumaria capreolata* in the DNBS model of mice colitis and intestinal epithelial CMT93 cells. Phytomedicine 23 (9), 901–913.

Cahlíková, L., Kucera, R., Host'álková, A., et al., 2012. Identification of pavinane alkaloids in the genera *Argemone* and *Eschscholzia* by GC-MS. Nat. Prod. Commun. 7 (10), 1279–1281.

Caliceti, C., Rizzo, P., Ferrari, R., et al., 2017. Novel role of the nutraceutical bioactive compound berberine in lectin-like OxLDL receptor 1-mediated endothelial dysfunction in comparison to lovastatin. Nutr. Metab. Cardiovasc. Dis. 27 (6), 552–563.

Chandra, P., Kishore, K., Ghosh, A.K., 2015. Evaluation of antisecretory, gastroprotective and in-vitro antacid capacity of *Fumaria indica* in rats. J. Environ. Biol. 36 (5), 1137–1142.

Chang, L., Hagel, J.M., Facchini, P.J., 2015. Isolation and characterization of O-methyltransferases involved in the biosynthesis of glaucine in *Glaucium flavum*. Plant Physiol. 169 (2), 1127–1140.

Chen, B.Z., Fang, Q.C., 1985. Chemical study on a traditional Tibetan drug *Hypecoum leptocarpum*. Yao Xue Xue Bao 20 (9), 658–661.

Chen, J., Lu, X., Lu, C., et al., 2016b. 13-Methyl-palmatrubine induces apoptosis and cell cycle arrest in A549 cells in vitro and in vivo. Oncol. Rep. 36 (5), 2526–2534.

Chen, P., Xia, C., Mei, S., et al., 2016a. Intra-articular delivery of sinomenium encapsulated by chitosan microspheres and photo-crosslinked GelMA hydrogel ameliorates osteoarthritis by effectively regulating autophagy. Biomaterials 81, 1–13.

Chen, R., Duan, R., Wei, Y., et al., 2015. Flavonol dimers from callus cultures of *Dysosma versipellis* and their in vitro neuraminidase inhibitory activities. Fitoterapia 107, 77–84.

Chen, W.F., Wu, L., Du, Z.R., et al., 2017a. Neuroprotective properties of icariin in MPTP-induced mouse model of Parkinson's disease: involvement of PI3K/Akt and MEK/ERK signaling pathways. Phytomedicine 25, 93–99.

Chen, Z., He, T., Zhao, K., et al., 2017b. Anti-metastatic activity of fangchinoline in human gastric cancer AGS cells. Oncol. Lett. 13 (2), 655–660.

Cheon, Y.H., Baek, J.M., Park, S.H., et al., 2015. *Stauntonia hexaphylla* (Lardizabalaceae) leaf methanol extract inhibits osteoclastogenesis and bone resorption activity via proteasome-mediateddegradation of c-Fos protein and suppression of NFATc1 expression. BMC Complement. Altern. Med. 15, 280.

Chilivery, R.K., Alagar, S., Darsini, T.P., 2016. In vitro anti-urolithiasis potentials of *Argemone mexicana* L. leaves. Curr. Clin. Pharmacol. 11 (4), 286–290.

Chlebek, J., Novák, Z., Kassemová, D., et al., 2016. Isoquinolinealkaloids from *Fumaria officinalis* L. and their biological activities related to Alzheimer's disease. Chem. Biodivers. 13 (1), 91–99.

Cho, J.H., Jung, J.Y., Lee, B.J., et al., 2017. *Epimedii Herba*: a promising herbal medicine for neuroplasticity. Phytother. Res.doi: 10.1002/ptr.5807.

Christina, A.J., Packia Lakshmi, M., Nagarajan, M., et al., 2002. Modulatory effect of *Cyclea peltata* Lam. on stone formation induced by ethylene glycol treatment in rats. Methods Find. Exp. Clin. Pharmacol. 24 (2), 77–79.

Dary, C., Bun, S.S., Herbette, G., et al., 2017. Chemical profiling of the tuber of *Stephania cambodica* Gagnep. (Menispermaceae) and analytical control by UHPLC-DAD. Nat. Prod. Res. 31 (7), 802–809.

Deng, Y., Long, L., Wang, K., et al., 2017. Icariside II, a broad-spectrum anti-cancer agent, reverses beta-amyloid-induced cognitive impairment through reducing inflammation and apoptosis in rats. Front. Pharmacol. 8, 39.

Dhama, K.D., Sachan, S., Khandia, R., et al., 2017. Medicinal and beneficial health applications of *Tinospora cordifolia* (Guduchi): a miraculous herb countering various diseases/disorders and its immunomodulatory effects. Recent Pat. Endocr. Metab. Immune Drug Discov.doi: 10.2174/1872214811666170301105101.

Ding, W., Li, Y., Li, G., et al., 2016. New 30-noroleanane triterpenoid saponins from *Holboellia coriacea* Diels. Molecules 21 (6), E734. doi: 10.3390/molecules21060734.

Doncheva, T., Yordanova, G., Vutov, V., et al., 2016. Comparative study of alkaloid pattern of four Bulgarian *Fumaria* species. Nat. Prod. Commun. 11 (2), 211–212.

Dong, P.L., Han, H., Zhang, T.Y., et al., 2014. P-glycoprotein inhibition increases the transport of dauricine across the blood-brain barrier. Mol. Med. Rep. 9 (3), 985–988.

Du, Y., Du, L., Cao, J., et al., 2017. Levo-tetrahydropalmatine inhibits the acquisition of ketamine-induced conditioned place preference by regulating the expression of ERK and CREB phosphorylation in rats. Behav. Brain Res. 317, 367–373.

Elango, G., Rahuman, A.A., Kamaraj, C., et al., 2011. Efficacy of medicinal plant extracts against malarial vector *Anopheles subpictus* Grassi. Parasitol. Res. 108 (6), 1437–1445.

Elango, G., Rahuman, A.A., Zahir, A.A., et al., 2010. Evaluation of repellent properties of botanical extracts against *Culex tritaeniorhynchus* Giles (Diptera: Culicidae). Parasitol. Res. 107 (3), 577–584.

Faison, S.L., Schindler, C.W., Goldberg, S.R., et al., 2016. l-tetrahydropalmatine reduces nicotine self-administration and reinstatement in rats. BMC Pharmacol. Toxicol. 17 (1), 49.

Fan, H., Shu, Q., Guan, X., et al., 2017. Sinomenine protects PC12 neuronal cells against H2O2-induced cytotoxicity and oxidative stress via a ROS-dependent up-regulation of endogenous antioxidant system. Cell. Mol. Neurobiol. 37 (8), 1387–1398.

Fateh, S., Dibazar, S.P., Daneshmandi, S., et al., 2015. Barberry's (*Berberis integerrima*) ingredients suppress T-cell response and shift immune responses toward Th2: an in vitro study. Future Sci. OA 1 (4), FSO49.

Forrester, M.B., 2017. Pediatric *Nandina domestica* ingestions reported to poison centers. Hum. Exp. Toxicol. 2017.doi: 10.1177/0960327117705429.

Fu, H., Koike, K., Zheng, Q., et al., 2001. Fargosides A-E, triterpenoid saponins from *Holboellia fargesii*. Chem. Pharm. Bull. 49 (8), 999–1002.

Fu, J.L., Tan, C.H., Lin, L.P., et al., 2007. Fibrarecisin, a novel triterpenoid from *Fibraurea recisa* with antitumor activity. Nat. Prod. Res. 21 (4), 351–353.

Fu, Y.C., Zhang, Y., Tian, L.Y., et al., 2016. Effects of allocryptopine on outward potassium current and slow delayed rectifier potassium current in rabbit myocardium. J. Geriatr. Cardiol. 13 (4), 316–325.

Fun, H.K., Salae, A.W., Razak, I.A., et al., 2011. Absolute configuration of fibaruretin B. Acta Crystallogr. Sect. E Struct. Rep. Online 67 (Pt 5), o1246–o1247.

Gadapuram, T.K., Murthy, J.S., Rajannagari, R.R., et al., 2013. Nephroprotective activity of *Cocculus hirsutus* leaf extract in 5/6 nephrectomized rat model. J. Basic Clin. Physiol. Pharmacol. 24 (4), 299–306.

Gala, V.C., John, N.R., Bhagwat, A.M., et al., 2016. Attenuation of quorum sensing-regulated behaviour by *Tinospora cordifolia* extract & identification of its active constituents. Indian J. Med. Res. 144 (1), 92–103.

Gao, H., Zhao, F., Chen, G.D., et al., 2009. Bidesmoside triterpenoid glycosides from *Stauntonia chinensis* and relationship to anti-inflammation. Phytochemistry 70 (6), 795–806.

Gao, S., Li, X., Ding, X., et al., 2017. Cepharanthine induces autophagy, apoptosis and cell cycle arrest in breast cancer cells. Cell. Physiol. Biochem. 41 (4), 1633–1648.

Ghosh, S., Tiwari, S.S., Kumar, B., et al., 2015. Identification of potential plant extracts for anti-tick activity against acaricide resistant cattle ticks, *Rhipicephalus* (*Boophilus*) microplus (Acari:Ixodidae). Exp. Appl. Acarol. 66 (1), 159–171.

Gong, L.L., Li, G.R., Zhang, W., et al., 2016. Akebia saponin D decreases hepatic steatosis through autophagy modulation. J. Pharmacol. Exp. Ther. 359 (3), 392–400.

Gong, S., Xu, D., Zou, F., et al., 2017. (-)-Curine induces cell cycle arrest and cell death in hepatocellular carcinoma cells in a p53-independent way. Biomed. Pharmacother. 89, 894–901.

Guerram, M., Jiang, Z.Z., Sun, L., et al., 2015. Antineoplastic effects of deoxypodophyllotoxin, a potent cytotoxic agent of plant origin, on glioblastoma U-87 MG and SF126 cells. Pharmacol. Rep. 67 (2), 245–252.

Guo, J.P., Pang, J., Wang, X.W., et al., 2006. In vitro screening of traditionally used medicinal plants in China against enteroviruses. World J. Gastroenterol. 12 (25), 4078–4081.

Hagel, J.M., Mandal, R., Han, B., et al., 2015b. Metabolome analysis of 20 taxonomically related benzylisoquinoline alkaloid-producing plants. BMC Plant Biol. 15, 220.

Hagel, J.M., Morris, J.S., Lee, E.J., et al., 2015a. Transcriptome analysis of 20 taxonomically related benzylisoquinoline alkaloid-producing plants. BMC Plant Biol. 15, 227.

Han, G.Q., Chang, M.N., Hwang, S.B., 1986. The investigation of lignans from *Sargentodoxa cuneata* (Oliv) RehdetWils. Yao Xue Xue Bao 21 (1), 68–70.

Hao, D.C., Gu, X., Xiao, P.G., 2015. Medicinal Plants: Chemistry, Biology and Omics, first ed. Elsevier-Woodhead, Oxford, 9780081000854.

He, J., Liu, Y., Kang, Y., et al., 2016. Identification of alkaloids in *Stephania hainanensis* by liquid chromatography coupled with quadrupole time-of-flight mass spectrometry. Phytochem. Anal. 27 (3–4), 206–216.

He, J.M., Mu, Q., 2015. The medicinal uses of the genus *Mahonia* in traditional Chinese medicine: an ethnopharmacological, phytochemical and pharmacological review. J. Ethnopharmacol. 175, 668–683.

Hilu, K.W., Black, C., Diouf, D., et al., 2008. Phylogenetic signal in matK vs. trnK: a case study in early diverging eudicots (angiosperms). Mol. Phylogenet. Evol. 48 (3), 1120–1130.

Hilu, K.W., Borsch, T., Muller, K., et al., 2003. Angiosperm phylogeny based on matK sequence information. Am. J. Bot. 90, 1758–1776.

Hoot, S.B., Magallón, S.A., Crane, P.R., 1999. Phylogeny of basal eudicots based on three molecular data sets: atpB, rbcL, and 18S nuclear ribosomal DNA sequences. Ann. Missouri Bot. Gard. 86, 1–32.

Hossen, F., Ahasan, R., Haque, M.R., et al., 2016. Crispene A, B, C and D, four new clerodane type furanoid diterpenes from *Tinospora crispa* (L). Pharmacogn. Mag. 12 (S1), 37–41.

Hu, S., Zhou, Q., Wu, W.R., et al., 2016a. Anticancer effect of deoxypodophyllotoxin induces apoptosis of human prostate cancer cells. Oncol. Lett. 12 (4), 2918–2923.

Hu, W., Wu, L., Qiang, Q., et al., 2016b. The dichloromethane fraction from *Mahonia bealei* (Fort) Carr. leaves exerts an anti-inflammatory effect both in vitro and in vivo. J. Ethnopharmacol. 188, 134–143.

Hu, X., Wang, S., Xu, J., et al., 2014. Triterpenoid saponins from *Stauntonia chinensis* ameliorate insulin resistance via the AMP-activated protein kinase and IR/IRS-1/PI3K/Akt pathways in insulin-resistant HepG2 cells. Int. J. Mol. Sci. 15 (6), 10446–10458.

Huang, L., Li, T., Zhou, H., et al., 2015. Sinomenine potentiates degranulation of RBL-2H3 basophils via up-regulation of phospholipase A2 phosphorylation by Annexin A1 cleavage and ERK phosphorylation without influencing on calcium mobilization. Int. Immunopharmacol. 28 (2), 945–951.

Huang, Y.L., Cui, S.Y., Cui, X.Y., et al., 2016. Tetrandrine, an alkaloid from *S. tetrandra* exhibits anti-hypertensive and sleep-enhancing effects in SHR via different mechanisms. Phytomedicine 23 (14), 1821–1829.

Hullatti, K.K., Gopikrishna, U.V., Kuppast, I.J., 2011. Phytochemical investigation and diuretic activity of *Cyclea peltata* leaf extracts. J. Adv. Pharm. Technol. Res. 2 (4), 241–244.

Hullatti, K.K., Sharada, M.S., 2010. Comparative phytochemical investigation of the sources of ayurvedic drug patha: a chromatographic fingerprinting analysis. Indian J. Pharm. Sci. 72 (1), 39–45.

Hwang, S.H., Kwon, S.H., Kim, S.B., et al., 2017. Inhibitory activities of *Stauntonia hexaphylla* leaf constituents on rat lens aldose reductase and formation of advanced glycation end products and antioxidant. Biomed. Res. Int., 4273257.

Iwasa, K., Doi, Y., Takahashi, T., et al., 2009. Enantiomeric separation of racemic 1-benzyl-N-methyltetrahydroisoquinolines on chiral columns and chiral purity determinations of the O-methylated metabolites in plant cell cultures by HPLC-CD on-line coupling in combination with HPLC-MS. Phytochemistry 70 (2), 198–206.

Jacob, J., Babu, B.M., Mohan, M.C., et al., 2017. Inhibition of proinflammatory pathways by bioactive fraction of *Tinospora cordifolia*. Inflammopharmacology. doi: 10.1007/s10787-017-0319-2.

Jayasinghe, U.L., Balasooriya, B.A., Hara, N., et al., 2005. Steroidal and triterpenoidal saponins from the fruits of *Diploclisia glaucescens*. Nat. Prod. Res. 19 (3), 245–251.

Jayasinghe, U.L., Hara, N., Fujimoto, Y., 2007. (2-Nitro ethyl)phenyl and cyanophenyl glycosides from the fruits of *Diploclisia glaucescens*. Nat. Prod. Res. 21 (3), 260–264.

Jayasinghe, U.L., Jayasooriya, C.P., Fujimoto, Y., 2002a. Oleanane glycosides from the leaves of *Diploclisia glaucescens*. Fitoterapia 73 (5), 406–410.

Jayasinghe, L., Jayasooriya, C.P., Oyama, K., et al., 2002b. 3-Deoxy-1beta, 20-dihydroxyecdysone from the leaves of *Diploclisia glaucescens*. Steroids 67 (7), 555–558.

Jayasinghe, U.L., Kumarihamy, B.M., Bandara, A.G., et al., 2003b. Antifeedant activity of some Sri Lankan plants. Nat. Prod. Res. 17 (1), 5–8.

Jayasinghe, L., Mallika Kumarihamy, B.M., et al., 2003a. A new ecdysteroid, 2-deoxy-5beta, 20-dihydroxyecdysone from the fruits of *Diploclisia glaucescens*. Steroids 68 (5), 447–450.

Jia, X.J., Li, X., Wang, F., et al., 2017. Berbamine exerts anti-inflammatory effects via inhibition of NF-κB and MAPK signaling pathways. Cell. Physiol. Biochem. 41 (6), 2307–2318.

Jiang, H., Zhang, G.J., Liu, Y.F., et al., 2017. Clerodane diterpenoid glucosides from the stems of *Tinospora sinensis*. J. Nat. Prod. 80 (4), 975–982.

Jiang, Y., Yu, L., Wang, M.H., 2015. N-trans-feruloyltyramine inhibits LPS-induced NO and PGE2 production in RAW 264.7 macrophages: involvement of AP-1 and MAP kinase signalling pathways. Chem. Biol. Interact. 235, 56–62.

Jiao, Y., Leebens-Mack, J., Ayyampalayam, S., et al., 2012. A genome triplication associated with early diversification of the core eudicots. Genome. Biol. 13 (1), R3.

Jung, S., Lee, S.H., Song, Y.S., et al., 2016. Effect of beverage containing fermented *Akebia quinata* extracts on alcoholic hangover. Prev. Nutr. Food Sci. 21 (1), 9–13.

Kamalakannan, S., Ananth, S., Murugan, K., et al., 2016. Bio fabrication of silver nanoparticle from *Argemone mexicana* for the control of *Aedes albopictus* and their antimicrobial activity. Curr. Pharm. Biotechnol. 17 (14), 1285–1294.

Kang, D.W., Moon, J.Y., Choi, J.G., et al., 2016. Antinociceptive profile of levo-tetrahydropalmatine in acute and chronic pain mice models: role of spinal sigma-1 receptor. Sci. Rep. 6, 37850.

Kawakami, Y., Nagai, Y., Nezu, Y., et al., 1987. Indonesian medicinal plants. I. New furanoditerpenes from *Arcangelisia flava* MERR. (2). Stereostructure of furanoditerpenes determined by nuclear magneticresonance analysis. Chem. Pharm. Bull. 35 (12), 4839–4845.

Khan, A.U., Gilani, A.H., 2015. Natural products useful in respiratory disorders: focus on side-effect neutralizing combinations. Phytother. Res., 10. 1002/ptr.5380.

Kim, D.R., Lee, J.E., Shim, K.J., et al., 2017a. Effects of herbal Epimedium on the improvement of bone metabolic disorder through the induction of osteogenic differentiation from bone marrow-derived mesenchymal stem cells. Mol. Med. Rep. 15 (1), 125–130.

Kim, G.J., Song, D.H., Yoo, H.S., et al., 2017b. Hederagenin supplementation alleviates the pro-inflammatory and apoptotic response to alcohol in rats. Nutrients 9 (1), E41. doi: 10.3390/nu9010041.

Kirana, H., Srinivasan, B.P., 2010. Effect of *Cyclea peltata* Lam. roots aqueous extract on glucose levels, lipid profile, insulin, TNF-alpha and skeletal muscle glycogen in type 2 diabetic rats. Indian J. Exp. Biol. 48 (5), 499–502.

Kodai, T., Horiuchi, Y., Nishioka, Y., et al., 2010. Novel cycloartane-type triterpenoid from the fruits of *Nandina domestica*. J. Nat. Med. 64 (2), 216–218.

Kölsch, A., Gleissberg, S., 2006. Diversification of CYCLOIDEA-like TCP genes in the basal eudicot families *Fumariaceae* and *Papaveraceae* s.str. Plant Biol. 8 (5), 680–687.

Kong, C.S., Kim, Y.A., Kim, H., et al., 2016. Evaluation of a furochromone from the halophyte *Corydalis heterocarpa* for cytotoxic activity against human gastric cancer (AGS) cells. Food Funct. 7 (12), 4823–4829.

Konoshima, T., Takasaki, M., Kozuka, M., et al., 1987. Studies on inhibitors of skin-tumor promotion, I. Inhibitory effects of triterpenes from *Euptelea polyandra* on Epstein-Barr virus activation. J. Nat. Prod. 50 (6), 1167–1170.

Kukula-Koch, W., Mroczek, T., 2015. Application of hydrostatic CCC-TLC-HPLC-ESI-TOF-MS for the bioguided fractionation of anticholinesterase alkaloids from *Argemone mexicana* L. roots. Anal. Bioanal. Chem. 407 (9), 2581–2589.

Kulkarni, R.R., Lee, W., Jang, T.S., et al., 2015. Caffeoyl glucosides from *Nandina domestica* inhibit LPS-induced endothelial inflammatory responses. Bioorg. Med. Chem. Lett. 25 (22), 5367–5371.

Laamech, J., El-Hilaly, J., Fetoui, H., et al., 2017. *Berberis vulgaris* L. effects on oxidative stress and liver injury in lead-intoxicated mice. J. Complement. Integr. Med. 14 (1).

Latheef, S.K., Dhama, K., Samad, H.A., et al., 2017. Immunomodulatory and prophylactic efficacy of herbal extracts against experimentally induced chicken infectious anaemia in chicks: assessing the viral load and cell mediated immunity. Virusdisease 28 (1), 115–120.

Lee, D.S., Keo, S., Cheng, S.K., et al., 2017a. Protective effects of Cambodian medicinal plants on tert-butylhydroperoxide-induced hepatotoxicity via Nrf2-mediated heme oxygenase-1. Mol. Med. Rep. 15 (1), 451–459.

Lee, J., Sohn, E.J., Yoon, S.W., et al., 2017b. Anti-metastatic effect of dehydrocorydaline on H1299 non-small cell lung carcinoma cells via inhibition of matrix metalloproteinases and B cell lymphoma 2. Phytother. Res. 31 (3), 441–448.

Lee, J.Y., Kim, K.J., Kim, J., et al., 2016. Anti-osteoclastogenic effects of isoquinoline alkaloids from the rhizome extract of *Sinomenium acutum*. Arch. Pharm. Res. 39 (5), 713–720.

Lee, Y., Jung, J.C., Ali, Z., et al., 2012. Anti-inflammatory effect of triterpene saponins isolated from *Blue cohosh* (*Caulophyllum thalictroides*). Evid. Based Complement. Alternat. Med., 798192, 2012.

Leyva-Peralta, M.A., Robles-Zepeda, R.E., Garibay-Escobar, A., et al., 2015. In vitro anti-proliferative activity of *Argemone gracilenta* and identification of some active components. BMC Complement. Altern. Med. 15, 13.

Li, C.H., Chen, C., Zhang, Q., et al., 2017b. Differential proteomic analysis of platelets suggested target-related proteins in rabbit platelets treated with *Rhizoma Corydalis*. Pharm. Biol. 55 (1), 76–87.

Li, D.H., Lv, Y.S., Liu, J.H., 2016a. Simultaneous determination of four active ingredients in *Sargentodoxa cuneata* by HPLC coupled with evaporative light scattering detection. Int. J. Anal. Chem., 8509858, 2016.

Li, H., Yuan, Y., Zhang, Y., 2017a. Icariin inhibits AMPK-dependent autophagy and adipogenesis in adipocytes in vitro and in a model of Graves' orbitopathy in vivo. Front. Physiol. 8, 45.

Li, H., Zhao, F.C., Yuan, X.D., 2015a. Determination of phenols and triterpenoid saponins in stems of *Sargentodoxa cuneata*. Zhongguo Zhong Yao Za Zhi 40 (10), 1865–1871.

Li, J., Feng, J., Luo, C., 2015b. Absolute configuration of podophyllotoxone and its inhibitory activity against human prostate cancer cells. Chin. J. Nat. Med. 13 (1), 59–64.

Li, J., Sun, H., Jin, L., et al., 2013. Alleviation of podophyllotoxin toxicity using coexisting flavonoids from *Dysosma versipellis*. PLoS One 8 (8), e72099.

Li, W., Wang, X., Zhang, H., et al., 2016d. Anti-ulcerogenic effect of cavidine against ethanol-induced acute gastric ulcer in mice and possible underlying mechanism. Int. Immunopharmacol. 38, 450–459.

Li, W., Zhang, L., Xu, L., et al., 2016c. Functional reversal of (-)-Stepholidine analogues by replacement of benzazepine substructure using the ring-expansion strategy. Chem. Biol. Drug Des. 88 (4), 599–607.

Li, X.A., Ho, Y.S., Chen, L., et al., 2016b. The protective effects of icariin against the homocysteine-induced neurotoxicity in the primary embryonic cultures of rat cortical neurons. Molecules 21 (11), E1557.

Liang, F., Han, Y., Gao, H., et al., 2015. Kaempferol identified by zebrafish assay and fine fractionations strategy from *Dysosma versipellis* inhibits angiogenesis through VEGF and FGF pathways. Sci. Rep. 5, 14468.

Liang, Y.H., Tang, C.L., Lu, S.Y., et al., 2016. Serum metabonomics study of the hepatoprotective effect of *Corydalis saxicola* Bunting on carbon tetrachloride-induced acute hepatotoxicity in rats by (1)H NMR analysis. J. Pharm. Biomed. Anal. 129, 70–79.

Liao, J., Lei, Y., Wang, J.Z., 2014. Chemical constituents of *Cocculus orbiculatus* var. mollis root. Zhong Yao Cai 37 (2), 254–257.

Lin, M., Xia, B., Yang, M., et al., 2013. Anti-ovarian cancer potential of two acidic polysaccharides from the rhizoma of *Menispermum dauricum*. Carbohydr. Polym. 92 (2), 2212–2217.

Ling, Y., Zhang, Q., Zhu, D.D., et al., 2016. Identification and characterization of the major chemical constituents in FructusAkebiae by high-performance liquid chromatography coupled with electrospray ionization-quadrupole-time-of-flight mass spectrometry. J. Chromatogr. Sci. 54 (2), 148–157.

Liscombe, D.K., Macleod, B.P., Loukanina, N., et al., 2005. Evidence for the monophyletic evolution of benzylisoquinoline alkaloid biosynthesis in angiosperms. Phytochemistry 66 (11), 1374–1393.

Liu, D., Li, S., Qi, J.Q., et al., 2016a. The inhibitory effects of nor-oleanane triterpenoid saponins from *Stauntonia brachyanthera* towards UDP-glucuronosyltransferases. Fitoterapia 112, 56–64.

Liu, Q., Mao, X., Zeng, F., et al., 2012. Effect of daurisoline on HERG channel electrophysiological function and protein expression. J. Nat. Prod. 75 (9), 1539–1545.

Liu, W., Liu, J., Yin, D., et al., 2015a. Influence of ecological factors on the production of active substances in the anticancer plant *Sinopodophyllum hexandrum* (Royle) T.S. Ying. PLoS One 10 (4), e122981.

Liu, X.L., Li, S., Meng, D.L., 2016b. Anti-gout nor-oleanane triterpenoids from the leaves of *Stauntonia brachyanthera*. Bioorg. Med. Chem. Lett. 26 (12), 2874–2879.

Liu, X.L., Wang, D.D., Wang, Z.H., et al., 2015b. Diuretic properties and chemical constituent studies on *Stauntonia brachyanthera*. Evid. Based Complement. Alternat. Med., 432419, 2015.

Liu, Y., Song, M., Zhu, G., et al., 2017. Corynoline attenuates LPS-induced acute lung injury in mice by activating Nrf2. Int. Immunopharmacol. 48, 96–101.

Lu, W.L., Ren, H.Y., Liang, C., et al., 2014c. *Akebia trifoliata* (Thunb.) Koidzseed extract inhibits the proliferation of human hepatocellular carcinoma cell lines via inducing endoplasmic reticulum stress. Evid. Based Complement. Alternat. Med., 192749, 2014.

Lu, X., Qiu, F., Pan, X., et al., 2014b. Simultaneous quantitative analysis of nine triterpenoid saponins for the quality control of *Stauntonia obovatifoliola* Hayata subsp. intermedia stems. J. Sep. Sci. 37 (24), 3632–3640.

Lu, X.R., Liu, S., Wang, M.Y., 2014a. Triterpenoids from *Stauntonia obovatifoliola* Hayata subsp. intermedia stems. Zhongguo Zhong Yao Za Zhi 39 (23), 4629–4636.

Lv, H.N., Zeng, K.W., Zhao, M.B., et al., 2017. Pyrrolo[2,1-a]isoquinoline and pyrrole alkaloids from *Sinomenium acutum*. J. Asian Nat. Prod. Res.doi: 10.1080/10286020.2017.1326910.

Ma, C., Du, F., Yan, L., et al., 2015. Potent activities of roemerine against Candida albicans and the underlying mechanisms. Molecules 20 (10), 17913–17928.

Ma, W.K., Li, H., Dong, C.L., et al., 2016. Palmatine from *Mahonia bealei* attenuates gut tumorigenesis in ApcMin/+ mice via inhibition of inflammatory cytokines. Mol. Med. Rep. 14 (1), 491–498.

Malik, T.A., Kamili, A.N., Chishti, M.Z., 2017. Breaking the resistance of *Escherichia coli*: antimicrobial activity of *Berberis lycium* Royle. Microb. Pathog. 102, 12–20.

Manosroi, A., Akazawa, H., Akihisa, T., 2015. In vitro anti-proliferative activity on colon cancer cell line (HT-29) of Thai medicinal plants selected from Thai/Lanna medicinal plant recipe database "MANOSROI III". J. Ethnopharmacol. 161, 11–17.

Mao, Z., Wang, X., Liu, Y., et al., 2017. Simultaneous determination of seven alkaloids from Rhizoma Corydalis Decumbentis in rabbit aqueous humor by LC-MS/MS: application to ocular pharmacokinetic studies. J. Chromatogr. B Analyt. Technol. Biomed. Life Sci. 1057, 46–53.

Matsuzaki, K., Murano, K., Endo, Y., et al., 2014. Nortriterpene saponins from *Akebia trifoliata*. Nat. Prod. Commun. 9 (12), 1695–1698.

Mehrara, M., Halakoo, M., Hakemi-Vala, M., et al., 2015. Antibacterial and antifungal activities of the endemic species *Glaucium vitellinum* Boiss and Buhse. Avicenna J. Phytomed. 5 (1), 56–61.

Meyer, A., Imming, P., 2011. Benzylisoquinoline alkaloids from the papaveraceae: the heritage of Johannes Gadamer (1867-1928). J. Nat. Prod. 74 (11), 2482–2487.

Mi, G., Liu, S., Zhang, J., 2017. Levo-tetrahydroberberrubine produces anxiolytic-like effects in mice through the 5-HT1A receptor. PLoS One 12 (1), e0168964.

Miyamae, Y., Nishito, Y., Nakai, N., et al., 2016. Tetrandrine induces lipid accumulation through blockade of autophagy in a hepatic stellate cell line. Biochem. Biophys. Res. Commun. 477 (1), 40–46.

Mo, J., Guo, Y., Yang, Y.S., 2007. Recent developments in studies of l-stepholidine and its analogs: chemistry, pharmacology and clinical implications. Curr. Med. Chem. 14 (28), 2996–3002.

Mo, Z.T., Li, W.N., Zhai, Y.R., et al., 2017. The effects of icariin on the expression of HIF-1α, HSP-60 and HSP-70 in PC12 cells suffered from oxygen-glucose deprivation-induced injury. Pharm. Biol. 55 (1), 848–852.

Morteza-Semnani, K., Mahmoudi, M., Heidar, M.R., 2006. Analgesic activity of the methanolic extract and total alkaloids of *Glaucium paucilobum*. Methods Find. Exp. Clin. Pharmacol. 28 (3), 151–155.

Morteza-Semnani, K., Saeedi, M., Hamidian, M., 2004. Anti-inflammatory and analgesic activity of the topical preparation of *Glaucium grandiflorum*. Fitoterapia 75 (2), 123–129.

Murakami, T., Oominami, H., Matsuda, H., et al., 2001. Bioactive saponins and glycosides. XVIII. Nortriterpene and triterpene oligoglycosides from the fresh leaves of Euptelea polyandra Sieb. et Zucc. (2): structures of eupteleasaponins VI, VI acetate, VII, VIII, IX, X, XI, and XII. Chem. Pharm. Bull. 49 (6), 741–746.

Nguyen-Pouplin, J., Tran, H., Tran, H., et al., 2007. Antimalarial and cytotoxic activities of ethnopharmacologically selected medicinal plants from South Vietnam. J. Ethnopharmacol. 109 (3), 417–427.

Nikolaev, S.M., Fedorov, A.V., Toropova, A.A., et al., 2014. Hepatoprotective effect of *Hypecoum erectum* extract on experimental D-galactosamine-induced damage of rat liver. Eksp. Klin. Farmakol. 77 (9), 18–22.

Orhan, I., Sener, B., Choudhary, M.I., et al., 2004. Acetylcholinesterase and butyrylcholinesterase inhibitory activity of some Turkish medicinal plants. J. Ethnopharmacol. 91 (1), 57–60.

Peng, C.Y., Liu, J.Q., Zhang, R., et al., 2014. A new alkaloid from the fruit of *Nandina domestica* Thunb. Nat. Prod. Res. 28 (15), 1159–1164.

Peng, Y., Chen, S.B., Liu, Y., et al., 2006. A pharmacophylogenetic study of the *Berberidaceae* (s.l.). Acta Phytotaxo. Sin. 44 (3), 241–257.

Philipov, S., Istatkova, R., Denkova, P., et al., 2009. Alkaloids from Mongolian species *Hypecoum lactiflorum* Kar.et Kir. Pazij. Nat. Prod. Res. 23 (11), 982–987.

Priya, L.B., Baskaran, R., Elangovan, P., et al., 2017. *Tinospora cordifolia* extract attenuates cadmium-induced biochemical and histological alterations in the heart of male Wistar rats. Biomed. Pharmacother. 87, 280–287.

Qiao, W., Wang, L., Ye, B., et al., 2015. Electrochemical behavior of palmatine and its sensitive determination based on an electrochemically reduced L-methionine functionalized graphene oxide modified electrode. Analyst 140 (23), 7974–7983.

Qiu, Y.L., Li, L., Hendry, T.A., et al., 2006. Reconstructing the basal angiosperm phylogeny: evaluating information content of mitochondrial genes. Taxon 55, 837–856.

Raja, R.D., Jeeva, S., Prakash, J.W., et al., 2011. Antibacterial activity of selected ethnomedicinal plants from South India. Asian Pac. J. Trop. Med. 4 (5), 375–378.

Rao, G.X., Zhang, S., Wang, H.M., et al., 2009. Antifungal alkaloids from the fresh rattan stem of *Fibraurea recisa* Pierre. J. Ethnopharmacol. 123 (1), 1–5.

Rao, S., Liu, S., Zou, L., et al., 2017. The effect of sinomenine in diabetic neuropathic pain mediated by the P2X3 receptor in dorsal root ganglia. Purinergic. Signal. 13 (2), 227–235.

Ren, K., Zhang, W., Wu, G., et al., 2016. Synergistic anti-cancer effects of galangin and berberine through apoptosis induction and proliferation inhibition in oesophageal carcinoma cells. Biomed. Pharmacother. 84, 1748–1759.

Ren, Y., Li, H.F., Zhao, L., Endress, P.K., 2007. Floral morphogenesis in *Euptelea* (Eupteleaceae Ranunculales). Ann. Bot. 100, 185–193.

Rücker, G., Mayer, R., Shin-Kim, J.S., 1991. Triterpenesaponins from the Chinese drug "Daxueteng" (Caulis sargentodoxae). Planta Med. 57 (5), 468–470.

Sakumoto, H., Yokota, Y., Ishibashi, G., et al., 2015. Sinomenine and magnoflorine, major constituents of Sinomeni caulis et rhizoma, show potent protective effects against membrane damage induced by lysophosphatidylcholine in rat erythrocytes. J. Nat. Med. 69 (3), 441–448.

Samita, F., Ochieng, C.O., Owuor, P.O., et al., 2017. Isolation of a new β-carboline alkaloid from aerial parts of *Triclisia sacleuxii* and its antibacterial and cytotoxicity effects. Nat. Prod. Res. 31 (5), 529–536.

Sauquet, H., Carrive, L., Poullain, N., et al., 2015. Zygomorphy evolved from disymmetry in *Fumarioideae* (Papaveraceae, Ranunculales): new evidence from an expanded molecular phylogenetic framework. Ann. Bot. 115 (6), 895–914.

Sedaghat, R., Taab, Y., Kiasalari, Z., et al., 2017. Berberine ameliorates intrahippocampal kainate-induced status epilepticus and consequent epileptogenic process in the rat: underlying mechanisms. Biomed. Pharmacother. 87, 200–208.

Shakya, A., Soni, U.K., Rai, G., et al., 2016. Gastro-protective and anti-stress efficacies of monomethyl fumarate and a *Fumaria indica* extract in chronically stressed rats. Cell. Mol. Neurobiol. 36 (4), 621–635.

Shafiee, A., Morteza-Semnani, K., 1998. Crabbine and other alkaloids from the aerial parts of Glaucium paucilobum. Planta. Med. 64 (7), 680.

Shang, W.Q., Chen, Y.M., Gao, X.L., et al., 2014. Phytochemical and pharmacological advance on Tibetan medicinal plants of Corydalis. Zhongguo Zhong Yao Za Zhi 39 (7), 1190–1198.

Shin, S., Son, D., Kim, M., et al., 2015. Ameliorating effect of *Akebiaquinata* fruit extracts on skin aging induced by advanced glycation end products. Nutrients 7 (11), 9337–9352.

Shine, V.J., Latha, P.G., Shyamal, S., et al., 2009. Gastric antisecretory and antiulcer activities of *Cyclea peltata* (Lam) Hook. f. & Thoms. in rats. J. Ethnopharmacol. 125 (2), 350–355.

Shine, V.J., Latha, P.G., Suja, S.N., et al., 2014. Ameliorative effect of alkaloid extract of *Cyclea peltata* (Poir) Hook. f. & Thoms. roots (ACP) on APAP/CCl4 induced liver toxicity in Wistar rats and invitro free radical scavenging property. Asian Pac. J. Trop. Biomed. 4 (2), 143–151.

Si, K., Liu, J., He, L., et al., 2010. Caulophine protects cardiomyocytes from oxidative and ischemic injury. J. Pharmacol. Sci. 113 (4), 368–377.

Sim, H.J., Kim, J.H., Lee, K.R., et al., 2013. Simultaneous determination of structurally diverse compounds in different *Fangchi* species by UHPLC-DAD and UHPLC-ESI-MS/MS. Molecules 18 (5), 5235–5250.

Singh, H., Kaur, T., Manchanda, S., et al., 2017. Intermittent fasting combined with supplementation with Ayurvedic herbs reduces anxiety in middle aged female rats by anti-inflammatory pathways. Biogerontology. doi: 10.1007/s10522-017-9706-8.

Singh, S., Verma, M., Malhotra, M., et al., 2016. Cytotoxicity of alkaloids isolated from *Argemone mexicana* on SW480 human colon cancer cell line. Pharm. Biol. 54 (4), 740–745.

Soltis, D.E., Smith, S.A., Cellinese, N., et al., 2011. Angiosperm phylogeny: 17 genes, 640 taxa. Am. J. Bot. 98, 704–730.

Soltis, D.E., Soltis, P.S., Chase, M.W., et al., 2000. Angiosperm phylogeny inferred from 18S rDNA, rbcL, and atpB sequences. Bot. J. Linn. Soc. 133, 381–461.

Soltis, P.S., Soltis, D.E., 2004. The origin and diversification of angiosperms. Am. J. Bot. 91 (10), 1614–1626.

Son, S.Y., Rhee, H.S., Lee, M.W., et al., 2014. Analysis of benzo[c]phenanthridine alkaloids in *Eschscholtzia californica* cell culture using HPLC-DAD and HPLC-ESI-MS/MS. Biosci. Biotechnol. Biochem. 78 (7), 1103–1111.

Sriwilaijareon, N., Petmitr, S., Mutirangura, A., et al., 2002. Stage specificity of *Plasmodium falciparum* telomerase and its inhibition by berberine. Parasitol. Int. 51 (1), 99–103.

Su, C.R., Chen, Y.F., Liou, M.J., et al., 2008. Anti-inflammatory activities of furanoditerpenoids and other constituents from *Fibraurea tinctoria*. Bioorg. Med. Chem. 16 (21), 9603–9609.

Su, C.R., Ueng, Y.F., Dung, N.X., et al., 2007. Cytochrome P3A4 inhibitors and other constituents of *Fibraurea tinctoria*. J. Nat. Prod. 70 (12), 1930–1933.

Su, Q., He, J., Wang, Z., et al., 2016. Intestinal anti-inflammatory effect of the rhizome extracts of *Menispermum dauricum* DC. on trinitrobenzene sulfonic acid induced ulcerative colitis in mice. J. Ethnopharmacol. 193, 12–20.

Su, Y.S., Fan, Z.X., Xiao, S.E., et al., 2017. Icariin promotes mouse hair follicle growth by increasing insulin-like growth factor 1 expression in dermal papillary cells. Clin. Exp. Dermatol. 42 (3), 287–294.

Subash-Babu, P., Alshammari, G.M., Ignacimuthu, S., et al., 2017. Epoxy clerodane diterpene inhibits MCF-7 human breast cancer cell growth by regulating the expression of the functional apoptotic genes Cdkn2A, Rb1, mdm2 and p53. Biomed. Pharmacother. 87, 388–396.

Subeki, Matsuura, H., Takahashi, K., et al., 2005. Antibabesial activity of protoberberine alkaloids and 20-hydroxyecdysone from *Arcangelisia flava* against Babesia gibsoni in culture. J. Vet. Med. Sci. 67 (2), 223–227.

Sun, D., Han, Y., Wang, W., et al., 2016d. Screening and identification of Caulis Sinomenii bioactive ingredients with dual-target NF-κB inhibition and β2- AR agonizing activities. Biomed. Chromatogr. 30 (11), 1843–1853.

Sun, D., Zhou, M., Ying, X., et al., 2014. Identification of nuclear factor-κB inhibitors in the folk herb Rhizoma Menispermi via bioactivity-based ultra-performance liquid chromatography/quadrupole time-of-flight mass spectrometry analysis. BMC Complement. Altern. Med. 14, 356.

Sun, M., Wang, J., Zhou, Y., et al., 2016c. Isotetrandrine reduces astrocyte cytotoxicity in neuromyelitis optica by blocking the binding of NMO-IgG to aquaporin 4. Neuroimmunomodulation 23 (2), 98–108.

Sun, Y., Moore, M.J., Zhang, S., et al., 2016a. Phylogenomic and structural analyses of 18 complete plastomes across nearly all families of early-diverging eudicots, including an angiosperm-wide analysis of IR gene content evolution. Mol. Phylogenet. Evol. 96, 93–101.

Sun, Y.J., Gao, M.L., Zhang, Y.L., et al., 2016b. Labdane diterpenes from the fruits of *Sinopodophyllum* emodi. Molecules 21 (4), 434.

Sun, Y.J., Pei, L.X., Wang, K.B., et al., 2015. Preparative isolation of two prenylated biflavonoids from the roots and rhizomes of *Sinopodophyllum* emodi by Sephadex LH-20column and high-speed counter-current chromatography. Molecules 21 (1), E10.

Sung, Y.Y., Kim, D.S., Kim, H.K., 2015. Akebiaquinata extract exerts anti-obesity and hypolipidemic effects in high-fat diet-fed mice and 3T3-L1 adipocytes. J. Ethnopharmacol. 168, 17–24.

Telang, N.T., Li, G., Katdare, M., et al., 2017. The nutritional herb *Epimedium grandiflorum* inhibits the growth in a model for the Luminal A molecular subtype of breast cancer. Oncol. Lett. 13 (4), 2477–2482.

Thakur, P., Chawla, R., Narula, A., et al., 2017. Protective effect of *Berberis aristata* against peritonitis induced by carbapenem-resistant Escherichia coli in a mammalian model. J. Glob. Antimicrob. Resist. 9, 21–29.

Thavamani, B.S., Mathew, M., Dhanabal, S.P., 2013. In vitro cytotoxic activity of *menispermaceae* plants against HeLa cell line. Anc. Sci. Life 33 (2), 81–84.

Thavamani, B.S., Mathew, M., Dhanabal, S.P., 2016. *Cocculus hirsutus*: molecular docking to identify suitable targets for hepatocellular carcinoma by in silico technique. Pharmacogn. Mag. 12 (S3), S350–352.

Thavamani, B.S., Mathew, M., Palaniswamy, D.S., 2014. Anticancer activity of *Cocculus hirsutus* against Dalton's lymphoma ascites (DLA) cells in mice. Pharm. Biol. 52 (7), 867–872.

Ti, H., Wu, P., Xu, L., et al., 2017. Anti-inflammatory neolignans from *Epimedium pseudowushanese*. Nat. Prod. Res. 31 (22), 2621–2628.

Tian, D.D., Zhang, R.X., Wu, N., et al., 2017. Tetrandrine inhibits the proliferation of human osteosarcoma cells byupregulating the PTEN pathway. Oncol. Rep. 37 (5), 2795–2802.

Tian, Y., Zhang, H.J., Tu, A.P., et al., 2005. Phenolics from traditional Chinese medicine *Sargentodoxa cuneata*. Yao Xue Xue Bao 40 (7), 628–631.

Toropova, A.A., Nikolaev, S.M., Razuvaeva, Y.G., et al., 2014. Effect of *Hypecoum erectum* extract on morphofunctional state of the liver in rats with tetracycline-associated hepatitis. Antibiot. Khimioter. 59 (9–10), 25–28.

Tsukiyama, M., Akaishi, T., Ueki, T., et al., 2007. The extract from *Nandina domestica* THUNBERG inhibits histamine- and serotonin-induced contraction in isolated guinea pig trachea. Biol. Pharm. Bull. 30 (11), 2063–2068.

Ueki, T., Akaishi, T., Okumura, H., et al., 2011. Biphasic tracheal relaxation induced by higenamine and nantenine from *Nandina domestica* Thunberg. J. Pharmacol. Sci. 115 (2), 254–257.

Ueki, T., Akaishi, T., Okumura, H., et al., 2012. Extract from *Nandina domestica* inhibits lipopolysaccharide-inducedcyclooxygenase-2 expression in human pulmonary epithelial A549 cells. Biol. Pharm. Bull. 35 (7), 1041–1047.

Vijayan, D., Cheethaparambil, A., Pillai, G.S., et al., 2014. Molecular authentication of *Cissampelos pareira* L. var. *hirsuta* (Buch.-Ham. Ex DC.) Forman, the genuine source plant of ayurvedic raw drug 'Patha', and its other source plants by ISSR markers. 3 Biotech. 4 (5), 559–562.

Vijayan, F.P., Rani, V.K., Vineesh, V.R., et al., 2007. Protective effect of *Cyclea peltata* Lam on cisplatin-induced nephrotoxicity and oxidative damage. J. Basic Clin. Physiol. Pharmacol. 18 (2), 101–114.

Vrancheva, R.Z., Ivanov, I.G., Aneva, I.Y., et al., 2016. Alkaloid profiles and acetylcholinesterase inhibitory activities of *Fumaria* species from Bulgaria. Z. Naturforsch. C. 71 (1–2), 9–14.

Wang, L., Ci, X., Lv, H., et al., 2016a. Isotetrandrine ameliorates tert-butyl hydroperoxide-induced oxidative stress through upregulation of heme oxygenase-1 expression. Exp. Biol. Med. 241 (14), 1568–1576.

Wang, J., Li, Y., Zu, X.B., et al., 2012. Dauricine can inhibit the activity of proliferation of urinary tract tumor cells. Asian Pac. J. Trop. Med. 5 (12), 973–976.

Wang, N., Liu, R., Liu, Y., et al., 2016c. Sinomenine potentiates P815 cell degranulation via upregulation of Ca2+ mobilization through the Lyn/PLCγ/IP3R pathway. Int. J. Immunopathol. Pharmacol. 29 (4), 676–683.

Wang, Q.H., Guo, S., Yang, X.Y., et al., 2017b. Flavonoids isolated from Sinopodophylli Fructus and their bioactivities against human breast cancer cells. Chin. J. Nat. Med. 15 (3), 225–233.

Wang, W., Cai, J., Tang, S., et al., 2016b. Sinomenine attenuates angiotensin II-induced autophagy via inhibition of P47-Phox translocation to the membrane and influences reactive oxygen species generationin podocytes. Kidney Blood Press Res. 41 (2), 158–167.

Wang, W., Lu, A.M., Ren, Y., et al., 2009. Phylogeny and classification of *Ranunculales* evidence from four molecular loci and morphological data. Persp. Plant Ecol.Evol.Syst. 11, 81–110.

Wang, J.B., Mantsch, J.R., 2012. l-tetrahydropalamatine: a potential new medication for the treatment of cocaine addiction. Future Med. Chem. 4 (2), 177–186.

Wang, W., Ortiz, R.D., Jacques, F.M., et al., 2017a. New insights into the phylogeny of Burasaieae *(Menispermaceae)* with the recognition of a new genus and emphasis on the southern Taiwanese and mainland Chinese disjunction. Mol. Phylogenet. Evol. 109, 11–20.

Wang, X.L., Liu, B.R., Chen, C.K., et al., 2011. Four new fluorenone alkaloids and one new dihydroazafluoranthene alkaloid from *Caulophyllum robustum* Maxim. Fitoterapia 82 (6), 793–797.

Wangchuk, P., Giacomin, P.R., Pearson, M.S., et al., 2016. Identification of lead chemotherapeutic agents from medicinal plants against blood flukes and whipworms. Sci. Rep. 6, 32101.

Wefferling, K.M., Hoot, S.B., Neves, S.S., 2013. Phylogeny and fruit evolution in *Menispermaceae*. Am. J. Bot. 100 (5), 883–905.

Wei, Y., Ma, C.M., Chen, D.Y., et al., 2008. Anti-HIV-1 protease triterpenoids from *Stauntonia obovatifoliola* Hayata subsp. intermedia. Phytochemistry 69 (9), 1875–1879.

Wei, Z.Z., Qin, Q.P., Chen, J.N., et al., 2016. Oxoisoaporphine as potent telomerase inhibitor. Molecules 21 (11), E1534.

Wu, F., Zheng, H., Yang, Z.T., et al., 2017b. Urinarymetabonomics study of the hepatoprotective effects of total alkaloids from *Corydalis saxicola* Bunting on carbon tetrachloride-induced chronic hepatotoxicity in rats using (1)H NMR analysis. J. Pharm. Biomed. Anal. 140, 199–209.

Wu, L., Du, Z.R., Xu, A.L., et al., 2017a. Neuroprotective effects of total flavonoid fraction of the *Epimedium koreanum* Nakai extract on dopaminergic neurons: in vivo and in vitro. Biomed. Pharmacother. 91, 656–663.

Xia, Y.G., Li, G.Y., Liang, J., et al., 2014a. A strategy for characterization of triterpene saponins in *Caulophyllum robustum* hairy roots by liquid chromatography with electrospray ionization quadrupoletime-of-flight mass spectrometry. J. Pharm. Biomed. Anal. 100, 109–122.

Xia, Y.G., Li, G.Y., Liang, J., et al., 2014b. Genus *caulophyllum*: an overview of chemistry and bioactivity. Evid. Based Complement. Alternat. Med., 684508, 2014.

Xia, Y.G., Liang, J., Li, G.Y., et al., 2015. Analysis of oligosaccharide sequences of trace *Caulophyllum robustum* saponins by direct infusion multiple-stage tandem mass spectrometry. J. Pharm. Biomed. Anal. 112, 106–115.

Xia, Y.G., Liang, J., Li, G.Y., et al., 2016. Energy-resolved technique for discovery and identification of malonyl-triterpene saponins in *Caulophyllum robustum* by UHPLC-electrospray Fourier transform mass spectrometry. J. Mass Spectrom. 51 (10), 947–958.

Xiao, P.G., Wang, L.W., Lv, S.J., et al., 1986. Statistical analysis of the ethnopharmacologic data based on Chinese medicinal plants by electronic computer I Magnoliidae. Chin. J. Integ. Trad. West. Med. 6 (4), 253–256.

Xiao, P.G., Xiao, W., Xu, L.J., et al., 2016. Coptids Rhizoma and Chinese herbal medicines which contain berberine-type alkaloids. Modern Chin. Med. 18 (11), 1381–1385.

Xool-Tamayo, J., Serrano-Gamboa, G., Monforte-González, M., et al., 2017. Development of newly sanguinarine biosynthetic capacity in in vitro rootless shoots of *Argemone mexicana* L. Mexican prickly poppy. Biotechnol. Lett. 39 (2), 323–330.

Xu, Q.L., Wang, J., Dong, L.M., et al., 2016. Two new pentacyclic triterpene saponins from the leaves of *Akebia trifoliata*. Molecules 21 (7), E962. doi: 10.3390/molecules21070962.

Yang, H., Yin, P., Shi, Z., et al., 2016b. Sinomenine, a COX-2 inhibitor, induces cell cycle arrest and inhibits growth of human colon carcinoma cells in vitro and in vivo. Oncol. Lett. 11 (1), 411–418.

Yang, J., Xiong, Q., Zhang, J., et al., 2014. The protective effect of *Stauntonia chinensis* polysaccharide on CCl4-induced acute liver injuries in mice. Int. J. Biomed. Sci. 10 (1), 16–20.

Yang, L., Li, D., Zhuo, Y., et al., 2016c. Protective role of liriodendrin in sepsis-induced acute lung injury. Inflammation 39 (5), 1805–1813.

Yang, Z., Wu, Y., Wu, S., et al., 2016a. A combination strategy for extraction and isolation of multi-component natural products by systematic two-phase solvent extraction-(13)C nuclear magneticresonance pattern recognition and following conical counter-current chromatography separation: podophyllotoxins and flavonoids from *Dysosma versipellis* (Hance) as examples. J. Chromatogr. A, 184–196, 2016.

Yi, L., Luo, J.F., Xie, B.B., et al., 2015. α7 Nicotinic acetylcholine receptor is a novel mediator of sinomenine anti-inflammation effect in macrophages stimulated by lipopolysaccharide. Shock 44 (2), 188–195.

Ying, C., Ning, W., Ying, L., et al., 2014. Anti-nociceptive and anti-inflammatory activities of the extracts of *Stauntonia chinensis*. Pak. J. Pharm. Sci. 27 (5), 1317–1325.

Yoo, M., Lee, S.J., Kim, Y.K., et al., 2016. Dehydrocorydaline promotes myogenic differentiation via p38 MAPK activation. Mol. Med. Rep. 14 (4), 3029–3036.

Yoshikawa, M., Murakami, T., Oomiinami, H., et al., 2000. Bioactive saponins and glycosides. XVI. Nortriterpene oligoglycosides with gastroprotective activity from the fresh leaves of *Euptelea polyandra* Sieb. et Zucc. (1): structures of Eupteleasaponins I, II, III, IV, V, and V acetate. Chem. Pharm. Bull. 48 (7), 1045–1050.

Yu, D.R., Ji, L.P., Wang, T., et al., 2017. Neuroprotective activity of two active chemical constituents from *Tinospora hainanensis*. Asian Pac. J. Trop. Med. 10 (2), 114–120.

Yu, L.L., Hu, W.C., Ding, G., et al., 2011. Gusanlungionosides A-D, potential tyrosinase inhibitors from *Arcangelisia gusanlung*. J. Nat. Prod. 74 (5), 1009–1014.

Yu, L.L., Li, R.T., Ai, Y.B., et al., 2014. Protoberberine isoquinoline alkaloids from *Arcangelisia gusanlung*. Molecules 19 (9), 13332–13341.

Yuan, X., Dou, Y., Wu, X., et al., 2017. Tetrandrine, an agonist of aryl hydrocarbon receptor, reciprocally modulates the activities of STAT3 and STAT5 to suppress Th17 cell differentiation. J. Cell. Mol. Med. 21 (9), 2172–2183.

Zeng, X., Wang, H., Gong, Z., et al., 2015. Antimicrobial and cytotoxic phenolics and phenolic glycosides from *Sargentodoxa cuneata*. Fitoterapia 101, 153–161.

Zhang, B., Huang, R., Hua, J., et al., 2016c. Antitumor lignanamides from the aerial parts of *Corydalis saxicola*. Phytomedicine 23 (13), 1599–1609.

Zhang, G., Ma, H., Hu, S., et al., 2016a. Clerodane-type diterpenoids from tuberous roots of *Tinospora sagittata* (Oliv) Gagnep. Fitoterapia 110, 59–65.

Zhang, H., Ren, Y., Tang, X., et al., 2015b. Vascular normalization induced by sinomenine hydrochloride results in suppressed mammary tumor growth and metastasis. Sci. Rep. 5, 8888.

Zhang, H.F., Wu, D., Du, J.K., et al., 2014b. Inhibitory effects of phenolic alkaloids of *Menispermum dauricum* on gastric cancer in vivo. Asian. Pac. J. Cancer. Prev. 15 (24), 10825–10830.

Zhang, L., Wang, X.Z., Li, Y.S., et al., 2016b. Icariin ameliorates IgA nephropathy by inhibition of nuclear factor kappa b/Nlrp3 pathway. FEBS Open Bio. 7 (1), 54–63.

Zhang, N., Lian, Z., Peng, X., 2017a. Applications of higenamine in pharmacology and medicine. J. Ethnopharmacol. 196, 242–252.

Zhang, Q., Luan, G., Ma, T., et al., 2015a. Application of chromatography technology in the separation of active alkaloids from *Hypecoum leptocarpum* and their inhibitory effect on fatty acid synthase. J. Sep. Sci. 38 (23), 4063–4070.

Zhang, S.L., Li, H., He, X., 2014a. Alkaloids from *Mahonia bealei* posses anti-H$^+$/K$^+$-ATPase and anti-gastrin effects on pyloric ligation-induced gastric ulcer in rats. Phytomedicine 21 (11), 1356–1363.

Zhang, Y., Wen, Y.L., Ma, J.W., et al., 2017b. Tetrandrine inhibits glioma stem-like cells by repressing β-catenin expression. Int. J. Oncol. 50 (1), 101–110.

Zhang, Y.M., Wang, N., Dai, B.L., et al., 2011. Effect of taspine derivatives on human liver cancer SMMC7721. Zhong Yao Cai 34 (7), 1094–1097.

Zhao, B., Chen, Y., Sun, X., et al., 2012a. Phenolic alkaloids from *Menispermum dauricum* rhizome protect against brain ischemia injury via regulation of GLT-1, EAAC1 and ROS generation. Molecules 17 (3), 2725–2737.

Zhao, C., Zhang, N., He, W., et al., 2014. Simultaneous determination of three major lignans in rat plasma by LC-MS/MS and its application to a pharmacokinetic study after oral administration of *Diphylleia sinensis* extract. Biomed. Chromatogr. 28 (4), 463–467.

Zhao, H., Song, L., Huang, W., et al., 2017. Total flavonoids of *Epimedium* reduce ageing-related oxidative DNA damage in testis of rats via p53-dependent pathway. Andrologia, 12756. doi: 10.1111/and.12756.

Zhao, J., Lian, Y., Lu, C., et al., 2012b. Inhibitory effects of a bisbenzylisoquinline alkaloid dauricine on HERG potassium channels. J. Ethnopharmacol. 141 (2), 685–691.

Zheng, J., Zhao, Y., Lun, Q., et al., 2017. *Corydalis edulis Maxim.*promotes insulin secretion via the activation of protein kinase Cs (PKCs) in mice and pancreatic β cells. Sci. Rep. 7, 40454.

Zhou, F.Z., Wang, X., Dai, A.Y., et al., 2016. Effects of *Sinopodophyllum hexundrum* on apoptosis in K562 cells. Nan Fang Yi Ke Da Xue Xue Bao 37 (2), 226–231.

Zhou, Z.G., Zhang, C.Y., Fei, H.X., et al., 2015. Phenolic alkaloids from *Menispermum dauricum* inhibits BxPC-3 pancreatic cancer cells by blocking of Hedgehog signaling pathway. Pharmacogn. Mag. 11 (44), 690–697.

CHAPTER

4

Drug Metabolism and Pharmacokinetic Diversity of Ranunculaceae Medicinal Compounds

Ranunculales Medicinal Plants. http://dx.doi.org/10.1016/B978-0-12-814232-5.00004-6

ABBREVIATIONS

AC	Aconitine
ADME/T	Absorption, distribution, metabolism, excretion, and toxicity
AUC	Area under the plasma concentration-time curve
BAC	Benzoylaconine
BCRP	Breast cancer resistance protein
BHA	Benzoylhypaconine
BMA	Benzoylmesaconine
CL	Total body clearance
Cmax	Maximum plasma concentration
CYP	Cytochrome P450
DDI	Drug–drug interaction
DMPK	Drug metabolism and pharmacokinetics
GFA	Guanfu base A
HDI	Herb–drug interaction
HHI	Herb-herb interaction
HLMs	Human liver microsomes
IC50	Half maximal inhibitory concentration of a substance
Km	The concentration of substrate that leads to half maximal velocity
MRP	Multidrug resistance-associated protein
MRT	Mean residence time
OCT	Organic cation transporter
Papp	Apparent permeability coefficient
P-gp	P glycoprotein
rhCYP	Recombinant human cytochrome P450
SC	Stratum corneum
SULT	Sulfotransferase
T1/2	Elimination half life
TCM	Traditional Chinese medicine
Tmax	Time to reach Cmax
UGT	Uridine diphosphate glucuronosyltransferase
UPLC/Q-TOF MS	Ultraperformance liquid chromatography-quadrupole time-of-flight mass spectrometry
Vd	Volume of distribution
Vmax	Maximum enzyme velocity

4.1 INTRODUCTION

Many Ranunculaceae genera, for example, *Ranunculus* (600 species), *Delphinium* (365), *Thalictrum* (330), *Clematis* (325) (Hao et al., 2013a), and *Aconitum* (300), are commonly used in traditional Chinese medicine (TCM) and worldwide ethnomedicine (Hao et al., 2013b). Forty-two genera and around seven hundred and twenty species are distributed throughout China, most of which are in the southwest mountainous region (Wu et al., 2003). Reports about plant systematics, phytochemistry, chemotaxonomy, and pharmacology of Ranunculaceae family (Xiao, 1980) are numerous. This family represents a promising model plant family for drug metabolism and pharmacokinetic (DMPK) studies of multicomponent medicinal herbs. However, DMPK characteristics of Ranunculaceae-derived medicinal compounds have not been summarized.

As far as we know, at least in mainland China, Korea, Hong Kong, Macau, and Taiwan, Ranunculaceae products are legally used in public hospitals, either alone or, more often, in combination with Western drugs/chemical drugs. Ten Ranunculaceae species are recorded in the main text of *China Pharmacopoeia* (CP), 2010 version, plus an additional 10 Ranunculaceae species in the CP appendix. Moreover, at least an additional 20 Ranunculaceae species are collected in local standards of China. These 40 Ranunculaceae species and the medicinal compounds thereof are frequently used in prescriptions at formal health settings. Data on herb-drug interaction (HDI) studies are relevant when the popularity of herb-drug combinations among health consumers is considered. ADME/T properties will be relevant in compounds, for example, berberine of *Coptis* and aconine of *Aconitum*, which are being used in public hospitals or have potential for medicinal utility. Black cohosh (*Cimicifuga racemosa* or *Actaea racemosa*) and goldenseal (*Hydrastis*) raise concerns of HDI. DMPK studies of other Ranunculaceae genera, for example, *Nigella*, *Delphinium*, *Aconitum*, *Trollius*, and *Coptis*, are also rapidly increasing and becoming more clinically relevant. In this chapter, the current knowledge, as well as the challenges around the DMPK-related issues in optimization of drug development and clinical practice of Ranunculaceae compounds, is highlighted. Exhaustive literature searches in PubMed, Google, and CNKI (http://cnki.net/) have been performed to outline the progress of DMPK studies of Ranunculaceae medicinal compounds during the last decade. Search terms "drug metabolism," "pharmacokinetic," "drug transporter," "absorption," "distribution," "excretion," and "toxicity" were used, combined with "Ranunculaceae" and the names of genera.

4.2 ABSORPTION OF RANUNCULACEAE COMPOUNDS

4.2.1 Absorption Via Gut

4.2.1.1 *Alkaloid of Ranunculoideae*

Intestinal absorption is a complex process of the transport of compounds from the apical side (AP) to the basolateral side (BL) of intestinal epithelia and may be influenced by multiple environmental and/or genetic factors (Fig. 4.1). Many Ranunculaceae herbs are conventionally orally used, thus it is essential to characterize the absorption process of the representative compounds. Flowers of *Trollius chinensis* (Jin Lian Hua in Chinese) have antiinflammatory and antimicrobial activities. Trolline and veratric acid (Fig. 4.2), isolated from the flowers of *T. chinensis*, are transported across Caco-2 cell monolayer in a concentration-dependent manner (Liu et al., 2014c). Trolline, a tetrahydroisoquinoline alkaloid, is transported at an apparent permeability coefficient (P_{app}) 10^{-6} cm/s with the $P_{appAP \to BL}/P_{appBL \to AP}$ ratio of more than 1.8 or less than 0.8, suggesting that it is moderately absorbed through an associative mechanism involving active and passive transport. Veratric acid, a phenolic acid, is transported at a P_{app} 10^{-5} cm/s with the $P_{app \, AP \to BL}/P_{app \, BL \to AP}$ ratio of close to 1.0, indicating that it is well absorbed, mainly through passive diffusion. These results are useful in chemically assessing the pharmacodynamic (PD) material basis of the flowers of *T. chinensis*.

Aconitine (AC) is a lethal alkaloid of the genus *Aconitum* and is abundant in some traditionally used medicinal herbs, such as *Aconitum carmichaeli* (Wu Tou in Chinese), *Aconitum*

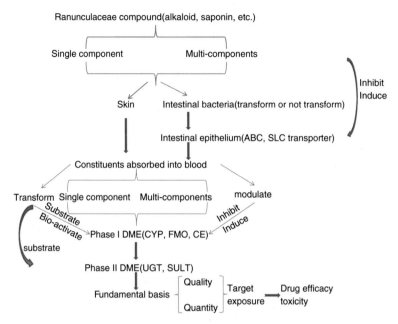

FIGURE 4.1   PK and PD mechanisms of Ranunculaceae medicinal compounds.

kusnezoffii, Aconitum brachypodum, and *Aconitum austroyunnanense* (Hao et al., 2013b). P-gly-coprotein (P-gp, MDR1, ABCB1) (Wang et al., 2006) is involved in the low and unpredictable bioavailability of AC regarding its oral use (Yang et al., 2013). The influx of AC through mono-layers of Caco-2 and MDCKII-MDR1 cells is less than its efflux, while the latter is dramatically reduced by the P-gp inhibitors, verapamil and cyclosporin A (CsA). The intestinal permeabil-ity of AC is enhanced, from 0.22×10^{-5} to 2.85×10^{-5} cm/s, by the verapamil coperfusion in rat intestines. Verapamil pretreatment increases the maximum plasma concentration (C_{max}) of oral AC in rats, from 39.43 to 1490.7 ng/mL, accompanying a sharp increase of the area under the plasma concentration-time curve (AUC) of AC. A common P-gp recognition mechanism might be used by AC and verapamil, and P-gp might curb the intestinal absorption of AC and mitigate its poisonousness to mammalians. Care should be taken when AC-containing for-mulations are administered, since drug-drug interaction (DDI) might be mediated by P-gp.

In Caco-2 cells, AC, mesaconitine (MA), and hypaconitine (HA) display decent absorbency with P_{app} values of more than 1×10^{-6} cm/s (Li et al., 2012). AC, MA, and HA, when mixed, exhibit better transport efficiency in the AP to BL than that in the reverse direction. Digoxin (DIG, a P-gp substrate) efflux to the AP side is deterred by these alkaloids. Verapamil inhibits the reverse transport of MA and HA. These alkaloids might be both P-gp inhibitors and their substrates, which interact to enhance their own bioavailability on concurrent use.

4.2.1.2 Alkaloid of Thalictroideae and Coptis

One advantage of the biopharmaceutics classification system of Chinese materia medica (CMMBCS) is expanding the classification research from single ingredients to multicom-ponents of Chinese herbs and from multicomponents research to holistic research of the

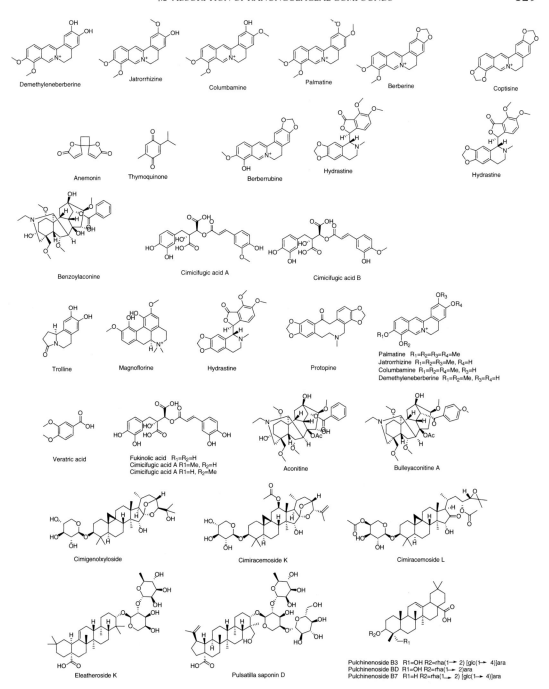

FIGURE 4.2 Examples of Ranunculaceae medicinal compounds mentioned in the text.

Chinese materia medica (Liu et al., 2017). The alkaloids of *Coptis* extract are used to explore the changes in solubility and intestinal permeability of single-components and multicomponents and to determine the biopharmaceutical classification of extract at the holistic level. The typical shake-flask method and high-performance liquid chromatography (HPLC) are used to study the solubility of the single ingredients of alkaloids. The alkaloids in the intestinal absorption are quantified in single-pass intestinal perfusion, and the permeability coefficient of *Coptis* extract is calculated by using the self-defined weight coefficient method.

The absorption of isoquinoline alkaloids berberine (BBR) and palmatine, the representative therapeutic ingredients of *Coptis* plants, in different regions of rat intestines varies significantly (Chen et al., 2011). Ileum has the highest P_{app} value of 50 mg/L BBR, trailed by duodenum, jejunum, and large intestines. Palmatine (50 mg/L) also has the highest P_{app} value in ileum, followed by colon, jejunum, and duodenum. The concentration of BBR and palmatine had distinct effects on their absorption rate constant (K_a) values, and their absorption was not passive diffusion. The other ingredients in Wuji pill (section below named Metabolism) and *Coptis* crude extract can promote the absorption of alkaloids.

Everted intestinal sac models are used to investigate the intestinal absorption of BBR, palmatine, coptisine, and epiberberine in Fuzheng Xiaozheng Fang (Xie et al., 2016). The absorption rate of four alkaloids is zero-order one. The absorption rate constant (K_a) of four alkaloids at low doses is significantly different from those of medium and high doses, but there is no significant difference of K_a between medium dose and high dose. There is both positive diffusion and passive absorption. The absorption amount of four alkaloids is slightly greater in ileum than in jejunum, but the difference is not significant.

Results in human colon cancer Caco-2 cells suggest that coptisine, jatrorrhizine, BBR, epiberberine, and palmatine interact to affect each other's absorption (Qi et al., 2013). The least square multilinear regression is used to obtain the interaction type and intensity (I) of different alkaloids (X, mg/mL) during absorption. For instance, the multiple regression equation of the impact of other alkaloids on coptisine is: $I_{coptisine} = 1.40X_{epiberberine} - 2.88X_{palmatine} + 0.62X_{epiberberine}X_{palmatine} + 0.48X_{jatrorrhizine}X_{BBR} + 0.48X_{jatrorrhizine}X_{BBR}X_{palmatine} - 0.75X_{epiberberine}X_{jatrorrhizine}X_{BBR}$, indicating that the most significant impact is the inhibition of palmatine, followed by the promotion of epiberberine, and the inhibition/promotion of the interaction between different alkaloids. Similarly, the predominant impact on jatrorrhizine absorption is the inhibition of coptisine and the promotion of the interaction between BBR and palmatine. BBR absorption is inhibited by palmatine but enhanced by the interaction between jatrorrhizine and palmatine. Palmatine absorption is inhibited by epiberberine but enhanced by the interaction between BBR and epiberberine, and coptisine and epiberberine. This regression method would be helpful in delineating crosstalk between multiple compounds during intestinal absorption.

BBR proves useful in many illnesses, for example, inflammation, infection, diabetes, and liver diseases, although its intestinal absorption is poor. Preparations having 0.5%, 1.5%, and 3.0% chitosan increased its $AUC_{0-36 h}$ values by 1.9, 2.2, and 2.5 times respectively (Chen et al., 2012). Chitosan might enhance the BBR paracellular pathway in the gut. In contrast, 2% and 3.3% chitosan hydrochloride failed to increase either C_{max} or $AUC_{0-36 h}$ of BBR, as the BBR chloride solubility is reduced in the presence of chitosan hydrochloride.

The poor bioavailability of BBR limits its development as a promising anticancer agent. Spray-dried mucoadhesive microparticle preparations of BBR, produced by dual-channel spray-gun technology, have dramatically increased gastrointestinal permeability in Caco-2 cell monolayer (Godugu et al., 2014). This technology might be useful for other Ranunculaceae compounds that are difficult to absorb.

4.2.1.3 *Saponin*

α-Hederin, a pentacyclic triterpene saponin, is abundant in *Nigella* (Liang et al., 2014), *Anemone*, *Pulsatilla*, and *Clematis*. In situ single-pass perfusion provides clues for the poor bioavailability of α-hederin (He et al., 2014). α-Hederin can be absorbed in each part of the intestines, while ileum has the highest K_a, followed by the colon, jejunum, and duodenum. Absorption parameters of α-hederin do not alter dramatically at different concentrations (75, 150, and 300 μg/mL) but rise when pH is increased. Disruption of the intestinal flora might affect the absorption of α-hederin. P-gp inhibitor does not significantly change K_a and permeability coefficient values, and α-hederin is not the substrate of P-gp. Saturation of absorption is not observed within the test concentration range of the drug, and passive diffusion might be predominant.

The K_a and P_{app} values of *Pulsatilla saponin* D (PSD) are highest in the colon (Rao et al., 2013), followed by the ileum, jejunum, and duodenum. These values are higher in basic conditions. PSD could undergo passive diffusion but may be a substrate of P-gp, and there is saturation of transporters. The K_a values of pulchinenosides B3 and BD and $P_{app}(P_{eff})$ values of B3 and B7 display significant differences in different intestinal segments (Liu et al., 2015). The highest absorption is in duodenum, followed by the jejunum, colon, and ileum. The *Pulsatilla* saponins exhibit oversaturation as the concentration increases over 0.05–2.5 g/L. The P_{app} values and saponin concentrations are not of linear correlation in the duodenum. The K_a and P_{app} values decline significantly when DIG is administered but increases when verapamil is used. The *Pulsatilla* saponins are not transported in a concentration-dependent manner, and the transporter protein might be involved in their transport.

In rats, calycosin-7-O-β-D-glucoside could enhance the absorption of prim-O-glucosylcimifugin and cimifugin and improve their bioavailability (Zhao et al., 2014).

Many transporters (e.g., PepT1, GLUT5, MRP2, and high-affinity glutamate transporter) are abundant in the small intestine and are less so in the large intestine of both humans and rats (Cao et al., 2006). On the contrary, some transporters (e.g., MDR1, MRP3, GLUT1, and GLUT3) display divergent expression contours in the mammalian duodenum and colon. Moreover, the intestinal expressions of CYP3A4/3A9 and UDP-glucuronosyltransferase (UGT) exhibit great differences between humans and rats, and their differential expression patterns are regionally dependent. Since intestinal drug absorption is similar in the small intestine of humans and rats, and the intestinal expression of metabolizing enzymes is species specific, rats can be utilized to forecast drug absorption in the human small intestine but not to envisage drug metabolism or oral bioavailability in humans.

4.2.2 Absorption Via Skin

Percutaneous absorption of alkaloids and anemonin is appreciated in some application contexts. (E)-2-isopropyl-5-methylcyclohexyl octadec-9-enoate (M-OA) significantly enhances the penetration of MA and HA through the skin (Zhao et al., 2011), while AC is not detected on the receiver side of side-by-side diffusion cells, and L-menthol does not influence the alkaloid permeation. M-OA extracts the stratum corneum (SC, horny layer) lipids to disrupt it and to desquamate the SC flake. SC lipid fluidization might change the protein conformation and facilitate drug absorption; therefore, M-OA could be included in transdermal preparations to improve permeation.

Anemonin, abundant in *Clematis*, *Anemone*, *Pulsatilla*, and *Ranunculus*, has antimicrobial, analgesic, and sedative effects. However, anemonin is highly irritated and not suitable for oral medication. The steady-state rate of anemonin penetrating through human skin is improved, up to 18.56 $\mu g/cm^2/h$, by ethanol and laurocapm (Ning et al., 2007), while the steady-state permeation rate of saturated water solution of anemonin is only 1.17 $\mu g/cm^2/h$. Ethanol and laurocapm might be used in the new transdermal preparation.

The transdermal osmosis processes of *A. brachypodum*'s liniment, gel, and patcher were investigated to provide the basis for selecting formulation and quality control (Lin et al., 2007). The three preparations have distinct characteristics of transdermal osmosis. The liniment follows dynamics zero-order process, while the gel and the patcher follow dynamics zero-order process of noncorroded drug systems. The relationship between their transdermal osmosis and release is well fitted with cubic equation. The microemulsion gel of *A. brachypodum* total alkaloids has more stable transdermal absorption, longer duration of action, and higher bioavailability than ordinary gel (Wu et al., 2016b). The transdermal bioavailability of bullatine A is similar between microemulsion gel and ordinary gel.

Aconitum tincture and unprocessed roots can be absorbed via the skin into the blood stream to result in deadly and nonlethal intoxication (Chan, 2012). Very high concentrations of *Aconitum* alkaloids can be absorbed along the diffusion gradient. The danger of intoxication increases if *Aconitum* alkaloids are in close contact with the damaged skin and epidermis (SC).

The *Aconiti Radix Cocta* gel and *Aconiti Radix Cocta* + *Paeoniae Radix Alba* gel were administered to mice (Li et al., 2016). The effects of *Paeoniae Radix Alba* on transdermal permeation of six aconitum alkaloids were investigated. T_{max} of the three monoester aconitum alkaloids in the medicine pair group is shorter, and C_{max} and AUC of benzoylaconine and benzoylhypaconine are increased. On the contrary, AUC of three diester-type alkaloids is reduced, with prolonged T_{max} of hypaconitine and lower C_{max}. The decreased toxicity and increased efficacy of aconitum alkaloids could be due to the combination of two herbs.

Microneedle-mediated transdermal delivery of nanostructured lipid carriers is used for alkaloids of *Aconitum sinomontanum* (AAS-NLCs-MN) (Guo et al., 2017), which shows benefits in eliminating paw swelling, decreasing inflammation and pain, and regulating immune function in adjuvant arthritis rats. AAS-NLCs-MN does not cause skin irritation in rabbits, and electrocardiograms of rats show improved arrhythmia. This approach provides safe transdermal delivery and improves the therapeutic efficacy through sustained release of AAS.

4.2.3 Absorption Via Mucosa

The main active saponins of Toutongning nasal sprays can be absorbed through the rat nasal mucosa (Wang et al., 2017). The absorption rate constant of high, middle, and low concentrations of prim-O-glucosylcimifugin is $(0.588 \pm 0.041) \times 10^{-3}$, $(0.547 \pm 0.023) \times 10^{-3}$, and $(0.592 \pm 0.063) \times 10^{-3}$ min^{-1} respectively, and that of 5-O-methylvisammioside is slightly lower. The absorption rate is not significantly different between different concentrations, and the absorption under pH 6 is better than that under other pH conditions. It is feasible to make nasal spray.

4.3 DISTRIBUTION

4.3.1 Thymoquinone

In pharmacokinetics (PK), distribution describes the reversible transfer of drugs from one location to another within the body. Plasma protein binding, physical volume of an organism, and removal rate are among the most fundamental factors that affect drug distribution. Binding between drug and plasma proteins reduces the drug's ultimate concentration in the tissues. Thymoquinone (THQ), extracted from *Nigella* essential oil and seeds, binds to human α_1-acid glycoprotein (AGP), which marginally enhances its thermal stability and changes the molten globulelike state to higher temperatures (Lupidi et al., 2012). Fluorescence quenching and molecular docking suggested that hydrophobic interactions and to a lesser extent hydrogen bonds are responsible for THQ binding.

Albumins are present in higher concentrations than glycoproteins and lipoproteins in the plasma, and they readily bind to other substances. Binding between THQ and human serum albumin (HSA) has no impact on the secondary structure of HSA (Lupidi et al., 2010). One mole THQ binds one mole HSA. The binding is driven by enthalpy and thus is spontaneous, and hydrophobic interactions stabilize the complex. The molecular modeling suggests that THQ binds to site I of HSA.

The binding of THQ to bovine serum albumin (BSA) and AGP were 94.5 ± 1.7% and 99.1 ± 0.1% respectively (El-Najjar et al., 2011). Cyst-34 was involved in the covalent binding between THQ and BSA. BSA protects against THQ-induced cell death; BSA preincubation abolished THQ's anticancer effects. Contrariwise, binding of THQ to AGP failed to change its anticancer activity. When THQ is pretreated with AGP before the BSA exposure, THQ still has the anticancer activity, indicating that AGP precluded the tie of THQ to BSA.

4.3.2 Isoquinoline Alkaloid

Fluorescence data reveals the presence of a single class of jatrorrhizine (one of the protoberberine alkaloids in *Coptis*) binding site on HSA (Li et al., 2005). The secondary structure of HAS is altered when jatrorrhizine is in aqueous solution. The standard enthalpy (ΔH^0) and entropy (ΔS^0) of the reaction are -10.89 kJ/mol and 56.267 J/mol•K respectively. Hydrophobic and electrostatic interactions are preeminent in the binding. Similar to THQ, jatrorrhizine binds to site I of HSA.

A drug must be distributed into interstitial and intracellular fluids to exert its effects once it is orally or intravenously (i.v.) used. In rats, oral BBR is swiftly distributed into the liver, kidneys, muscle, etc. (Tan et al., 2013). BBR's tissue concentrations are higher than the plasma concentration 4 h after medication. BBR kept moderately steady in the liver, heart, brain, etc. The phase I metabolites of BBR, for example, thalifendine (M1), berberrubine (M2), and jatrorrhizine (M4), are simply detected in liver and kidney. M2 is the most abundant liver metabolite, followed by M1 and M4.The plasma has much less BBR metabolites than liver, indicated by much less AUC_{0-t}. The distribution landscape of BBR partially explains its versatile bioactivities, such as lipid-modulating, antidiabetic, antimicrobial, and antiinflammatory activities. Theoretically, the blood concentration of herb compound would be lower, provided that it is more distributed in the target organ and vice versa. Despite the drug category, the plasma concentration should reach a certain level, at which the enough drug concentration in the target organ can be maintained. Regardless of plasma concentration or tissue concentration, there should be additive effect of effective forms (original compounds and metabolites) (Xu et al., 2014).

4.3.3 Diterpenoid Alkaloid

Data on the tissue distribution of the *Aconitum* alkaloids in poisoning are valuable to illuminate numerous actions of alkaloids. In all three autopsy cases (Niitsu et al., 2013), there were more *Aconitum* alkaloids (jesaconitine, MA, and AC) in the liver and kidney than in the heart and cerebrum. There were fewer alkaloids in the cerebrum than in the blood. In the corpse that died of acute aconite intoxication (Liu et al., 2009a), the content of aconite was highest in the urine, followed by bile, gastric content, heart blood, etc., and aconite was not detected in the brain. Urine, bile, and blood are the top samples to quantify aconite in the severe poisoning. AC, benzoylaconine, and aconine distributed throughout all rat organs (Zhang et al., 2016b), and the distribution rate of AC was slower than that of the other two.

In mice, Mdr1a deficiency significantly enhanced the analgesic effect of AC and exacerbated its toxicity by upregulating its distribution to the brain and decreasing its plasma elimination rate (Zhu et al., 2017). Mdr1a dysfunction may cause severe AC poisoning.

4.3.4 Others

Orientin, one of active flavonoid glycosides in Ranunculaceae genera such as *Trollius* and *Ranunculus*, is rapidly allocated and excreted within 1.5 h following tail vein injection (Li et al., 2008a). Orientin mainly distributed into rat liver, lung, and kidney while the blood-brain barrier prevented it from entering the brain. Orientin did not have long-standing buildup in rat organs. Compared to alkaloids, both distributions and eliminations of three flavonoids/glycosides in rats were fast (Table 4.1) (Li et al., 2008a,b).

4.4 METABOLISM

4.4.1 Metabolism Via Gut Flora

4.4.1.1 Alkaloid

Gut flora (microbiota) consists of a complex of microorganism species that live in the alimentary tracts of animals and is the largest reservoir of microorganisms communal to humans. A TCM formula, Wuji pill, consists of *Coptis chinensis* (CC), *Tetradium ruticarpum* (Wu Zhu Yu in Chinese), and *Paeonia lactiflora* (Bai Shao in Chinese), and the proportion of the three herbs can be adjusted in terms of the individual disease status. Wuji pill or CC extract was coincubated with fresh human excrement under anaerobic conditions for 24 h to simulate in vivo drug metabolism via gut flora (Men et al., 2013). Metabolism of BBR was positively correlated with doses, whereas metabolism of palmatine was negatively correlated with doses in CC extracts. Compound compatibility (interaction) of Wuji pill accelerated the metabolism of low-dose BBR, which was positively correlated with the doses of *Tetradium* and *Paeonia*; the metabolism

TABLE 4.1 Examples of PK Parameters of Ranunculaceae Compounds Determined in Human and Animal Studies

Compound/mixture	Subject	C_{max}	$T_{1/2}$	AUC	Volume of distribution	Total clearance	Absolute bioavailability	T_{max}	MRT	Ref.
Higenamine	Healthy Chinese volunteers	15.1–44.0 ng/mL	0.133 h (0.107–0.166 h)	5.39 ng·h/mL (3.2–6.8 ng h/mL)	48 L (30.8–80.6 L)	249 L/h (199–336 L/h)	ND	ND	ND	Feng et al. (2012)
Guanfu base A	Dogs	ND	$T_{1/2}\pi$ 0.07 h, $T_{1/2}\alpha$ 1.5 h, $T_{1/2}\beta$ 13.5 h	61.43 µg·h/mL	Central compartment volume 0.37 L/kg	Plasma clearance 0.14 L/kg/h	ND	ND	ND	Wu et al. (2002)
AC	Rats	8.72 ± 5.32 ng/mL (oral)	322.12 ± 70.46 min (single oral); 80.98 ± 6.40 min (iv)	3297.23 ± 1007.64 (single oral); 3201.89 ± 338.91 min·ng/mL (iv)	0.632 ± 0.332 L/kg (iv)	0.317 ± 0.144 L/h/kg (iv)	8.24 ± 2.52% (single oral)	30.08 ± 9.73 min (single oral)	322.12 ± 70.46 min (single oral); 116.86 ± 9.23 min (iv)	Tang et al. (2012)
AC	Dogs	14.13 ± 0.75 ng/mL	272.81 ± 5.36 min	5513.8 ± 75.52 ng· min/mL	ND	ND	ND	70 ± 8.66 min	ND	Xiao et al. (2014)
MA	Dogs	45.42 ± 1.76 ng/mL	374.89 ± 5.70 min	21638 ± 144.02 ng· min/mL	ND	ND	ND	70 ± 8.66 min	ND	
HA	Dogs	43.18 ± 1.49 ng/mL	291.52 ± 13.94 min	18890.6 ± 455.49 ng· min/mL	ND	ND	ND	70 ± 8.66 min	ND	
BAC	Dogs	23.61 ± 0.92 ng/mL	456.0 ± 11.69 min	13427.6 ± 415.72 ng· min/mL	ND	ND	ND	70 ± 8.66 min	ND	
BMA	Dogs	123.86 ± 5.43 ng/mL	255.05 ± 9.12 min	35382.6 ± 1025.77 ng· min/mL	ND	ND	ND	70 ± 8.66 min	ND	
BHA	Dogs	52.27 ± 2.41 ng/mL	339.18 ± 3.13 min	22038.9 ± 521.39 ng· min/mL	ND	ND	ND	70 ± 8.66 min	ND	
AC	Rats	2.1 ± 1.28 ng/mL (reflux 1 h)	7.7 ± 4.92 h (reflux 1 h)	8.58 ± 1.45 ng·h/mL (reflux 1 h)	ND	ND	ND	0.33 ± 0.13 h	ND	Liu et al. (2014b)
HA	Rats	7.47 ± 3.2 ng/mL	5.16 ± 1.7	36.53 ± 11.75	ND	ND	ND	0.33 ± 0.13	ND	
MA	Rats	6.15 ± 3.75	10.64 ± 3.48	23.89 ± 4.79	ND	ND	ND	0.42 ± 0.2	ND	
BAC	Rats	7.88 ± 4.19	13.82 ± 3.1	41.99 ± 13.7 (reflux 3 h)	ND	ND	ND	0.19 ± 0.04	ND	
BHA	Rats	3.34 ± 1.33	13.32 ± 3.62	32.18 ± 10.85	ND	ND	ND	0.25 ± 0.13	ND	
BMA	Rats	22.44 ± 15.31	14.58 ± 5.48	99.28 ± 35.74	ND	ND	ND	0.28 ± 0.11	ND	
Fuziline	Rats	72.1 ± 28.9 ng/mL(po)	5.0 ± 1.9 (po)/6.3 ± 2.6 (iv)h	595.0 ± 229.5 (po)/733.1 ± 239.9 (iv) ng·h/mL	14663 ± 3727.4 (po)/2522.1 ± 1886.7 (iv) mL/kg	1745.6 ± 818.1 (po)/305.1 ± 146.3 (iv) mL/kg/h	21.1 ± 7.0%	2.8 ± 0.7 h	11 ± 4.1 (po)/5.1 ± 1.6 (iv) h	Sun et al. (2013)
Fuzi extract	Rats	3.24 ± 0.4 ng/mL (single); 2.61 ± 0.98 ng/ mL (multiple)	217.88 ± 86.08 min (single); 383.83 ± 96.6 min (multiple)	588.47 ± 101.52 (single); 1105.56 ± 42.91 min·ng/ mL (multiple)	ND	ND	4.72 ± 2.66% (single)	58.00 ± 21.68 min (single); 20 ± 8.66 min (multiple)	150.99 ± 59.66 min (single); 265.95 ± 66.98 min (multiple)	Tang et al. (2012)
Berberine	Rats	1176.6 ± 341.8 ng/mL	11.9 ± 3 h	631.1 ± 77.6 ng·h/mL	ND	ND	ND	0.083 h	ND	Liu et al. (2009)

(Continued)

TABLE 4.1 Examples of PK Parameters of Ranunculaceae Compounds Determined in Human and Animal Studies (*cont.*)

Compound/mixture	Subject	C_{max}	$T_{1/2}$	AUC	Volume of distribution	Total clearance	Absolute bioavailability	T_{max}	MRT	Ref.
Demethylated berberine	Rats	37.2 ± 3.8 ng/mL	19 ± 5 h	198.1 ± 53 ng·h/mL	ND	ND	ND	0.625 ± 0.25 h	ND	Shi et al. (2012)
Demethylenated berberine	Rats	19.4 ± 5.6 ng/mL	1.2 ± 0.2 h	149 ± 45.2 ng·h/mL	ND	ND	ND	1.25 ± 0.5 h	ND	
Jatrorrhizine	Rats	3.3 ± 0.8 ng/mL	14.9 ± 1.2 h	21.8 ± 1.8 ng·h/mL	ND	ND	ND	0.625 ± 0.25 h	ND	
Jatrorrhizine	Rats	ND	8.5 ± 2.6 h (0.1 mg/kg), 10.6 ± 5.4 h (0.3 mg/kg), 8.9 ± 2.2 h (3 mg/kg)	9.6 ± 3.6 μg·h/L (0.1 mg/kg), 32.1 ± 13.4 μg·h/L (0.3 mg/kg), 308.9 ± 85.7 μg·h/L (3 mg/kg)	188.9 ± 121.7 L/kg (0.1 mg/kg), 149.9 ± 74.4 L/kg (0.3 mg/kg), 137 ± 57.5 L/kg (3 mg/kg)	11.6 ± 3.8 L/h/kg (0.1 mg/kg), 10.6 ± 3.9 L/h/kg (0.3 mg/kg), 10.3 ± 2.8 L/h/kg (3 mg/kg)	ND	ND	5.7 ± 2.3 h (0.1 mg/kg), 8.3 ± 4.2 h (0.3 mg/kg), 8.8 ± 1.4 h (3 mg/kg)	
Thymoquinone	Rabbits	ND	Absorption: 217 min, elimination: 63.43 ± 10.69 min (iv), 74.61 ± 8.48 min (po), $T_{1/2}\alpha$ ~8.9 min, $T_{1/2}\beta$ ~86.6 min	ND	700.90 ± 55.01 mL/kg (Vss/iv), 5, 109.46 ± 196.08 mL/kg (po)	7.19 ± 0.83 mL/kg/min (iv), 12.30 ± 0.30 mL/min/kg (po)	~58% lag time ~23 min	ND	ND	Alkharfy et al. (2015) 146
Thymoquinone-loaded nanostructured lipid carriers	Rabbits	4811.33 ± 55.52 ng/mL	4.493 ± 0.015 h	26821.61 ± 9.40 ng·h/mL	4.32 ± 0.34 L/kg	0.71 ± 0.031 L/h/kg	ND	3.96 ± 0.19 h	ND	Abdelwahab et al. (2013) 147
Orientin	Rats	ND	α, 1.48 ± 0.14 min, β, 7.22 ± 0.87 min, γ, 25.74 ± 3.05 min	513.7 ± 19.8 mg·min/L	Central compartment volume 0.234 ± 0.006 L/kg	0.04 ± 0.001 L/kg/min	ND	ND	15.38 ± 0.15 min	Li et al. (2008a)
Orientin	Rats	ND	α, 1.88 ± 0.227 min, β, 11.88 ± 0.46 min, γ, 64.497 ± 9.217 min	2856.3 ± 215.8 mg·min/L	0.016 ± 0.001 L/kg	0.002 L/kg/min	ND	ND	15.85 ± 0.59 min	Li et al. (2008b)
Orientin-2''-O-β-L-galactopyranosyl	Rats	ND	α, 1.72 ± 0.58 min, β, 10.38 ± 5.748 min, γ, 36.88 ± 3.71 min	3600.5 ± 106.1 mg·min/L	0.018 ± 0.007 L/kg	0.002 L/kg/min	ND	ND	31.45 ± 0.87 min	
Vitexin	Rats	ND	α, 1.216 ± 0.424 min, β, 9.99 ± 1.05 min, γ, 45.637 ± 7.657 min	586.5 ± 60.45 mg·min/L	0.015 ± 0.006 L/kg	0.003 L/kg/min	ND	ND	14.25 ± 0.48 min	
Cimicifugoside H-1	Rats	4.05–17.69 pmol/mL	1.1 h	ND	ND	15.7 mL/kg/h	1.86%–6.97%	0.46–1.28 h	ND	Gai et al. (2012)
23-Epi-26-deoxyactein	Rats	90.93–395.7 pmol/mL	2.5 h	ND	ND	0.48 mL/kg/h	26.8%–48.5%	2.00–4.67 h	ND	
Cimigenolxyloside	Rats	407.1–1180 pmol/mL	5.7 h	ND	ND	0.24 mL/kg/h	238%–319%	14.67–19.67 h	ND	
25-O-Acetylcimigenoside	Rats	21.56–45.09 pmol/mL	4.2 h	ND	ND	1.13 mL/kg/h	32.9%–48%	8.08–14.27 h	ND	

Compound	Species	C_{max}	T_{max}	AUC	V_d	CL	F	MRT	$t_{1/2}$	References
Cimicifugoside H-1(Cim A)	Rats	1219 ± 153.3 pmol/mL	1.13 ± 0.53 h	484.9 ± 203.4 pmol·h/mL	26.7 ± 21.1 mL/kg	15.7 ± 7.37 mL/kg/h	ND	ND	0.39 ± 0.16 h	
23-Epi-26-deoxyactein(Cim B)	Rats	9106 ± 692.9 pmol/mL	2.50 ± 0.53 h	17100 ± 4183 pmol·h/mL	1.77 ± 0.79 mL/kg	0.48 ± 0.14 mL/kg/h	ND	ND	2.54 ± 0.68 h	
Cimigenolxyloside (Cim C)	Rats	6824 ± 2267 pmol/mL	5.69 ± 1.29 h	31820 ± 1741 pmol·h/mL	1.98 ± 0.44 mL/kg	0.24 ± 0.01 mL/kg/h	ND	ND	9.74 ± 1.07 h	
25-O-Acetylcimigenoside (Cim D)	Rats	4682 ± 943 pmol/mL	4.17 ± 0.34 h	7924 ± 3026 pmol·h/mL	6.75 ± 2.25 mL/kg	1.13 ± 0.42 mL/kg/h	ND	ND	2.86 ± 0.38 h	
Nigella A	Rats	ND	2.138 ± 1.088 h, 2.175 ± 0.82 h, 2.838 ± 1.616 h (10, 20, 30 mg/kg iv)	38.77 ± 5.3, 77.6 ± 18.9, 128.1 ± 48.6 µg·h/mL (10, 20, 30 mg/kg iv)	0.8 ± 0.382, 0.838 ± 0.345, 1.051 ± 0.6 L/kg (10, 20, 30 mg/kg iv)	0.263 ± 0.038, 0.27 ± 0.058, 0.261 ± 0.088 L/h/kg (10, 20, 30 mg/kg iv)	ND	ND	1.169 ± 0.235 h, 1.191 ± 0.347 h, 1.093 ± 0.28 h (10, 20, 30 mg/kg iv)	Hu et al. (2014) 156
Raddeanin A	Rats	28.8 ± 2.6 (iv)/2.5 ± 0.7 (ip) µg/mL	2.6 ± 0.4 (iv)/2 ± 0.5 (ip) h	25.6 ± 7.6 (iv)/15.3 ± 5.4 (ip) µg·h/mL	0.11 ± 0.01 (iv)/0.15 ± 0.03 (ip) L/kg	31.8 ± 10.4 (iv)/55.3 ± 17.3 (ip) mL/kg/h	ND	2 h(ip)	1.9 ± 0.5 (iv)/4.4 ± 3.6 (ip) h	Luan et al. (2013)
Raddeanin A	Rats	11 ± 1.87 µg/L (po)	5.88 ± 3.24 (po)/7.12 ± 1.07 (iv) h	0.15 ± 0.058 (po)/42.9 ± 12.9 (iv) mg·h/L	55 ± 12.8 (po)/0.257 ± 0.091 (iv) L/kg	7.74 ± 3.69 (po)/0.025 ± 0.008 (iv) L/h/kg	0.30%	5 ± 2 h (po)	6.97 ± 2.64 (po)/7.96 ± 1.35 (iv)	Liu et al. (2013b)
Clematichinenoside AR	Rats	59.73 ± 25.6 (8 mg/kg) 197.57 ± 61.8 (32 mg/kg) ng/mL	4.10 ± 2.36 (8 mg/kg) 3.50 ± 1.94 (32 mg/kg) h	358.17 ± 135.8 (8 mg/kg) 1041.57 ± 322.78 (32 mg/kg) ng·h/mL	ND	ND	ND	3.83 ± 2.07 (8 mg/kg) 1.83 ± 0.75 (32 mg/kg) h	4.54 ± 1.31 (8 mg/kg) 6.00 ± 1.77 (32 mg/kg)h	Wang et al. (2012) 157
Pulchinenoside B3	Rats	338 ± 93.7(po)/120 ± 28 (iv) ng/mL	12.8 ± 8.7 (po)/0.432 ± 0.3 (iv) h	731 ± 328 (po)/31.5 ± 7.4 (iv) µg·h/L	ND	ND	1.16%	0.33 ± 0.1 h (po)	18.8 ± 9.4 h (iv)	Liu et al. (2013a)
Pulchinenoside BD	Rats	36.6 ± 24.8 (po)/16.1 ± 6.3 (iv) ng/mL	18.5 ± 9.6 (po)/2.23 ± 0.5 (iv) h	172 ± 72.5 (po)/7.36 ± 1.4 (iv)	ND	ND	1.17%	0.367 ± 0.1 h (po)	27.3 ± 10.2 h (iv)	
Pulchinenoside B7	Rats	103 ± 104 (po)/73.9 ± 21.7 (iv) ng/mL	16.3 ± 10.0 (po)/0.565 ± 0.2 (iv) h	473 ± 258.5 (po)/42.7 ± 12.1 (iv)	ND	ND	0.55%	0.5 ± 0.2 h (po)	24.7 ± 12.6 h (iv)	
Pulchinenoside B10	Rats	48.2 ± 24.7 (po)/19 ± 6.7 (iv) ng/mL	7.03 ± 2.1 (po)/0.346 ± 0.1 (iv) h	191 ± 68.2(po)/10.0 ± 2.6 (iv)	ND	ND	0.96%	0.5 ± 0.2 h (po)	19.1 ± 6.5 h (iv)	
Pulchinenoside B11	Rats	30.2 ± 12.8 (po)/15.3 ± 6.8 (iv) ng/mL	12.5 ± 8.3 (po)/0.34 ± 0.2 (iv) h	174 ± 107.3 (po)/3.49 ± 1.3 (iv)	ND	ND	2.50%	0.5 ± 0.2 h (po)	21.5 ± 11.3 h (iv)	
Hederacolchiside E	Rats	0.07, 0.13, 0.36 µg/mL (oral 100, 200, 400 mg/kg)	31.1 ± 37.2/19.0 ± 1 8.5/28.9 ± 19.9 h (oral 100, 200, 400 mg/kg)	0.56 ± 0.1/1.27 ± 0.27/6.4 6 ± 4.1 µg·h/mL (oral 100, 200, 400 mg/kg)	ND	ND	ND	0.38 ± 0.14/5.69 ± 4.13/11.5 ± 9.1 h (oral 100, 200, 400 mg/kg)	9.46 ± 0.61/10.1 ± 0.41/16.1 ± 2.3 h (oral 100, 200, 400 mg/kg)	Yoo et al. (2008)
Pulsatilla saponin D	Rats	179.6 ± 35.4(po)/2458.7 ± 380.3 (iv) ng/mL	758.8 ± 65.6(po)/120.5 ± 49.3 (iv) min	3663 ± 890(po)/10768.8 ± 1837.8 (iv) ng·min/mL	ND	9.54 ± 1.8 mL/mg/min (iv)	2.83%	20 ± 13.5(po)/5 (iv) min	431.4 ± 54(po)/78.41 ± 13 (iv) min	Ouyang et al. (2015)

AC, Aconitine; BAC, benzoylaconine; BHA, benzoylhypaconine; BMA, benzoylmesaconine; HA, hypaconitine; ip, intraperitoneal; iv, intravenous; MA, mesaconitine; ND, not determined; po, oral.

of high-dose BBR was lowered, which was negatively correlated with the dose of *Tetradium*. In contrast, both acceleration of the metabolism of high-dose palmatine and retardation of the metabolism of low-dose palmatine were observed, and the metabolic rates under these two conditions were negatively correlated with the doses of *Tetradium* and *Paeonia*. It seems that Wuji pill has a balancing effect. Additionally, different proportions of the three herbs are involved in differential metabolism rates of BBR and palmatine.

The anaerobic intestinal bacteria transformed AC into various lipoaconitines (Wang et al., 2013), such as 8-O-oleoylbenzoylaconine and 8-O-palmitoylbenzoylaconine, via esterification, which is in favor of bioavailability. *Pinelliae Rhizoma, Fritillariae thunbergii* and *Fritillariae Cirrhosae Bulbus* inhibited the intestinal bacterial biotransformation of diester-alkaloids (Xin et al., 2015), implying that these three Chinese herbal medicines increased the toxicity of Wu-Tou-Tang. The scientific connotation of TCM's "18 incompatible medicaments" theory from the point of intestinal bacterial metabolism might be valid.

4.4.1.2 Saponin

Many types of reactions of phytochemicals, such as hydrolysis, oxidation, reduction, isomerization, nitrogen-oxygen exchange, and polymerization, can be catalyzed by gut microbiota. The sample of PSD incubated in rat intestinal microflora was analyzed by ultraperformance liquid chromatography/hybrid triple quadrupole linear ion trap mass spectrometry (UPLC-Q-trap-MS) (Ouyang et al. 2014a). Seven metabolites were identified, including hederagenin 3-O-α-L-rhamnopyranosyl-(1→2)-α-L-arabinosyl, hederagenin 3-O-α-L-glucopyranosyl-(1→4)- α-L-arabinosyl, hederagenin, hydroxylated PSD, methylated PSD, and dehydrogenated PSD.

Gut microflora was anaerobically incubated with anemoside B4 at 37°C for 24, 48, 72, and 96 h, respectively (Wan et al., 2017). Ten metabolites were detected and identified by UPLC/TOF-MS, including the products of oxygenation and deglycosylation reactions. Microflora in the large intestine had more comprehensive metabolic pathways, possibly due to the more diverse bacteria and greater number of them.

Acteoside was anaerobically incubated with rat intestinal flora in vitro (Zhang et al., 2016a). Its metabolites were 3,4-dihydroxyphenyl acid; caffeic acid; and 3-(3'-hydroxyphenyl) propionic acid.

4.4.2 Cytochrome P450s

4.4.2.1 Substrate

Cytochrome P450s (CYPs) are responsible for various types of metabolic reactions of Ranunculaceae compounds, not only hydroxylation. Recently, more Ranunculaceae compounds are found to be substrates, inhibitors, or inducers of CYPs (Fig. 4.1; Tables 4.2–4.4). CYP3A in human intestine microsomes catalyzes phase I metabolism of diester diterpene alkaloids (DDAs) and monoester-diterpene alkaloids (MDAs), and the product has increased polarity (Zhang et al., 2017). The intestinal metabolic profiles of DDAs are more various than those of MDAs. No glucuronide metabolites are detected, similar to the human liver microsome (HLM) metabolism. Eight new metabolites from DDAs have been identified in microsome incubations by UPLC-Q-TOF-HRMS/MS.

TABLE 4.2 Eaxmples of Ranunculaceae Compounds as the Substrates of DMEs/Transporters in Human and Animal Studies

Compound type	Herbal source	Phytochemicals	In vitro model	DME/transporter	Metabolic reaction	Km (µM)	Vmax (pmol/min/mg protein)	Ref.
Diterpenoid alkaloid	Aconitum	Aconitine	HLMs/rhCYP	3A4/5, 2D6	N-deethylation	ND	ND	Tang et al. (2011)
				3A5	Dehydrogenation	ND	ND	
				3A5, 2D6	Demethylation, hydroxylation	ND	ND	
		Hypaconitine	HLMs/rhCYP	3A4/5, 2C19, 2D6, 2E1	Demethylation, dehydrogenation, hydroxylation, didemethylation	ND	ND	Ye et al. (2011b)
		Mesaconitine	HLMs	3A4/5	Demethylation	ND	ND	Ye et al. (2011a)
		Benzoylaconine	HLMs/rhCYP	3A4/5	Dehydrogenation	ND	ND	Ye et al. (2013a)
				3A4/5, 2D6	Demethylation	ND	ND	
				3A4	Didemethylation or Deethylation	ND	ND	
		Benzoylhypaconine	HLMs/rhCYP	3A4/5	Dehydrogenation, demethylation, hydroxylation, didehydrogenation	ND	ND	
		Benzoylmesaconine	HLMs/rhCYP	3A4/5	Dehydrogenation, hydroxylation	ND	ND	
				3A4/5, 2C8	Demethylation	ND	ND	
		Mesaconitine	RLMs	3A, 2C, 2D	Demethylation, deacetylation, dehydrogenation, hydroxylation	ND	ND	Bi et al. (2013)
	Aconitum bulleyanum	Bulleyaconitine A	RLMs	3A, 2C, 2D6, 2E1, 1A2	Demethylation, deacetylation, dehydrogenation, deacetylation, hydroxylation	ND	ND	Bi et al. (2015)

TABLE 4.2 Eaxmples of Ranunculaceae Compounds as the Substrates of DMEs/Transporters in Human and Animal Studies (*cont.*)

Compound type	Herbal source	Phytochemicals	In vitro model	DME/transporter	Metabolic reaction	Km (μM)	Vmax (pmol/min/mg protein)	Ref.
Isoquinoline alkaloid	*Coptis*	Berberine	rhCYP	1A2	10-Demethylation	100 ± 8.74	5.4 ± 0.094 pmol/min/nmol CYP	Li et al. (2011a,b) 160
				2D6	10-Demethylation	31.9 ± 1.23	2.7 ± 0.0052 pmol/min/nmol CYP	
				3A4	10-Demethylation	27.8 ± 0.79	0.024 ± 0.0021 pmol/min/nmol CYP	
			rhCYP	1A2	Demethylenation	125.8 ± 3.57	4.8 ± 0.54 pmol/min/nmol CYP	
				2D6	Demethylenation	12.0 ± 0.68	2.9 ± 0.19 pmol/min/nmol CYP	
				3A4	Demethylenation	27.1 ± 1.03	0.59 ± 0.012 pmol/min/nmol CYP	
		Berberine	HLMs	2D6, 1A2	O-demethylation	2.69 nmol/mL	1.51 nmol/mg/h	Chen et al. (2013c) 40
		Berberine	RLMs	CYP	ND	0.243 ± 0.004 mmol/L	1.714 ± 0.029 μmol/min/g	Li et al. (2014)
		Coptisine	RLMs	CYP	ND	1.082 ± 0.043 mmol/L	8.976 ± 0.351 μmol/min/g	
		Epiberberine	RLMs	CYP	ND	0.611 ± 0.003 mmol/L	1.383 ± 0.006 μmol/min/g	
		Palmatine	RLMs	CYP	ND	0.681 ± 0.002 mmol/L	1.101 ± 0.002 μmol/min/g	
		Jatrorrhizine	RLMs	CYP	ND	1.244 ± 0.051 mmol/L	7.694 ± 0.32 μmol/min/g	
		Jatrorrhizine	HLMs/rhCYP	1A2	Demethylation	191.1 ± 22.3 (HLM),113 ± 13 (rCYP)	106.7 ± 3.9 (HLM), 1.14 ± 0.04 (pmol/min/pmol rCYP)	Zhou et al. (2013)
		Jatrorrhizine	RLMs	3A1/2, 2D2	Demethylation	55.2 ± 2.9	136.7 ± 1.5	Shi et al. (2012)

Substrate	System	Enzyme	Reaction			Reference
Demethylated jatrorrhizine	HLMs	UGT1A1, 1A3, 1A7, 1A8, 1A9, 1A10	Glucuronidation	281.4 ± 41.4	302.2 ± 17.5	Zhou et al. (2013)
	rhUGT	UGT1A1	Glucuronidation	230.7 ± 12.9	666.1 ± 13.5	
		UGT1A3	Glucuronidation	501.7 ± 40.4	280.3 ± 11.2	
		UGT1A7	Glucuronidation	509.3 ± 105.6	115.5 ± 11.9	
		UGT1A8	Glucuronidation	252.3 ± 22.2	195.2 ± 6.5	
		UGT1A9	Glucuronidation	213.7 ± 19.8	39.7 ± 1.3	
		UGT1A10	Glucuronidation	165.9 ± 6.1	757.6 ± 8.9	
Demethylated jatrorrhizine	RLMs	UGT1A1, 1A3	Glucuronidation	144.4 ± 8.9	1105 ± 18.9	Shi et al. (2012)
Berberine	RLMs	3A1/2	Demethylation	26.8	1.74 μmol/mg/min	Liu et al. (2009a,b)
Berberine	RLMs	2B	Demethylenation	60.1	1.25 μmol/mg/min	
Demethylated berberine	RLMs	UGT1A1, 2B1	Glucuronidation	72.8	0.61 μmol/mg/min	
Demethylenated berberine	RLMs	UGT1A1, 2B1	Glucuronidation	10.6	4.89 μmol/mg/min	
Epiberberine	RLMs	CYP	Demethylation, hydrogenation	ND	ND	Yang et al. (2014)
		UGT	Glucuronidation	ND	ND	
Berberine	Transfected MDCK cells	OCT2	NA	1.01 ± 0.17	92.7 ± 3.3	Sun et al. (2014)
	Transfected MDCK cells	OCT3	NA	2.17 ± 0.2	44.7 ± 1.2	

(Continued)

TABLE 4.2 Eaxmples of Ranunculaceae Compounds as the Substrates of DMEs/Transporters in Human and Animal Studies (*cont.*)

Compound type	Herbal source	Phytochemicals	In vitro model	DME/transporter	Metabolic reaction	Km (µM)	Vmax (pmol/min/mg protein)	Ref.
			Transfected MDCKII cells	OCT1	NA	14.8	685 ± 124	Nies et al. (2008)
			Transfected MDCKII cells	OCT2	NA	4.4	194 ± 25	
			Transfected MDCKII cells	P-gp	NA	ND	ND	
Triterpene saponin	Pulsatilla chinensis	Pulsatilla saponin D	SD rats	3A, UGT, SULT	Isomerization, deglycosylation, deoxidation, dehydrogenation, oxidation, methylation, glucuronidation, demethylation, sulfation, dehydroxylation	ND	ND	Ouyang et al. (2014b)
	Pulsatilla, Clematis, Anemone, Nigella	α-Hederin	SD rats	CYP, UGT	Isomerization, glucuronidation, demethylation, hydrogenation	ND	ND	Liang et al. (2014)

NA, Not applicable; *ND*, not determined.

TABLE 4.3 Examples of Ranunculaceae Compounds as the Inhibitors of DMEs/Transporters in Human and Animal Studies

Compound type	Herbal source	Herbal medicine/ phytochemical	Ezyme source	DME/trans- porter	IC50/Ki	Mode of inhibition	Ref.
Alkaloid	Goldenseal	Berberine	HLM	2E1	K_i 18 µM	Mixed type	Raner et al. (2007)
		Hydrastine	HLM	2E1	K_i 2.8 µM	Mixed type	
		Canadine	HLM	2E1	K_i 17 µM	Mixed type	
		Mixture	16 Healthy volunteers	3A	ND	ND	Gurley et al. (2008a)
		Hydrastine, berberine	16 Healthy volunteers	2D6	ND	ND	Gurley et al. (2008b)
	Goldenseal	Mixture	RLM	1A2	IC_{50} 15.65 µg/mL	ND	Yamaura et al. (2011)
		Mixture	RLM	2D6	IC_{50} 7.35 µg/mL	ND	
		Mixture	RLM	2E1	IC_{50} 4.32 µg/mL	ND	
		Mixture	RLM	3A	IC_{50} 52.07 µg/mL	ND	
	Coptis, goldenseal	Berberine	Eight-week-old male C57BL/6 mice	3a11, 3a25	ND	Gene expression	Guo et al. (2011)
			Eight-week-old male C57BL/6 mice	2d22	ND	ND	
		Berberine	Primary mouse hepatocyte	Cyp1a1, 1a2, 2e1, 3A4(Cyp3a11), Cyp4a10, 4a14	ND	Gene expression	Chatuphon- prasert et al. (2012)
		Berberine	Streptozotocin- induced diabetic mice	2e1	ND	Gene expression	
				3a11, 4a10, 4a14	ND	Gene expression	
		Berberine	Healthy male subjects	2D6	ND	ND	Guo et al. (2012)
		Berberine	Healthy male subjects	2C9	ND	ND	

(Continued)

TABLE 4.3 Examples of Ranunculaceae Compounds as the Inhibitors of DMEs/Transporters in Human and Animal Studies (*cont.*)

Compound type	Herbal source	Herbal medicine/ phytochemical	Enzyme source	DME/trans-porter	IC50/Ki	Mode of inhibition	Ref.
		Berberine	Healthy male subjects	3A4	ND	ND	Chen et al. (2013c)
		Berberine	HLM	2D6, 2E1	ND	ND	
		Berberine	Transfected MDCK cells	OCT2	IC$_{50}$ 0.371 (5HT) 0.438 (NE) 8.0 (MPP$^+$) μM	ND	Sun et al. (2014)
			Transfected MDCK cells	OCT3	IC$_{50}$ 0.225 (5HT) 0.566 (NE) 9.9 (MPP$^+$) μM	ND	
			Transfected MDCKII cells	OCT1	ND	ND	Nies et al. (2008)
			Transfected MDCKII cells	OCT2	ND	ND	
		Berberine	Rats	P-gp	ND	Possibly competitive	Qiu et al. (2009)
		Berberine	E.coli membrane-expressing human CYP1	1A1.1, 1B1.1	K_i 44 ± 16 nm / IC50 94 ± 8 nm (1B1.1), K_i 679 ± 106 nm / IC$_{50}$ 1.38 ± 0.12 μM (1A1.1)	Noncompetitive	Lo et al. (2013)
		Berberine	E.coli membrane expressing human CYP1	1B1.1	IC$_{50}$33.9 ± 1.7 μM	ND	
		Berberine	E.coli membrane expressing human CYP1	1B1.3 (V432L), 1B1.4 (N453S)	IC$_{50}$ 71 ± 4 nm (1B1.3), 65 ± 7 nm (1B1.4)	ND	
		Berberine	E.coli membrane-expressing human CYP1	1A2	IC$_{50}$ > 60 μM (WT), 45.8 ± 7.2 μM (T223 N)	ND	

Class	Herb	Compound	Enzyme source	CYP	Kinetic parameters	Inhibition type	References
		Palmatine	E.coli membrane-expressing human CYP1	1A1.1, 1B1.1	K_i 12.77 ± 1.33 μM/IC$_{50}$ 8.71 ± 0.55 μM (1A1.1), K_i 5.64 ± 0.41 μM/IC$_{50}$ 37.2 ± 4.2 μM (1B1.1)	Mixed (1A1.1), competitive (1B1.1)	Yang et al. (2014)
		Jatrorrhizine	E. coli membrane-expressing human CYP1	1A1.1, 1B1.1	K_i 4.98 ± 0.39 μM/IC$_{50}$ 2.17 ± 0.08 μM (1A1.1), K_i 0.47 ± 0.05 μM/IC$_{50}$ 1.71 ± 0.08 μM (1B1.1)	Mixed type	
	Coptis	Epiberberine	RLM	2D6	IC$_{50}$ 35.22 μM	ND	Wei et al. (2013)
	Coptis	Mixture	RLM	1A2, 2D6	ND	ND	Li et al. (2011a,b)
	Black cohosh	Protopine	HLM	2D6	K_i 78 nm	Competitive	
		Allocryptopine	HLM	2D6	K_i 122 nm	Competitive	
Alkaloid	Aconitum	Radix Aconite preparata	RLM	2C, 2D	IC$_{50}$ 19.40 μg/L(2C), 21.60 μg/L(2D)	ND	Bi et al. (2014)
Alkaloid		Radix Aconite	RLM	3A, 2C, 2D	IC$_{50}$ 61.16 μg/L(3A), 59.34 μg/L(2C), 38.65 μg/L(2D)	ND	
Triterpene saponin	Black cohosh	Mixture	HLM	2D6, 3A4, FMO	IC$_{50}$ 0.52 mg/mL	Noncompetitive	Gorman et al. (2013)
	Black cohosh	Mixture	HLM	Carboxylesterase	K_i 1.62 mg/mL, IC$_{50}$ 1.69 mg/mL (high K_m), 4.74 mg/mL (low K_m)	Competitive	
		Mixture	Rhcyp	2C19	IC$_{50}$ 0.37 μg/mL	ND	Ho et al. (2011)

(Continued)

TABLE 4.3 Examples of Ranunculaceae Compounds as the Inhibitors of DMEs/Transporters in Human and Animal Studies (*Cont.*)

Compound type	Herbal source	Herbal medicine/phytochemical	Enzyme source	DME/trans-porter	IC50/Ki	Mode of inhibition	Ref.
		23-*O*-acetylsheng-manol-3-α-L-arabinopyranoside (23R)	HLM	3A4	IC$_{50}$ 2.3 ± 0.2 μM	Competitive	Li et al. (2011a,b)
		Cimiracemoside K	HLM	3A4	IC$_{50}$ 3.8 ± 0.5 μM	Competitive	
		23-*O*-acetylsheng-manol-3-β-D-xylopyranoside	HLM	3A4	IC$_{50}$ 2.7 ± 0.2 μM/K_i 1.7 μM	Competitive	
		Cimiracemoside O	HLM	3A4	IC$_{50}$ 5.1 ± 0.4 μM	Competitive	
		Cimiracemoside L	HLM	3A4	IC$_{50}$ 2.4 ± 0.2 μM/K_i 1.1 μM	Competitive	
		Cimicifugoside M	HLM	3A4	IC$_{50}$ 4.3 ± 0.3 μM	Competitive	
		7-β-hydroxycimigenol aglycone	HLM	3A4	64.9% ± 1.1% inhibition (10 μM)	Competitive	
		25-*O*-acetyl-7-β-hydroxycimi-genol-3-*O*-β-xylopyranoside	HLM	3A4	61.8% ± 0.2% inhibition (10 μM)	Competitive	
Phenolic acid	Black cohosh	Fukinolic acid	Rhcyp	1A2, 3A4, 2C9, 2D6	IC$_{50}$ 1.8 (1A2) 7.2 (3A4) 7.1 (2C9) 5.4 (2D6) μM	ND	Huang et al. (2010)
		Cimicifugic acid A	Rhcyp	1A2, 3A4, 2C9, 2D6	IC$_{50}$ 7.2 (1A2) 9.7 (3A4) 8.3 (2C9) 9.0 (2D6) μM	ND	
		Cimicifugic acid B	Rhcyp	1A2, 3A4, 2C9, 2D6	IC$_{50}$ 7.35 (1A2) 9.8 (3A4) 12.5 (2C9) 12.6 (2D6) μM	ND	
Others	*Nigella sativa*	Mixture	RLM	2C11 (human 2C9)	ND	Gene expression	Korashy et al. (2015)

ND, Not determined.

TABLE 4.4 Examples of Ranunculaceae Compounds as the Inducers of CYPs/Transporters in Human and Animal Studies

Compound type	Herbal sources	Phytochemicals	Experimental model	Target enzyme	Transactivation mechanism	References
Alkaloid	*Coptis, Hydrastis*	Berberine	HepG2	3A4	PXR	Liu et al. (2011)
		Berberine	Mice	Cyp1a2	Gene expression	Guo et al. (2011)
		Berberine	Primary mouse hepatocyte	Cyp2b9, 2b10	Increase the gene expression	Chatuphonprasert et al. (2012)
	Coptis alkaloid extract	Alkaloid	High lipid diet-induced hyper-lipidemic rats	7A1	Upregulate gene expression of PPARα, down-modulation of the FXR mRNA expression	Cao et al. (2012)
	Aconiti Laterlis Radix extract	Mixture	Rat	3A4	ND	Zhang et al. (2012)
	CC + *Scutellaria baicalensis* (HQ)	Mixture	RLM	2D6, 3A4	ND	Wei et al. (2013)
	CC + cinnamon	Mixture	Db/db mice	GLUT4	AMP-activated protein kinase (AMPK)	Hu et al. (2013)
Multiple	*Nigella sativa*	Mixture	Beagle dogs	3A4	ND	Al-Mohizea et al. (2015)
	Black cohosh	Mixture	Mice	2B, 3A	ND	Yokotani et al. (2013)
	Black cohosh	Mixture	Mice	3a11	mouse PXR	Pang et al. (2011)

CC, *Coptis chinensis*; *FXR*, farnesoid X receptor; *GLUT4*, solute carrier family 2 glucose transporter type 4; *ND*, not determined; *PPAR*, peroxisome proliferator-activated receptor; *PXR*, pregnane X receptor.

HLMs and recombinant CYPs were used to explore the metabolic process of AC (Tang et al., 2011). Six CYP-transformed metabolites were found. The CYP3A inhibitor strongly inhibited AC metabolism; the CYP2C9, 2C8, and 2D6 inhibitors mildly inhibited AC transformation; while the 2C19, 1A2, and 2E1 inhibitors failed to inhibit AC metabolism. The recombinant CYP3A5 and 2D6 were responsible for the hydroxylation and di-demethylation of AC, 3A4/5 was responsible for the dehydrogenation, while 3A4/5 and 2D6 transformed AC to demethyl-ACand N-deethyl-AC.

In male HLMs, MA underwent demethylation, dehydrogenation, hydroxylation, and demethylation-dehydrogenation (Ye et al. 2011a). Recombinant CYP3A4/5 was responsible for the generation of MA metabolites, while the contributions of 2C8, 2C9, and 2D6 were marginal. The metabolic reactions of MA in rat liver microsomes (RLMs) included demethylation, deacetylation, dehydrogenation, and hydroxylation (Bi et al., 2013). CYP3A metabolized MA, while 2C and 2D were also involved in metabolic reactions of MA, and CYP1A2 and 2E1 did not have any contribution to MA metabolism in RLMs.

Bulleyaconitine A (BLA) of *Aconitum bulleyanum* is a traditional antiinflammatory drug in China and Southeast Asia. In RLMs, BLA underwent deacetylation, demethylation, hydroxylation, and dehydrogenation-deacetylation (Bi et al., 2015). CYP3A and 2C transformed BLA, while 2D6 and 2E1 also had contributions. In contrast, 1A2 was only involved in one metabolite (M11, N-deethyl-BLA). C19 diterpenoid alkaloid crassicauline A is eliminated in rats predominantly by metabolism under toxic dosage, and the hydroxylation at C-15 might be a potential bioactivation pathway, producing toxic deoxyjesaconitine in both rats and humans (Fan et al., 2017).

Hydrolyzing diester-diterpene alkaloids into monoester-diterpene ones mitigates *Aconitum* poisonousness. The processed *Aconitum* (Zhi Fu Zi in Chinese) contains MDAs, for example, benzoylaconine (BAC), benzoylhypaconine (BHA), and benzoylmesaconine (BMA). CYPs are an integral part of the defense mechanisms that minimize the harmful effects of toxic compounds. Seven, eight, and nine metabolites were detected by high-resolution mass spectrometry (MS) for BAC, BMA, and BHA, respectively (Ye et al., 2013b). Dehydrogenation, demethylation, hydroxylation, demethylation-dehydrogenation, and didemethylation were the core metabolic pathways, and the toxicity of metabolites was significantly lower than that of AC, MA, and HA metabolites. CYP3A4/5 is responsible for the metabolism of BAC, BMA, and BHA.

CYP2D6 and 1A2 were responsible for 75.25% and 23.32% of the BBR metabolite demethyleneberberine (M1) and 46.89% and 8.67% of M2 (thalifendine or berberrubine) (Chen et al., 2013a). The major metabolic pathway of BBR in pooled HLMs is O-demethylation.

As a promising gastric prokinetic drug, jatrorrhizine is a major metabolite of BBR after oral administration (Zhou et al., 2014). In HLMs, demethyleneberberine (demethylated product) was identified as the phase I metabolite of jatrorrhizine (Zhou et al., 2013). The enzyme kinetics for demethylation was described by the Michaelis-Menten equation. CYP1A2 catalyzed demethylation, which was inhibited significantly by furafylline.

After i.v. administration in rats, blood jatrorrhizine concentrations displayed a biphasic decay, dose-independent elimination, and a relatively large distribution volume (V_d) (Shi et al., 2012). Similar to humans, the demethylated metabolites were predominant in rats. The enzyme kinetics in RLMs follows the Michaelis-Menten equation. CYP3A1/2 and 2D2 catalyzed demethylation in RLMs.

4.4.2.2 *Inhibitor and Inducer*

Protoberberine alkaloids, such as BBR, palmatine, and jatrorrhizine, were inhibitors of CYP1A1*1 and 1B1*1, which catalyzed 7-ethoxyresorufin O-deethylation (EROD) (Lo et al., 2013), and 1A2*1 was not inhibited. BBR noncompetitively inhibited EROD activities, while the inhibition of palmatine and jatrorrhizine was either competitive or mixed type. Compared to other protoberberines, BBR was the most potent and selective inhibitor of CYP1B1*1. Compared with BBR, palmatine and jatrorrhizine showed less selectivity between CYP1A1 and 1B1 inhibition. CYP1B1*1 activities toward 7-methoxyresorufin and 7-ethoxycoumarin were strongly constrained by BBR, which only marginally inhibited benzo(a)pyrene hydroxylation. The polymorphic variants, CYP1B1*3 (V432L) and 1B1*4 (N453S), were also strongly inhibited by BBR. A mutation of Asn228 to Thr in CYP1B1*1 abrogated BBR inhibition, while a reversal mutation of Thr223 to Asn in CYP1A2*1 augmented the inhibition. Molecular docking results suggested that Asn228 and Gln332 might be important for the selective inhibition of CYP1B1 by BBR. The hydrogen-bonding interaction between the methoxy group of BBR and Asn228 of CYP1B1*1 is essential. The inhibitory outcomes of BBR on 1B1 activities are substrate dependent.

Frequent administration of BBR inhibited CYP2D6, 2C9, and 3A4 in healthy male subjects (Guo et al., 2012). In contrast, CYP2C19 and 1A2 were not affected. In primary mouse liver cells, BBR concentration dependently inhibited the induction of Cyp1a1, 1a2, 2e1, 3a11, 4a10, and 4a14 gene expression by their archetypal stimulators (Chatuphonprasert et al., 2012). Nonetheless, BBR on its own enhanced the gene expression of Cyp2b9 and 2b10. The hepatic transcript levels of Cyp1a1, 2b9, 2b10, 3a11, 4a10, and 4a14 were upregulated in streptozotocin-induced diabetic mice. Remarkably, BBR (1, 5, or 10 µM) alone suppressed the isoniazid-induced expression of Cyp2e1 (an adverse reaction-related enzyme) and upregulated Cyp3a11, 4a10, and 4a14 transcripts to normal level. BBR modulates the CYPs by either inhibition or augmentation of CYP levels. The capacity of BBR to restore the expression of Cyp2e1, 3a, and 4a to normal levels might be advantageous to diabetic patients. However, an HDI is possible as any BBR-containing formulation would certainly result in prominent CYP3A4 inhibition-based interactions.

BBR is abundant in both *Coptis* and *Hydrastis*. Although hydrastine is present at lower concentrations than BBR in ethanolic *Hydrastis canadensis* extracts, it is a more potent CYP2E1 inhibitor than BBR and another *Hydrastis* alkaloid canadine (Table 4.3) (Raner et al., 2007). Moreover, *Coptis* alkaloid extract, containing BBR, magnoflorine, columbamine, jatrorrhizine, epiberberine, coptisine, and palmatine, dose-dependently upregulated the gene expression of CYP7A1 (cholesterol 7α-hydroxylase) in the livers of hyperlipidemic rats (Cao et al., 2012). Therefore, DDIs should be considered when BBR and relevant alkaloids are administered.

Three phenolic acids of black cohosh (fukinolic acid, cimicifugic acids A and B) sturdily inhibited recombinant human CYP1A2, 2D6, 2C9, and 3A4 (IC_{50} 1.8–12.6 µM) (Huang et al., 2010). In in vitro assays using recombinant human CYPs, black cohosh was found to inhibit CYP2C19 (Ho et al., 2011), which should be tested by appropriate in vivo studies. In mice, black cohosh dose-dependently increased the hepatic weight, total CYP amount, and CYP 2B and 3A activities (Yokotani et al., 2013). However, the induction of Cyp3a11 is hepatic-specific and only mouse pregnane X receptor (PXR) is involved (Pang et al., 2011), not the human PXR. Thus, further in vivo studies on whether the incidence of HDI in patients having black cohosh is mediated by human CYP3A4 are warranted.

The aqueous extract of *Coptis* caused downregulation of rat CYP2E1, 3A9, 2C23, 1A2, 2C70, 2A1, 4A14, and 7A1 and upregulation of rat CYP2D3, 2B2, 2J3, 2D26, 2C7, and 2C6 (Yang et al., 2016), while the other 15 CYPs remained unchanged.

4.4.2.3 Probe

The CYP3A subfamily is the predominant group of hepatic CYPs and participates in the transformation of more drugs, including Ranunculaceae herbal medicine, compared to other CYP subfamilies. Bufalin 5β-hydroxylation was specifically catalyzed by CYP3A4, instead of CYP3A5 and 3A7; therefore, bufalin could be a biotransformation probe substrate (Ge et al., 2013). The well-depicted probe reaction can be used to quantify the genuine catalytic activities of CYP3A4 from different sources toward diterpenoid alkaloids; isoquinoline alkaloids; and other substrates, inhibitors, and/or inducers of Ranunculaceae (Fig. 4.3). The highly selective and specific fluorescent probe substrates of drug metabolizing enzymes (DMEs) and drug transporters can be designed for direct interaction between Chinese materia medica

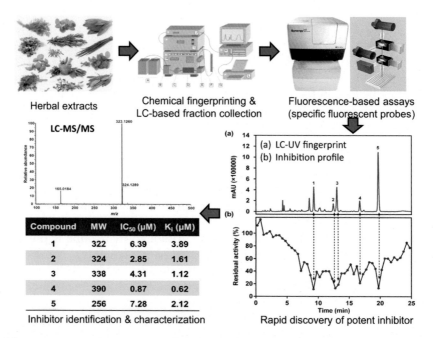

FIGURE 4.3 **Screening and discovery of DME/transporter inhibitors via fluorescent probe.** Ultrafast liquid chromatography (UFLC)-UV fingerprinting combined with the inhibition profile of each fraction can be used to find naturally occurring inhibitors of a given drug-metabolizing enzyme from the crude extract of herbals. Among the reported selective probes of human enzymes, fluorescent probes have attracted increasing attention because of their inherent advantages, such as being highly sensitive, nondestructive, easily conducted, as well as being applicable to high-throughput screening or determination. The fluorescent probe substrate is designed based on the substrate preference and/or favorable metabolic reaction type of DME/transporter. The target DME is able to selectively catalyze the activity-based fluorescent probe and generate the fluorescent product. Harnessing long wavelength fluorescent probes can minimize the background interference of the complex biological system and sample. The preparations of plant tissues/cells can be used as the DME/transporter source, and the rapid quantification of the effects of herbal crude extracts on the target DME/transporter can be fulfilled.

(CMM) extract and its chemical constituents and DME/transporter within the context of high-throughput screening. Screenings of both inhibitors and inducers could be performed based on such a probe.

4.4.3 Herb-Drug Interaction

HDIs are significant safety concerns in clinic. Many literatures have reported that herbs could interact with several clinical narrow therapeutic index drugs, including methotrexate (Liu et al., 2014d; Wang et al., 2014), anticoagulants (e.g., warfarin), immunosuppressant drugs (e.g., tacrolimus and cyclosporin), anti-HIV agents (e.g., indinavir and saquinavir), cardiovascular drugs (e.g., digoxin), and anticancer agents (e.g., docetaxel, irinotecan, and imatinib). Experimental determination of the absorption and disposition properties of herbal medicine, especially TCM constituents, is attracting more research groups worldwide (Wu et al., 2012). Multicomponent herbs are subject to sequential metabolism, concurrent metabolism, and multiple metabolism in vivo. Data on the interaction between Ranunculaceae herbal medicine and DMEs/transporters are accumulating. It should be highlighted that the HDI is a double-edged sword, given that the mild HDI could alleviate the metabolic clearance of the coadministered drugs and increase their AUC and half-life ($T_{1/2}$), which might be good for their in vivo therapeutic effects, especially those with a relatively wide therapeutic window (Meng and Liu, 2014; Wang and Liu, 2014).

Herbal supplements are broadly used in cancer patients, but how they affect the chemotherapy is frequently undisclosed. Black cohosh was a stronger inhibitor than St. John's wort and ginger root extract, for both CYP- and carboxylesterase (CE)-mediated biotransformation of tamoxifen and irinotecan, respectively (Gorman et al., 2013). Eight triterpene glycosides of black cohosh proved to be competitive CYP3A4 inhibitors with IC_{50} values ranging from 2.3 to 5.1 μM (Li et al., 2011a), while the alkaloids protopine and allocryptopine of black cohosh, also abundant in other Ranunculaceae genera (e.g., *Coptis*, *Thalictrum*, and *Aquilegia*), were competitive CYP2D6 inhibitors.

The microsomal CE catalyzes the irinotecan bioactivation to SN-38, a topoisomerase I inhibitor. However, the role of multiple CE isozymes is not known. A highly selective ratiometric fluorescent probe of human CE1 has been developed for in vitro monitoring and cellular imaging (Liu et al., 2014e). Two highly selective fluorescent probes can be used for the detection of hCE2 (Feng et al., 2014a,b). These innovative specific probes are highly valuable for real-time monitoring of hCE activity in complex biological systems and provide novel solutions for HDI studies (Fig. 4.3).

Of note is that BBR-containing Ranunculaceae herbs are involved in HDIs. For instance, goldenseal (*H. canadensis*) sizably inhibited CYP3A (Gurley et al., 2008a). $AUC_{0-\infty}$(107.9 vs. 175.3 ng•h/mL), $T_{1/2}$(2.01 vs. 3.15 h), and C_{max} (50.6 vs. 71.2 ng/mL) of midazolam increased, while total body clearance (CL) of midazolam decreased (1.26 vs. 0.81 L/h/kg). The simultaneous intake of goldenseal and CYP3A substrates may result in noteworthy HDIs. Goldenseal, but not the black cohosh extracts, significantly inhibited (~50%) CYP2D6 activity (Gurley et al., 2008b). Goldenseal inhibited CYP2E1 most potently, followed by 1A2, 2D6, and 3A (Yamaura et al., 2011). Since CYP2E1 metabolizes acetaminophen (APAP) to the highly active intermediate, goldenseal could ameliorate APAP-induced acute liver failure.

Various amounts of BBR did not significantly alter the hepatic function of mice (Guo et al., 2011), and repeated use of the lower doses of BBR for 2 weeks had no influence on the gene expression of more than 20 main Cyps. However, the highest dose of BBR (300 mg/kg) downregulated Cyp3a11 and 3a25 expression by 67.6% and 87.4%, respectively, while Cyp1a2 (for 7-ethoxyresorufin O-dealkylation) mRNA was increased by 43.2%, and Cyp3a11 (for testosterone 6β-hydroxylation) and 2d22 (for dextromethorphan O-demethylation) activities decreased by 67.9% and 32.4%, respectively. The gene expression and enzyme activity of Cyp2a4 (for testosterone15α-hydroxylation), 2b10, and 2c29 (both for testosterone 16β-hydroxylation) were not altered. Lower dose BBR might not result in DDIs. However, high-dose BBR may reduce Cyp activities and cause DDIs.

Compared with ciprofloxacin alone, comedication of BBR (50 mg/kg) and ciprofloxacin significantly decreased C_{max} of ciprofloxacin (Hwang et al., 2012). The pretreatment of BBR (50 mg/kg/day) and BBR-containing Huang-Lian-Jie-Du-Tang (HR; 1.4 g/kg/day) significantly decreased C_{max} and $AUC_{0\to\infty}$ of ciprofloxacin, as compared with the control group. P-gp and OCT (organic cation transporter) could be involved in reduced oral bioavailability of ciprofloxacin by BBR and HR.

Jiao-Tai-Wan (JTW), consisting of CC and *Cinnamomum cassia*, efficiently guarded the pancreatic islet morphology, enhanced the activation of hepatic AMP-activated protein kinase (AMPK), and upregulated the expression of glucose transporter 4 (GLUT4) in white fat and skeletal muscle (Hu et al., 2013). Thus, BBR-involved DDIs might also be mediated by transporter superfamily members.

Nigella-related HDIs are also highlighted in recent studies. For instance, *Nigella sativa* dose-dependently inhibited the gene and enzyme expressions of rat CYP2C11 (Korashy et al., 2015), thus reducing the amount of 4-hyroxy-tolbutamide, a tolbutamide metabolite, in vitro. The inhibitory effect of *Nigella* on rat CYP2C11 was stronger than that of *Trigonella foenum-graecum* and *Ferula asafoetida*, which could result in the undesirable effect of CYP2C11 substrates.

CYP3A4 and to a lesser extent CYP2C9-mediated metabolism of sildenafil could be impacted by *Nigella* (Hyland et al., 2001). Oral administration of *N. sativa* resulted in reduction of $AUC_{0-\infty}$, C_{max}, and $T_{1/2}$ as compared with the control (Al-Mohizea et al., 2015). Concurrent use of *Nigella* alters the PK of sildenafil, which might result in a decrease in sildenafil bioavailability. In rabbits, the concurrent use of *N. sativa* significantly decreased the C_{max} and $AUC_{0-\infty}$ of CsA, a commonly used immunosuppressant (Al-Jenoobi et al., 2013). The aqueous extract of *N. sativa* dose-dependently inhibited sodium-dependent glucose transport through rat jejunum (Meddah et al., 2009). On the contrary, methanol and hexane extracts of *Nigella* seeds enhanced amoxicillin availability in both in vivo and in vitro studies (Ali et al., 2012). *Nigella* might increase intestinal absorption of amoxicillin.

Does AC inhibit/induce CYP3A? In one study, the production of 1-(2-pyrimidinyl) piperazine (PP) and 6'-hydroxybuspirone from the probe substrate buspirone (BP) did not change, suggesting that the rat CYP3A activity was not impacted by the single and repeated use of 0.125 mg/kg AC (Zhu et al., 2013). In RLMs, one-week AC pretreatment did not affect CYP3A protein levels. Therefore, the authors claimed that AC does not inhibit or induce CYP3A in rats and might not lead to CYP3A-associated DDI in the liver. However, in another study, multiple AC exposure (0.125 mg/kg) increased the $AUC_{0-\infty}$ of BP by 110% (Lijun et al., 2014), and the amounts of 1-PP and 6'-OH-BP were increased by 229% and decreased by 95%, respectively. Single/multiple AC exposure did not alter the first-pass (intestinal and hepatic) CYP3A activity when using oral BP as a probe in rats. Nonetheless, whether multiple AC exposure prominently changes the production of BP metabolites warrants further in vivo studies.

Most studies highlighted the impact of herbal medicine on Western drugs, but not vice versa. The possible reasons are: (1) the constituents of herbal medicine are too complicated and their effects are versatile; therefore, it is challenging to select the appropriate PK and PD markers of herbal medicine; and (2) the bioactivity and systemic exposure of the single ingredients of herbal medicine is very often moderate, and there is a lack of rapid and strong potency. In HDI studies, special attention should be given to the effects of dose, regimen, and mode of medication, since empirically more HDIs occur only under high-dose or long-term administration.

To date, information is still lacking for the main CYP and UGT enzymes in the less-studied medicinal plants. Information of P-gp and other drug transporters is also limited (section below, Phase III: Drug Transporter). The role of another phase I DME flavin-containing monooxygenase (Hao et al., 2007; Hao and Xiao, 2011) in HDIs remains elusive. ABC transporter and solute carrier (SLC) (Hao et al., 2013d) superfamilies have many other transporters, besides P-gp, OCT, and GLUT, which await future investigations. To date, there are very few systematic methods for quantitative forecast of the scale and probability of herb-drug interactions. Physiologically based pharmacokinetic (PBPK) modeling could be used to increase prediction correctness of possible HDIs (Brantley et al., 2014), but the premise is that abundant in vitro information for building such quantitative relationships is available in the near future (Azam et al., 2014). The DDI prediction of Ranunculaceae shares the same challenges and problems as that of other CMM. For instance, very often there is more than one inhibitor of the same DME. The integrative effects of the sum of the multiple weak inhibitions are sometimes considerably high. It also should be noted that HDIs are always complex, as each herb contains many ingredients that may simultaneously interact with multiple targets. To date, it is not realistic to predict the potential HDI by IVIVE (in vitro–in vivo extrapolation) and PBPK modeling, as many key parameters, for example, the plasma concentration of each component, the unbound fraction, the inhibitory activity of metabolites, and the half-life of each inhibitor, are absent.

4.4.4 Herb-Herb Interaction: *Aconitum* Related

Contrary to most HDIs, in most circumstances the HHI within a TCM formula is preferred by professionals to achieve synergistic therapeutic effects of different herbs. Each herb in the formula exerts unique and complimentary effects as Jun (monarch drug), Chen (minister), Zuo (adjuvant), or Shi (courier). Different herbal ingredients may regulate either the same or different targets in various pathways (Hao and Xiao, 2014) and consequently work together in anagonistic and synergistic ways. Most HDIs are based on the interactions between herbs and DMEs/drug transporters. Analogously, it is plausible to infer that ingredients in one herb could regulate DMEs and transporters to modulate systemic exposure of ingredients of the other herbs. Many herb-herb interactions (HHIs) are beneficial to minimize the adverse effects of toxic ingredients and enhance pharmacological potency of agents.

4.4.4.1 *Fu Zi*

TCM herb pairs consist of only two herbs and are the simplest form of TCM formula. Fu Zi, commonly used in TCM herb pair/formula, is made from the lateral roots of *Aconitum* plants, as the main root (Chuan Wu in Chinese) is more toxic and brings about more adverse reactions in clinical settings (Jaiswal et al., 2014). Processed Fu Zi (Heishunpian, HSP) has much less toxic diester diterpene alkaloid than raw Fu Zi and is safer in clinical use. Rats were fed with HSP decoction (Tang in Chinese), Zhufu (ZF, HSP + *Atractylodes macrocephala*) decoction, or Gancaofuzi (GF, HSP + *Radix Glycyrrhizae*) decoction, respectively (Xin et al., 2011). UPLC/Q-TOF MS was used to determine HA concentration in the plasma. It was found that the absorption of HA from ZF decoction was lower than that from HSP decoction, while the absorption of HA from GF decoction was higher than that from HSP decoction. Some components in ZF decoction might limit HA absorption, whereas some components in GF decoction may promote HA absorption; therefore, the dissimilar drug efficacy between ZF decoction and GF decoction may stem from the differential absorption of HA into plasma.

GF is a typical acid-base herb pair. In everted gut sac permeability experiments, when GF was used, the permeability of HA was highest in ileum (Zhang et al., 2013a), and its uptake was enhanced by P-gp inhibitors. In situ single-pass gut perfusion suggested the active transport mechanism of AC, HA, and MA, all of which are P-gp substrates. Some alkaloid molecules could bind glycyrrhizic acid or other molecules of Gan Cao during boiling (decoction process). In rats, the first order absorption of the three alkaloids in GF could follow the two-compartment model with lag time. When GF was orally administered, the three toxic diester diterpenoid alkaloids could dissolve slowly in the gut and then absorb into circulation with prolonged mean residence time (MRT) and large absorption amounts, thus shunning dose dumping in single use of Fu Zi. Such an interaction mechanism could also apply to other acid-base herb pairs.

On the other hand, the MRT and AUC of glycyrrhetic acid in rats having GF were 27.6 h and 122.8 µg·h/mL respectively (Gao et al., 2004), which is meaningfully greater than those in rats having glycyrrhizic acid (15.0 h and 40.9 µg·h/mL, respectively), suggesting the increased effect of glycyrrhetic acid on GF medication, as well as the complex interaction between *Aconitum* alkaloids and other herbal medicines.

Rats were orally administered with either decoction of *Radix Aconiti Laterlis* (1.5 g/kg; Fu Zi), blend decoction of Fu Zi and *Radix Glycyrrhizae* (Gan Cao) that decocted separately, or decoction of Fu Zi and *Radix Glycyrrhizae* that decocted together (Shen et al., 2011). C_{max} and AUC of AC, MA, and HA were decreased on combined use of two herbal medicines. With MRT, $T_{1/2}$ was prolonged but T_{max} did not change significantly when two herbs were combined. The effect of herb pair was more prominent when two herbs were decocted together than when decocted separately. In rats, single dose *Radix Aconiti Laterlis* extract (0.5 g/kg) decreased buspirone hydrochloride (CYP3A4 specific substrate) $AUC_{0-2\,h}$ by 47% (Zhang et al., 2012), and increased CL by 22%. Compared to the saline treatment, the combined use of Fuzi and *Radix Glycyrrhizae* extract has no effect on CYP3A4. Fuzi extract might induce CYP3A4 while *Radix Glycyrrhizae* extract abolished this effect in vivo.

Shen-fu (Fu Zi + *Panax ginseng*) injectable powder is powerful in the treatment of heart failure and cerebral infarction. *P. ginseng* proved to decrease the systemic exposure of the toxic diester diterpene alkaloids of Fu Zi (Zhang et al., 2008). *Herba Ephedrae* (Ma Huang in Chinese) and *Radix Aconiti Lateralis* are combined in TCM for treating colds and rheumatic arthralgia. The plasma alkaloids of Ma Huang + Fu Zi, except methylephedrine (from Ma Huang), BMA, and BHA, showed slower elimination (longer MRT and $T_{1/2}$) than those of single herbs (Song et al., 2015), although the C_{max} and AUC values were smaller. Buildup of alkaloids might result from repeated drug ingestion. Drug monitoring may be warranted for the innocuous intake of the Mahuang-Fuzi decoction.

In TCM, *Rhizoma Zingiberis* (Gan Jiang in Chinese) is used with Fuzi as the herb pair to reduce deadliness and improve efficacy. Compared with the Fuzi group, both $T_{1/2}$ and AUC_{0-t} of the toxic AC and HA declined with statistical significance (Peng et al., 2013), while $T_{1/2}$, AUC_{0-t}, and C_{max} of BAC and BHA increased in Fuzi-Ganjiang group, which suggests that Gan Jiang could help eliminate AC and HA and augment the uptake of the less toxic MDAs. *Glycyrrhiza uralensis* (GU, *Radix Glycyrrhizae*) is traditionally combined with these two herbs in the formula Sini decoction (SND). Compared with Fuzi group, T_{max}, C_{max}, k (elimination rate constant), AUC_{0-24}, and $AUC_{0-\infty}$ of HA decreased in SND groups (Zhang et al., 2015a), and $T_{1/2}$ and MRT increased. Compared with those in Fuzi group, C_{max}, AUC_{0-24}, and CL of MDAs increased and $T_{1/2}$ declined in Fuzi + Ganjiang, Fuzi + GU, and SND groups (Zhang et al., 2015b). Gan Jiang (minister) and GU (adjuvant/courier) of SND help maintain toxic alkaloids within a moderate range and thus maximize the potency of Fu Zi (king).

Fuzi diterpene alkaloids' PK were greatly influenced by Dahuang, which may account for the compatibility mechanism of effect enhancing and toxicity reducing (Li et al., 2017). Dahuang Fuzi decoction (DFD) consists of *Radixet Rhizoma Rhei* (Da Huang), Fu Zi, and *Radixet Rhizoma Asari* (Xi Xin). After intake, the AUC_{0-t}, $AUC_{0-\infty}$ and C_{max} of MDAs and HA of DFD were much less than those of the Fuzi extract group (Liu et al., 2014a). V_d and CL values of BHA, BMA, BAC, and HA increased. $T_{1/2}$ and MRT_{0-t} values of BHA, BMA, and BAC in the DFD group were much longer than those of the Fuzi extract. The T_{max} of toxic HA increased significantly in the DFD group compared with that in the Fuzi group. These PK parameters rationalize the combination of three herbs in DFD. An optimal ratio of herbs results in an optimal ADME (absorption, distribution, metabolism, excretion) that maximizes the drug efficacy.

Fuzi Xiexin Tang (FXT) is composed of Fu Zi, Da Huang, CC, and *Scutellaria baicalensis*. Maceration method is used for oral administration in ancient China, while in modern clinical practice decoction method is adopted for preparation. Different preparative methods resulted in significant differences in exposure and PK features of alkaloids, flavones, and anthraquinones from FXT, especially protoberberine alkaloids (from *Coptis*) (Zhang et al., 2014a). Concentrations of MDAs (from Fu Zi) were below the detection limit in rat plasma after administration of FXT due to the presence of the other three herbs. Maceration could decrease the absorption of flavones while increasing the absorption of anthraquinones. C_{max} of emodin and rhein were increased by 3.1 and 10.3-fold respectively, while eliminations of these two constituents were 8.0 and 19.0-fold slower, respectively, after administration of macerated FXT. However, how Fu Zi interacts with other herbs during maceration is not clear. Bioavailability of both flavones and anthraquinones increased after oral use of macerated FXT, especially emodin and rhein increasing as much as 13.5 and 20.7-fold. As mentioned previously, HHIs considerably influence the exposure of *Aconitum* alkaloids.

The alkaloids of *H. Ephedrae-Radix Aconiti Lateralis* were widely distributed in the heart, liver, spleen, lung, kidney, and brain (Ren et al., 2017). Lower bioavailability and higher clearance of some alkaloids were observed in the rats orally administered with the medicine pair, but hypaconitine had a longer residence time and lower clearance. The ephedra and aconitum alkaloids were mainly excreted in urine and feces, respectively. Compared with single-herb extracts, the use of medicine pairs led to a prolonged residence time and delayed elimination of aconitum alkaloids, increasing the risk of drug accumulation.

4.4.4.2 *Chuan Wu*

Chuan Wu is more toxic than Fu Zi, and how other herbs of Chuan Wu–containing formula neutralize its toxicity warrant detailed investigation. The metabolic fingerprint of Chuan Wu (*Radix Aconite*, RA) +*Radix Paeoniae Alba* (RPA; Bai Shao in Chinese) and its effect on CYPs were investigated using ultra-performance liquid chromatography-mass spectrometry (UPLC-MS)/MS and cocktail probe substrates (Bi et al., 2014). Couse of RPA of different proportions could alleviate RA's inhibition on CYP3A, 2D, 2C, and 1A2 of rat liver, while it did not influence RA's inhibition of CYP2E1. Compared with RA decoction alone, the intensity of diester diterpene alkaloids decreased significantly, and the MDAs significantly increased in the metabolic fingerprints of codecoctions of RA and RPA. These results suggest that coadministration of RPA reduced the systemic exposure of toxic alkaloid and enhanced the drug efficacy.

4.4.5 Herb-Herb Interaction: *Coptis* Related

CC and *S. baicalensis* (*Radix Scutellariae*, RS) are among the most popular TCM prescriptions. CC observably inhibited CYP2D6 and 1A2 in RLMs (Wei et al., 2013), while RS alone remarkably inhibited CYP1A2, 2E1, and 2C9. The combination of CC and RS at the ratio of 1:1 inhibited CYP1A2 but remarkably activated CYP2D6 and 3A4, which might minimize bioactivation of toxic metabolites and facilitate detoxification of toxic alkaloid. However, at the ratio of 2: 1, CYP1A2 and 2C9 were inhibited in vitro.

Compared to RS alone, RS + CC decreased AUC and C_{max} of baicalin and wogonoside, two active flavonoids of RS (Shi et al., 2009). CC, with antimicrobial activity, decreased the hydrolysis of baicalin and wogonoside by intestinal flora. CC constituents, for example, BBR, might conjugate with flavonoids and decrease transport of baicalein to the basolateral side of intestinal epithelia, thus decreasing the bioavailability of baicalin and wogonoside.

JTW is comprised of CC and Cinnamon (Rou Gui in Chinese) granules. Compared to CC alone, higher plasma concentration of BBR, shorter T_{max}, longer $T_{1/2}$, and lower CL were obtained when JTW was used in healthy male volunteers (Huang et al., 2011; Chen et al., 2013b). Reciprocally, CC increased the relative bioavailability of cinnamic acid of Cinnamon (Chen et al., 2008).

Zuojinwan, consisting of CC and *Evodia rutaecarpa* (Wu Zhu Yu in Chinese) powder (6:1, g/g), is used in TCM for the treatment of gastrointestinal disorders. Multiple peaks were observed on the plasma concentration time curve after oral use of Zuojinwan (Yan et al., 2011), while there was only one peak after administration of CC alone. Possible reasons are: (1) Drug redistribution between tissues and plasma, reabsorption in kidney, and/or enterohepatic circulation (Deng et al., 2008); (2) CC and *E. rutaecarpa* decreases the dissolution rate of herb powder reciprocally; and (3) metabolic conversion between different alkaloids. Compared with single herbs, the mean plasma concentration of dehydroevodiamine increased and that of coptisine (from CC) decreased after combining. In RLMs, rutaecarpine inhibited the hepatic metabolism of coptisine, epiberberine, berberine, palmatine, and jatrorrhizine (Xue et al., 2014). The half inhibitory concentration (IC_{50}) was greater than 50 μM, suggesting that rutaecarpine had a weak inhibition on *Coptis* alkaloids. However, differences of the inhibition constant (K_i) were statistically significant. Inhibition of berberine was greater than jatrorrhizine, followed by palmatine, epiberberine, and coptisine. One should be cautious to extrapolate these in vitro results back to in vivo.

CC is often used in more complex TCM formulas. For instance, Gegenqinlian decoction consists of Ge Gen (pueraria, root of kudzu vine), RS, CC, and *Radix Glycyrrhizae*. Flavonoids, alkaloids, and triterpene saponins are key therapeutic components of this TCM formula. In rat single pass gut perfusion, pueraria and *Radix Glycyrrhizae* promote the absorption of jatrorrhizine, BBR, and palmatine (An et al., 2012). The antiinflammatory Sanhuang Xiexin Tang, composed of Da Huang, RS, and CC, displayed distinct PK characteristics when prepared by decoction and maceration respectively (Zhang et al., 2013b), especially the protoberberine alkaloids (BBR, palmatine, jatrorrhizine, and coptisine) thereof.

Wuji pill consists of CC, *T. ruticarpum* (*E. rutaecarpa*, Wu Zhu Yu) and *P. lactiflora* (*Radix Paeoniae Alba*, Bai Shao), and the proportion of three herbs varies according to different patients. The representative bioactive ingredients, that is, BBR, palmatine (both from CC), evodiamine, rutaecarpine (both from *Tetradium*), and paeoniflorin (from *Paeonia*), in rat liver were quantified after oral administration of Wuji pill/single herb at 2 h time point (Zhang et al., 2014b). Compared to the single herb of the same dosage, the bioactive ingredients have distinct concentrations in different combinations of three herbs. CC was positively correlated with evodiamine concentration when low or high dose of *T. ruticarpum* was administered. *T. ruticarpum* was negatively correlated with BBR concentration when low dose CC was administered, but it was positively correlated with middle dose CC. *P. lactiflora* was negatively correlated with palmatine concentration when middle dose CC was administered. The HHIs could explain the differential concentration of each ingredient in rat liver. The combination 12 (CC): 6 (*Tetradium*): 6 (*Paeonia*) maximized the concentration of each ingredient in rat liver. It was argued that BBR, abundant in CC, *Thalictrum*, and *Hydrastis*, induced CYP3A4 in HepG2 cells (Liu et al., 2011) and inhibited CYP2E1 in HLMs (Raner et al., 2007), which, however, cannot explain all the observations. *P. lactiflora* (minister) and *T. ruticarpum* (adjuvant/courier) are beneficial in maintaining toxic alkaloids within a moderate range and thus maximize the potency of CC (king).

Major therapeutic components of RS + CC extract, for example, baicalin and BBR, were uptook into the rat blood (Jiang et al., 2014). The normal and type II diabetic rat plasma had similar metabolite classes, as did the urine samples. Nevertheless, the type II diabetic rat plasma had much higher concentrations of baicalin and methylated BBR than normal samples, whereas the trend is reversed in the urine, which helps keep a high plasma drug concentration and could be helpful in handling type II diabetes.

Almost every TCM formula has historically undergone long-term tests in both therapeutic efficacy and side effects, even if no evidence-based medicine data is available for many formulas. Notwithstanding, HHI predictions remain challenging. Network pharmacology might be helpful in such a prediction (Hao and Xiao, 2014), but bioavailability and other PK properties of one herbal compound must be taken into account when inferring whether it is more druglike. The establishment of recommendations for quantitative forecasts of clinically significant HHIs could be possible in the PBPK modeling framework. Interactions between some TCM formula constituents and the other constituents, which could be the probe substrates of the DME/transporter of interest, could be simulated. The proof-of-concept medical research should be performed to confirm the prediction. In short, PK takes precedence in contemporary plant-based drug development.

4.4.6 Phase II Drug Metabolizing Enzymes

4.4.6.1 Substrate

The phase II DMEs can work on both the absorbed Ranunculaceae constituents and the phase I metabolites. The glucuronidation of demethylated BBR (M1) was much slower than that of demethylenated BBR (M2, CYP catalysis product) (Liu et al., 2009a,b). Both M1 and M2 could be glucuronidated by UGT1A1 and 2B1, and M2 glucuronidation was mainly catalyzed by UGT1A1.

Jatrorrhizine (an isoquinoline alkaloid of *Coptis*) glucuronide was a phase II metabolite in HLMs (Zhou et al., 2013). The UGT kinetics followed the Michaelis-Menten equation. The recombinant UGT1A1, 1A3, 1A7, 1A8, 1A9, and 1A10 catalyzed jatrorrhizine glucuronidation, which was hindered by quercetin, 1-naphthol, and silibinin. Similar to humans, glucuronidated alkaloids were the main metabolites in rats (Shi et al., 2012), and glucuronidation in RLMs was catalyzed by UGT1A1 and 1A3. Identifying a selective UGT probe is formidable due to the significant overlapping substrate specificity displayed by the enzyme. Lv et al. found that UGT1A1-catalyzed NCHN (N-3-carboxy propyl-4-hydroxy-1,8-naphthalimide)-4-O-glucuronidation generated a single fluorescent product (Lv et al., 2015), which can be used for sensitive measurements of UGT1A1 activities in human liver preparations, as well as for rapid screening of UGT1A1 modulators from variable enzyme sources. Jiang et al. found that desacetylcinobufagin (DACB) 3-O- and 16-O-glucuronidation are isoform-specific probe reactions for UGT1A4 and 1A3, respectively (Jiang et al., 2015). DACB, the well-characterized fluorescent probe, can be used to simultaneously determine the catalytic activities of O-glucuronidation mediated by UGT1A3 and 1A4 of various enzyme sources. These fluorescent probe substrates could be useful in studying phase II metabolism of Ranunculaceae compounds and the relevant HDIs.

Neither phase I nor phase II metabolites of 23-epi-26-deoxyactein, the most abundant triterpene glycoside of black cohosh, were detected in clinical samples or in vitro (van Breemen et al., 2010).

Higenamine was used in rabbits by i.v. bolus, p.o. route, and i.v. infusion (Lo and Chen, 1996). After urine samples were hydrolyzed with β-glucuronidase, urinary concentrations of higenamine were prominently increased. MA was given via intragastric (i.g.) infusion in rats, and urine metabolites were analyzed (Chen et al., 2010). MA and its metabolites, hypo-MA glucuronic acid conjugate, 10-hydroxy-MA, 1-O-demethyl MA, deoxy-MA, and hypo-MA, were found in the rat urine. UGT is involved in phase II metabolism of *Aconitum* alkaloids.

Two phase I metabolites of guanfu base A (GFA; *Aconitum coreanum*), guanfu base I (GFI), and guanfu alcohol-amine (AA) were detected in rat urine (A et al., 2002). Phase II conjugates, glucuronide and sulfate conjugates of GFA and GFI, were isolated and tentatively identified by hydrolysis with glucuronidase or sulfatase. The polarity of the metabolites is higher, and they are less potent than the parent drug. Sulfate conjugates of jatrorrhizine were identified in rats after i.v. administration (Shi et al., 2012). Besides deglycosylation, dehydrogenation, and hydroxylation, sulfation was also the major metabolic transformation of PSD in rat plasma (Ouyang et al., 2014a,b). However, the responsible UGT and sulfotransferase (Hao et al., 2010) have not been identified.

4.4.6.2 Inhibitor and Inducer

BBR, epiberberine, coptisine, and jatrorrhizine significantly inhibited rat liver microsome UGT activity (Zhao et al., 2016), and epiberberine had the strongest inhibition. UGT1A1 activity was slightly inhibited by jatrorrhizine, with IC_{50} about 227 $\mu mol \cdot L^{-1}$, and coptisine and magnoflorine significantly activated UGT1A1. BBR, coptisine, jatrorrhizine, and palmatine significantly inhibited mice liver microsome UGT activity, and six alkaloids significantly activated UGT1A1. In vivo, mice UGTs were significantly activated by BBR, and UGT1A1 was significantly activated by jatrorrhizine.

4.4.7 Phase III: Drug Transporter

4.4.7.1 ABC Transporter

Drug transport through lipid membrane is sometimes called phase III metabolism and can be mediated by drug transporters. AC, MA, and HA are very poisonous, while their hydrolysates, for example, MDAs, aconine, and mesaconine, are noticeably less lethal. Efflux transporters, for example, P-gp, breast cancer resistance protein (BCRP), and multidrug resistance-associated protein isoform 2 (MRP2), are the integral part of defence mechanisms and indispensable in deadliness avoidance (Hao et al., 2011). The authors performed the bidirectional transport assays of alkaloids in the presence or absence of P-gp (CsA and verapamil), BCRP (Ko143), and MRP2 (MK571) inhibitors (Ye et al., 2013a,b). The efflux ratio (Er) of AC in Caco-2 cells was higher than those of MA and HA; MDAs had an Er of around 4, and aconine and mesaconine had Er values of 1. The Er values of AC, MA, and HA in parental MDCKII cells were expressively lower than those in MDR1-MDCKII and BCRP-MDCKII cells, where P-gp and BCRP are overexpressed respectively. Inhibition studies suggest that P-gp and BCRP participated in the transport of AC, MA, and HA, while MRP2 could carry AC, MA, HA, and MDAs.

Aconitum alkaloids increased P-gp expression in LS174T and Caco-2 cells in the order AC > benzoylaconine > aconine (Wu et al., 2016a). Nuclear receptors were involved in the induction of P-gp. AC and benzoylaconine increased the P-gp transport activity, intracellular ATP level, and mitochondrial mass. Exposure to AC decreased the toxicity of vincristine and doxorubicin toward the cells. In vivo, AC significantly upregulated the P-gp protein levels in the jejunum, ileum, and colon of FVB mice and protected them against acute AC toxicity.

The involvement of ABC transporters in the secretion of BBR in *Thalictrum minus* cells (Terasaka et al., 2003) implies that BIAs could also be the substrates of human ABC transporters. Rifampin and clarithromycin regulate DIG transport, while black cohosh or goldenseal did not affect DIG (Gurley et al., 2006, 2007), suggesting that they are not powerful regulators of P-gp in vivo.

The effects of BBR on the PK of DIG, CsA (a dual substrate of P-gp and CYP3A), carbamazepine (CYP3A substrate), and its metabolite were studied in rats (Qiu et al., 2009). A 14-dBBR pretreatment caused a dose-dependent rise of DIG in AUC and C_{max} in the i.g.-medicated rats. The 14-d BBR pretreatment also substantially increased AUC and C_{max} of i.g.-administered CsA. BBR dose-dependently enhanced bioavailability of DIG and CsA by suppression of gut P-gp. In addition, the inhibition of hepatic P-gp by BBR may result in declined biliary elimination of CsA. BBR did not significantly change CYP3A activity.

4.4.7.2 Solute Carrier Family

Organic cation transporter 2 (OCT2, SLC22A2) and 3 (OCT3, SLC22A3) are weak-affinity, high-capacity transporters in the brain, liver, and kidney. In the tail suspension test and forced swim test, BBR exerted antidepressant-like activity by elevating serotonin/norepinephrine/dopamine (5-HT/NE/DA) concentration in the brains of mice. OCT inhibition by BBR could enhance serotonergic and noradrenergic effects in mouse brain synaptosomes (Sun et al., 2014). In transfected MDCK cells, BBR is a powerful inhibitor of human OCT2 and OCT3, as its IC_{50} values for 5-HT/NE uptake inhibition are below 1 μM. BBR is also a substrate of hOCT2 and hOCT3. The higher capacity (V_{max}) and higher apparent binding affinity (K_m) for hOCT2, which led to ~4-fold higher transport efficiency (V_{max}/K_m) than hOCT3, suggest that hOCT2 is more crucial than hOCT3 in BBR uptake. Future OCT-related HDI studies are warranted.

OCT1 and OCT2 were unambiguously expressed on the basolateral membrane of human liver cells and tubular epithelia of kidneys, respectively. The basolateral membrane of MDCKII transfectants had both transporters (Nies et al., 2008). The affinity between BBR and OCT2 is higher than that between BBR and OCT1. The transport of the cations tetraethylammonium and 1-methyl-4-phenylpyridinium by MDCK-OCT1 and MDCK-OCT2 transfectants was inhibited by BBR. In polarized cells, BBR transfer from the basolateral to the apical compartments was much faster in MDCK-OCT1/P-gp double transfectants than in MDCK-OCT1 or MDCK-P-gp single ones. The MDCK-OCT1/P-gp double transfectants could be used to recognize other cationic substrates, suppressors of OCT1 and P-gp, and potential HDIs.

In the study of multicomponent herb drug metabolism, simultaneous determination of multiple compounds/metabolites should be pursued, and QAMS (quantitative analysis of multicomponent with single marker) (Wang et al., 2015) method is recommended. Multicomponent drug metabolism should be characterized continuously along both time and space axes. Quantitative determination should be combined with qualitative methods, given that it is not realistic to quantify every component.

4.5 TOXICITY

4.5.1 Alkaloid

Toxicity of diterpenoid alkaloids of *Aconitum* and *Delphinium* is a major concern in drug development research. *Aconitum* alkaloid poisoning could be due to contamination of herbs by aconite roots (Chan, 2016). Five *Aconitum* poisoning cases were reported (Fujita et al., 2007). Patient 1 had ventricular tachycardia and ventricular fibrillation, while toxic symptoms of other four patients were relatively mild. $T_{1/2}$ of AC varied between 5.8 and 15.4 h in these cases, and other alkaloids had $T_{1/2}$ close to that of the main alkaloid in each patient. $T_{1/2}$ of the major alkaloid in patient 1 was much longer than those of the other patients, and the AUC and MRT values were much higher in patient 1. The severity of deadly symptoms in *Aconitum* poisoning could be featured by the alkaloid toxicokinetic parameters.

Delphinium, similar to *Aconitum*, contains various types of alkaloids, which are useful in folk medicine and important for natural product-based drug development. Sheep taking in low larkspur (*Delphinium*) had no symptoms of poisoning (Welch et al., 2013). When sheep simultaneously took in death camas (*Zigadenus*) and low larkspur, the heart rate, exercise-induced muscle fatigue, and blood zygacine kinetics did not change dramatically, suggesting that low larkspur has no influence on the toxicity of death camas in sheep.

Delphinium nuttallianum and *Delphinium andersonii* were fed to 10 cattle (Green et al., 2013). The concentrations of the alkaloids in the two species were distinct. The C_{max} of serum alkaloid and AUC values of 16-deacetylgeyerline and geyerline/nudicauline were also distinct between the two groups. $T_{1/2}$ of the alkaloid was similar in the two species, suggesting that the excretion rates of norditerpene alkaloids of these species in cattle are comparable. The individual alkaloid composition of the plant decides the *Delphinium* toxicity.

Both N-(methylsuccinimido) anthranoyllycoctonine (MSAL) type and 7,8-methylenedioxylycoctonine (MDL) type norditerpenoid alkaloids might be responsible for much of the *Delphinium* toxicity in cattle (Green et al., 2011). Cattle that have consumed larkspur will excrete 99% of methyllycaconitine (MLA; MSAL-type) and deltaline (MDL-type) from circulation within 6 days (Green et al., 2009). In mice, a normal biphasic redistribution and excretion pattern could delineate the MLA elimination, with elimination constant of 0.0376 and $T_{1/2}$ of 18.4 min (Stegelmeier et al., 2003). Clearance rates of other tissues were alike. MLA is quickly allocated and eliminated. In mice, alkaloid poisoning affects the brain, manifesting as dyspnea, "explosive" muscle twitches, and spasms.

There are more mildly poisonous MDL-type alkaloids than MSAL-type ones in most *Delphinium barbeyi* and *Delphinium occidentale* populations (Welch et al., 2012), and MDL-type alkaloids aggravate the toxicity of the MSAL-type ones. These results increase knowledge about the toxicity of *Delphinium* in mammalians.

Some renal toxic alkaloids are from *Tripterygium regelii*, *Stephania tetrandra*, *Strychnos nux-vomica*, and *A. carmichaeli* (Xu et al., 2016).

4.5.2 Others

Intravenous formulations of saponins are developed due to their poor intestinal absorption. During blood collection, hemolysis was seen after i.v. administration (0.15 mg/kg) of *Pulsatilla chinensis* saponins (Liu et al., 2013a). The hemolysis disappeared after about 1 h. The greater the dose, the more severe the hemolysis, and the rats died when the i.v. dose of 1 mg/kg was implemented, demonstrating that a safe dose range would be an important issue for pulchinenoside anticancer therapy.

Renal toxicity components from TCMs include aristolochic acids (AAS), alkaloids, anthraquinones, and others (Xu et al., 2016). TCM renal toxicity is most commonly caused by AAS and some alkaloids. AAS mainly come from *Aristolochia contorta*, *Aristolochia manshuriensis*, *Clematis chinensis*, etc.

4.6 PHARMACOKINETICS AND PHARMACODYNAMICS

4.6.1 *Aconitum* Alkaloid

PK tackles how an organism works on drugs, while PD highlights the effects of drugs on an organism. Diterpenoid alkaloids, especially those isolated from various *Aconitum* and *Delphinium* species, display extensive bioactivities (Hao et al., 2013b). PK and PD studies should be performed in developing these natural products into clinically useful drugs. For instance, the plasma concentration of fuziline (15α-hydroxyneoline) in rats following i.v. and i.g. administrations was determined by HPLC/-MS (Sun et al., 2013), and fuziline showed desirable absolute bioavailability (21.1% ± 7.0%). Higenamine, isolated from *Aconitum* root, has cardioactive effects (Hao et al., 2013b). In the PD model, a simple direct effect model with baseline could describe the correlation between the heart rates and the blood higenamine concentrations (Feng et al., 2012). Higenamine has desirable PK and PD features in human subjects. Species differences of guanfu-base APK behavior was found between humans and dogs (Wu et al., 2002), especially $T_{1/2}$ of the slow distribution phase (α) and the terminal elimination phase (β). Chronic administration of AC gradually decreased AC concentration and increased BAC and aconine concentration in organs and blood (Wada et al., 2005), implying the increased AC metabolism. Accordingly, the arrhythmias became less frequent with time and repeated use of AC. The PK and PD results provide important information for future clinical studies of *Aconitum* alkaloids.

The absolute bioavailability (F%) after the intake of 0.5 mg/kg AC and Fuzi extract (0.118 mg/kg AC) in rats was 8.24% ± 2.52% and 4.72% ± 2.66%, respectively (Tang et al., 2012). The T_{max} of AC and Fuzi extract is 30.08 ± 9.73 min and 58.00 ± 21.68 min respectively, suggesting a very fast absorption. AC was excreted speedily with a short $T_{1/2}$ (i.v., 80.98 ± 6.40 min) and a low protein binding (23.9%–31.9%). All PK parameters were not distinct between the single and multiple doses of AC. Nevertheless, the absorption of AC after a single dose was much slower than that of repeated ingestions of Fuzi extract (T_{max}: 58.0 vs. 20.0 min), and a single dose was followed by a smaller AUC.

In beagle dogs, the optimal PK model for the six *Aconitum* alkaloids (AC, MA, HA, and three MDAs) was the one-compartment model (Table 4.1) (Xiao et al., 2014). The absorption and elimination rates of six alkaloids are close to each other. Three extraction methods (Liu et al., 2014b), that is, ultrasonic extraction, 1 h reflux extraction, and 3 h reflux extraction, were used to obtain *A. kusnezoffii* root extract. After oral administration of crude extract, double-absorption peaks were observed on the concentration–time plots of the six alkaloids, which might be caused by enterohepatic circulation, delayed gastric emptying, and/or differential absorption within different gut portions. The DDAs exhibited faster absorption and elimination than MDAs, and the absorption of both DDA and MDA after intake of ultrasonic extract and 3 h reflux extract was expressively slower than that after 1 h reflux extract.

4.6.2 *Coptis* Alkaloid

The systemic exposures of the *Coptis* alkaloids are extremely low after oral administration (Ma and Ma, 2013). The alkaloids may exert their systemic activities via tissue distribution and/or metabolites, or by modulating targets in the gut. The drug transporters and DMEs involved in the in vivo process have been documented. However, significant differences between the blood and tissue exposure make it difficult to find suitable PK markers of the alkaloids in blood, and the dose-systemic exposure-response relationships of the alkaloids have not been determined. Derivatives or formulations of the alkaloids should be designed to obtain optimal PK features and improve the oral bioavailability and efficacy.

Significant differences in the PK behaviors, such as C_{max}, AUC_{0-t}, apparent V_d, and apparent CL, of BBR and palmatine were noticed between nondiseased and postinflammation irritable bowel syndrome (PI-IBS) in model rats (Gong et al., 2014). In diseased rats, C_{max} and AUC_{0-t} values were much higher, whereas apparent V_d and apparent CL were much less. Second peaks at 3 h for BBR and 4 h for palmatine after i.g. administration of the CC extract were observed in the disease group, implying that the change of PK behavior plays an important role in drug efficacy. It is essential to scrutinize the PK of the Ranunculaceae compounds in various pathological conditions.

Compared with single alkaloid and their mixture, T_{max} and $T_{1/2ka}$ of BBR and jatrorrhizine in *Coptidis Rhizoma* powder group were shorter when orally administered (Wei et al., 2015), and C_{max}, AUC_{inf}, AUC_{last}, and V_L/F were significantly increased. Compared with the control group, blood glucose levels were decreased significantly in the powder group since the 18th day and were decreased in BBR or BBR + jatrorrhizine (BJ) group since the 36th day. The free fatty acid values in all treatment groups were decreased significantly. Triglyceride, HDL, and LDL values in the powder group, LDL values in the BBR group, and HDL values in the BJ group were improved significantly. *Coptidis Rhizoma* powder showed better PK characteristics and excellent activity of lowering blood glucose and lipid. Its oral administration is recommended in the treatment of type II diabetes.

4.6.3 Saponin and Others

THQ, the active constituent of *N. sativa* seeds, is a compound with relatively slower absorption and rapid elimination following oral administration (Table 4.1) (Alkharfy et al., 2015). THQ exhibited the biphasic decline on the concentration–time curve after i.v. administration. The initial rapid decline represents rapid drug distribution phase with binding to both plasma and tissue. However, the V_d at steady state was relatively small (0.7 L/kg), which can be explained by high THQ binding to plasma proteins (>99%) due to its lipophilicity.

Nanostructured lipid carriers (NLCs), comprised of lipids and surfactants, are promising colloidal drug transporters. In rabbits, the T_{max}, C_{max}, and elimination $T_{1/2}$ of THQ-loaded NLCs were 3.96 h, 4811.33 ng/mL, and 4.493 h, respectively (Abdelwahab et al., 2013), indicating that THQ-loaded NLC could be used extravascularly.

The purified extracts of *T. chinensis* flowers contain multiple flavonoids and their glycosides, which display antiinflammatory and antifebrile activities. Various orientin glycosides could be transformed into orientin in rabbits, which resulted in the increase of orientin AUC (Li et al., 2007).

Cimicifuga foetida has antiinflammatory, antipyretic, and analgesic effects. Tetracyclic triterpene saponins are the primary active constituents of *Cimicifuga* plants (Hao et al., 2013c,d). After *C. foetida* extract intake, systemic exposure to three *C. foetida* saponins, that is, 23-O-acetylshengmanol (Cim B), cimigenolxyloside(C), and 25-O-acetylcimigenoside(D), was more significant than that to cimicifugoside H-1(A), despite the prevalence of Cim A (21.2%) in the extract (Gai et al., 2012). Significantly different clearance and transformation from Cim A to Cim C partially accounts for the differential exposure to the four cimicifugosides.

The five pulchinenosides of *P. chinensis* exhibited rapid absorption, quick elimination, and significant double-peak on the plasma concentration–time curve (Liu et al., 2013a). The first peak occurred within 30 min and the second between 8–12 h. The first absorption probably represents direct absorption, while enterohepatic circulation may contribute to the second peak. However, the oral bioavailability, comparable to that of PSD (2.83%) (Ouyang et al., 2015), is quite low (0.55%–2.5%) due to the unfavorable molecular size ($\geq$733.5 Da) of pentacyclic triterpene saponins, poor gut absorption, and/or extensive metabolism after absorption (Ouyang et al., 2014a,b).

Hederacolchiside E is a neuroactive oleanolic-acid saponin, abundant in *Pulsatilla koreana* extract. C_{max} and AUC of hederacolchiside E after *P. koreana* extract was orally administered at 100, 200, and 400 mg/kg suggested nonlinear PK pattern in rats (Yoo et al., 2008). Unlike five pulchinenosides of *P. chinensis*, hederacolchiside E exhibited slow absorption and sluggish elimination, and no double peaks were observed.

Anemone is evolutionarily closer to *Pulsatilla* than to other Ranunculaceae genera and is also characterized by abundant triterpene saponin with therapeutic efficacy. The three hydrophilic sugar moieties of raddeanin A, a oleanane-type saponin from *Anemone raddeana*, provide hydrogen bonding possibility and polar surface region (Luan et al., 2013), which, along with its large molecular weight (>500 Da), could account for reduced membrane penetrability and low absolute bioavailability (0.295%) (Liu et al., 2013b). Many saponins undergo quick and wide-ranging biliary elimination via active transport (Yu et al., 2012), which could result in short $T_{1/2}$, low-systemic exposure and small V_d.

Nigella A, a potential anticancer triterpene saponin from seeds of *Nigella glandulifera*, presented dose-dependent PK behavior in rats after i.v. administration (Hu et al., 2014). The residence time of *Nigella* A was short, and MRT, V_d, and CL were not significantly different with regard to dose or gender. Similar to other Ranunculaceae saponins (Wang et al., 2012; Luan et al., 2013), *Nigella* A showed low oral bioavailability, possibly due to extensive gut metabolism and poor gut absorption of saponins. The anticancer activity of *Nigella* A, if any, would only be achieved by i.v. administration.

Rats are the most commonly used animal for in vivo TCM PK studies (Table 4.1). Unlike human subjects, rats can be fed with high-dose herb decoction so that more original compounds and metabolites could be identified by metabolomics techniques, given that compounds having similar structures have similar metabolic pathways. Based on the collated structure information of metabolites, the appropriate LC-MS protocol could be established to monitor metabolites and their PK behavior after using decoction of clinical dose. Metabolomics focuses on the dynamic change of endogenous substances under the systemic exposure of xenobiotics (Lan and Jia, 2010), while PK delineates the dynamic change of xenobiotic compounds and their in vivo metabolites. Provided that the combinatorial use of PK and state-of-the-art metabolomics is implemented, the plant metabolome could be linked to the human/animal metabolome, and the gap between multicomponent agents and molecular pharmacology could be bridged.

PK trends of most Ranunculaceae compounds could not be fitted with compartmental models properly (Table 4.1). The PK-PD combined model is an effective tool for studying the quantitative relationship between the drug dose and drug effect, which could shed light on the three-dimensional relationship of "time-plasma concentration-drug effect" and could be of great help in formulation development, mechanism study, clinical drug optimization, and rational drug use. It is essential to launch the PK-PD combined model for multicomponent herb drugs.

4.7 CONCLUSION AND PROSPECT

The clinical utility of Ranunculaceae-derived medicinal compounds has been validated by traditional uses of thousands of years and current evidence-based medicine studies. DMPK studies of plant-based natural products are an indispensable part of comprehensive medicinal plant exploration, which could facilitate conservation and sustainable utilization of Ranunculaceae pharmaceutical resources, as well as new chemical entity development with improved DMPK parameters. HHI of Ranunculaceae herb-containing TCM formula could significantly influence the in vivo PK behavior of compounds thereof, which may partially explain the complicated therapeutic mechanism of TCM formulas. Yet, little information is available on the relationship between plasma/tissue drug concentration and pharmacological outcome. The absorption, distribution, and excretion data of many Ranunculaceae compounds are lacking, and the in vivo metabolism data are much less available than the in vitro data.

The vibrant concentration contour of Ranunculaceae compounds, resulting from dynamic absorption and liver and intestinal microbial transformation, and the human metabolic reaction outline in both healthy and diseased statuses, should be integrated to discuss the holdup problem during the therapeutic appraisal of multiconstituent medicines (e.g., Ranunculaceae plant extract), resulting in the direct clarification of the healing and toxic mechanisms of these compounds. In this sense, a new concept "precision pharmacokinetics/pharmacodynamics" could be put forward and the proof-of-concept interrogations would follow. This concept would define crosstalk between drug and recipient by underlying molecular causes and other factors in addition to conventional signs and markers. This concept also highlights the interindividual variation and agrees with the fundamental ideas of personalized medicine.

Recent years have seen the progress in revealing the ADME/T of Ranunculaceae compounds. Nevertheless, there is a lack of DMPK studies of important medicinal genera *Aquilegia*, *Thalictrum*, and *Clematis*. Fluorescent probe compounds could be promising substrate (Fig. 4.3), inhibitor, and/or inducer in future DMPK studies of Ranunculaceae compounds. A better understanding of the important HDI/HHIs, bioavailability, and metabolomics aspects of Ranunculaceae compounds will illuminate future natural product-based drug design and a more detailed examination of personalized drug disposition.

References

A, J.Y., Wang, G.J., Liu, X.Q., et al., 2002. Study on the metabolites of guanfu base A hydrochloride in rat urine by high performance liquid chromatograph-mass spectrum. Yao Xue Xue Bao 37, 283–287.

Abdelwahab, S.I., Sheikh, B.Y., Taha, M.M., et al., 2013. Thymoquinone-loaded nanostructured lipid carriers: preparation, gastroprotection, in vitro toxicity, and pharmacokinetic properties after extravascular administration. Int. J. Nanomed. 8, 2163–2172.

Ali, B., Amin, S., Ahmad, J., et al., 2012. Bioavailability enhancement studies of amoxicillin with *Nigella*. Indian J. Med. Res. 135, 555–559.

Al-Jenoobi, F.I., Al-Suwayeh, S.A., Muzaffar, I., et al., 2013. Effects of *Nigella sativa* and *Lepidium sativum* on cyclosporine pharmacokinetics. Biomed. Res. Int. 2013, 953520.

Alkharfy, K.M., Ahmad, A., Khan, R.M., et al., 2015. Pharmacokinetic plasma behaviors of intravenous and oral bioavailability of thymoquinone in a rabbit model. Eur. J. Drug Metab. Pharmacokinet. 40 (3), 319–323.

Al-Mohizea, A.M., Ahad, A., El-Maghraby, G.M., et al., 2015. Effects of *Nigella sativa, Lepidium sativum* and Trigonella foenum-graecum on sildenafil disposition in beagle dogs. Eur. J. Drug Metab. Pharmacokinet. 40 (2), 219–224.

An, R., Zhang, H., Zhang, Y.Z., et al., 2012. Intestinal absorption of different combinations of active compounds from *Gegenqinlian decoction* by rat single pass intestinal perfusion in situ. Yao Xue Xue Bao 47, 1696–1702.

Azam, Y.J., Machavaram, K.K., Rostami-Hodjegan, A., et al., 2014. The modulating effects of endogenous substances on drug metabolising enzymes and implications for inter-individual variability and quantitative prediction. Curr. Drug Metab. 15, 599–619.

Bi, Y., Zheng, Z., Pi, Z., et al., 2014. The metabolic fingerprint of the compatibility of *Radix Aconite* and *Radix Paeoniae* Alba and its effect on CYP450 enzymes. Acta Pharm. Sin. 49, 1705–1710.

Bi, Y., Zhuang, X., Zhu, H., et al., 2015. Studies on metabolites and metabolic pathways of bulleyaconitine A in rat liver microsomes using LC-MS^n combined with specific inhibitors. Biomed. Chromatogr. 29 (7), 1027–1034.

Bi, Y.F., Liu, S., Zhang, R.X., et al., 2013. Metabolites and metabolic pathways of mesaconitine in rat liver microsomal investigated by using UPLC-MS/MS method in vitro. Yao Xue Xue Bao 48, 1823–1828.

Brantley, S.J., Gufford, B.T., Dua, R., et al., 2014. Physiologically based pharmacokinetic modeling framework for quantitative prediction of an herb-drug interaction. CPT Pharmacometrics Syst. Pharmacol. 3, e107.

Cao, X., Gibbs, S.T., Fang, L., et al., 2006. Why is it challenging to predict intestinal drug absorption and oral bioavailability in human using rat model. Pharm. Res. 23, 1675–1686.

Cao, Y., Bei, W., Hu, Y., et al., 2012. Hypocholesterolemia of Rhizoma Coptidis alkaloids is related to the bile acid by up-regulated CYP7A1 in hyperlipidemic rats. Phytomedicine 19, 686–692.

Chan, T.Y., 2012. Aconite poisoning following the percutaneous absorption of *Aconitum* alkaloids. Forensic Sci. Int. 223, 25–27.

Chan, T.Y., 2016. Aconitum alkaloid poisoning because of contamination of herbs by aconite roots. Phytother. Res. 30 (1), 3–8.

Chatuphonprasert, W., Nemoto, N., Sakuma, T., et al., 2012. Modulations of cytochrome P450 expression in diabetic mice by berberine. Chem. Biol. Interact. 196, 23–29.

Chen, G., Lu, F., Wang, F., et al., 2008. Effects of *Rhizoma coptidis* on relative bioavailability of cinnamic acid in *Cinnamomum cassia*. Chin. Pharm. J. 43, 696–698.

Chen, G., Lu, F., Xu, L., et al., 2013b. The anti-diabetic effects and pharmacokinetic profiles of berberine in mice treated with Jiao-Tai-Wan and its compatibility. Phytomedicine 20, 780–786.

Chen, J.L., Zhang, Y.L., Dong, Y., et al., 2013a. Enzyme reaction kinetics, metabolic enzyme phenotype, and metabolites of berberine. Chin. Trad. Herb. Drug 44, 3334–3340.

Chen, J.L., Zhang, Y.L., Dong, Y., et al., 2013c. CYP450 enzyme inhibition of berberine in pooled human liver microsomes by cocktail probe drugs. Zhongguo Zhong Yao Za Zhi 38, 2009–2014.

Chen, P.P., Zhao, N., Xu, X.L., et al., 2010. Analysis on the metabolites of mesaconitine in the rat urine by liquid chromatography and electrospray ionization mass spectrometry. Yao Xue Xue Bao 45, 1043–1047.

Chen, W., Fan, D., Meng, L., et al., 2012. Enhancing effects of chitosan and chitosan hydrochloride on intestinal absorption of berberine in rats. Drug Dev. Ind. Pharm. 38, 104–110.

Chen, Y., Yang, Q., Zou, L., et al., 2011. Studies on intestinal absorption of alkaloids in *Coptis chinensis* by in situ single-pass perfused rat intestinal model. Zhongguo Zhong Yao Za Zhi 36, 3523–3527.

Deng, Y., Liao, Q., Li, S., et al., 2008. Simultaneous determination of berberine, palmatine and jatrorrhizine by liquid chromatography-tandem mass spectrometry in rat plasma and its application in a pharmacokinetic study after oral administration of coptis-evodia herb couple. J. Chromatogr. B Analyt. Technol. Biomed. Life Sci. 863, 195–205.

El-Najjar, N., Ketola, R.A., Nissilä, T., et al., 2011. Impact of protein binding on the analytical detectability and anticancer activity of thymoquinone. J. Chem. Biol. 4, 97–107.

Fan, X., Yin, S.S., Li, X.J., et al., 2017. Hydroxylation metabolisms of crassicauline A in rats under toxic dose. Eur. J. Drug. Metab. Pharmacokinet. 42 (5), 857–869.

Feng, L., Liu, Z.M., Hou, J., et al., 2014a. A highly selective fluorescent ESIPT probe for the detection of human carboxylesterase 2 and its biological applications. Biosens. Bioelectron. 65C, 9–15.

Feng, L., Liu, Z.M., Xu, L., et al., 2014b. A highly selective long-wavelength fluorescent probe for the detection of human carboxylesterase 2 and its biomedical applications. Chem. Commun. 50, 14519–14522.

Feng, S., Jiang, J., Hu, P., et al., 2012. A phase I study on pharmacokinetics and pharmacodynamics of higenamine in healthy Chinese subjects. Acta Pharmacol. Sin. 33, 1353–1358.

Fujita, Y., Terui, K., Fujita, M., et al., 2007. Five cases of aconite poisoning: toxicokinetics of aconitines. J. Anal. Toxicol. 31, 132–137.

Gai, Y.Y., Liu, W.H., Sha, C.J., et al., 2012. Pharmacokinetics and bioavailability of cimicifugosides after oral administration of *Cimicifuga foetida* L. extract to rats. J. Ethnopharmacol. 143, 249–255.

Gao, Q.T., Chen, X.H., BiK, S., 2004. Comparative pharmacokinetic behavior of glycyrrhetic acid after oral administration of glycyrrhizic acid and Gancao-Fuzi-Tang. Biol. Pharm. Bull. 27, 226–228.

Ge, G.B., Ning, J., Hu, L., et al., 2013. A highly selective probe for human cytochrome P4503A4: isoformselectivity, kinetic characterization and its applications. Chem. Commun. 49, 9779–9781.

Godugu, C., Patel, A.R., Doddapaneni, R., et al., 2014. Approaches to improve the oral bioavailability and effects of novel anticancer drugs berberine and betulinicacid. PLoS One 9, e89919.

Gong, Z., Chen, Y., Zhang, R., et al., 2014. Pharmacokinetics of two alkaloids after oral administration of rhizomacoptidis extract in normal rats and irritable bowel syndrome rats. Evid. Based Complement. Alternat. Med. 2014, 845048.

Gorman, G.S., Coward, L., Darby, A., et al., 2013. Effects of herbal supplements on the bioactivation of chemotherapeutic agents. J. Pharm. Pharmacol. 65, 1014–1025.

Green, B.T., Welch, K.D., Gardner, D.R., et al., 2009. Serum elimination profiles of methyllycaconitine and deltaline in cattle following oral administration of larkspur (*Delphinium barbeyi*). Am. J. Vet. Res. 70, 926–931.

Green, B.T., Welch, K.D., Gardner, D.R., et al., 2011. A toxicokinetic comparison of norditerpenoid alkaloids from *Delphinium barbeyi* and *D. glaucescens* in cattle. J. Appl. Toxicol. 31, 20–26.

Green, B.T., Welch, K.D., Gardner, D.R., et al., 2013. A toxicokinetic comparison of two species of low larkspur (*Delphinium* spp.) in cattle. Res. Vet. Sci. 95, 612–615.

Guo, T., Zhang, Y., Li, Z., et al., 2017. Microneedle-mediated transdermal delivery of nanostructured lipid carriers for alkaloids from *Aconitum sinomontanum*. Artif. Cells Nanomed. Biotechnol. 12, 1–11.

Guo, Y., Chen, Y., Tan, Z.R., et al., 2012. Repeated administration of berberine inhibits cytochromes P450 in humans. Eur. J. Clin. Pharmacol. 68, 213–217.

Guo, Y., Pope, C., Cheng, X., et al., 2011. Dose-response of berberine on hepatic cytochromes P450 mRNA expression and activities in mice. J. Ethnopharmacol. 138, 111–118.

Gurley, B.J., Barone, G.W., Williams, D.K., et al., 2006. Effect of milk thistle (*Silybummarianum*) and black cohosh (*Cimicifuga racemosa*) supplementation on digoxin pharmacokinetics in humans. Drug Metab. Dispos. 34, 69–74.

Gurley, B.J., Swain, A., Barone, G.W., et al., 2007. Effect of goldenseal (*Hydrastis canadensis*) and kava kava (*Piper methysticum*) supplementation on digoxin pharmacokinetics in humans. Drug Metab. Dispos. 35, 240–245.

Gurley, B.J., Swain, A., Hubbard, M.A., et al., 2008a. Supplementation with goldenseal (*Hydrastis canadensis*), but not kava kava (*Piper methysticum*), inhibits human CYP3A activity in vivo. Clin. Pharmacol. Ther. 83, 61–69.

Gurley, B.J., Swain, A., Hubbard, M.A., et al., 2008b. Clinical assessment of CYP2D6-mediated herb-drug interactions in humans: effects of milk thistle, black cohosh, goldenseal, kava kava, St. John's wort, and Echinacea. Mol. Nutr. Food Res. 52, 755–763.

Hao, D.C., Feng, Y., Xiao, R., et al., 2011. Non-neutral nonsynonymous single nucleotide polymorphisms in human ABC transporters: the first comparison of six prediction methods. Pharmacol. Rep. 63, 924–934.

Hao, D.C., Gu, X.J., Xiao, P.G., et al., 2013a. Chemical and biological research of *Clematis* medicinal resources. Chin. Sci. Bull. 58, 1120–1129.

Hao, D.C., Gu, X.J., Xiao, P.G., et al., 2013b. Recent advances in the chemical and biological studies of *Aconitum* pharmaceutical resources. J. Chin. Pharm. Sci. 22, 209–221.

Hao, D.C., Gu, X.J., Xiao, P.G., et al., 2013c. Recentadvance in chemical and biological studies on Cimicifugeae pharmaceutical resources. Chin. Herb. Med. 5, 81–95.

Hao, D.C., Sun, J., Furnes, B., et al., 2007. Allele and genotype frequencies of polymorphic FMO3 gene in two genetically distinct populations. Cell Biochem. Funct. 25, 443–453.

Hao, D.C., Xiao, B., Xiang, Y., et al., 2013d. Deleterious nonsynonymous single nucleotide polymorphisms in human solute carriers: the first comparison of three prediction methods. Eur. J. Drug Metab. Pharmacokinet. 38, 53–62.

Hao, D.C., Xiao, P.G., 2011. Prediction of sites under adaptive evolution in flavin-containing monooxygenases: selection pattern revisited. Chin. Sci. Bull. 56, 1246–1255.

Hao, D.C., Xiao, P.G., 2014. Network pharmacology: a Rosetta stone for traditional Chinese medicine. Drug Dev. Res. 75, 299–312.

Hao, D.C., Xiao, P.G., Chen, S., et al., 2010. Phenotype prediction of nonsynonymous single nucleotide polymorphisms in human phase II drug/xenobiotic metabolizing enzymes: perspectives on molecular evolution. Sci. China Life Sci. 53, 1252–1262.

He, M., Liang, Q., Ouyang, H., et al., 2014. In vivo intestinal absorption characteristics of α-hederin in rats. Chin. Trad. Herb. Drug 45, 807–812.

Ho, S.H., Singh, M., Holloway, A.C., et al., 2011. The effects of commercial preparations of herbal supplements commonly used by women on the biotransformation of fluorogenic substrates by human cytochromes P450. Phyto-ther. Res. 25, 983–989.

Hu, N., Yuan, L., Li, H.J., et al., 2013. Anti-diabetic activities of Jiaotaiwan in db/dbmice by augmentation of AMPK protein activity and upregulation of GLUT4 expression. Evid. Based Complement. Alternat. Med., 180721.

Hu, X., Liu, X., Gong, M., et al., 2014. Development and validation of liquid chromatography-tandem mass spectrometry method for quantification of a potential anticancer triterpene saponin from seeds of *Nigella glandulifera* in rat plasma: application to a pharmacokinetic study. J. Chromatogr. B Analyt. Technol. Biomed. Life Sci. 967, 156–161.

Huang, Y., Jiang, B., Nuntanakorn, P., et al., 2010. Fukinolic acid derivatives and triterpene glycosides from black cohosh inhibit CYP isozymes, but are not cytotoxic to Hep-G2 cells in vitro. Curr. Drug Saf. 5, 118–124.

Huang, Z., Lu, F., Dong, H., et al., 2011. Effects of cinnamon granules on pharmacokinetics of berberine in *Rhizoma Coptidis* granules in healthy male volunteers. J. Huazhong Univ. Sci. Technol. Med. Sci. 31, 379–383.

Hwang, Y.H., Cho, W.K., Jang, D., et al., 2012. Effects of berberine and hwangryunhaedok-tang on oral bioavailability and pharmacokinetics of ciprofloxacin in rats. Evid. Based Complement. Alternat. Med. 2012, 673132.

Hyland, R., Roe, E.G., Jones, B.C., et al., 2001. Identification of the cytochrome P450 enzymes involved in the N-demethylation of sildenafil. Br. J. Clin. Pharmacol. 51, 239–248.

Jaiswal, Y., Liang, Z., Ho, A., et al., 2014. Distribution of toxic alkaloids in tissues from three herbal medicine *Aconitum* species using laser micro-dissection, UHPLC-QTOF MS and LC-MS/MS techniques. Phytochemistry 107, 155–174.

Jiang, L., Liang, S.C., Wang, C., et al., 2015. Identifying and applying a highly selective probe to simultaneously determine the O-glucuronidation activity of human UGT1A3 and UGT1A4. Sci. Rep. 5, 9627.

Jiang, S., Xu, J., Qian, D.W., et al., 2014. Comparative metabolites in plasma and urine of normal and type 2 diabetic rats after oral administration of the traditional Chinese scutellaria-coptis herb couple by ultra performance liquid chromatography-tandem mass spectrometry. J. Chromatogr. B Analyt. Technol. Biomed. Life Sci. 965, 27–32.

Korashy, H.M., Al-Jenoobi, F.I., Raish, M., et al., 2015. Impact of herbal medicines like *Nigella sativa*, *Trigonella foenum-graecum*, and *Ferula asafoetida*, on cytochrome P450 2C11 gene expression in rat liver. Drug Res. 65 (7), 366–372.

Lan, K., Jia, W., 2010. An integrated metabolomics and pharmacokinetics strategy for multi-component drugs evaluation. Curr. Drug Metab. 11, 105–114.

Li, D., Wang, Q., Xu, L., et al., 2008b. Pharmacokinetic study of three active flavonoid glycosides in rat after intravenous administration of *Trollius ledebourii* extract by liquid chromatography. Biomed. Chromatogr. 22, 1130–1136.

Li, D., Wang, Q., Yuan, Z.F., et al., 2008a. Pharmacokinetics and tissue distribution study of orientin in rat by liquid chromatography. J. Pharm. Biomed. Anal. 47, 429–434.

Li, J., Gödecke, T., Chen, S.N., et al., 2011a. In vitro metabolic interactions between black cohosh (*Cimicifuga racemosa*) and tamoxifen via inhibition of cytochromes P450 2D6 and 3A4. Xenobiotica 41, 1021–1030.

Li, N., Tsao, R., Sui, Z., et al., 2012. Intestinal transport of pure diester-type alkaloids from an aconite extract across the Caco-2 cell monolayer model. Planta Med. 78, 692–697.

Li, X., Huo, T., Qin, F., et al., 2007. Determination and pharmacokinetics of orientin in rabbit plasma by liquid chromatography after intravenous administration of orientin and *Trollius chinensis* Bunge extract. J. Chromatogr. B Analyt. Technol. Biomed. Life Sci. 853, 221–226.

Li, X.L., Wu, L., Wu, W.G., et al., 2016. Effects of combined administration of Paeoniae Radix Alba on local pharmacokinetics of six aconite alkaloids by skin microdialysis in vivo. China J. Chin. Mater. Med. 41 (5), 948–954.

Li, Y., He, W., Liu, J., et al., 2005. Binding of the bioactive component jatrorrhizine to human serum albumin. Biochim. Biophys. Acta 1722, 15–21.

Li, Y., Li, Y.X., Zhao, M.J., et al., 2017. The effects of *Rheum palmatum* L. on the pharmacokinetic of major diterpene alkaloids of *Aconitum carmichaelii* Debx. in rats. Eur. J. Drug Metab. Pharmacokinet. 42 (3), 441–451.

Li, Y., Ren, G., Wang, Y.X., et al., 2011b. Bioactivities of berberine metabolites after transformation through CYP450 isoenzymes. J. Transl. Med. 9, 62.

Li, Z., Xue, B., Zhang, Y., et al., 2014. Comparison on in vitro hepatic metabolic characteristics of five kinds of alkaloids from *Coptis chinensis*. Chin. Trad. Herb. Drug 45, 532–535.

Liang, Q., He, M., Ouyang, H., et al., 2014. Analysis on in vivo metabolites of α-hederin in rats by UPLC-MS/MS. Chin. Trad. Herb. Drug 45, 1883 1888.

Lijun, Z., Linlin, L.U., Enshuang, G., et al., 2014. The influences of aconitine, an active/toxic alkaloid from *Aconitum*, on the oral pharmacokinetics of CYP3A probe drug buspirone in rats. Drug Metab. Lett. 8, 135–144.

Lin, Y.P., Zhao, Y., Zhang, Y.P., et al., 2007. Comparative study on transdermal osmosis in vitro of *Aconitum brachypodium* liniment, gel and patcher. Zhongguo Zhong Yao Za Zhi 32, 203–206.

Liu, J., Li, Q., Yin, Y., et al., 2014b. Ultra-fast LC-ESI-MS/MS method for the simultaneous determination of six highly toxic *Aconitum* alkaloids from *Aconiti kusnezoffii radix* in rat plasma and its application to a pharmacokinetic study. J. Sep. Sci. 37, 171–178.

Liu, L., Wu, X., Wang, R., et al., 2014c. Absorption properties and mechanism of trolline and veratric acid and their implication to an evaluation of the effective components of the flowers of *Trollius chinensis*. Chin. J. Nat. Med. 12, 700–704.

Liu, Q., Wang, C., Meng, Q., et al., 2014d. MDR1 and OAT1/OAT3 mediate the drug–drug interaction between puerarin and methotrexate. Pharm. Res. 31, 1120–1132.

Liu, W., Shen, M., Qin, Z.Q., 2009a. Distribution of aconitum alkaloids in the corpse died of acute aconite intoxication. Fa Yi Xue Za Zhi 25, 176–178.

Liu, X., Li, H., Song, X., et al., 2014a. Comparative pharmacokinetics studies of benzoylhypaconine, benzoylmesaconine, benzoylaconine and hypaconitine in rats by LC-MS method after administration of *Radix AconitiLateralis Praeparata* extract and Dahuang Fuzi Decoction. Biomed. Chromatogr. 28, 966–973.

Liu, Y., Hao, H., Xie, H., et al., 2009b. Oxidative demethylenation and subsequent glucuronidation are the major metabolic pathways of berberine in rats. J. Pharm. Sci. 98, 4391–4401.

Liu, Y., Ma, B., Zhang, Q., et al., 2013b. Development and validation of a sensitive liquid chromatography/tandem mass spectrometry method for the determination of raddeanin A in rat plasma and its application to a pharmacokinetic study. J. Chromatogr. B Analyt. Technol. Biomed. Life Sci. 912, 16–23.

Liu, Y., Song, Y., Xu, Q., et al., 2013a. Validated rapid resolution LC-ESI-MS/MS method for simultaneous determination of five pulchinenosides from *Pulsatilla chinensis* (Bunge) Regel in rat plasma: application to pharmacokinetics and bioavailability studies. J. Chromatogr. B Analyt. Technol. Biomed. Life Sci. 942–943, 141–150.

Liu, Y., Yin, X.W., Wang, Z.Y., et al., 2017. Study on biopharmaceutics classification system for Chinese materia medica of extract of Huanglian. China J. Chin. Mater. Med. 42 (21), 4127–4134.

Liu, Y.H., Mo, S., Bi, H., et al., 2011. Regulation of human pregnane X receptor and its target gene cytochrome P450 3A4 by Chinese herbal compounds and a molecular docking study. Xenobiotica 41, 259–280.

Liu, Y.L., Song, Y., Guan, Z., et al., 2015. Intestinal absorption of pulchinenosides from *Pulsatilla chinensis* in rats. Zhongguo Zhong Yao Za Zhi 40, 543–549.

Liu, Z.M., Feng, L., Ge, G.B., et al., 2014e. A highly selective ratiometric fluorescent probe for in vitro monitoring and cellular imaging of human carboxylesterase 1. Biosens. Bioelectron. 57, 30–35.

Lo, C.F., Chen, C.M., 1996. Pharmacokinetics of higenamine in rabbits. Biopharm. Drug Dispos. 17, 791–803.

Lo, S.N., Chang, Y.P., Tsai, K.C., et al., 2013. Inhibition of CYP1 by berberine, palmatine, and jatrorrhizine: selectivity, kinetic characterization, and molecular modeling. Toxicol. Appl. Pharmacol. 272, 671–680.

Luan, X., Guan, Y.Y., Wang, C., et al., 2013. Determination of Raddeanin A in rat plasma by liquid chromatography-tandem mass spectrometry: application to a pharmacokinetic study. J. Chromatogr. B Analyt. Technol. Biomed. Life Sci. 923–924, 43–47.

Lupidi, G., Camaioni, E., Khalifé, H., et al., 2012. Characterization of thymoquinone binding to human α-acid glycoprotein. J. Pharm. Sci. 101, 2564–2573.

Lupidi, G., Scire, A., Camaioni, E., et al., 2010. Thymoquinone, a potential therapeutic agent of *Nigella sativa*, binds to site I of human serum albumin. Phytomedicine 17, 714–720.

Lv, X., Ge, G.B., Feng, L., et al., 2015. An optimized ratiometric fluorescent probe for sensing human UDP-glucuronosyltransferase 1A1 and its biological applications. Biosens. Bioelectron. 72, 261–267.

Ma, B.L., Ma, Y.M., 2013. Pharmacokinetic properties, potential herb-drug interactions and acute toxicity of oral *Rhizoma coptidis* alkaloids. Expert Opin. Drug Metab. Toxicol. 9, 51–61.

Meddah, B., Ducroc, R., El Abbes Faouzi, M., et al., 2009. *Nigella sativa* inhibits intestinal glucose absorption and improves glucose tolerance in rats. J. Ethnopharmacol. 121, 419–424.

Men, W., Chen, Y., Yang, Q., et al., 2013. Study on metabolism of Coptis chinensis alkaloids from different compatibility of Wuji Wan in human intestinal flora. Zhongguo Zhong Yao Za Zhi 38, 417–421.

Meng, Q., Liu, K.X., 2014. Pharmacokinetic interactions between herbal medicines and prescribed drugs: focus on drug metabolic enzymes and transporters. Curr. Drug Metab. 15, 791–807.

Nies, A.T., Herrmann, E., Brom, M., et al., 2008. Vectorial transport of the plant alkaloid berberine by double-transfected cells expressing the human organic cation transporter 1 (OCT1, SLC22A1) and the efflux pump MDR1 P-glycoprotein (ABCB1). Naunyn. Schmiedebergs Arch. Pharmacol. 376, 449–461.

Niitsu, H., Fujita, Y., Fujita, S., et al., 2013. Distribution of Aconitum alkaloids in autopsy cases of aconite poisoning. Forensic Sci. Int. 227, 111–117.

Ning, Y.M., Rao, Y., Liang, W.Q., 2007. Influence of permeation enhancers on transdermal permeation of anemonin. Zhongguo Zhong Yao Za Zhi 32, 393–396.

Ouyang, H., Guo, Y., He, M., et al., 2014a. Identification of metabolites of *Pulsatilla* saponin D in intestinal microflora of rats in vitro by UPLC-Q-trap-MS. Chin. Trad. Herb. Drug 45, 523–526.

Ouyang, H., Guo, Y., He, M., et al., 2015. A rapid and sensitive LC-MS/MS method for the determination of *Pulsatilla* saponin D in rat plasma and its application in a rat pharmacokinetic and bioavailability study. Biomed. Chromatogr. 29, 373–378.

Ouyang, H., Zhou, M., Guo, Y., et al., 2014b. Metabolites profiling of *Pulsatilla saponin* D in rat by ultra performance liquid chromatography-quadrupole time-of-flight mass spectrometry (UPLC/Q-TOF-MS/MS). Fitoterapia 96, 152–158.

Pang, X., Cheng, J., Krausz, K.W., et al., 2011. Pregnane X receptor-mediated induction of Cyp3a by black cohosh. Xenobiotica 41, 112–123.

Peng, W.W., Li, W., Li, J.S., et al., 2013. The effects of *Rhizoma Zingiberis* on pharmacokinetics of six *Aconitum* alkaloids in herb couple of *Radix Aconiti Lateralis-Rhizoma Zingiberis*. J. Ethnopharmacol. 148, 579–586.

Qi, J., Guo, T., Li, H., et al., 2013. Absorption of multi-components from *Coptidis Rhizoma* in Caco-2 monolayer model and their interactions. Chin. Trad. Herb. Drug 44, 1801–1806.

Qiu, W., Jiang, X.H., Liu, C.X., et al., 2009. Effect of berberine on the pharmacokinetics of substrates of CYP3A and P-gp. Phytother. Res. 23, 1553–1558.

Raner, G.M., Cornelious, S., Moulick, K., et al., 2007. Effects of herbal products and their constituents on human cytochrome P450 (2E1) activity. Food Chem. Toxicol. 45, 2359–2365.

Rao, X., Gong, M., Yin, S., et al., 2013. Study on in situ intestinal absorption of *Pulsatilla saponin* D in rats. Chin. Trad. Herb. Drug 44, 3515–3520.

Ren, M., Song, S., Liang, D., et al., 2017. Comparative tissue distribution and excretion study of alkaloids from *Herba Ephedrae-Radix Aconiti Lateralis* extracts in rats. J. Pharm. Biomed. Anal. 134, 137–142.

Shen, H., Zhu, L.Y., Yao, N., et al., 2011. The effect of the compatibility of *Radix Aconiti Laterlis* and radix glycyrrhizae on pharmacokinatic of aconitine, mesaconitine and hypacmitine in rat plasma. Zhong Yao Cai 34, 937–942.

Shi, R., Zhou, H., Liu, Z., et al., 2009. Influence of coptis Chinensis on pharmacokinetics of flavonoids after oral administration of radix *Scutellariae* in rats. Biopharm. Drug Dispos. 30, 398–410.

Shi, R., Zhou, H., Ma, B., et al., 2012. Pharmacokinetics and metabolism of jatrorrhizine, a gastric prokinetic drug candidate. Biopharm. Drug Dispos. 33, 135–145.

Song, S., Tang, Q., Huo, H., et al., 2015. Simultaneous quantification and pharmacokinetics of alkaloids in *Herba Ephedrae-Radix Aconiti Lateralis* extracts. J. Anal. Toxicol. 39, 58–68.

Stegelmeier, B.L., Hall, J.O., Gardner, D.R., et al., 2003. The toxicity and kinetics of larkspur alkaloid, methyllycaconitine, in mice. J. Anim. Sci. 81, 1237–1241.

Sun, J., Zhang, F., Peng, Y., et al., 2013. Quantitative determination of diterpenoid alkaloid Fuziline by hydrophilic interaction liquid chromatography (HILIC)-electrospray ionization mass spectrometry and its application to pharmacokinetic study in rats. J. Chromatogr. B Analyt. Technol. Biomed. Life Sci. 913–914, 55–60.

Sun, S., Wang, K., Lei, H., et al., 2014. Inhibition of organic cation transporter 2 and 3 may be involved in the mechanism of the antidepressant-like action of berberine. Prog. Neuropsychopharmacol. Biol. Psychiatry 49, 1–6.

Tan, X.S., Ma, J.Y., Feng, R., et al., 2013. Tissue distribution of berberine and its metabolites after oral administration in rats. PLoS One 8, e77969.

Tang, L., Gong, Y., Lv, C., et al., 2012. Pharmacokinetics of aconitine as the targeted marker of Fuzi (*Aconitum carmichaeli*) following single and multiple oral administrations of Fuzi extracts in rat by UPLC/MS/MS. J. Ethnopharmacol. 141, 736–741.

Tang, L., Ye, L., Lv, C., et al., 2011. Involvement of CYP3A4/5 and CYP2D6 in the metabolism of aconitine using human liver microsomes and recombinant CYP450 enzymes. Toxicol. Lett. 202, 47–54.

Terasaka, K., Sakai, K., Sato, F., et al., 2003. *Thalictrum minus* cell cultures and ABC-like transporter. Phytochemistry 62, 483–489.

van Breemen, R.B., Liang, W., Banuvar, S., et al., 2010. Pharmacokinetics of 23-epi-26-deoxyactein in women after oral administration of a standardized extract of black cohosh. Clin. Pharmacol. Ther. 87, 219–225.

Wada, K., Nihira, M., Hayakawa, H., et al., 2005. Effects of long-term administrations of aconitine on electrocardiogram and tissue concentrations of aconitine and its metabolites in mice. Forensic Sci. Int. 148, 21–29.

Wan, J.Y., Zhang, Y.Z., Yuan, J.B., et al., 2017. Biotransformation and metabolic profile of anemoside B4 with rat small and large intestine microflora by ultra-performance liquid chromatography-quadrupole time-of-flight tandem mass spectrometry. Biomed. Chromatogr. 31 (5), doi: 10.1002/bmc.3873.

Wang, C., Liu, K.X., 2014. The drug-drug interaction mediated by efflux transporters and CYP450 enzymes. Yao Xue Xue Bao 49, 590–595.

Wang, C.Q., Jia, X.H., Zhu, S., et al., 2015. A systematic study on the influencing parameters and improvement of quantitative analysis of multi-component with single marker method using notoginseng as research subject. Talanta 134, 587–595.

Wang, D., Li, F., Li, P., et al., 2012. Validated LC-MS/MS assay for the quantitative determination of clematichinenoside AR in rat plasma and its application to a pharmacokinetic study. Biomed. Chromatogr. 26, 1282–1285.

Wang, L., Wang, C., Peng, J., et al., 2014. Dioscin enhances methotrexate absorption by down-regulating MDR1 in vitro and in vivo. Toxicol. Appl. Pharmacol. 277, 146–154.

Wang, R., Yuan, M., Yang, X., et al., 2013. Intestinal bacterial transformation—a nonnegligible part of Chinese medicine research. J. Asian Nat. Prod. Res. 15, 532–549.

Wang, T.F., Wang, J.P., Pan, J.H., et al., 2017. Nasal resorption of prim-*O*-glucosylcimifugin and 5-*O*-methylvisammioside in rats. China J. Chin. Mater. Med. 42 (9), 1772–1776.

Wang, Y., Hao, D.C., Stein, W.D., et al., 2006. A kinetic study of Rhodamine 123 pumping by P-glycoprotein. Biochim. Biophys. Acta 1758, 1671–1676.

Wei, L.Y., Zhang, Y.J., Wei, B.H., et al., 2013. Effect of comparability of Coptis chinensis and *Scutellaria baicalensis* on five sub-enzymatic activities of liver microsomes in rats. Zhongguo Zhong Yao ZaZhi 38, 1426–1429.

Wei, S.C., Xu, L.J., Zou, X., et al., 2015. Pharmacokinetic and pharmacodynamic characteristics of berberine and jateorhizine in *Coptidis Rhizoma* powder and their monomeric compounds in type 2 diabetic rats. China J. Chin. Mater. Med. 40 (21), 4262–4267.

Welch, K.D., Green, B.T., Gardner, D.R., et al., 2012. The effect of 78-methylenedioxylycoctonine-type diterpenoid alkaloids on the toxicity of tall larkspur (Delphinium spp.) in cattle. J. Anim. Sci. 90, 2394–2401.

Welch, K.D., Green, B.T., Gardner, D.R., et al., 2013. The effect of low larkspur (*Delphinium* spp) co-administration on the acute toxicity of death camas (Zigadenus spp.) in sheep. Toxicon 76, 50–58.

Wu, J., Lin, N., Li, F., et al., 2016a. Induction of P-glycoprotein expression and activity by *Aconitum alkaloids*: implication for clinical drug-drug interactions. Sci. Rep. 6, 25343.

Wu, J.J., Ai, C.Z., Liu, Y., et al., 2012. Interactions between phytochemicals from traditional Chinese medicines and human cytochrome P450 enzymes. Curr. Drug Metab. 13, 599–614.

Wu, M.S., Wang, G.J., Cai, X.H., et al., 2002. Determination of guanfu base A hydrochloride in plasma by LC-MS method and its pharmacokinetics in dogs. Yao Xue Xue Bao 37, 551–554.

Wu, Y.M., Chen, X.L., Liu, W., et al., 2016b. Pharmacokinetics of bullatine A in Aconitum brachypodum total alkaloids gel in transdermal delivery. Zhongguo Zhong Yao Za Zhi 41 (8), 1530–1534.

Wu, Z.Y., Lu, A.M., Tang, Y.C., et al., 2003. The Families and Genera of Angiosperms in China: A Comprehensive Analysis. Science Press, Beijing.

Xiao, P.G., 1980. A preliminary study of the correlation between phylogeny, chemical constituents and pharmaceutical aspects in the taxa of Chinese Ranunculaceae. Acta Phytotax. Sin. 18, 142–153.

Xiao, R.P., Lai, X.P., Zhao, Y., et al., 2014. Pharmacokinetic study of six aconitine alkaloids in aconitilateralis radix praeparata in beagle dogs. Zhong Yao Cai 37, 284–287.

Xie, X.K., Yan, L.H., Zhu, J.J., et al., 2016. Absorption characteristics of alkaloids in Fuzheng Xiaozheng Fang by rat everted intestinal sac models. China J. Chin. Mater. Med. 41 (11), 2144–2148.

Xin, Y., Liu, Z.Q., Liu, S.Y., 2015. Study on effects of incompatible herbal medicine on intestinal bacterial metabolism of Wu-Tou-Tang by ESI-MS. Zhong Yao Cai 38 (8), 1728–1731.

Xin, Y., Pi, Z., Song, F., et al., 2011. Comparison of intracorporal absorption of hypaconitine in *Heishunpian decoction* and its compound recipe decoction by ultra performance liquid chromatography-quadrupole time-of-flight mass spectrometry. Se Pu 29, 389–393.

Xu, F., Yang, D., Shang, M., et al., 2014. Effectiveforms, additive effect, and toxicities scattering effect of pharmaco-dynamic substances of TCMs—some reflections evoked by the study on the metabolic disposition of traditional Chinese medicines. World Sci. Tech./Modern Trad. Chin. Med. Mat. Med. 16, 688–703.

Xu, X.L., Yang, L.J., Jiang, J.G., 2016. Renal toxic ingredients and their toxicology from traditional Chinese medicine. Expert Opin. Drug Metab. Toxicol. 12 (2), 149–159.

Xue, B.J., Li, Z.H., Zhang, Y., et al., 2014. Inhibition of rutaecarpine on in vitro hepatic metabolism of five *Coptis* alkaloids in rats. Chin. Trad. Herb. Drug 45, 1293–1296.

Yamaura, K., Shimada, M., Nakayama, N., et al., 2011. Protective effects of goldenseal (*Hydrastis canadensis* L.) on acetaminophen-induced hepatotoxicity through inhibition of CYP2E1 in rats. Pharmacognosy Res. 3, 250–255.

Yan, R., Wang, Y., Shen, W., et al., 2011. Comparative pharmacokinetics of dehydroevodiamine and coptisine in rat plasma after oral administration of single herbs and Zuojinwan prescription. Fitoterapia 82, 1152–1159.

Yang, C., Zhang, T., Li, Z., et al., 2013. P-glycoprotein is responsible for the poor intestinal absorption and low toxicity of oral aconitine: in vitro, in situ, in vivo and in silico studies. Toxicol. Appl. Pharmacol. 273, 561–568.

Yang, X., Wang, Q.H., Wang, M., et al., 2016. iTRAQ technology combined with 2D-LC-MS/MS to analyze effect of *Coptidis Rhizoma* on cytochrome P450 isoenzyme expression. China J. Chin. Mater. Med. 41 (4), 731–736.

Yang, X.Y., Ye, J., Sun, G.X., et al., 2014. Identification of metabolites of epiberberine in rat liver microsomes and its inhibiting effects on CYP2D6. Zhongguo Zhong Yao Za Zhi 39, 3855–3859.

Ye, L., Tang, L., Gong, Y., et al., 2011a. Characterization of metabolites and human P450 isoforms involved in the microsomal metabolism of mesaconitine. Xenobiotica 41, 46–58.

Ye, L., Wang, T., Yang, C., et al., 2011b. Microsomal cytochrome P450-mediated metabolism of hypaconitine, an active and highly toxic constituent derived from *Aconitum* species. Toxicol. Lett. 204, 81–91.

Ye, L., Yang, X., Yang, Z., et al., 2013a. The role of efflux transporters on the transport of highly toxic aconitine, mesaconitine, hypaconitine, and their hydrolysates, as determined in cultured Caco-2 and transfected MDCKII cells. Toxicol. Lett. 216, 86–99.

Ye, L., Yang, X.S., Lu, L.L., et al., 2013b. Monoester-diterpene *Aconitum* alkaloid metabolism in human liver microsomes: predominant role of CYP3A4 and CYP3A5. Evid. Based Complement. Alternat. Med. 2013, 941093.

Yokotani, K., Chiba, T., Sato, Y., et al., 2013. Effect of three herbal extracts on cytochrome P450 and possibility of interaction with drugs. Shokuhin Eiseigaku Zasshi 54, 56–64.

Yoo, H.H., Lee, S.K., Lim, S.Y., et al., 2008. LC-MS/MS method for determination of hederacolchiside E, a neuroactive saponin from *Pulsatilla koreana* extract in rat plasma for pharmacokinetic study. J. Pharm. Biomed. Anal. 48, 1425–1429.

Yu, K., Chen, F., Li, C., 2012. Absorption, disposition, and pharmacokinetics of saponins from Chinese medicinal herbs: what do we know and what do we need to know more? Curr. Drug Metab. 13, 577–598.

Zhang, F., Tang, M.H., Chen, L.J., et al., 2008. Simultaneous quantitation of aconitine, mesaconitine, hypaconitine, benzoylaconine, benzoylmesaconine and benzoylhypaconine in human plasma by liquid chromatography-tandem mass spectrometry and pharmacokinetics evaluation of "SHEN-FU" injectable powder. J. Chromatogr. B Analyt. Technol. Biomed. Life Sci. 873, 173–179.

Zhang, G., Zhu, L., Zhou, J., et al., 2012. Effect of aconitilaterlis radix compatibility of glycyrrhizae radix on CYP3A4 in vivo. Zhongguo Zhong Yao Za Zhi 37, 2206–2209.

Zhang, H., Liu, M., Zhang, W., et al., 2015b. Comparative pharmacokinetics of three monoester-diterpenoid alkaloids after oral administration of *Acontium carmichaeli* extract and its compatibility with other herbal medicines in Sini Decoction to rats. Biomed. Chromatogr. 29 (7), 1076–1083.

Zhang, H., Sun, S., Zhang, W., et al., 2016b. Biological activities and pharmacokinetics of aconitine, benzoylaconine, and aconine after oral administration in rats. Drug Test Anal. 8 (8), 839–846.

Zhang, J.M., Liao, W., He, Y.X., et al., 2013a. Study on intestinal absorption and pharmacokinetic characterization of diesterditerpenoid alkaloids in precipitation derived from fuzi-gancao herb-pair decoction for its potential interaction mechanism investigation. J. Ethnopharmacol. 147, 128–135.

Zhang, M., Peng, C.S., Li, X.B., 2017. Human intestine and liver microsomal metabolic differences between C19-diester and monoester diterpenoid alkaloids from the roots of *Aconitum carmichaelii* Debx. Toxicol. In Vitro 45 (Pt 3), 318–333.

Zhang, Q., Ma, Y.M., Wang, Z.T., et al., 2013b. Differences in pharmacokinetics and anti-inflammatory effects between decoction and maceration of Sanhuang Xiexin Tang in rats and mice. Planta Med. 79, 1666–1673.

Zhang, Q., Ma, Y.M., Wang, Z.T., et al., 2014a. Pharmacokinetics difference of multiple active constituents from decoction and maceration of Fuzi Xiexin Tang after oral administration in rat by UPLC-MS/MS. J. Pharm. Biomed. Anal. 92, 35–46.

Zhang, R.J., Chen, Y., Gong, Z.P., et al., 2014b. Research on bioactive ingredients in rat liver after oral administration of different combinations of Wujipill. Zhongguo Zhong Yao Za Zhi 39, 1695–1703.

Zhang, W., Zhang, H., Sun, S., et al., 2015a. Comparative pharmacokinetics of hypaconitine after oral administration of pure hypaconitine, *Aconitum carmichaelii* extract and Sini Decoction to rats. Molecules 20, 1560–1570.

Zhang, W.X., Yang, B., Hu, Y.M., et al., 2016a. Effect of rat intestinal flora in vitro on metabolites of acteoside. China J. Chin. Mater. Med. 41 (8), 1541–1545.

Zhao, L., Fang, L., Li, Y., et al., 2011. Effect of (E)-2-isopropyl-5-methylcyclohexyl octadec-9-enoate on transdermal delivery of *Aconitum* alkaloids. Drug Dev. Ind. Pharm. 37, 290–299.

Zhao, X.L., Liu, L., Di, L.Q., et al., 2014. Studies on effects of calycosin-7-O-β-D-glucoside on prim-O-glucosylcimifugin and cimifugin in vivo pharmacokinetics. China J. Chin. Mater. Med. 39 (23), 4669–4674.

Zhao, Y.Y., Miao, P.P., Miao, Q., et al., 2016. In vitro and In vivo effects of six Coptidis alkaloids on liver microsomes UGTs and UGT1A1 activities in rats and mice. China J. Chin. Mater. Med. 41 (2), 309–313.

Zhou, H., Shi, R., Ma, B., et al., 2013. CYP450 1A2 and multiple UGT1A isoforms are responsible for jatrorrhizine metabolism in human liver microsomes. Biopharm. Drug Dispos. 34, 176–185.

Zhou, Y., Cao, S., Wang, Y., et al., 2014. Berberine metabolites could induce low density lipoprotein receptor upregulation to exert lipid-lowering effects in human hepatoma cells. Fitoterapia 92, 230–237.

Zhu, L., Wu, J., Zhao, M., et al., 2017. Mdr1a plays a crucial role in regulating the analgesic effect and toxicity of aconitine by altering its pharmacokinetic characteristics. Toxicol. Appl. Pharmacol. 320, 32–39.

Zhu, L., Yang, X., Zhou, J., et al., 2013. The exposure of highly toxic aconitine does not significantly impact the activity and expression of cytochrome P450 3A in rats determined by a novel ultra performance liquid chromatography-tandem mass spectrometric method of a specific probe buspirone. Food Chem. Toxicol. 51, 396–403.

Drug Metabolism and Disposition Diversity of Ranunculales Phytometabolites

Ranunculales Medicinal Plants. http://dx.doi.org/10.1016/B978-0-12-814232-5.00005-8

ABBREVIATIONS

ADME/T Absorption, distribution, metabolism, excretion, and toxicity
AUC Area under the plasma concentration-time curve
BBR Berberine
BCRP Breast cancer resistance protein
BIA Benzylisoquinoline alkaloid
CDL Corydaline
CL Total body clearance
C_{max} Maximum plasma concentration
CYP Cytochrome P450
DDI Drug–drug interaction
DHC Dehydrocorydaline
DMPK Drug metabolism and pharmacokinetics
ER Estrogen receptor
HDI Herb–drug interaction
HLMs Human liver microsomes
IC_{50} Half maximal inhibitory concentration of a substance
K_m The concentration of substrate that leads to half maximal velocity
MRP Multidrug resistance-associated protein
MRT Mean residence time
OCT Organic cation transporter
P_{app} Apparent permeability coefficient
P-gp P glycoprotein
rhCYP Recombined human cytochrome P450
SULT Sulfotransferase
$T_{1/2}$ Elimination half life
TCM Traditional Chinese medicine
THP Tetrahydropalmatine
T_{max} Time to reach C_{max}
UGT Uridine diphosphate glucuronosyltransferase
UPLC/Q-TOF MS Ultra performance liquid chromatography-quadrupole time-of-flight mass spectrometry
V_d Volume of distribution
V_{max} Maximum enzyme velocity

5.1 INTRODUCTION

Ranunculales, an order of angiosperm plants, consists of the families Ranunculaceae (buttercup family) (Hao et al., 2015a,e), Berberidaceae (barberry family), Menispermaceae, Lardizabalaceae, Circaeasteraceae, Papaveraceae (poppy family), and Eupteleaceae (Fig. 5.1). Ranunculales belongs to the basal eudicots and is evolutionarily older than other eudicots. Various Ranunculales species have been employed in traditional Chinese medicine (TCM), traditional oriental medicine, and Western folk medicine (Datta et al., 2014). Light is being shed on widely known Ranunculales members, for example, blue cohosh, poppies, barberries, and buttercups, by advanced omics techniques and from the systems perspective. Ranunculales contributes plenty of botanical extracts with therapeutic efficacy and/or health-promoting effects.

Botanical extracts of Ranunculales, different from conventional Western supplements, are intricate mixtures of many bioactive compounds. Berberidaceae has about 17 genera

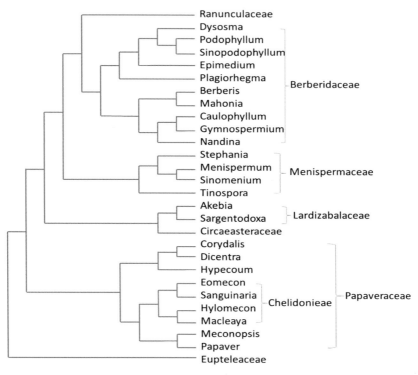

FIGURE 5.1 Ranunculales genera mentioned in the text and their phylogenetic relationship. The genera *Eschscholtzia* and *Glaucium* are closer to *Papaver* than to Chelidonieae in morphology, while *Fumaria* is closer to *Corydalis* than to *Papaver*. However, the exact position of these three genera is not known due to lack of molecular data.

and 650 species. For instance, the TCM plant *Epimedium* sp. (Berberidaceae), known as an herbal "Viagra" for men, contains various prenyl-flavonoids (Fig. 5.2). Menispermaceae and Papaveraceae have about 72 (450 species) and 44 (770 species) genera respectively. *Stephania* (Menispermaceae) and *Corydalis* (Papaveraceae) produce multiple classes of alkaloids. The simple isoquinolines are present in *Papaver* (Hao et al., 2015b) and *Corydalis* (Papaveraceae) and *Thalictrum* (Ranunculaceae) (Hao et al., 2015c). Benzylisoquinoline alkaloids (BIAs) are representative Ranunculales secondary metabolites stemming from tyrosine (Hagel et al., 2015b). Seven types of BIAs are present in Ranunculales. Benzyltetrahydroisoquinolines, for example, demethylcoclaurine and papaverine (PAP), are found in *Aconitum* (Ranunculaceae) (Hao et al., 2013a, 2015c) and *Papaver*. Bis-benzyltetrahydroisoquinolines, for example, tetrandrine, trilobine, and dauricine, are found in Menispermaceae and *Thalictrum*. Aporphine and isoaporphine, for example, magnoflorine and stephanine, are present in Ranunculaceae (Hao et al., 2015e) and Menispermaceae. Morphinanes, such as morphine, codeine, and sinomenine, are present in Menispermaceae and Papaveraceae. Protoberberine and berberine (BBR), such as tetrahydrocoptisine, tetrahydropalmatine (THP), and jatrorrhizine (Fig. 5.2), are abundant in Ranunculaceae, Berberidaceae, and Papaveraceae. Protopine and phenanthridine alkaloids are abundant in Ranunculaceae and Chelidonieae (Papaveraceae) (Hao et al., 2015b).

Protoberberine alkaloids

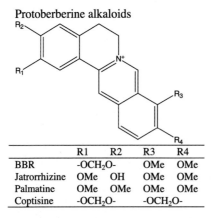

	R1	R2	R3	R4
BBR	-OCH$_2$O-		OMe	OMe
Jatrorrhizine	OMe	OH	OMe	OMe
Palmatine	OMe	OMe	OMe	OMe
Coptisine	-OCH$_2$O-		-OCH$_2$O-	

Bi-isobenzylquinoline alkaloid: berbamine

Aporphine alkaloids

Magnoflorine   Sinomenine

Menispermaceae alkaloids

Cepharanthine

FIGURE 5.2 Examples of Ranunculales phytometabolites with medicinal utility.

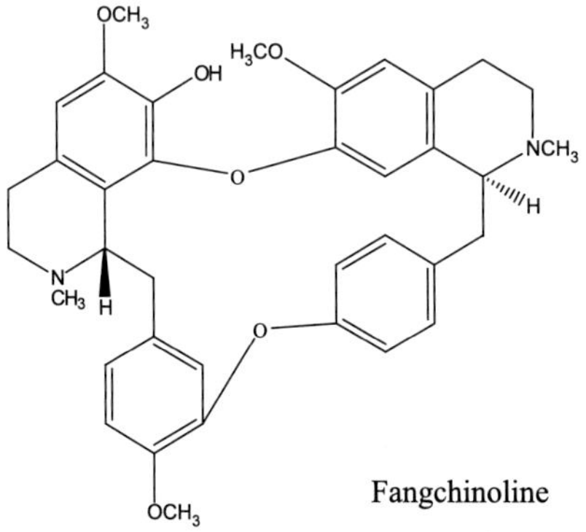

Fangchinoline

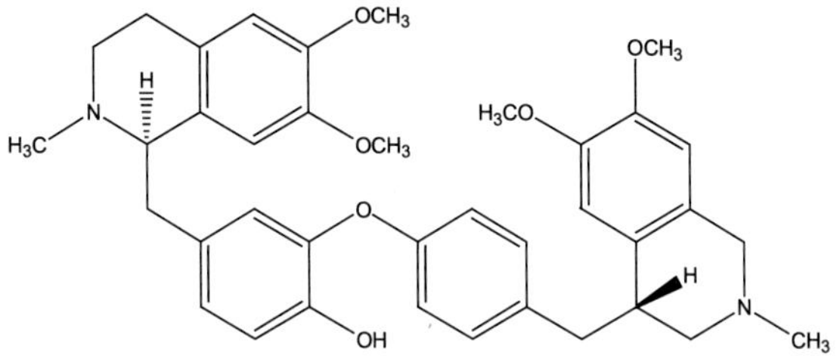

Dauricine

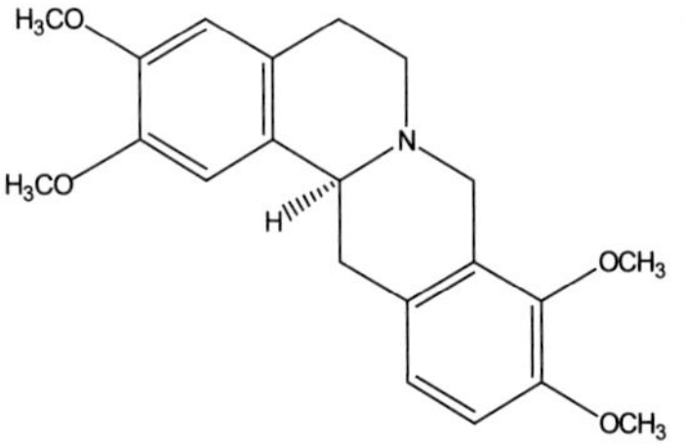

l-THP

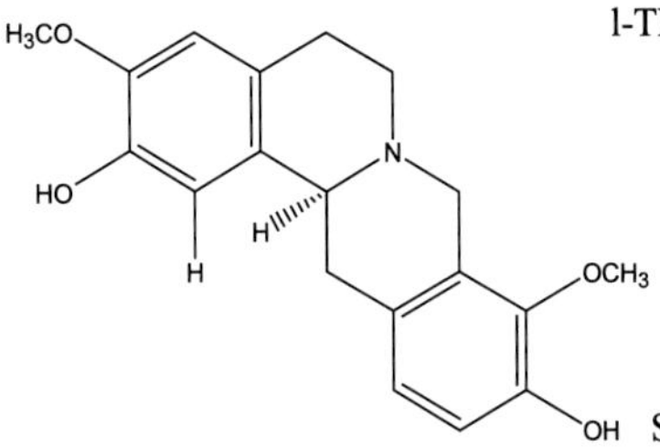

Stepholidine

Papaveraceae alkaloids

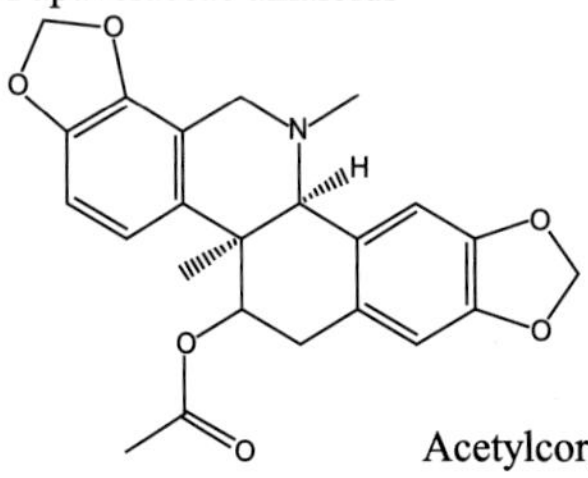

Acetylcorynoline

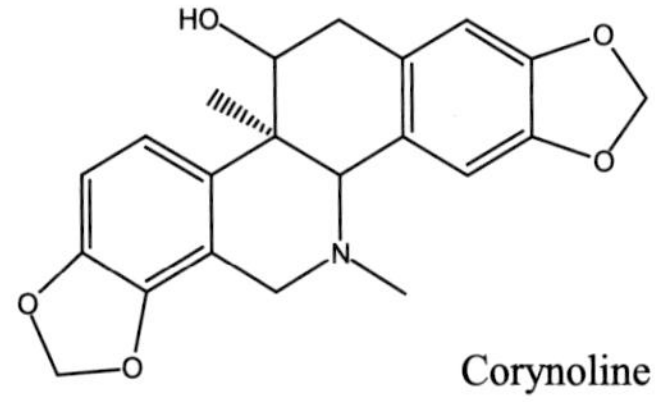

Corynoline

FIGURE 5.2   (*Cont.*)

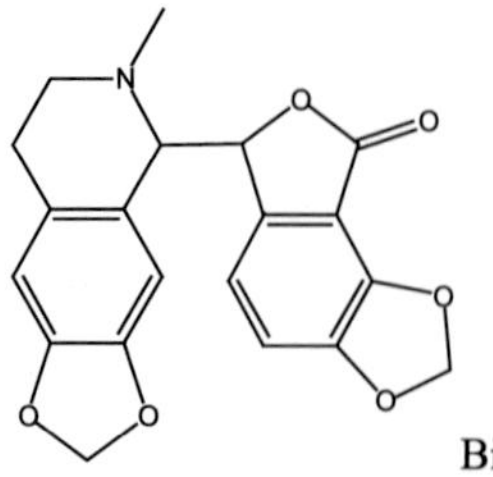

Bicuculline

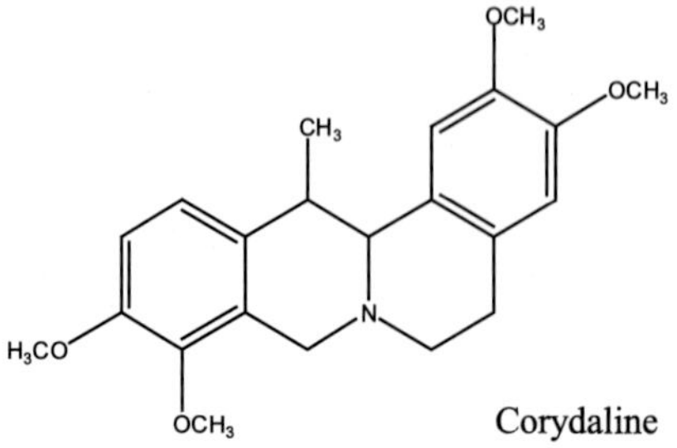

Corydaline

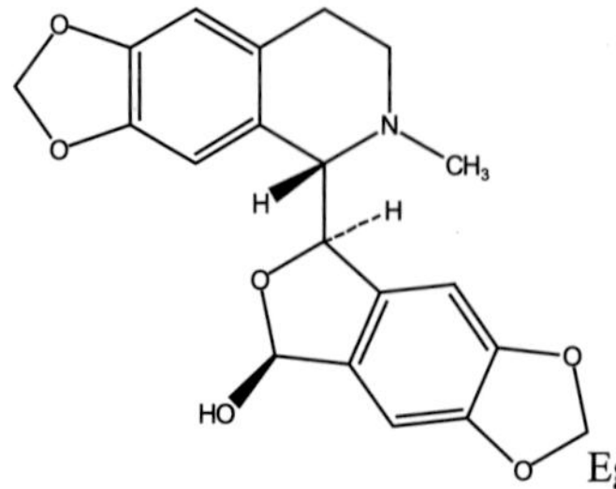

Egenine

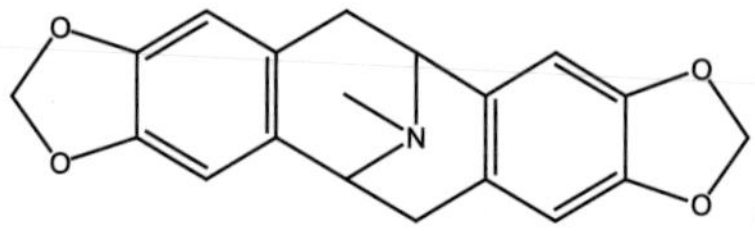

Californine

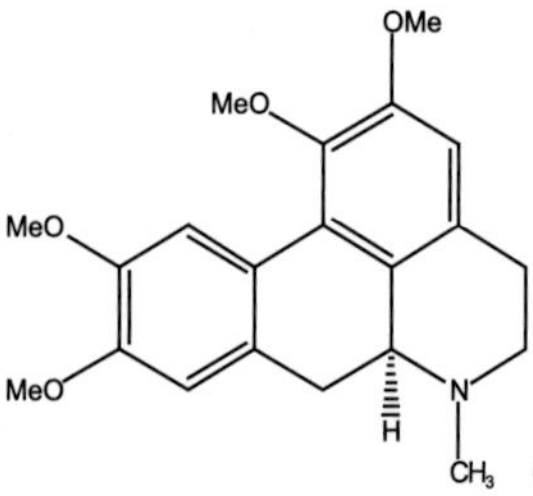

Glaucine

Quaternary benzo[c]phenanthridine alkaloids

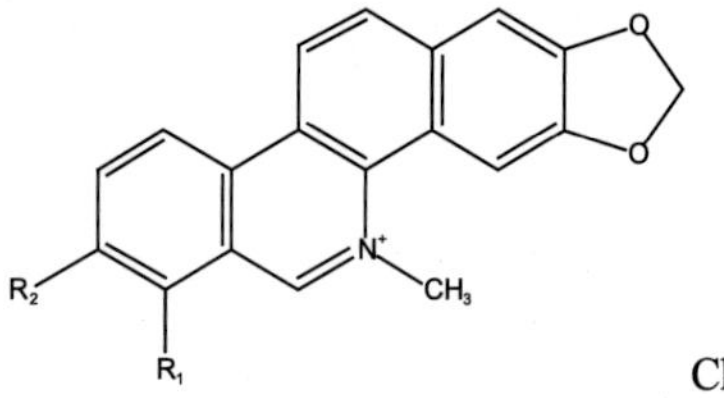

Chelerythrine(R1=R2=OCH$_3$),
Sanguinarine(R1+R2=OCH$_2$O)

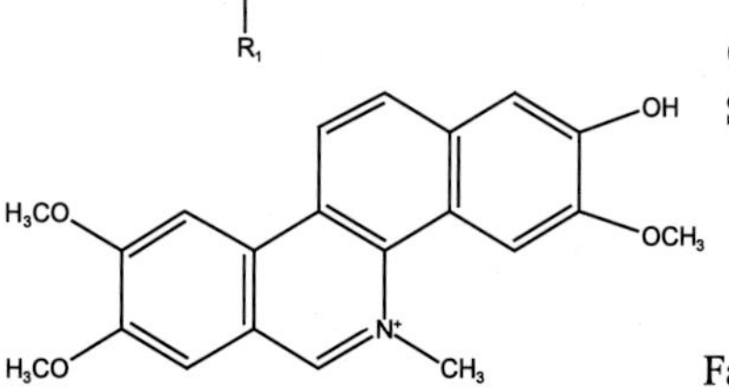

Fagaronine

FIGURE 5.2   (*Cont.*)

Piperidine alkaloids:

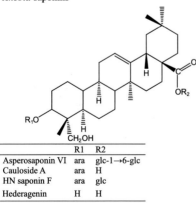

O-acetylbaptifolin

Anagyrine

Lupanine

Caulophyllumine B

Saponins

	R1	R2	R3	R4
Cauloside C	2-O-β-D-glucopyranosyl	OH	H	H
Cauloside D	H	OH	H	4-O-(6-deoxy-α-L-mannopyranosyl)-β-D-glucopyranosyl
Cauloside G	2-O-β-D-glucopyranosyl	OH	H	4-O-(6-deoxy-α-L-mannopyranosyl)-β-D-glucopyranosyl

Akebia saponins

	R1	R2
Asperosaponin VI	ara	glc-1→6-glc
Cauloside A	ara	H
HN saponin F	ara	glc
Hederagenin	H	H

FIGURE 5.2 *(Cont.)*

Epimedium flavonoids

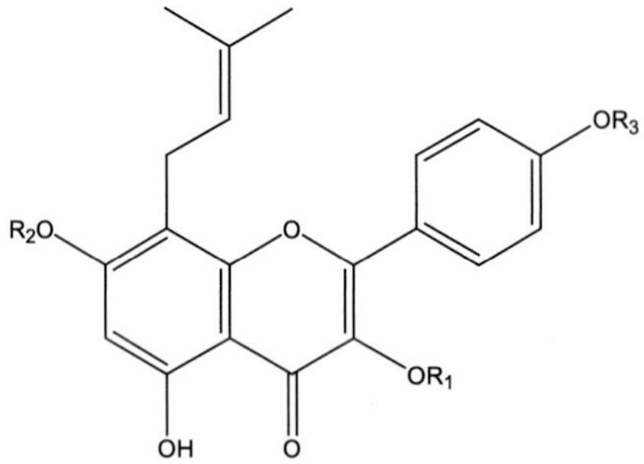

	R1	R2	R3
EpiA	rha(2-1)glc	glc	CH$_3$
EpiB	rha(2-1)xyl	glc	CH$_3$
EpiC	rha(2-1)rha	glc	CH$_3$
icariin	rha	glc	CH$_3$
BaoI	rha	H	CH$_3$
SagA	rha(2-1)glc	H	CH$_3$
SagB	rha(2-1)xyl	H	CH$_3$
2''-O-rha-icaII	rha(2-1)rha	H	CH$_3$
IcaI	H	glc	CH$_3$
Icaritin	H	H	CH$_3$

FIGURE 5.2   *(Cont.)*

Various BIAs exhibit significant bioactivities, and BIA-abundant botanicals have long been used in traditional therapy. In recent decades, BIA biosynthetic pathways of some important Ranunculales compounds, for example, morphine, sanguinarine (Hao et al., 2015b), and BBR, have been elucidated in a few model plants (Hagel et al., 2015b). The transcriptome data sets of 20 nonmodel Ranunculales species, which produce a wide variety of alkaloids, flavonoids, saponins, etc., characterize a gold mine of biocatalyst diversity appreciated by medicinal chemists, phytochemists, and researchers of synthetic biology (Hagel et al., 2015b).

Standardization of the therapeutic reagents from the medicinal plant complex system is challenging, while appraising the bench and bedside efficacy of the product is labor intensive. Most Ranunculales medicinal plants are not domesticated, thus it is indispensible to tackle the botanical taxonomy and to authenticate samples by combining the phylogenetic analysis, chemotaxonomy, and therapeutic efficacy (Hao et al., 2015b; Hao and Xiao, 2015). The evolutionarily closely related species often possess similar chemical profiles and consequently relevant metabolic patterns/therapeutic efficacy (Hao et al., 2014, 2015d). The pharmacokinetic/pharmacodynamics (PK/PD) endeavors constitute a fundamental part of TCM tonic efficacy evaluations. Here, we document and capture the global research efforts in drug metabolism (DM) and disposition of the myriad bioactive compounds in Ranunculales extracts, informing the potential drug–drug interactions (DDIs) and capricious PK/PD behaviors for safety and efficacy of clinical use. Exhaustive literature searches in PubMed, Google, and CNKI (http://cnki.net/) were performed to outline the research trends in DMPK studies of Ranunculales medicinal phytometabolites during the recent decade. Search terms "drug metabolism," "pharmacokinetic," "drug transporter," "absorption," "distribution," "excretion," and "cytochrome p450" were used, combined with "Ranunculales" and the names of each family and genera.

5.2 ABSORPTION

5.2.1 Flavonoids

Research on the processing (Pao Zhi in Chinese) mechanisms of TCM is a critical issue to the modernization of TCM. Presently, chemical and pharmacology methods are used to explore the underlying mechanisms of improved drug efficacy, attenuated toxicity, delayed drug effect, and differential use of raw and cooked herbal drugs. New research avenues of biotransformation, intestinal absorption, pharmacokinetics, and metabolomics are available in TCM processing mechanisms (Sun et al., 2014c). In-depth studies should be performed on absorption and metabolism from a systems perspective, highlighting the integration of multiple disciplines and data statistics.

Flavonoids are ubiquitous in Berberidaceae, Menispermaceae, and Papaveraceae, such as *Epimedium*, *Dysosma*, and *Sinopodophyllum*. *Epimedium* (Yin Yang Huo in Chinese) is recorded in China *Pharmacopoeia* and is traditionally used in kidney yang-invigorating, bone-strengthening, and rheumatic-relieving (Chinese Pharmacopoeia Commission, 2015). The genus *Epimedium* is highlighted for its efficacy in sexual dysfunction and osteoporosis. More than 260 phytometabolites have been identified in *Epimedium* (Li et al., 2015a), most of which belong to flavonoids. During *Epimedium* processing, heating changes the contents of major flavonoids and more absorption-amenable flavonoids, for example, icariin and baohuoside (icariside) I, are available (Tan et al., 2009; Sun et al., 2014b), thus enhancing the therapeutic efficacy of *Epimedium* reagents. The duodenum and jejunum are the major absorption sites of *Epimedium* flavonoids. The excipient suet oil has profound effects on in vivo PK characteristics of icariside I in rats (Qian et al., 2012a). In rats intragastrically administrated with extract solutions, the C_{max} and area under the plasma concentration-time curve (AUC) of raw *Epimedium* products, heating products, and suet oil-processed products were significantly different (processed products > heating products > raw products). Suet oil could promote the oral absorption of icariside I and increase its bioavailability. Fatty acids are abundant in suet oil, which, along with the natural biosurfactant bile salt, could be the base of the spontaneously arranged medicine transport structure to facilitate the absorption of prenyl-flavonoids (Sun et al., 2014e). The nanomicelles consisting of circinal-icaritin, suet oil, and sodium deoxycholate were verified using transmission electron microscopy (Jiang et al., 2015a). It is proposed that the TCM-processing mechanism might be based on the coupled effects of the chemical constituent transformation and intestinal absorption barrier (Sun et al., 2014c). Correspondingly, a research paradigm is put forward: chemical constituent alteration after Pao Zhi → biotransformation → gut metabolism of in vivo and in vitro models → bowel absorption/transport → PK plus PD → mechanismof drug efficacy (Sun et al., 2014d).

The optimized self-emulsifying medicine transport system has oleic acid as the hydrophobic phase, Tween-80 as the nonionic surfactant, and PEG400 as the cosurfactant in the proportion of 2:4:4 (Jiang and Mo, 2010), the dissolution of which in water was more than 85% in 25 min, while that of the *Epimedium* flavonoids capsule was less than 50% in 60 min. The application of the optimized formulation of the self-emulsifying drug delivery system could significantly increase the solubility of *Epimedium* flavonoids in the water and improve its bioavailability. Adding the phospholipid could improve the bowel absorption of icariside II (Jin et al., 2012). A microemulsion-based delivery system enhanced absorption of epimedium flavonoids by real-time enzymolysis and mucoadhesive strategy.

The gut flora or enzymes could hydrolyze *Epimedium* flavonoids into secondary glycosides or aglycon, thus facilitating their absorption and enhancing the antiosteoporoticaction (Chen et al., 2008; Zhou et al., 2015). In osteoporosis rats, the hydrolysis rates of icariin and epimedin A, B, and C were reduced by 21%, 24%, 8%, and 31% respectively when they were incubated with duodenal enzymes for 60 min, while the hydrolysis rates of four compounds were decreased by 13%, 9%, 7%, and 47% respectively when treated with jejunum enzymes. The apparent permeability coefficient (P_{app}) and elimination of the four flavonoids in the four segments (duodenum, jejunum, ileum, and colon) were reduced dramatically as compared with healthy rats. The chief metabolites of the four flavonoids were not dissimilar between osteoporosis rats and normal rats, despite either the gut perfusion or enzyme incubation. The ovariectomized rats had significantly reduced amount and activity of enzymes, which decreased the gut transformation and absorption of 7-glucose-containing flavonoids.

5.2.2 Alkaloids

5.2.2.1 *Berberidaceae*

Mahoniae Caulis is officially listed in China Pharmacopoeia as *Mahonia bealei* and *M. fortunei* (Berberidaceae). In rats the gut absorption of total alkaloids from *Mahoniae Caulis* (TAMC) was much more than that of a sole molecule or a combination of molecules (jatrorrhizine, palmatine, and BBR) (Sun et al., 2014a). The promotion of absorption by the bicyclic monoterpenoids (borneol or camphor) was greater than that by the monocyclic monoterpenes (menthol or menthone), and the promotion by compounds with a hydroxyl group (borneol or menthol) was greater than that by those with a carbonyl group (camphor or menthone). The P_{app} of TAMC was increased to 1.8-fold by verapamil (P-glycoprotein inhibitor) but was reduced to one-half by thiamine. The absorption-rate constant (K_a) and P_{app} of TAMC were not altered by probenecid and pantoprazole. The intestinal absorption of TAMC might be passive diffusion, and the small bowel was the predominant absorption site. TAMC might be transported by P-glycoprotein (P-gp, MDR1, ABCB1) and organic cation transporter [OCT, an solute carrier transporter (SLC) family member], rather than breast cancer resistance protein (BCRP, ABCG2) and multidrug resistance protein (MRP, ABCC). Compared with a single compound and a mixture of compounds, TAMC was more easily absorbed into the circulation.

5.2.2.2 *Menispermaceae*

Sinomenine can be absorbed via skin and might be useful in antiinflammatory and analgesic therapy (Zhang et al., 2010). The PK characteristic followed the one-order rate and two-compartment model. Sinomenine hydrochloride (SMH) is a weak base, and its solubility and partition coefficiently on pH. The permeation enhancer azone was most effective in facilitating SMH absorption (Li et al., 2010). Compared with the patch, the spray formulation allowed for higher accumulative permeated SMH and shorter permeated time. The skin SMH transport system had a comparable antiinflammatory efficiency as the oral ingestion. The C_{max} and AUC of the oral drug were higher than those of topical formulations, implying that the systemic side effects caused by the latter might be negligible.

5.2.2.3 *Papaveraceae*

d-corydaline (CDL) and THP are stable in the human intestinal flora incubation system (Liu et al., 2013b). In the Caco-2 monolayer, the P_{app} magnitudes of both CDL and THP were 1×10^{-5} cm/s in the bidirectional transport, which were similar to that of propranolol. The transport of CDL and THP was concentration dependent between 0 and 180 min and is mainly passive diffusion. The K_a and effective permeability (P_{eff}) of THP were similar at 8, 16, and 32 µg/mL in perfusion or in four segments of the rat gut (Wu et al., 2007). The absorption of (−)-THP (L-THP) and THP in the jejunum was significantly different. Eleven prototype components were detected in the rodent blood after ingesting *Corydalis* rhizome extracts (Hong et al., 2012b). Three isomers, with the m/z 356.18, were differentiated as THP, glaucine, and corybulbine.

Corydalis yanhusuo (Yuan Hu in Chinese) is recorded in China Pharmacopoeia and is traditionally used in blood-activating and pain-relieving (Chinese Pharmacopoeia Commission, 2015). Ten alkaloids, that is, columbamine, fumaricine, THP, 13-methyl-dehydrocorydalmine, dehydrocorybulbine, 13-methyl-palmatrubine, unknown, palmatine, BBR, and dehydrocorydalmine, were found in the rat blood after ingesting *C. yanhusuo* (Cheng et al., 2009), which is the prototype contained in the crude drug. The glucuronide metabolites were found in the circulation. The prototype compounds and their metabolites might be in charge of the therapeutic efficacy. The serum pharmacochemistry should be further probed to shed light on the pharmacology and the underlying machinery of Yuan Hu. Oligochitosan promotes the intestinal absorption of protoberberine alkaloids in *Corydalis saxicola* (Yan Huang Lian in TCM) total alkaloids (Li et al., 2015b).

The representative BIAs PAP, laudanosine (LAU), and cepharanthine (CEP) are used against cardiovascular and cerebral vascular diseases. Their P_{app} is almost identical to that of propranolol (Ma and Yang, 2008), a transcellular transport marker. On the other hand, the efflux transport of PAP is 1.45 times higher than its influx transport, while $P_{app\ AP \to BL}/P_{app\ BL \to AP}$ values of LAU and CEP are 1.07 and 1.19, respectively, suggesting that the efflux transport might not be involved in their absorption in Caco-2 cell monolayers. A good correlation is found between the P_{app} value and apparent distribution coefficient (Log D) at pH 7.35 for the three alkaloids. PAP, LAU, and CEP can be absorbed across intestinal epithelial cells, and they are completely absorbed compounds. The oil/water partition coefficient plays a key role in their transmembrane permeation.

In rabbits, quaternary ammonium alkaloids of *Rhizoma Corydalis Decumbentis* show worse corneal permeability than tertiary amine alkaloids (Mao et al., 2017). Borneol dose dependently enhances corneal penetration of these alkaloids into aqueous humor.

5.2.3 Saponins

Triterpene saponins are common in Berberidaceae (e.g., *Caulophyllum* and *Gymnospermium*) and Lardizabalaceae (e.g., *Akebia*). *Akebia* is recorded in the ancient *Shen Nong Ben Cao Jing* (The Classic of Herbal Medicine) and traditionally used in diuresis-inducing and strangurtia (Miao, 2011). The different methods can present a drug as different solid forms, which have distinct bioavailability (Wang et al., 2015d). *Akebia* saponin D (ASD, asperosaponin VI) is useful in combating against lower backache, fatty liver, traumatic hematoma, bone rupture, and even Alzheimer's disease. The microcrystalline form of ASD (ASD-2) has poorer water solubility than the amorphous ASD (ASD-1), and it shows lower dissolution than ASD-1 at the end of the 3 h test period. However, the $AUC_{0-20 h}$ of ASD-2 is much larger than ASD-1. ASD-2 is cost effective and can be prepared in large-scale.

5.3 DISTRIBUTION

5.3.1 Flavonoids

G-002M, isolated from *Podophyllum* (*Sinopodophyllum*) *hexandrum* rhizome, contains a flavonoid, a lignin, and its glucoside and displays strong radioprotective activity (Dutta et al., 2012). In mice, G-002M was distributed extensively and enriched in lung, liver, upper small bowel, and kidney. The kidney had the maximal retention of G-002M, indicating that it was eliminated mainly via the renal path. Histology of lung, liver, jejunum, and kidney was not altered after G-002M use. G-002M is of high stability, multitissue distribution, longer $t_{1/2}$, and no toxicity, thus it is a promising protective reagent against lethal radiation.

5.3.2 Alkaloids

In rats, sinomenine was distributed broadly (concentrations high to low) in kidney, liver, lung, spleen, heart, etc. (Liu et al., 2005a). The organ concentrations were reduced noticeably after 1.5 h of oral dosing. Sinomenine accumulated in the liver and kidney, the major metabolism and elimination sites. In the albumin solution, the protein-binding rate of sinomenine was more than 60%, while α_1-acid-glycoprotein (AGP) bound much less drug (~33%) due to its acidic property. When used intraperitoneally, sinomenine (e.g., 0.5–2.0 µg/g) was found in the brain, where it exerted neuroprotective activity on H_2O_2-induced damage in PC12 cells (Long et al., 2010).

Sinomenium (Menispermaceae) is recorded in the renowned *Ben Cao Gang Mu* (Compendium of Materia Medica) of the 16th century (Li, 2015). Sinomenine, of the genera *Sinomenium* and *Stephania*, shows strong antirheumatic activity, and it must go into the synovial compartment. The intramuscular injection is a faster approach than oral administration in terms of the time going into blood and synovial fluid (SF) (Yan et al., 2015). Sinomenine penetrates into SF equally well via two routes, and electroporation, a transdermal delivery method, dramatically increases the C_{SF}/C_{Plasma} between 90 and 480 min as compared with systemic dosing.

L-stepholidine, isolated from *Stephania*, is a kind of dopamine receptor antagonist with intriguing characteristics. In rats, the amounts of [³H] stepholidine bound to plasma protein, liver, and kidney homogenates are around 37%, 31%, and 30% respectively (Zhang et al., 1990). After an intravenous (iv) dose of [¹⁴C] stepholidine in mice, the drug rapidly distributes to kidneys, liver, brain, salivary glands, Harder's glands, heart blood, and muscle in 2 min. The highest radioactivities are found in kidneys and stomach mucosa at 30 min. Since L-stepholidine can penetrate the blood-brain barrier (BBB) easily (Natesan et al., 2008), if the hasty excretion is not a serious flaw, it may be useful for schizophrenia.

The binding between chiral compounds and blood proteins is frequently stereoselective, which influences bioactivities and PK contours (Shen et al., 2013). The human serum albumin

(HSA), AGP, and lipoproteins are vital proteins in circulation. THP stereoselectively binds to HSA, AGP, and other plasma proteins (Sun et al., 2010). The protein binding of (+)-THP is more expressive than that of (−)-THP, which is not so in rat blood. The affinity of HSA and AGP to (+)-THP is notably higher than those of (−)-THP. Only Site I of HSA binds (−)-THP, while both Sites I and II bind (+)-THP. The AGP binding drugs diminish the binding between THP and AGP but do not affect the binding between THP and proteins in the human blood.

Six rat brain regions display distinct PK contours of the THP enantiomers (Hong et al., 2006). The AUC and C_{max} of the (−)-THP are much greater than those of the (+)-enantiomer, and the striatum has the highest C_{max}, followed by the hippocampus, diencephalon, plasma, brain stem, cerebellum, and cortex. The tissue distribution of the two enantiomers is substantially different in all organs, apart from the lung. Both enantiomers are most enriched in the liver. The (−)/(+)-enantiomer ratios in different brain areas and other organs correspond to that of the blood.

The total alkaloids of *C. yanhusuo* (TAC) consist of more than 100 phytometabolites (Wang et al., 2010). About 40 compounds go into the rat circulation and ~20 compounds penetrate the BBB. Four alkaloids (protopine, glaucine, THP, and CDL) possessing profound antinociceptive activities accumulate in the striatum to exert synergistic effects in attenuating the formalin-induced nociception. The comprehensive 2-D biochromatography can be used to screen the therapeutic ingredients of *C. yanhusuo* and provide clues for recognizing the machinery by which *C. yanhusuo* regulates nociception.

Identifying brain penetration potential molecules in complex herbal mixtures at the early stage of drug development calls for a high-throughput analytical approach (Fig. 5.3). A parallel artificial membrane permeability assay (PAMPA) of the BBB shows good performance in predictive power and is phytochemically selective. It is an exclusive screening tool in selecting brain-penetrable phytometabolites (Könczöl et al., 2013). After simple modifications of the assay design, PAMPA-BBB filtered samples, for example, extracts of *Corydalis cava*, can be directly used in nuclear magnetic resonance (NMR) and liquid chromatography-mass spectrometry (LC-MS). THP and CDL are found to go through the BBB.

The Chelidonieae alkaloids chelerythrine and sanguinarine strongly bind bovine serum albumin (BSA) and HSA (Sedo et al., 2002). Albumin substantially reduces the inhibitory effects of chelerythrine, sanguinarine, and fagaronine against amino peptidase N activity, which is dose dependent within 50–500 μg/mL BSA.

The alkaloids morphine and codeine were found in all oral fluid (OF) samples 30 min after poppy seed (15.7 mg morphine) ingestion, which remained positive for 0.5–13 h (Concheiro et al., 2015). C_{max} of OF morphine and codeine were 177 and 32.6 μg/L, and T_{max} was 0.5–1 h and 0.5–2.5 h respectively. The detection window after the second seed ingestion was prolonged at least 1 day for morphine and 18 h for codeine.

5.3.3 TCM Formula

Yuanhu Zhitong (YZ) prescription, a TCM formula composed of *Corydalis Rhizoma* (CR) and *Angelicae Dahuricae Radix* (ADR), is used to alleviate various pains, including headaches, spastic pain, and traumatic swelling pain. In mice, peak brain levels for the four CR alkaloids were detected about 15 min after CR administration (Gao et al., 2014), suggesting that they enter the brain rapidly without lag time. AUC_{brain} to AUC_{plasma} ratios for coptisine, palmatine, dehydrocorydaline (DHC), and THP were 0.18, 0.38, 0.26, and 0.20 in the CR

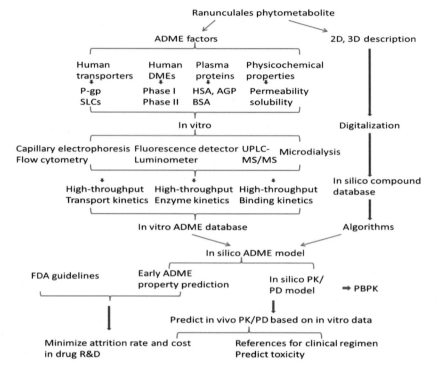

FIGURE 5.3   **Drug metabolism and disposition diversity of Ranunculales phytometabolites revealed from a systems perspective.**

group, respectively. In the YZ group, the ratios of the four alkaloids were 0.13, 0.29, 0.26, and 0.17, respectively. The AUC_{brain} to AUC_{plasma} ratios between the two groups were not significantly different. Multiple components could go across the BBB to the brain, substantiating the beneficial neuropharmacological effects of YZ.

5.4 METABOLISM

5.4.1 Gut, Microflora, and Others

The intestinal microflora and the host cells synergistically preserve gut homeostasis. Drugs and xenobiotics are cometabolized by the microbiota and intestinal epithelia, and the mRNA levels of drug-metabolizing enzymes (DMEs), transporters, and nuclear receptors are influenced by the microbes (Swanson, 2015). *Streptococcus* sp., *Enterococcus* sp., and *Blautia* sp., isolated from human intestines, react with icariin under anaerobic conditions (Wu et al., 2016a). The metabolites icariside II, icaritin, and desmethylicaritin, but not icariside I, are produced. The *Streptococcus* and *Enterococcus* strains hydrolyze only the glucose moiety of icariin, and icariside II is the only metabolite. However, the *Blautia* strain metabolizes icariin further to desmethylicaritin via icariside II and icaritin. It is inferred that most icariin is quickly transformed to icariside II before absorption in the human intestine;

therefore, the pharmacokinetics of icariin metabolites should be highlighted to ameliorate the discrepancy between the in vitro bioassay and pharmacological effects. In osteoporotic rats the hydrolysis of icariin and epimedin A, B, and C incubated with intestinal flora for 60 min is decreased by 19%, 26%, 19%, and 14%, respectively (Zhou et al., 2015), compared with that in normal rodents. The main metabolites of the four *Epimedium* flavonoids after microflora incubation are not different in the two groups, but the amount and activity of intestinal flora are distinct between ovariectomized and healthy rats, and the gut absorption and hydrolysis of 7-glucose-containing flavonoids are weakened by the pathological conditions.

ASD is totally degraded in the large intestine contents of rats in 8 h (Gong et al., 2014). ASD is very stable in the artificial gastric juice, intestinal juice, and gastric contents of rats, implying that it might be absorbed as the prototype. ASD is easily degraded by coliform bacteria but stable in acidic environments and with the presence of digestive enzymes.

Various glucosidases, for example, glucocerebrosidase, lactase-phlorizin hydrolase (LPH), β-glucosidase, and pyridoxine glucoside hydrolase, are functional in the mammalian gut (Chen et al., 2011). In the rat duodenum and jejunum, icariin and epimedin A, B, and C are hydrolyzed quickly before absorption, while baohuoside I is absorbed without transformation. The LPH inhibitor gluconolactone significantly inhibits the hydrolysis of diglycosides and triglycosides. Icariin and epimedin A, B, and CC are converted to baohuoside I, sagittatoside A, sagittatoside B, and 2″-O-rhamnosylicariside II, respectively. The rat gut enzymes transform flavonoids faster than the gut flora (Zhou et al., 2013a). The gut flora transforms icariin the fastest, followed by epimedin B, A, and C and baohuoside I. The gut enzymes also transform icariin the fastest, followed by epimedin A, C, and Band baohuoside I. The gut enzymes or microflora convert icariin into three metabolites; epimedin A, B, and C into four; and baohuoside I into one product. For the same compound, the metabolic pathways via intestinal flora or enzymes are the same. Gut enzymes such as LPH play a more important role in transforming prenylated flavonoids as compared with gut microflora.

Zebrafish transform icariin and epimedin C by removing glucose rather than rhamnose (Wei et al., 2012), which is similar to mammalian models. Little is known about the transformation of the less abundant epimedin A in mammalians. Zebrafish metabolize epimedin A, icariin, and epimedin C in a similar way. Zebrafish, like the mammalian models, are a powerful tool in deciphering transformation pathways of photometabolites, which is not expensive, needs much less compounds, is easily arranged, and of high efficiency. The deglycosylation metabolites and one hydroxylation metabolite of ASD have been identified in zebrafish (Wei et al., 2013). The results are highly inconsistent with rat metabolism of ASD. Zebrafish metabolic models may be used in the quick prediction of metabolism of even trace amounts of compound, providing an alternative for mainstream models.

Penicillium janthinellum AS 3.510 metabolizes L-THP to three compounds (Li et al., 2006a): L-corypalmine, L-corydalmine, and 9-O-desmethyl-L-THP.

5.4.2 Cytochrome P450s

5.4.2.1 *Berberidaceae*

After ingestion of *Epimedium* extract, icariin can be metabolized, possibly via cytochrome P450s (CYPs), to demethylicaritin in Sprague Dawley (SD) rats (Wong et al., 2009), whereas no demethylicaritin is detected in the human serum (Shen et al., 2007), suggesting a difference in the species. Treatment with multiple doses of icaritin has inhibitory effects on rat CYP1A2, 2C9, and 3A4 enzyme activities (Table 5.1) (Liang and Zheng, 2014), and icaritin has no inductive or inhibitory effect on CYP2E1. Caution should be taken when icaritin is coadministered with CYP1A2, 2C9, or 3A4 substrates, which may cause therapy failure and herb–drug interactions (HDIs). In mice, icariin has no effect on Cyp2b10 and 3a11 mRNA levels (Xu et al., 2014) but has surprisingly enhanced Cyp4a14 expression (Table 5.2).

Ethanol extract of *Epimedium koreanum* directly inhibits CYP1A2 (IC_{50} 121.8 µg/ml, K_i 110.7 ± 36.8 µg/ml) and CYP2B6 (IC_{50} 59.5 µg/ml, K_i 18.1 ± 2.9 µg/ml) (Zhong et al., 2017). For CYP2C9, 2C19, 2D6, 2E1, and 3A4, only negligible effects are observed. Time-dependent and irreversible inhibition by *E. koreanum* is observed for CYP1A2 (K_I 32.9 ± 18.4 µg/ml, k_{inact} 0.031 ± 0.006 min^{-1}). Ethanol extract of *E. koreanum* (1.5–150 µg/ml) does not change the activity or mRNA expressions of CYP3A4, 1A2, 2C19, and 2B6.

Both quinolizidine alkaloids and BIAs are found in the genus *Caulophyllum* (Berberidaceae). The alkaloid fraction of blue cohosh (*Caulophyllum thalictroides*) strongly inhibits CYP 2C19, 3A4, 2D6, and 1A2 (>80% inhibition at 100 µg/mL) (Madgula et al., 2009). Four lysine-derived alkaloids (O-acetylbaptifolin, caulophyllumine B, anagyrine, and lupanine) also inhibit these CYPs (IC_{50} 2.5–50 µM) (Table 5.1). A mixture of alkaloids inhibits four CYPs more substantially than a single alkaloid. The saponins caulosides C and D also deter CYP3A4 at the highest test concentration. Blue cohosh, when used in a dietary supplement, may cause DDIs if the coadministered drugs or herbs are CYP substrates.

Aqueous extracts of *Dysosma versipellis* (100 µg/mL) inhibit CYP3A4 by more than 85% (Ashour et al., 2017). The methanol extracts of *M. bealei* inhibit the enzyme by more than 50%.

5.4.2.2 THP

After oral administration of THP enantiomers, more than 30 potential metabolites were found in rat urine (Zhang et al., 2014a), and the metabolic pathways were involved in demethylation, oxidation, glucuronide conjugation, sulfation, etc. (Table 5.3). The tridesmethylation metabolite and didesmethylation, coupled with the oxidation metabolite, were found only in (+)-THP-treated rats. The phase I metabolites of THP dominated over THP and the phase II metabolites in urinary excretion, and there was competition between THP enantiomers. The urinary excretion of THP was stereoselective, and (−)-THP/metabolites were favorably excreted.

THP is an achiral alkaloid that has analgesic and many other bioactivities. (+)-THP is transformed better by liver microsomes of rodents, dogs, and primates, and the intrinsic clearance (CL_{int}) of (+)-THP is much higher than that of (−)-THP. In RLMs (rat liver microsomes), dexamethasone (Dex) and β-naphthoflavone (β-NF) enhance THP conversion. The CL_{int} of (+)-THP is much higher than that of (−)-THP in Dex-induced RLM but not in a β-NF-induced one. CYP3A1/2 and 1A2 are responsible for metabolizing THP in RLM. CYP3A1/2 favors (+)-THP, while CYP1A2 favors (−)-THP. (+)-THP is preferentially metabolized by HLM (Sun et al., 2013a). THP enantiomers do not substantially inhibit the activity of various CYP isoforms; however, (−)-THP significantly inhibited the activity of CYP2D6. In HLM, CYP3A4/5. and 1A2 are in charge of metabolizing THP. CYP1A2 prefers (+)-THP, while CYP3A4/5 favors both enantiomers. DDI might result from the (−)-THP inhibition of CYP2D6.

TABLE 5.1 Examples of Ranunculales Compounds as the Inhibitors of DMEs/Transporters in Human and Animal Studies

Herbal source	Herbal medicine/phytochemical	Enzyme source	Dme/transporter	IC50/Ki (μM)	Mode of inhibition	References
Alkaloid						
Caulophyllum thalictroides	Alkaloidal fraction	cDNA coexpressed enzymes	CYP2C19, 3A4, 2D6, 1A2	IC50, 2 (2C19), 4 (2D6), 8 (3A4), 20 (1A2) μg/mL	ND	Madgula et al. (2009)
	Equimolar mixture of alkaloids		CYP2C19, 3A4, 2D6, 1A2	IC50, 10 (3A4), 4 (2D6), 4 (1A2), 0.5 (2C19) μM	ND	
	Caulophyllumine B		CYP2C19, 3A4, 2D6, 1A2	IC50, 2.5 (2C19), 20 (2D6), 9 (1A2), 50 (3A4) μM	ND	
	O-acetlybaptifolin		CYP2C19, 3A4, 2D6, 1A2	IC50, 40 (3A4), 50 (2D6), 35 (1A2), 12 (2C19) μM	ND	
	Anagyrine		CYP2C19, 3A4, 2D6, 1A2	IC50, 40 (3A4), >100 (2D6), 18 (1A2), 3 (2C19) μM	ND	
	Lupanine		CYP2C19, 3A4, 2D6, 1A2	IC50, >100 (3A4), 10 (2D6), 60 (1A2), 11 (2C19) μM	ND	
	N-methylcytisine		3A4	32% inhibition at 100 μM	ND	
Stephania	Sinomenine	HLM	CYP2C19	Decrease CYP2C19 activity by 69%	ND	Yao et al. (2007)
	Tetrandrine	MCF7/DOX	P-gp	ND	Inhibit P-gp efflux, reduce gene expression	Shan et al. (2013)
	Tetrandrine	HepG2	3A4	IC50, 3.9 μM	ND	
Tinospora cordifolia	Nonpolar fraction, berberine, jatrorrhizine, palmatine	*Escherichia coli* expressing human CYP3A4	3A4	IC50 13.06 ± 1.38, 6.25 ± 0.30, 15.18 ± 1.59, 15.53 ± 1.89 μg/mL	ND	Patil et al. (2014)
Eschscholtzia californica	Californine	RLM	2D1, 2C11	Ki, 750.3 ± 173.5 μM	Substrate inhibition	Paul et al. (2004)

(Continued)

TABLE 5.1 Examples of Ranunculales Compounds as the Inhibitors of DMEs/Transporters in Human and Animal Studies (*cont.*)

Herbal source	Herbal medicine/phytochemical	Enzyme source	Dme/transporter	IC50/Ki (μM)	Mode of inhibition	References
Menispermoideae, papaveraceae	Protopine	RLM	2D1, 2C11	Ki, 2297.0 ± 532.6 μM	Substrate inhibition	Zhao et al. (2015)
	Tetrahydropalmatine	cDNA coexpressed enzymes	2D6	IC50, 3.04 ± 0.26 μM, Ki 1.17 μM	Competitive	
		cDNA coexpressed enzymes	1A2	IC50, >280 μM, Ki 244.8 μM	Competitive	
		cDNA coexpressed enzymes	3A4	IC50, 41.5 ± 3.8 μM, Ki 52.2 μM	Competitive-noncompetitive	
	Neferine (as a control)	cDNA coexpressed enzymes	1A2	IC50, >320 μM, Ki 343.75 μM	Competitive-noncompetitive	
		cDNA coexpressed enzymes	3A4	IC50, 25.1 ± 3.5 μM	Allosteric inhibition	
		cDNA coexpressed enzymes	2D6	IC50, 73.4 ± 16.8 μM, Ki 86.37 μM	Noncompetitive-uncompetitive	
	Berberine	cDNA coexpressed enzymes	2D6	IC50, 7.40 ± 0.36 μM, Ki 3.17 μM	Competitive-noncompetitive	
		cDNA coexpressed enzymes	1A2	IC50, 73.2 ± 5.5 μM, Ki 19.15 μM	Competitive-noncompetitive	
		cDNA coexpressed enzymes	3A4	IC50, 48.9 ± 9.0 μM, Ki 18.47 μM	Competitive	
	(−)-Tetrahydropalmatine	HLM	2D6	Ki 6.42 ± 0.38 μM	ND	

Papaveraceae	Glaucine	Multiple drug-resistant MCF-7/adriamycin	P-gp, MRP1	Reversal ratio 3.25 (12.5 μM), 4.63 (25 μM), 7.87 (50 μM)	Suppress gene expression	Lei et al. (2013)
	Corydaline	HLM	CYP2C19, 2C9	Ki 1.7, 7.0 mM	ND	Ji et al. (2011)
		HLM	UGT1A1, 1A9	Ki 57.6, 37.3 mM	ND	
	Corydalis tuber extract	HLM	CYP2D6	Ki 3.7 μg/mL	Competitive inhibition	Ji et al. (2012)
		HLM	UGT1A1	IC50, 290 μg/mL	ND	
	DA-9701 (Corydalis tuber + Pharbittidis semen)	HLM	CYP2D6	Ki 6.3 μg/mL	Noncompetitive inhibition	
		HLM	UGT1A1	IC50, 188 μg/mL	ND	
Saponin						
C. thalictroides	Caulosides C and D	cDNA coexpressed enzymes	3A4	43% and 35% inhibition at 100 μM	ND	Madgula et al. (2009)
Flavonoid						
Epimedium	Icaritin	Rat	1A2, 2C9, 3A4	ND	ND	Liang and Zheng (2014)
	Icariin	Recombinant human UGT	UGT1A3	IC50 12.4 ± 0.1 μM	ND	Cao et al., 2012
	Icariside II	Recombinant human UGT	UGT1A4, 1A7, 1A9, 2B7	IC50 2.9 ± 0.1 (1A4), 2.8 ± 0.1 (1A7), 2.4 ± 0.1 (1A9), 12.5 ± 0.1 (2B7) μM	ND	
	Icaritin	Recombinant human UGT	UGT1A7, 1A9	IC50 0.3 (1A7), 1.5 ± 0.1 (1A9) μM	ND	
	Icariin	HEK293 overexpressing transporter	OATP1B3, 2B1	IC50 3.0 ± 1.3 (1B3), 6.4 ± 1.9 (2B1) μM	ND	
	Icaritin	Multiple drug-resistant HepG2/adriamycin	P-gp	Reversal fold 1.65 (1 μM), 2.5 (15 μM), 7.18 (30 μM)	Decrease gene expression	

ND, Not determined.

TABLE 5.2 Examples of Ranunculales Compounds as the Inducers of DMEs/Transporters in Human and Animal Studies

Compound type	Herbal sources	Phytochemicals	Experimental model	Target enzyme	Transactivation mechanism	References
Alkaloid	Corydalis yan-husuo	Total alkaloid extract 30 and 150 mg/kg	Rat	CYP2E1, 3A1	Upregulate gene expression	Yan et al. (2014)
		Total alkaloid extract 150 mg/kg	Rat	CYP1A2, 2C11	Upregulate gene expression	
	Fumaria officinalis	Protopine, allocryp-topine	Primary human hepatocytes, HepG2 cells	CYP1A1, 1A2	Increase gene expression	Vrba et al. (2011)
Flavonoid	Epimedium	Icariin	Mice	Cyp4a14	Increase gene expression	Xu et al. (2014)
				Ugt2b1, 2b5, 2b36	Increase gene expression	

TABLE 5.3 Examples of Ranunculales Alkaloids as the Substrates of DMEs/Transporters in Human and Animal Studies

Herbal source	Phytochemicals	In vitro/ in vivo model	DME/ transporter	Metabolic reaction	Km (µM)	Vmax (pmol/min/ mg protein)	References
Stephania	Sinomenine	Human, rat, HLM	CYP	N-demethylation	ND	ND	Yao et al. (2007)
Menispermum dauricum	Dauricine	CD-1 mice	CYP3A4/5	Dehydrogenation	ND	ND	Jin et al. (2010)
Rhizoma Corydalis	(+)-Tetrahydropalmatine	Rat	CYP	N-demethylation, didesmethylation, trides- methylation, oxidation	ND	ND	Zhang et al. (2014)
	(−)-Tetrahydropalmatine	Rat	CYP	N-demethylation, oxidation	ND	ND	
	(+)-Tetrahydropalmatine	Rat	UGT	Glucuronide conjugation	ND	ND	
	(−)-Tetrahydropalmatine	Rat	UGT	Glucuronide conjugation	ND	ND	
	(+)-Tetrahydropalmatine	Rat	SULT	Sulfation	ND	ND	
	(−)-Tetrahydropalmatine	Rat	SULT	Sulfation	ND	ND	
	Tetrahydropalmatine	HLM	3A4/5 and CYP1A2	ND	ND	ND	

(Continued)

TABLE 5.3 Examples of Ranunculales Alkaloids as the Substrates of DMEs/Transporters in Human and Animal Studies (*cont.*)

Herbal source	Phytochemicals	In vitro/vivo model	DME/transporter	Metabolic reaction	Km (μM)	Vmax (pmol/min/mg protein)	References
	Tetrahydro-palmatine	Mouse, dog, monkey LM	CYP	ND	(−)-THP, 52.6 ± 3.5 (mouse), 63.5 ± 1.5 (dog), 33.9 ± 3.3 (monkey), (+)-THP, 31.2 ± 2.8 (mouse), 47.8 ± 7.7 (dog), 22.4 ± 2.9 (monkey)	(−)-THP, 1657 ± 54 (mouse), 481 ± 36 (dog), 1443 ± 120 (monkey), (+)-THP, 2810 ± 137 (mouse), 1467 ± 253 (dog), 1600 ± 131 (monkey)	Zhao et al. (2015)
		RLM	CYP3A1/2 and CYP1A2	ND	30.4 ± 2.3 ((−)-THP), 41.7 ± 4.7 ((+)-THP)	878 ± 13 ((−)-THP), 3200 ± 435 ((+)-THP)	
Menisper-moideae, papaveraceae	Tetrahydro-palmatine	cDNA coex-pressed enzymes	2D6	N-demethylation, didesmethyl-ation, trides-methylation, oxidation	16.49 ± 4.07	19.27 ± 5.18 pmol/min/pmol	
		cDNA coex-pressed enzymes	1A2	N-demethylation, didesmethyl-ation, trides-methylation, oxidation	112.59 ± 7.56	37.59 ± 4.07 pmol/min/pmol	
		cDNA coex-pressed enzymes	3A4	N-demethylation, didesmethyl-ation, trides-methylation, oxidation	150.10 ± 18.13	169.49 ± 22.82 pmol/min/pmol	
Ranuncu-laceae, berberi-daceae, meni-sper-moideae	Berberine	cDNA coex-pressed enzymes	2D6	Demethylation, demethylena-tion	10.40 ± 0.58	14.16 ± 1.44 pmol/min/pmol	

Corydalis bugeana	Corynoline	1A2	cDNA coexpressed enzymes	Demethylation, demethylenation	158.18 ± 24.87	23.26 ± 4.83 pmol/min/pmol	Mao et al. (2015)
		3A4	cDNA coexpressed enzymes	Demethylation, demethylenation	1235.69 ± 350.94	625.0 ± 208.33 pmol/min/pmol	
		CYP2C9, 3A4, 2C19	RLM, recombinant CYP, rat	Oxidation	ND	ND	
		?	RLM	N-acetylcysteine conjugation	ND	ND	
		Glutathione S-transferase	Rat	Glutathione conjugation	ND	ND	
		γ-Glutamyltranspeptidase	Rat	Glu removal to produce Cys-Gly conjugate	ND	ND	
C. yanhu-suo	Alkaloids	UGTs	Rat	Glucuronide conjugation	ND	ND	Cheng et al. (2009)
E. californica	Californine	CYP3A2, 1A2, 2D1	RLM	N-demethylation	4.5 ± 4.7 (high affinity), 161.3 ± 16.7 (low affinity)	22.9 ± 13.7 (high), 311.8 ± 39.4 (low) (peak area ratio/min/mg protein)	Paul et al. (2004)
		2D1, 2C11	RLM	Demethylenation	4.6 ± 0.4	79.6 ± 1.7 (peak area ratio/min/mg protein)	
		UGT, SULT	Rat	Glucuronide conjugation, sulfation	ND	ND	Paul and Maurer (2003)
	Protopine	2D1, 2C11	RLM	Demethylenation	6.2 ± 0.3	152.4 ± 1.5 (peak area ratio/min/mg protein)	Paul et al. (2004)
		UGT, SULT	Rat	Glucuronide conjugation, sulfation	ND	ND	Paul and Maurer (2003)

(Continued)

TABLE 5.3 Examples of Ranunculales Alkaloids as the Substrates of DMEs/Transporters in Human and Animal Studies (*cont.*)

Herbal source	Phytochemicals	In vitro/vivo model	DME/transporter	Metabolic reaction	Km (μM)	Vmax (pmol/min/mg protein)	References
Glaucium flavum	Glaucine	HLM	CYP	2-O-DM-glaucine	81 ± 21	0.35 ± 0.04 pmol/min/pmol	Meyer et al. (2013a)
			CYP	9-O-DM-glaucine	25 ± 9	0.1 ± 0.01 pmol/min/pmol	
			CYP	N-DM-glaucine	70 ± 23	0.83 ± 0.07 pmol/min/pmol	
		Recombinant CYP	1A2	2-O-DM-glaucine	46 ± 8	0.18 ± 0.01 pmol/min/pmol	
		Recombinant CYP	3A4	2-O-DM-glaucine	64 ± 23	0.35 ± 0.05 pmol/min/pmol	
		Recombinant CYP	1A2	9-O-DM-glaucine	55 ± 10	0.27 ± 0.02 pmol/min/pmol	
		Recombinant CYP	2C19	9-O-DM-glaucine	67 ± 14	0.3 ± 0.02 pmol/min/pmol	
		Recombinant CYP	2D6	9-O-DM-glaucine	96 ± 20	1.21 ± 0.11 pmol/min/pmol	
		Recombinant CYP	2D6	N-DM-glaucine	102 ± 40	0.29 ± 0.05 pmol/min/pmol	
		Recombinant CYP	3A4	N-DM-glaucine	140 ± 21	1.92 ± 0.14 pmol/min/pmol	
		Rat	UGT, SULT	Glucuronide conjugation, sulfation	ND	ND	Meyer et al. (2013b)
Poppy	Morphine	Adult male volunteer	UGT	Glucuronide conjugation	ND	ND	Chen et al. (2014)

ND, Not determined.

THP inhibits CYP2D6 more significantly than BBR (Zhao et al., 2012a). The inhibition of CYP3A4 and 1A2 by THP, neferine, and BBR is much weaker. The THP inhibition of CYP1A2 and 2D6 is mechanism based (Zhao et al., 2015), as is the BBR inhibition of CYP2D6. The inhibition of CYP1A2 by neferine and BBR is nonmechanistic. The significant inhibition of CYP3A4 by these alkaloids is mechanism based. THP and BBR show reversible/irreversible inhibition of CYP3A4 and 2D6. The H-bond and π-bond connections are found between alkaloids and particular amino acids of CYP1A2, 2D6, and 3A4, which might be related to the recognized HDIs.

5.4.2.3 Menispermaceae

Tetrandrine moderately inhibits CYP3A4, which participates in the biotransformation of BBR (Shan et al., 2013). Tetrandrine can improve the pharmacokinetics of BBR. The nonpolar fraction of *Tinospora cordifolia* (Menispermaceae) shows significant CYP3A4 inhibition (Table 5.1) (Patil et al., 2014). Major constituents of the nonpolar fraction are BBR, jatrorrhizine, and palmatine. The *T. cordifolia* stem extract contains protoberberine alkaloids, which might interact with CYP3A4 substrates, such as cancer chemotherapy drugs. CYP2C19 is inhibited by 50 μM sinomenine in human microsomes (Table 5.1) (Yao et al., 2007), but in vivo sinomenine at therapeutic concentrations fosters the metabolism of mephenytoin, a CYP2C19 substrate.

T. cordifolia extract has significantly higher IC_{50} value than the positive inhibitors against CYP3A4, 2D6, 2C9, and 1A2 (Bahadur et al., 2016). *T. cordifolia* may not cause any adverse effects when consumed along with other xenobiotics.

5.4.2.4 Papaveraceae

In RLM, corynoline has three oxidative metabolites (Mao et al., 2015). M1 and M2 are two isomers of catechol derivatives, and M3 is a dicatechol. These three metabolites are excreted in rat urine. Corynoline is also activated to orthoquinone derivatives. In RLM, corynoline is also transformed to *N*-acetylcysteine conjugates (M4–M7). M4 and M5 are from M1, M6 from M2, and M7 from M3. However, M4–M7 are not found in rat rurine; alternatively, the cysteinylglycine conjugates (M8–M10) are found, possibly transformed from glutathione conjugates. Corynoline is activated by CYPs 2C9, 3A4, and 2C19. The metabolic investigation benefits the understanding of corynoline-prompted cytotoxicity and CYP inhibition. The liver microsomal incubation of govaniadine, isolated from *Corydalis govaniana*, results in new *O*-demethylated, dihydroxylated and monohydroxylated compounds (Marques et al., 2016).

In HLMs the CYP2C19-mediated S-mephenytoin-4'-hydroxylatoin and CYP2C9-mediated diclofenac 4-hydroxylation are inhibited by CDL (Ji et al., 2011). CDL also suppresses the CYP3A-mediated midazolam hydroxylation in a dose-dependent mode. CDL potentially causes HDIs in vivo due to strong inhibition of CYP2C19 and 2C9. DA-9701, consisting of *Corydalis* tuber and *Pharbitidis* semen, is used to treat functional dyspepsia in Korea (Ji et al., 2012). The DA-9701 inhibition of CYP2D6 is noncompetitive, but DA-9701 may not be an effective CYP2D6 inhibitor. Further clinical investigations are needed to assess the in vivo possibility of the in vitro inhibition.

Californine undergoes *N*-demethylation and/or demethylation with successive catechol-*O*-methylation of one of the hydroxy groups (Paul and Maurer, 2003), while protopine is subjected to the demethylation of the 2,3-methylenedioxy group before catechol-O-methylation. UDP glucuronosyltransferases (UGTs) and/or sulfotransferases (SULTs) catalyzes the conjugation of phenolic hydroxy metabolites. CYP3A2 is responsible to the californine N-demethylation, and CYP1A2 and 2D1 also participates in this transformation (Table 5.3) (Paul et al., 2004). The demethylenation of californine and protopine is catalyzed byCYP2D1 and 2C11, and CYP1A2 and 3A2 play a minor role. The substrate inhibition of CYPs shown by californine demethylenation and protopine demethylenation is negligible.

Glaucine, a main isoquinoline alkaloid of *Glaucium flavum* (Papaveraceae), underwent O- and N-demethylation to generate four isomers in rats. The typical Michaelis-Menten equation explains the kinetic behavior (Meyer et al., 2013a). In the liver, CYP3A4 and 1A2 are responsible for 2-O-demethylation, and CYP1A2, 2D6, and 2C19 are in charge of 9-O-demethylation. CYP3A4 is also responsible for N-demethylation. As glaucine is metabolized via three initial steps and different CYPs are involved in the hepatic clearance of glaucine, a clinically relevant interaction with single CYP inhibitors is highly impossible.

Yanhusuo (*C. yanhusuo*) is used in TCM for chest pain, epigastric pain, and dysmenorrhea. Its alkaloid ingredients, for example, THP, inhibit CYPs in vitro (Yan et al., 2014). The hepatic CYP2E1 and 3A1 activities are increased by the total alkaloid extract (TAE) (Table 5.2). The TAE 6–30 mg/kg significantly upregulates gene expression levels of CYP2E1 and 3A1 of the rat liver, lung, and gut. In rats the activities and gene expression levels of CYP1A2 and 2C11 are increased by 150 mg/kg TAE. In contrast, CYP2D1 is not induced. Coadministration of prescriptions containing Yanhusuo might be involved in a potential HDI via the induction of CYP2E1 and 3A1 enzymes.

The EtOH extract and fractions of *Eschscholzia californica* show strong time-dependent inhibition of CYP 3A4, 2C9, and 2C19 (Manda et al., 2016), and reversible inhibition of CYP 2D6. The alkaloids escholtzine and allocryptopine time-dependently inhibit CYP 3A4, 2C9, and 2C19 (IC$_{50}$ shift ratio > 2), while protopine and allocryptopine reversibly inhibit CYP 2D6. The pregnane X receptor was significantly activated (>2-fold) by the EtOH extract, basic CHCl$_3$ fraction, and alkaloids (except protopine), resulting in an increased expression of mRNA and activity of CYP 3A4 and 1A2.

5.4.2.5 CYP Induction

In HepG2 cells the CYP1A1 gene expression level is significantly upregulated by protopine and allocryptopine (Vrba et al., 2011). In human hepatocytes, CYP1A1 and 1A2 gene expression levels are dose-dependently increased by these alkaloids, but their effects are less significant than those of 2,3,7,8-tetrachlorodibenzo-p-dioxin, an archetypal CYP1A inducer. The induction of CYP1A1 expression by alkaloids might not be related to the activation of the aryl hydrocarbon receptor. The CYP1A protein or activity levels are not increased, thus the reagents containing protopine and/or allocryptopine might be safe.

In mice, icariin has no effect on Cyp2b10 and 3a11 mRNA level (Xu et al., 2014), but surprisingly it enhances Cyp4a14 expression (Table 5.2).

5.4.2.6 TCM Formula

Contrary to most HDIs, in most circumstances the herb–herb interaction within a TCM formula is preferred by professionals to achieve synergistic therapeutic effects of different herbs. In the YZ formula group the AUCs of the four alkaloids in both mouse plasma and brain are significantly increased as compared with those of the CR group (Gao et al., 2014). Remarkable reductions of CL are found in the YZ group, which might be due to the inhibitory effects of the ADR extract on CYP enzymes. ADR inhibits activities of CYP1A2 (Yi et al., 2009), and coumarins in the ADR (imperatorin, oxypeucedanin, byakangelicol, and byakangelicin) are involved in the metabolic inhibition of CYP3A4 (Guo et al., 2001). The ADR and coumarins therefore potentially influence the pharmacokinetics of CR alkaloids.

5.4.2.7 CYP Probe

The CYP3A subfamily is the principal group of liver CYPs and participates in the transformation of 50%–60% of drugs, including Ranunculales herbal medicine. Bufalin 5β-hydroxylation was specifically catalyzed by CYP3A4, instead of CYP3A5 and 3A7; therefore, bufalin could be a biotransformation probe substrate (Ge et al., 2013). The well-depicted probe reaction can be used to quantify the genuine activities of CYP3A4 from various sources toward alkaloids, flavonoids, and other substrates, inhibitors, and/or inducers of Ranunculales. Gomisin A, with a dibenzocyclooctadiene skeleton, is an isoform-specific probe for the selective detection of human CYP3A4 in HLMs and cells (Wu et al., 2016b). Gomisin A has been successfully applied to selectively monitor the modulation of CYP3A4 activities by the inducer rifampin in hepG2 cells, which reflects the level change of CYP3A4 mRNA expression.

Analogously, a ratiometric two-photon fluorescent probe N-(3-carboxy propyl)-4-methoxy-1,8-naphthalimide that allows for selective and sensitive detection of CYP1A was developed (Dai et al., 2015). The probe can be used to real-time monitor the enzyme activity of CYP1A in complex biological systems and has the potential for fast screening of CYP1A regulators based on tissue preparations. The highly selective and specific fluorescent probe of DMEs and drug transporters can be designed for direct interaction between Chinese materia medica extract/its chemical constituents and DME/transporters within the context of high-throughput screening. Screenings of both inhibitors and inducers could be performed based on such a probe.

5.4.3 Phase II DMEs

5.4.3.1 Flavonoid

After oral administration of icariin, 19 metabolites were detected and identified in rat blood (Qian et al., 2012b). The glucuronide conjugates of icariin and the aglycone were abundant in their circulation. Forty-three metabolites of epimedin B found in rat feces, bile, urine, and blood (Cui et al., 2013b) were generated via deglycosylation, dehydrogenation, hydrogenation, demethylation, hydroxylation, glucuronidation, glycosylation, etc. The deglycosylation of 7-O-glucosides in the bowel and glucuronidation in the liver were major metabolic pathways of epimedin B.

In rats, glucuronide conjugates of flavonoids were the major plasma compounds after oral administration of *Epimedii wushanensis herba* (EWH) (Wang et al., 2015c). Due to the saturation effects of gut conjugation, when high doses of EWH were used, the cycling of flavonoid conjugates back to the gut lumen were be strikingly declined (Silberberg et al., 2006). When the highest dose of icaritin was used, the elimination was delayed, due to different phase II enzymes being activated (Wong et al., 2009). The SULTs, rather than the UGTs, could be favored at higher doses. Since the sulfo-conjugates are more hydrophilic, they may be less effectively excreted by the kidneys (Liu and Hu, 2007). Similar depot effects were found in ERE-luc transgenic mice, where a single physiological dose of the isoflavone was subjected to prolonged elimination and induced strong estrogen receptor (ER)/ERE-driven activity in organs such as the liver, brain, and testis for more than 24 h (Montani et al., 2008).

Icariin strongly inhibited UGT1A3 (Table 5.1) (Cao et al., 2012), while the intestinal metabolites of icariin had a distinct inhibition profile. Icariside II potently inhibited UGT1A4, 1A7, 1A9, and 2B7, and icaritin potently inhibited UGT1A7 and 1A9. The in vivo inhibition of gut UGT1A3, 1A4, and 1A7 is possible after ingestion of icariin. In mice, icariin did not suppress Ugt1 family members (Xu et al., 2014). The highest dose (320 mg/kg) of icariin marginally induced Ugt2b1, 2b5, and 2b36.

5.4.3.2 Alkaloid

After oral administration of BBR of 100 mg/kg/day for three consecutive days in rats, 97 metabolites were identified (Wang et al., 2017a), including 68 in urine, 45 in plasma, 44 in bile, and 41 in feces. Demethylation, demethylenation, reduction, hydroxylation, and subsequent glucuronidation, sulfation, and methylation were the major metabolic pathways of BBR in vivo.

In rats, O-demethylation, hydroxylation, dihydroxylation, glucuronidation of O-demethyl DHC, sulfation of O-demethyl DHC, and dihydroxylation of dehydro-DHC were the major metabolic pathways of DHC (Guan et al., 2017). In HLMs, UGT1A1, and 1A9 were modestly inhibited by CDL (Ji et al., 2011). UGT1A1 was marginally inhibited by the Corydalis-containing DA-9701 and Corydalis tuber extract (Ji et al., 2012). Glaucine undergoes O- and N-demethylation, hydroxylation, N-oxidation, glucuronidation, and/or sulfation (Meyer et al., 2013b). Among 21 phase II metabolites, 6 were sulfo-conjugates and 15 glucuronides.

Forty-four metabolites (including 13 phase I and 31 phase II metabolites) of ambinine, a major hexahydrobenzo[c]phenanthridine alkaloids of Corydalis ambigua var. amurensis, and the parent drug were identified in the plasma, bile, urine and feces of rats (Liu et al., 2016a). Demethylation, sulfation, and glucuronidation were the major metabolic pathways of ambinine in vivo.

Papaver somniferum is used in TCM as an astringent and pain killer. Morphine and morphine-3-glucuronide were detected in the urine up to 24 h after the ingestion of poppy seeds (Chen et al., 2014). A glucuronide metabolite (ATM4G), a marker of "street" heroin administration, was not identified in the urine. The 6-monoacetylmorphine (MAM) was not found in the volunteers' urine samples.

Identifying a selective UGT probe was challenging due to the significant overlapping substrate specificity exhibited by the enzyme. UGT1A1 specifically transformed N-3-carboxy propyl-4-hydroxy-1,8-naphthalimide-4-O-glucuronidation to a single fluorescent metabolite (Lv et al., 2015), which is useful in sensing UGT1A1 in human liver and is also powerful in fast screening of UGT1A1 regulators. The desacetylcinobufagin (DACB) 3-O- and 16-O-glucuronidation are specifically catalyzed by UGT1A4 and 1A3, respectively (Jiang et al., 2015b). DACB fluorescent probes can be used to concurrently determine UGT1A3 and 1A4 of numerous enzyme sources. These fluorescent probes could be useful in studying phase II metabolism of Ranunculales compounds and the relevant HDIs.

5.4.3.3 Saponin

Seven nor-oleanane triterpenoid saponins, isolated from Stauntonia brachyanthera (Lardizabalaceae), showed inhibitions toward UGT1A1, 1A3, and 1A10 (Liu et al., 2016); the inhibitions of UGT1A10, 1A1, and 1A3 was best fit to noncompetitive and competitive types, respectively.

5.4.4 Phase III: Drug Transporter

5.4.4.1 Flavonoid

The efflux via apical transporters attenuated the absorption of baohuoside I (Chen et al., 2008). The inhibitors of BCRP and MRP2 dramatically decreased the efflux of baohuoside I, and the P-gp inhibitor verapamil inhibited the efflux of icariin.

In HepG2/adriamycin (Adr) cells, icaritin downregulated the P-gp gene expression and significantly augmented the intracellular buildup of Adr (Sun et al., 2013b). Icaritin, a strong multidrug resistance (MDR) reversal drug, could be used in cancer chemotherapy.

The icariin/hydroxypropyl-β-cyclodextrin (HP-β-CD) complex facilitated the gut absorption of icariin as compared with the icariin/β-CD complex (Zhang et al., 2012). The solubilizing effect and/or P-gp inhibitory effect could explain the augmentation effect. β-CD did not affect P-gp, and HP-β-CD suppressed the ATPase activity of P-gp instead of altering the membrane fluidity.

The SLC family members transport extremely diverse solutes, including charged and uncharged organic compounds (Hao et al., 2013b). Icariin significantly inhibits organic anion transporting polypeptide 1B3 and 2B1 (Table 5.1) (Li et al., 2014a). The competition between icariin and other SLC substrates may interfere with multidrug therapy.

5.4.4.2 Alkaloid

The elimination of sinomenine in the bile was slow (Tsai and Wu, 2003). It took 20–40 min to reach peak concentration after iv dosing. The bile AUC was significantly decreased when P-gp inhibitor cyclosporin A was coadministered, and the bile/plasma ratio was significantly decreased. Contrarily, the CYP inhibitor proadifen increased the plasma and bile AUCs, and the plasma-to-bile distribution ratio was not significantly changed. Sinomenine might undergo active hepatobiliary elimination via P-gp, and CYP could participate in its metabolism.

The k_a and P_{eff} of THP increases dramatically when P-gp inhibitor verapamil is coperfused with THP (Wu et al., 2007), while those of l-THP are not influenced by verapamil. The intestinal absorption of THP is a passive diffusion process and without a special absorption region. The stereoselective absorption difference may result from a stereoselective combination of P-gp with (+)-THP (d-THP).

Dauricine is able to pass the BBB, and P-gp plays an important role in the transportation of dauricine across the BBB (Dong et al., 2014).

P-gp and MRP1 (MDR-associate protein 1) are inhibited by glaucine, but the ATPase activity of the transporters is activated by glaucine (Lei et al., 2013). Glaucine, a transporter substrate, competitively suppresses P-gp and MRP1. The expression of ABC transporter genes s inhibited by glaucine. The resistance of MCF-7/Adr to adriamycin and mitoxantrone is reversed efficiently by glaucine.

Tetrandrine and fangchinoline, bisbenzylisoquinoline alkaloids from *Stephania tetrandra*, can reverse MDR by inhibiting P-gp activity in multidrug-resistant human cancer cells (Sun and Wink, 2014). The efflux of P-gp substrates rhodamine-123 and BBR is suppressed by tetrandrine of *Stephania*, and the uptake of P-gp substrates is increased by tetrandrine in MCF7/DOX cells and Caco-2 cells (Table 5.1) (Shan et al., 2013). The expression of P-gp is seriously reduced by tetrandrine in Caco-2 cells. Tetrandrine dose-dependently increases the average C_{max} and $AUC_{0-24\,h}$ of coadministered BBR in mice. In diabetic mice the hypoglycemic effect of BBR is enhanced by tetrandrine. Tetrandrine decreases cell resistance to phenytoin and valproate and decreases seizure rate, thus the treatment efficacy of refractory epilepsy is improved. Tetrandrine reduces the expression of P-gp at mRNA and protein levels in vivo.

5.4.5 Drug–Drug Interaction/Herb–Drug Interaction

Both *Epimedium sagittatum* and sildenafil, a phosphodiesterase 5 inhibitor, are used to treat erectile dysfunction (Hsueh et al., 2013). A high dose of *E. sagittatum* extract significantly decreases AUC of sildenafil (10 mg/kg). There is significant HDI between *E. sagittatum* and sildenafil at various daily doses, and coadministration of the two drugs in clinical settings has to be avoided.

Tight junctions (TJs) are the structures formed between Caco-2 cells to mediate the paracellular transport of molecules (van Breemen and Li, 2005). Sinomenine HCl decreases the transepithelial electrical resistance of cell monolayers (Lu et al., 2010). Sinomenine expressively enhances the apical-to-basolateral movement of cimetidine (P-gp substrate), vitamin C, rutin, luteolin, and insulin and reduces the reverse movement of cimetidine. Sinomenine augments the temporary opening of TJs and/or inhibits the active efflux of the P-gp substrate. *Paeonia lactiflora* root and *Sinomenium acutum* are used in combination in TCM to combat against rheumatoid arthritis. Sinomenine greatly increases the bioavailability of paeoniflorin in rat models (Liu et al., 2005c).

The expression of claudin-1 is inhibited by sinomenine (Li et al., 2013), which leads to TJ opening and increased paracellular transport. After sinomenine is removed, the claudin-1 increases and the TJ closes. PKC-α is activated by sinomenine and is moved from the cytosol to the membrane. In rats the absolute bioavailability of octreotide is increased by sinomenine, which are related to the key role of claudin-1 in sinomenine-elicited reversible TJ opening and PKC activation.

In schizophrenic patients, those who were treated with *Akebia Caulis* were more likely to suffer from a poor prognosis (OR = 14.6, 95% CI 2.50–31.87) (Zhang et al., 2011a). The side effects in the schizophrenic population could result from the simultaneous herbal and antipsychotic treatments. Potential HDIs need to be further evaluated.

5.5 EXCRETION AND ELIMINATION

5.5.1 Urinary Excretion

5.5.1.1 Berberidaceae

Desmethylicaritin, icariside II, and icaritin, the metabolites of icariin, were observed in a urine sample (Liu et al., 2005b). Icaritin and desmethylicaritin, the weak xenoestrogen, significantly increased proliferation of MCF-7 cells, but icariin had no such effect.

Compared with icariin control solution, icariin stealth solid lipid nanoparticles (Ica-SSLN) had seven times longer $t_{1/2\beta}$ and four times larger AUC (Liu et al., 2012). Ica-SSLN significantly increased the renal excretion of icariin.

5.5.1.2 *Papaveraceae*

THP and tetrahydroberberine, following oral use of *Rhizoma corydalis*, were detected in rat urine (Zhang et al., 2011b). In rats, the cumulative urinary excretion over 4 days of (−)-THP was much more than that of (+)-THP (Hong et al., 2010). The collective biliary elimination over 1 day of (−)-THP was slightly higher than that of (+)-THP. The average cumulative quantity of (−)-THP was much higher than that of (+)-THP in urine and bile. However, the (−)-/(+)-THP concentration ratio in rat plasma was much higher than that in urine and bile. The excretion of THP enantiomers was stereoselective but did not reflect the chiral PK aspects in the circulation. (−)-THP was favorably eliminated in the rat urine and bile. THP and its *O*-demethylation metabolites (e.g., 10-*O*-demethyl-THP) were found in rat urine.

Six metabolites were detected in rats, three of which (l-corydalmine, l-corypalmine, and 9-*O*-desmethyl-L-THP) can also be generated through microbial transformation (Li et al., 2006a). The other three metabolites were 2-*O*-desmethyl-L-THP and two di-*O*-demethylated L-THPs. Renal excretion was not the main excretion pathway of corynoline and acetycorynoline in humans (Liu et al., 2016c).

Could poppy seed ingestion produce positive results in the heroin marker assay? Uncooked poppy seeds with known morphine and codeine content were ingested to investigate urine opiate pharmacokinetics (Smith et al., 2014). Among 391 urine specimens after 32 h following the ingestion of 45 g of poppy seeds, morphine was found in 26.6% (2000 µg/L cutoff) and 83.4% (300 µg/L) samples, respectively. The morphine concentration varied from <300 to 7522 µg/L. As early as 6.6 h, morphine was detected in a urine sample at 2000 µg/L cutoff. Codeine never exceeded above 2000 µg/L, and 20.2% of urine samples had codeine of more than 300 µg/L. These data are useful in understanding urine opiate results. Neither 6-acetyl-codeine (AC) and 6-MAM nor ATM4G (potential street heroin marker acetylated-thebaine-4-metabolite glucuronide), but morphine and codeine could be detected in urine samples following poppy seed ingestion (Maas et al., 2017). Neither papaverine nor noscapine could be observed, even after consumption of poppy seeds containing up to 9.8 µg of papaverine and up to 37 µg of noscapine. In urine samples with suspicion of preceding heroin consumption, ATM4G was detected in 9 of 43 cases, while 6-AC and 6-MAM was found in only 7 urine samples. ATM4G should be measured additionally when requiring discrimination of street heroin consumption from poppy seed intake.

5.5.2 Hepatobiliary Excretion

BBR, abundant in *Berberis aristata* (Berberidaceae) and *Coptis* (Ranunculaceae), exhibits a linear PK profile between 10 and 20 mg/kg (Tsai and Tsai, 2004). BBR is excreted through the hepatobiliary route against a concentration gradient. The coadministration of BBR and CsA (P-gp inhibitor) or quinidine (OCT and P-gp inhibitor) greatly reduces the BBR in bile, indicating that the active BBR efflux could be affected by P-gp and OCT.

Fourteen metabolites of icariin, including demethylicariin, icariside I-3-O-glucuronide, demethylicariside II, demethylicariside II-7-O-glucuronide, and dehydroxyicaritin-glucuronide, were detected in rat bile, which is the main excretion pathway for icariin and its metabolites (Cheng et al., 2016).

5.6 TOXICITY AND SAFETY

5.6.1 Alkaloid

5.6.1.1 Berberidaceae

Nicotine and nicotine-like alkaloids, such as anabasine, n-methylcytisine, coniine, cytisine, n-methylconiine, and γ-coniceine, present in *C. thalictroides*, are absorbed easily and quickly and broadly distributed (Schep et al., 2009). These alkaloids easily penetrate the BBB and the placenta and are abundant in breast milk. They are metabolized mainly in the liver and excreted rapidly through the kidney. The acute intoxication symptoms usually have two phases. The early cholinergic stimulation leads to abdominal pain, hypertension, tachycardia, tremors, etc. (Brown, 2017). Hypotension, bradycardia, and dyspnea can be observed in the delayed inhibitory phase, followed by coma and respiratory failure. Nicotine and nicotine-like alkaloids may result in serious poisoning.

5.6.1.2 Menispermaceae

Dauricine (Fig. 5.2) is abundant in the roots of *Menispermum dauricum*. Mouse lung microsomal enzymes transform dauricine to a quinone methide (Jin et al., 2010), and ketoconazole (CYP3A4 inhibitor) inhibits its formation. Pulmonary injury can be caused by dauricine transformation in CD-1 mice, which might be dependent on CYP3A. The electrophilic metabolite undoubtedly plays a key role in dauricine-induced lung toxicity. Some renal toxic alkaloids are derived from *S. tetrandra* (Xu et al., 2016a). However, Fang-Ji-Huang-Qi-Tang extract is a high-safety-index Chinese medicine for antinociceptive and antiinflammatory application when *Radix S. tetrandra* is correctly used in it (Lin et al., 2015).

An antihypertensive herbal recipe containing *Tinospora crispa* did not cause death or any toxic signs in mice or rats (Charoonratana et al., 2017). The oral LD_{50} in mice was more than 5.0 g/kg. Some hematological and serum biochemical values of treated rats were significantly different from those of the control group; however, all values were within the normal ranges. No damage to liver or kidney was observed in the treatment group. Jinqing granules, consisting of *Radix Tinosporae* and *Canarii fructus*, can be used to prevent and treat gastric ulcer. In Sprague-Dawley rats, they were orally administered for 30 days at 8 g/kg or less, which is safe (Zhou et al., 2017), but higher doses might not be safe.

5.6.1.3 Papaveraceae

The minimum effective concentrations of *Argemone* oil that produce significant frequencies of chromosome aberration and sister chromatid exchange are 0.1 and 0.01 ml/kg, respectively (Ghosh and Mukherjee, 2016). *Argemone* oil and sanguinarine induce an insignificant increase of average generation time, indicating that they are noncytotoxic in the concentrations tested. *Argemone* oil is genotoxic even at low concentrations and its usage should be checked.

The aqueous extract (tea) of *E. californica* and its main alkaloid californidine does not affect CYPs, P-gp, or the pregnane X receptor (Manda et al., 2016). The tea might be safer than the EtOH extract.

The safety of standardized *Macleaya cordata* extract was assessed in an 84-day dietary study in dairy cows (Wang et al., 2017b). Concentrations of sanguinarine and chelerythrine in milk samples collected on day 84 were below the detection limit of HPLC-MS/MS. *M. cordata* extract, when used at up to 10,000 mg/animal/day, is well tolerated by dairy cows.

5.6.2 Flavonoid and Lignin

Podophyllotoxin, a lignin, is the main toxic ingredient of *Dysosma pleiantha* (Bajiaolian) rhizome. Bajiaolian is the fifth highest cause of poisoning among herbal medicine in Taiwan (Karuppaiya and Tsay, 2015). Since the therapeutic and toxic doses are very close, Bajiaolian poisoning is frequently reported. Early diagnosis of *Dysosma* poisoning is difficult because physicians are unfamiliar with this medicine's multiple clinical presentations in different stages of intoxication. Acute *D. versipellis* poisoning causes multiorgan pathological changes (Xu et al., 2013). There is a positive correlation between the toxic effect and dosage. The target tissues and organs are brain (neurons), heart, liver, and kidney.

The water extract of *E. koreanum* had no genotoxicity in bacterial reverse mutation, mammalian chromosomal aberration and in vivo micronuclei formation (Hwang et al., 2017).

5.7 PHARMACOKINETICS AND PHARMACODYNAMICS

5.7.1 Flavonoids

The effectiveness and safety of *Herba Epimedii*, its most abundant ingredient icariin, and other compounds have been validated by recent PK and toxicity studies (Zhai et al., 2013). Epimedin C (1 mg/kg, iv), a prenylflavonoid, displayed rapid distribution and slow elimination (Lee et al., 2014). The oral bioavailability of epimedin C in the pure compound was higher than that in *Herba Epimedii*, implying that other components of the herb could decrease the oral bioavailability of epimedin C.

Five prenylflavonoids of a standardized *Epimedium* extract show dose-dependent nonlinear AUC escalations (Wong et al., 2009). There were two consecutive PK phases. In the early phase, the maximal concentration of icariin and icariside II was found at 30–60 min. In the late phase, icariside I, icaritin, and desmethylicaritin reached their peak concentrations at 8 h. The metabolites icariside I, icaritin, and desmethylicaritin exerted estrogenic activity in *Epimedium*-treated rats. The glucuronidase/sulphatase treatment might increase the amount of icaritin, thus prolonging estrogenic activity. The depot effect led to larger AUCs at the three doses and thus bridged between the pharmacokinetics of flavonoids and the dynamics of their bioactivities.

Few studies correlate PK parameters with the PD marker. The *Epimedium* and extracts from differentially processed *Epimedium* influenced PK parameters distinctly in kidney-yang deficient mice (Cui et al., 2013a). The increased plasma superoxide dismutase was used as the PD indicator, and the pharmacological effect method was used to reveal the relationship between time and equivalent body dose. The synergistic effects of multiple components in TCM/formula and their dynamic alteration can thus be investigated. The suet oil-processed *Epimedium* showed higher C_{max} and larger AUC than heated *Epimedium*, and the latter was followed by crude *Epimedium*. The effect of promoting absorption of oil-processed *Epimedium* was stronger than that of the heated product.

In male SD rats, the ER α activity was substantially increased by the ingestion of *Epimedium brevicornum* (Yap et al., 2007). *E. brevicornum* extract intensifies the estrogenic activity in the serum, and clinical studies are warranted to assess whether it is useful in the estrogen replacement therapy.

5.7.2 Alkaloids

5.7.2.1 *Berberidaceae*

Berbamine (BBM, Fig. 5.2), a bisbenzylisoquinoline alkaloid of *Berberis amurensis*, was extensively distributed in the organs (Liu et al., 2013a). The $t_{1/2}$ 496.1 min indicates that BBM is eliminated slowly in rats. The higher dose led to the larger AUC and higher plasma concentration, indicating the linear PK process of BBM. In rats, the C_{max} and AUC_{0-t} of 8-cetyl-berberine (BBR-C16) were 2.8 and 12.9 times higher than those of BBR (Hu et al., 2017), and the relative bioavailability of BBR to 8-BBR-C16 was 7.7%. The total excretion amount of 8-BBR-C16 via urine and feces was 76.9% and that of BBR was only 20.5%. In rats, 8-BBR-C16 was subjected to phase I demethylation and phase II glucuronidation or sulfation.

The absolute bioavailability of berberrubine was 31.6% (Wang et al., 2015b). Berberrubine is rapidly absorbed and widely distributed in various tissues, with the highest level in the liver, followed by the kidney, spleen, and heart. The concentration of berberrubine in various tissues could be predicted by a BP-ANN model.

5.7.2.2 *Menispermaceae*

Sinomenine, abundant in the family Menispermaceae, is traditionally used to alleviate inflammation associated with rheumatoid arthritis. Eight healthy male volunteers ingested an 80 mg sinomenine HCl tablet (Yan et al., 1997). A two-compartment open model with first-order elimination could explain the serum concentration—time curve. The AUC and C_{max} after a single dose of sinomenine were much less than those after a multiple-dose treatment (Su et al., 2012), with an accumulation factor of 2.49 ± 0.77. In beagle dogs, the absolute bioavailability of sinomenine is low (Chen et al., 2009), and the elimination of sinomenine is fast. On the contrary, around 80% bioavailability was attained after a single oral dose of 90 mg/kg sinomenine in rats. The C_{max}, AUC, CL, and V_d of sinomenine alone are distinct from those in the *S. acutum* extract (Zhang et al., 2014c), while the MRT, t_{max}, and $t_{1/2}$ were similar in the two groups. The absorption of sinomenine might be deterred by other compounds of the *S. acutum* extract, which may consequently result in variable tonic and detoxification effects.

Cepharanthine could be extensively transformed in the human liver, and the absolute bioavailability was only 6%–9%, as shown in Hao et al. (2010). Cepharanthine was speedily distributed into tissues, but its elimination was relatively slow, corresponding to the long $t_{1/2}$. In rabbits, the PK profile of crebanine, an aporphine alkaloid of *Stephania*, could be outlined by a two-compartment open model (Ma et al., 2007). The male and female rabbits had similar PK profiles. Crebanine injection was subjected to a fast and extensive disposition in rabbits.

Dauricine exhibits antiarrhythmic effect by prolonging the action potential. In 12 healthy Chinese volunteers, dauricine showed two peaks of plasma concentration and was quickly eliminated after 6 h (Liu et al., 2010). The T_{max} corresponded to the first peak. The second peak was also observed in rats and dogs. At most, three fold variability of AUC and C_{max} among subjects was found, which agrees with other PK studies (Seaber et al., 1997; Yates et al., 2002).

5.7.2.3 *Papaveraceae*

Compared with the THP monomer, the T_{max} of THP in *Corydalis* and vinegar-processed *Corydalis* is shorter (Dou et al., 2007), thus enhancing the pain-relieving effect. In rats, DHC alone is absorbed faster and excreted slower than the DHC of *C. yanhusuo* (Li et al., 2014b). Other components of *C. yanhusuo* might be P-gp inhibitors, and other alkaloids of *C. yanhusuo* may compete with DHC in transporter-mediated processes, CYP-mediated metabolism, and binding to blood proteins.

Corynoline and acetylcorynoline of *Corydalis bungeana* show numerous bioactivities, for example, sedation, antileptospira, and hepatoprotection (Yang et al., 2014a). The two alkaloids are of slow absorption. The excretion of corynoline might be delayed by other components of the *C. bungeana* extract in rat plasma. The oral bioavailability of the alkaloid in the extract is higher than that of the respective alkaloid. Rapid elimination is observed after iv administration of acetylcorynoline (Wen et al., 2014). After rats were exposed to *Rhizoma Corydalis Decumbentis* (RCD, Xia Tian Wu in TCM) extract, THP and palmatine underwent fast absorption within 2 h (Ma et al., 2009). The elimination of palmatine might be slow because of the longer $t_{1/2}$ and MRT (about 10 h). The T_{max} of the six RCD alkaloids ranged from 1 to 2 h following oral administration (Liao et al., 2014), suggesting moderately rapid absorption. The concentration-time profiles of jatrorrhizine, BBR, palmatine, and THP displayed multiple peaks, which might be attributed to distribution, reabsorption and enterohepatic circulation.

Corydalis decumbens, listed in China Pharmacopoeia, is clinically used for the treatment of paralytic stroke, headache, rheumatic arthritis, and sciatica in China. The C_{max}, AUC, and bioavailability of protopine and THP in CDAs-SFE (an alkaloid extract)/HCl are significantly higher than those in CDAs-SFE and in CDAs-SFE/HP-β-CD (Wu et al., 2013). In CDAs-SFE/HP-β-CD, AUC and bioavailability of bicuculline (toxic compound) are higher than those in CDAs-SFE and CDAs-SFE/HCl. CDAs-SFE/HCl is the best formulation among the three formulations for the alkaloid extract of *C. decumbens*.

Levo-THP is one of the main active alkaloids isolated from *Rhizoma corydalis* (Wang et al., 2012). L-THP can quickly pass through the BBB into striatum, where it elicits profound effects on the dopaminergic system. The larger $t_{1/2}$ value in the striatum is beneficial to L-THP-induced anti-nociceptive effects.

The AUC of THP in spontaneously hypertensive rats (SHR) is 81.44 ± 45.0 μg h/mL (Hong et al., 2012a), significantly higher than that in normotensive rats (44.06 ± 19.6 μg h/mL). The $t_{1/2}$ and MRT in SHR are much longer than those in healthy SD rats, indicating slow elimination of THP in SHR. It is very important to investigate the PK properties of drugs in pathological conditions.

The absolute bioavailability of four *C. saxicola* alkaloids, dehydrocavidine, coptisine, dehydroapocavidine, and tetradehydroscoulerine, is very low (10.8%–24.5%) (Li et al., 2006b). Four alkaloids are quickly eliminated after iv administration. Only 7.6%, 3.4%, 13.7%, and 18.1% of dehydrocavidine, coptisine, dehydroapocavidine, and tetradehydroscoulerine respectively are excreted in the urine following iv dosing, which is much higher than those after oral administration. These alkaloids might be metabolized and/or they might be eliminated via other routes, for example, bile and feces.

The five alkaloids of *Chelidonium majus* (Papaveraceae), that is, protopine, chelidonine, coptisine, sanguinarine, and chelerythrine (Fig. 5.2), are rapidly absorbed after administration, with the T_{max} ranging from 1.17 to 1.92 h(Zhou et al., 2013b). The C_{max} of sanguinarine is about five to ten times higher than those of the other four alkaloids. Coptisine has the longest $t_{1/2}$(~7.03 h), followed by sanguinarine (~4.72 h). Sanguinarine has the largest AUC of the five alkaloids. Sanguinarine has approximately sevenfold larger AUC than coptisine despite administration of threefold lower doses of sanguinarine than coptisine. Although chelerythrine has strong structural similarity to sanguinarine, the total exposure of chelerythrine is significantly lower. The increased systemic exposure and $t_{1/2}$ with decreased peak plasma concentration after administration of the *C. majus* extract may be associated with enhanced efficacy and reduced toxicity (Zhou et al., 2013a). The extracts of *C. majus* exhibit relatively low toxicity compared with sanguinarine (Mazzanti et al., 2009).

5.7.3 Other Phytometabolites

In rats the oral bioavailability of columbin, an antiinflammatory furanoditerpen from *Radix Tinosporae* (Menispermaceae), is only 3.18% (Shi et al., 2007), indicating that columbin has poor absorption or underwent extensive first-pass metabolism.

ASD and its metabolites have different pharmacological effects. In rats, no significant gender difference is found in the pharmacokinetics of ASD and its metabolite hederagenin (M1) (Liu et al., 2013). There is no accumulation of ASD and M1 after multiple administrations of ASD (0.09 g/kg). After oral administration of ASD, the double peaks of ASD are observed in biliary and urinary excretion rate-time curves. M1 is detected in the feces at 6 h after oral use. The average plasma protein binding of ASD is 92.9% in rats. In rats, T_{max} of the metabolites cauloside A, HN saponin F, and hederagenin are 9.33 ± 2.49, 7.33 ± 0.47, and 12.33 ± 2.36 h, respectively (Liu et al., 2014). In both rats and beagle dogs, the double peak was observed in the plasma concentration-time curve of ASD (Liu et al., 2014; Shakya et al., 2012).

5.7.4 TCM Pair

Epimedin A, B, C, and icariin are the major drug efficacy compounds of Chuankezhi injection, consisting of *Radix Morindae Officinalis* and *Herba Epimedii*. After intermuscular administration of Chuankezhi injection, the plasma concentration of four flavonoid glycosides rapidly arose to peaks at about 10 min and then quickly declined in rats (Xu et al., 2016b). Their mean elimination half-life ($t_{1/2z}$) was 0.60, 0.62, 0.47, and 0.49 h. Serum concentration of four epimedium flavonoids in Chuankezhi injection was low, and their absorption and elimination were similar in rats.

5.7.5 TCM Formula

Fangji Huangqi Tang (FHT), used in the treatment of chronic glomerulonephritis, is composed of *S. tetrandra* (Menispermaceae), *Astragalus membranaceus* (Huang Qi), *Atractylodes macrocephala*, and *Glycyrrhiza uralensis*. It is not unusual to find the double plasma concentration peaks of tetrandrine and fangchinoline (Fig. 5.2) in rats after administration of

S. tetrandra extracts or FHT (Wang et al., 2015a). Enterohepatic circulation might explain this phenomenon, which extends the therapeutic time period of drugs. Compared with *S. tetrandra*, the substantially longer MRT and $t_{1/2}$ of tetrandrine in FHT are observed. Tetrandrine in FHT presented better absorption with relatively larger AUC, as well as slower elimination. Fangchinoline in FHT shows a similar tendency as tetrandrine. The oral use of FHT might lead to an improved bioavailability of the alkaloids.

The C_{max} ratio ($-/+$) of THP is 2.91, 1.38, and 1.19, and the $AUC_{0-\infty}$ ratio ($-/+$) of THP is 2.84, 1.50, and 1.35 in rats after dosed with rac-THP, extracts of *Rhizoma corydalis*, and the TCM formula Yuanhu-Baizhi (YB), respectively (Hong et al., 2008). The mean $AUC_{0-\infty}$ and C_{max} of (+)-THP dosed with YB extracts are significantly higher than those with rac-THP and *Rhizoma corydalis* extracts. The mean $AUC_{0-\infty}$ and T_{max} of rac-THP dosed with YB extracts are considerably higher than those with two other reagents. The stereoselectivity in pharmacokinetics of THP enantiomers in rats is decreased when dosed in plant form, while the increased $AUC_{0-\infty}$ of rac-THP after YB treatment rationalizes the drug combination of *Rhizoma corydalis* and *Radix angelicae dahuricae*.

Deng-yan granule (DYG), consisting of *Herba Erigerontis Breviscapi*, *Rhizoma C. yanhusuo*, and *Radix Astragali Mongolici*, is a widely used TCM formula for the treatment of coronary heart disease (Zhang et al., 2014b). Scutellarin and THP are the main active constituents in *Herba Erigerontis Breviscapi* and *Rhizoma C. yanhusuo*, respectively. Scutellarin and THP are detected in rat plasma 5 min after oral administration of DYG, indicating immediate absorption. Following oral administration of DYG, double peaks are observed in both scutellarin and THP, which may be relevant to the distribution, reabsorption, and/or enterohepatic circulation.

The TCM formula Shuanghua Baihe tablet (SBT) is composed of 10 crude herbs and contains corynoline, BBR, and various other ingredients. Corynoline alone shows a low bioavailability and a high elimination rate (Liu et al., 2016b). The $t_{1/2}$ is prolonged about 3-fold, and the C_{max} and AUC_{0-12} of corynoline are increased by 46.5% and 34.2%, respectively, when the same dosage of corynoline was administered in SBT. Compared with the corynoline alone, the C_{max} and AUC_{0-12} are increased by 11.1-fold and 5.0-fold respectively, in the rats treated with corynoline + BBR. The oral administration of the SBT prolonged the $t_{1/2}$ of corynoline and increased its bioavailability. BBR played an important role in the DDI of corynoline and other ingredients of SBT, and the influence of other coexisting compounds in SBT on the PK profiles of corynoline might not be trivial.

Three monoterpene glycosides (paeoniflorin, alibiflorin, and oxypaeoniflorin) and four alkaloids (THP, CDL, DHC, and BBR) are the main active ingredients of *Radix Paeoniae Rubra* extract (RPE) and *C. yanhusuo* extract (CYE) in the TCM formula Huo Luo Xiao Ling Dan (HLXLD) (Ai et al., 2015). Remarkable differences in PK properties of the analytes between the herbal formula and single herb groups, and between normal and arthritic groups, were found. Paeoniflorin, alibiflorin, and oxypaeoniflorin exhibits a similar absorption and metabolic route in the plasma of normal and arthritic rats after oral administration of HLXLD or RPE. The PK behaviors of THP, CDL, DHC, and BBR are also similar. Multiple peaks of alkaloids are observed, possibly due to reabsorption and/or enterohepatic circulation. The $AUC_{0-\infty}$ and C_{max} of four alkaloids in the HLXLD group are significantly decreased as compared with those of the CYE group, suggesting that their absorption might be inhibited. The T_{max} of all analytes increased after oral administration of HLXLD, indicating the slower absorption. Some ingredients of HLXLD could increase the $AUC_{0-\infty}$ and C_{max} of RP compounds to exert synergistic antiinflammatory activity and decrease the absorption of CY to attenuate the alkaloids' drastic action and side effects. The prolonged elimination of THP could increase the effective time and achieve sustained release. Such studies help explain the therapeutic efficacy of TCM formulas.

Upon HLXLD administration, the absorption of the four alkaloids is increased in arthritic rats, while the distribution and elimination are slower (Ai et al., 2015). The mechanisms of absorption and metabolism might be different between normal and arthritic rats. Some metabolic pathways (microfloras, enzymes, transporters) may be affected in arthritic rats. For instance, arthritis might cause a great decrease in microflora or enzyme activity that could transform alkaloids to their corresponding glucuronide and/or sulfate conjugates. Arthritis might result in a decrease of the efflux of the active ingredients via inhibiting the efflux pump. Therefore, the bioavailability of the four alkaloids in the arthritic group is higher than that in the normal group.

5.8 CONCLUSIONS AND PROSPECT

The structural complexity of Ranunculales natural products differentiates them from common synthetic drugs, allowing them to access a biological target space that is beyond the enzyme active sites and receptors targeted by conventional small molecule drugs. Naturally occurring Ranunculales flavonoids and alkaloids, in particular, exhibit a wide variety of unusual and potent bioactivities (Desgrouas et al., 2014; Li et al., 2015a; Yang et al., 2014b). Many of these compounds penetrate cells by passive diffusion, and some, like the *Epimedium* flavonoids, are orally bioavailable. In evaluating the drug-likeness of Ranunculales phytometabolites, their DMPK properties must be taken into account (Hao and Xiao, 2014). Because of their size, diversity, and complexity, Ranunculales compounds occupy a noticeable chemical space in the plant-based drug discovery that may provide useful scaffolds for modulating more challenging biological targets such as protein–protein interactions and allosteric binding sites. This review focuses on Ranunculales natural products from a systematic ADME/T perspective, outlining what we know and don't know about their metabolism and disposition properties, and unearthing trends that might be useful in the design of novel rule-breaking molecules.

DMPK studies of various alkaloids of non-Ranunculaceae Ranunculales plants predominate over those of saponins and flavonoids, which is reminiscent of the high abundance of alkaloids in Thalictroideae of Ranunculaceae (Hao et al., 2015e). In the subfamily Ranunculoideae, only the tribe Delphinieae has abundant diterpenoid alkaloids, and correspondingly the DMPK studies of these bioactive compounds are overwhelming (Hao et al., 2015a). However, the DMPK behavior of various phytometabolites is more related to the class that the compounds belong to rather than the taxonomic group with which they are affiliated. According to the pharmacophylogeny (Hao et al., 2014; Hao and Xiao, 2015; Hao et al., 2015d), the evolutionarily related taxa quite often possess similar chemical profiles, and closely related compounds often display both similar bioactivities (Zou et al., 2016) and comparable DMPK profiles, which are supported by the in vitro, ex vivo, and in vivo data summarized in this review.

The effectiveness and safety of Herba Epimedii, its most abundant ingredient icariin, and other compounds, have been validated by recent PK and toxicity studies (Zhai et al., 2013). The semisynthetic modulation on these flavonoids could be performed to obtain a series of derivatives with overlapping or nonoverlapping bioacitivities and ADME/T properties. The structure-activity relationship analysis (Zou et al., 2016) and DMPK prediction could bring novel insights into structure modification. Not only new leads with improved therapeutic efficacy and ADME/T properties, but also the novel and highly selective inhibitor/inducer/substrate of DMEs/transporters could be generated by such modifications.

Since species-specific differences in DMPK exist, data obtained from animal studies may not be sufficient to predict DM reactions in humans. The relationship between compound structure and PK behavior, especially cell permeability and metabolic clearance, in classic and novel Ranunculales phytometabolites has not been studied systematically, and the generality of flavonoids, alkaloids, and saponins as orally bioavailable scaffolds remains an open question. Also, there is a lack of knowledge about ADME/T properties of many compounds, especially those of the little investigated genera, such as *Vancouveria*, *Cocculus*, *Sargentodoxa*, and Circaeasteraceae (Fig. 5.1), in which the blockbuster compound comparable to paclitaxel might exist.

The concept, research blueprint, and methodology of systems biology are integrated into the recent applications and research progress of the related methods with regard to pharmaceutical studies, taking physiologically based pharmacokinetics, and physiologically-based pharmacodynamics (PBPD) studies as the examples(Shi et al., 2003; Yang et al., 2007) (Fig. 5.3). However, few Ranunculales studies correlate PK parameters with the PD marker. Given that the PK data of various alkaloids have been accumulated, it would be timely to call for more correlation studies integrating PK and PD markers. Although every TCM formula is a complex system, some significant PK data have been reported for Ranunculales compound-containing ones, for example, FHT, YB (Hong et al., 2008), DYG, etc., providing DMPK evidence for their salient therapeutic efficacy. Systems biology is a kind of systems theory with the reductionism as its solid foundation. TCM is a complex system under the guidance of holistic and dialectic philosophy of ancient China; however, it is challenged by many fundamental questions with the implementation of systems biology. The study of the ADME/T properties of herbal medicines would be the basis of the integrative study of TCM and systems biology, which bridges the studies of molecular level and organismic level. The summary of the fruitful trials of the worldwide groups in the previously mentioned fields suggests that the early ADME/T study of TCM is a commanding device in elucidating the riddle of TCM at the level of systems biology.

Omics-based methodologies (genomics, epigenomics, transcriptomics, proteomics, microbiomics, and metabolomics) (Hagel et al., 2015a, b; Swanson, 2015) and next-generation sequencing should be used to enhance the reliability of DMPK prediction. It is unlikely that a single in vitro system will be capable of mimicking the complex interactions in human tissues. The fluorescent probes (Ma et al., 2017) may bridge the gap between conventional models and in vivo clinical studies in humans and provide a consistent basis for DMPK assay development.

References

Ai, Y., Wu, Y., Wang, F., et al., 2015. A UPLC-MS/MS method for simultaneous quantitation of three monoterpene glycosides and four alkaloids in rat plasma: application to a comparative pharmacokinetic study of Huo Luo Xiao Ling Dan and single herb extract. J. Mass Spectrom. 50 (3), 567–577.

Ashour, M.L., Youssef, F.S., Gad, H.A., et al., 2017. Inhibition of cytochrome P450 (CYP3A4) activity by extracts from 57 plants used in traditional Chinese medicine (TCM). Pharmacogn. Mag. 13 (50), 300–308.

Bahadur, S., Mukherjee, P.K., Milan Ahmmed, S.K., et al., 2016. Metabolism-mediated interaction potential of standardized extract of *Tinospora cordifolia* through rat and human liver microsomes. Indian J. Pharmacol. 48 (5), 576–581.

Brown, A.C., 2017. Heart toxicity related to herbs and dietary supplements: online table of case reports. Part 4 of 5. J. Diet. Suppl. 5, 1–40.

Cao, Y.F., He, R.R., Cao, J., et al., 2012. Drug-drug interactions potential of icariin and its intestinal metabolites via inhibition of intestinal UDP-glucuronosyltransferases. Evid. Based Complement. Alternat. Med. 2012, 395912.

Charoonratana, T., Puntarat, J., Vinyoocharoenkul, S., et al., 2017. Innocuousness of a polyherbal formulation: a case study using a traditional Thai antihypertensive herbal recipe in rodents. Food Chem. Toxicol. doi: 10.1016/j.fct.2017.07.052.

Chen, P., Braithwaite, R.A., George, C., et al., 2014. The poppy seed defense: a novel solution. Drug Test. Anal. 6 (3), 194–201.

Chen, Y., Wang, J., Jia, X., et al., 2011. Role of intestinal hydrolase in the absorption of prenylated flavonoids present in Yinyanghuo. Molecules 16 (2), 1336–1348.

Chen, Y., Zhao, Y.H., Jia, X.B., et al., 2008. Intestinal absorption mechanisms of prenylated flavonoids present in the heat-processed *Epimedium koreanum* Nakai (Yin Yanghuo). Pharm. Res. 25 (9), 2190–2199.

Chen, W., Zhou, Y., Kang, J., et al., 2009. Study on pharmacokinetics and absolute bioavailability of sinomenine in beagle dogs. Zhongguo Zhong Yao Za Zhi 34 (4), 468–471.

Cheng, T., Sheng, T., Yi, Y., et al., 2016. Metabolism profiles of icariin in rats using ultra-high performance liquid chromatography coupled with quadrupole time-of-flight tandem mass spectrometry and in vitro enzymatic study. J. Chromatogr. B Analyt. Technol. Biomed. Life Sci. 1033–1034, 353–360.

Cheng, X.Y., Shi, Y., Sun, H., et al., 2009. Identification and analysis of absorbed components in rat plasma after oral administration of active fraction of *Corydalis yanhusuo* by LC-MS/MS. Yao Xue Xue Bao 44 (2), 167–174.

Chinese Pharmacopoeia Commission, 2015. Pharmacopoeia of the People's Republic of China. China Medical Science Press, Beijing.

Concheiro, M., Newmeyer, M.N., da Costa, J.L., et al., 2015. Morphine and codeine in oral fluid after controlled poppy seed administration. Drug Test. Anal. 7 (7), 586–591.

Cui, L., Sun, E., Qian, Q., et al., 2013a. Comparative study on effect of crude and different processed products of epimedium on pharmacokinetics characteristics in mice. Zhongguo Zhong Yao Za Zhi 38 (10), 1614–1617.

Cui, L., Sun, E., Zhang, Z., et al., 2013b. Metabolite profiles of epimedin B in rats by ultraperformance liquid chromatography/quadrupole-time-of-flight mass spectrometry. J Agric. Food Chem. 61 (15), 3589–3599.

Dai, Z.R., Ge, G.B., Feng, L., et al., 2015. A highly selective ratio metric two-photon fluorescent probe for human cytochrome p450 1A. J. Am. Chem. Soc. 137 (45), 14488–14495.

Datta, S., Mahdi, F., Ali, Z., et al., 2014. Toxins in botanical dietary supplements: blue cohosh components disrupt cellular respiration and mitochondrial membrane potential. J. Nat. Prod. 77 (1), 111–117.

Desgrouas, C., Taudon, N., Bun, S.S., et al., 2014. Ethnobotany, phytochemistry and pharmacology of *Stephania rotunda* Lour. J. Ethnopharmacol. 154 (3), 537–563.

Dong, P.L., Han, H., Zhang, T.Y., et al., 2014. P-glycoprotein inhibition increases the transport of dauricine across the blood-brain barrier. Mol. Med. Rep. 9 (3), 985–988.

Dou, Z.Y., Sun, W., Mi, X.L., et al., 2007. Comparison of pharmacokinetics of tetrahydropalmatine monomer and extractive of corydalis and corydalis processed with vinegar. Zhong Yao Cai 30 (12), 1499–1501.

Dutta, A., Verma, S., Sankhwar, S., et al., 2012. Bioavailability, antioxidant and non toxic properties of a radioprotective formulation prepared from isolated compounds of *Podophyllum hexandrum*: a study in mouse model. Cell. Mol. Biol. 58S, OL1646–OL1653.

Gao, Y., Hu, S., Zhang, M., et al., 2014. Simultaneous determination of four alkaloids in mice plasma and brain by LC-MS/MS for pharmacokinetic studies after administration of *Corydalis Rhizoma* and Yuanhu Zhitong extracts. J. Pharm. Biomed. Anal. 92, 6–12.

Ge, G.B., Ning, J., Hu, L., et al., 2013. A highly selective probe for human cytochrome P450 3A4: isoform selectivity, kinetic characterization and its applications. Chem. Commun. 49, 9779–9781.

Ghosh, I., Mukherjee, A., 2016. Argemone oil induces genotoxicity in mice. Drug Chem. Toxicol. 39 (4), 407–411.

Gong, C.Y., Xiao, W., Wang, Z.Z., et al., 2014. Stability of akebiasaponin D in gastrointestinal contents of rats. Zhongguo Zhong Yao Za Zhi 39 (12), 2311–2313.

Guan, H., Li, K., Wang, X., et al., 2017. Identification of metabolites of the cardioprotective alkaloid dehydrocorydaline in rat plasma and bile by liquid chromatography coupled with triple quadrupole linear ion trap mass spectrometry. Molecules 22 (10), E1686.

Guo, L.Q., Taniguchi, M., Chen, Q.Y., et al., 2001. Inhibitory potential of herbal medicines on human cytochrome P450-mediated oxidation: properties of umbelliferous or citrus crude drugs and their relative prescriptions. Jpn. J. Pharmacol. 85, 399–408.

Hagel, J.M., Mandal, R., Han, B., et al., 2015a. Metabolome analysis of 20 taxonomically related benzylisoquinoline alkaloid-producing plants. BMC Plant Biol. 15, 220.

Hagel, J.M., Morris, J.S., Lee, E.J., et al., 2015b. Transcriptome analysis of 20 taxonomically related benzylisoquinoline alkaloid-producing plants. BMC Plant Biol. 15, 227.

Hao, D.C., Ge, G.B., Xiao, P.G., et al., 2015a. Drug metabolism and pharmacokinetic diversity of ranunculaceae medicinal compounds. Curr. Drug Metab. 16 (4), 294–321.

Hao, D.C., Gu, X.J., Xiao, P.G., 2013a. Recent advances in the chemical and biological studies of *Aconitum* pharmaceutical resources. J. Chin. Pharm. Sci. 22 (3), 209–221.

Hao, D.C., Gu, X.J., Xiao, P.G., 2015b. Medicinal Plants: Chemistry, Biology and Omics, first ed. Elsevier-Woodhead, Oxford, ISBN 9780081000854.

Hao, D.C., Gu, X., Xiao, P.G., 2015c. Chemistry, biology, and phylogenetic analysis of *Thalictrum* pharmaceutical resources. Lishizhen Med. Mat. Med. Res. 26 (7), 1731–1733.

Hao, G., Liang, H., Li, Y., et al., 2010. Simple, sensitive and rapid HPLC-MS/MS method for the determination of cepharanthine in human plasma. J. Chromatogr. B Analyt. Technol. Biomed. Life Sci. 878 (28), 2923–2927.

Hao, D.C., Xiao, P.G., 2014. Network pharmacology: a Rosetta Stone for traditional Chinese medicine. Drug Dev. Res. 75 (5), 299–312.

Hao, D.C., Xiao, P.G., 2015. Genomics and evolution in traditional medicinal plants: road to a healthier life. Evol. Bioinform. 11, 197–212.

Hao, D.C., Xiao, P.G., Liu, M., et al., 2014. Pharmaphylogeny vs. pharmacophylogenomics: molecular phylogeny, evolution and drug discovery. Yao Xue Xue Bao 49 (10), 1387–1394.

Hao, D.C., Xiao, P.G., Liu, L.W., 2015d. Essentials of pharmacophylogeny: knowledge pedigree, epistemology and paradigm shift. China J. Chin. Mater. Med. 40 (17), 3335–3342.

Hao, D.C., Xiao, P.G., Ma, H.Y., et al., 2015e. Mining chemodiversity from biodiversity: pharmacophylogeny of medicinal plants of *Ranunculaceae*. Chin. J. Nat. Med. 13 (7), 507–520.

Hao, D.C., Xiao, B., Xiang, Y., et al., 2013b. Deleterious nonsynonymous single nucleotide polymorphisms in human solute carriers: the first comparison of three prediction methods. Eur. J. Drug Metab. Pharmacokinet. 38 (1), 53–62.

Hong, Z., Cai, G., Ma, W., et al., 2012a. Rapid determination and comparative pharmacokinetics of tetrahydropalmatine in spontaneously hypertensive rats and normotensive rats. Biomed. Chromatogr. 26 (6), 749–753.

Hong, Z., Fan, G., Le, J., et al., 2006. Brain pharmacokinetics and tissue distribution of tetrahydropalmatine enantiomers in rats after oral administration of the racemate. Biopharm. Drug Dispos. 27 (3), 111–117.

Hong, Z., Le, J., Lin, M., et al., 2008. Comparative studies on pharmacokinetic fates of tetrahydropalmatine enantiomers in different chemical environments in rats. Chirality 20 (2), 119–124.

Hong, Z., Wen, J., Zhang, Q., et al., 2010. Study on the stereoselective excretion of tetrahydropalmatine enantiomers in rats and identification of in vivo metabolites by liquid chromatography-tandem mass spectrometry. Chirality 22 (3), 355–360.

Hong, Z., Zhao, L., Wang, X., et al., 2012b. High-performance liquid chromatography-time-of-flight mass spectrometry with adjustment of fragmentor voltages for rapid identification of alkaloids in rat plasma after oral administration of rhizoma *Corydalis* extracts. J. Sep. Sci. 35 (13), 1690–1696.

Hu, Y., Fan, S., Liao, X., et al., 2017. Pharmacokinetics, excretion of 8-cetylberberine and its main metabolites in rat urine. J. Pharm. Biomed. Anal. 132, 195–206.

Hsueh, T.Y., Wu, Y.T., Lin, L.C., et al., 2013. Herb-drug interaction of *Epimedium sagittatum* (Sieb. et Zucc) maxim extract on the pharmacokinetics of sildenafil in rats. Molecules 18 (6), 7323–7335.

Hwang, Y.H., Yang, H.J., Yim, N.H., et al., 2017. Genetic toxicity of *Epimedium koreanum* Nakai. J. Ethnopharmacol. 198, 87–90.

Ji, H.Y., Liu, K.H., Jeong, J.H., et al., 2012. Effect of a new prokineticagent DA-9701 formulated with Corydalis tuber and *Pharbitidis semen* on cytochrome P450 and UDP-glucuronosyltransferase enzyme activities in human liver microsomes. Evid. Based Complement. Alternat. Med. 2012, 650718.

Ji, H.Y., Liu, K.H., Lee, H., et al., 2011. Corydaline inhibits multiple cytochrome P450 and UDP-glucuronosyltransferase enzyme activities in human liver microsomes. Molecules 16 (8), 6591–6602.

Jiang, J., Li, J., Zhang, Z., et al., 2015a. Mechanism of enhanced antiosteoporosis effect of circinal-icaritin by self-assembled nanomicelles in vivo with suet oil and sodium deoxycholate. Int. J. Nanomed. 10, 2377–2389.

Jiang, L., Liang, S.C., Wang, C., et al., 2015b. Identifying and applying a highly selective probe to simultaneously determine the O-glucuronidation activity of human UGT1A3 and UGT1A4. Sci. Rep. 5, 9627.

Jiang, Y.N., Mo, H.Y., 2010. Preparation of self-emulsifying soft capsule and its pharmacokinetic in rats for epimedium flavonoids. Zhong Yao Cai 33 (5), 767–771.

Jin, H., Dai, J., Chen, X., et al., 2010. Pulmonary toxicity and metabolic activation of dauricine in CD-1 mice. J. Pharmacol. Exp. Ther. 332 (3), 738–746.

Jin, X., Zhang, Z.H., Sun, E., et al., 2012. Preparation of icariside II-phospholipid complex and its absorption across Caco-2 cell monolayers. Pharmazie 67 (4), 293–298.

Karuppaiya, P., Tsay, H.S., 2015. Therapeutic values, chemical constituents and toxicity of Taiwanese *Dysosma pleiantha*—a review. Toxicol. Lett. 236 (2), 90–97.

Könczöl, A., Müller, J., Földes, E., et al., 2013. Applicability of a blood-brain barrier specific artificial membrane permeability assay at the early stage of natural product-based CNS drug discovery. J. Nat. Prod. 76 (4), 655–663.

Lee, C.J., Wu, Y.T., Hsueh, T.Y., et al., 2014. Pharmacokinetics and oral bioavailability of epimedin C after oral administration of epimedin C and Herba Epimedii extract in rats. Biomed. Chromatogr. 28 (5), 630–636.

Lei, Y., Tan, J., Wink, M., et al., 2013. Anisoquinoline alkaloid from the Chinese herbal plant *Corydalis yanhusuo* WT. Wang inhibits P-glycoprotein and multidrug resistance-associate protein 1. Food Chem. 136 (3–4), 1117–1121.

Li, S.Z., 2015. Ben Cao Gang Mu. Traditional Chinese Medicine. Classics Press, Beijing, (Ming Dynasty).

Li, Z., Cheung, F.S., Zheng, J., et al., 2014a. Interaction of the bioactive flavonol, icariin, with the essential human solute carrier transporters. J. Biochem. Mol. Toxicol. 28 (2), 91–97.

Li, Y., Duan, Z., Tian, Y., et al., 2013. A novel perspective and approach to intestinal octreotide absorption: sinomenine-mediated reversible tight junction opening and its molecular mechanism. Int. J. Mol. Sci. 14 (6), 12873–12892.

Li, C.R., Li, Q., Mei, Q., et al., 2015a. Pharmacological effects and pharmacokinetic properties of icariin, the major bioactive component in *Herba Epimedii*. Life Sci. 126, 57–68.

Li, Q.Y., Li, K.T., Sun, H., et al., 2014b. LC-MS/MS determination and pharmacokinetic study of dehydrocorydaline in rat plasma after oral administration of dehydrocorydaline and *Corydalis yanhusuo* extract. Molecules 19 (10), 16312–16326.

Li, X., Li, X., Zhou, Y., et al., 2010. Development of patch and spray formulations for enhancing topical delivery of sinomenine hydrochloride. J. Pharm. Sci. 99 (4), 1790–1799.

Li, X.Y., Xie, H., Lu, T.L., et al., 2015b. Study on effect of oligochitosan in promoting intestinal absorption of protoberberine alkaloids in extracts from *Corydalis saxicola* total alkaloids. Zhongguo Zhong Yao Za Zhi 40 (9), 1812–1816.

Li, L., Ye, M., Bi, K., et al., 2006a. Liquid chromatography-tandem mass spectrometry for the identification of L-tetrahydropalmatine metabolites in *Penicillium janthinellum* and rats. Biomed. Chromatogr. 20 (1), 95–100.

Li, H.L., Zhang, W.D., Liu, R.H., et al., 2006b. Simultaneous determination of four active alkaloids from a traditional Chinese medicine *Corydalis saxicola* Bunting, (Yanhuanglian) in plasma and urine samples by LC-MS-MS. J. Chromatogr. B Analyt. Technol. Biomed. Life Sci. 831 (1–2), 140–146.

Liang, D.L., Zheng, S.L., 2014. Effects of icaritin on cytochrome P450 enzymes in rats. Pharmazie 69 (4), 301–305.

Liao, C., Chang, S., Yin, S., et al., 2014. A HPLC-MS/MS method for the simultaneous quantitation of six alkaloids of *Rhizoma Corydalis Decumbentis* in rat plasma and its application to a pharmacokinetic study. J Chromatogr. B Analyt. Technol. Biomed. Life Sci. 944, 101–106.

Lin, Y.C., Chang, C.W., Wu, C.R., 2015. Anti-nociceptive, anti-inflammatory and toxicological evaluation of Fang-Ji-Huang-Qi-Tang in rodents. BMC Complement. Altern. Med. 15, 10.

Liu, Z.Q., Chan, K., Zhou, H., et al., 2005a. The pharmacokinetics and tissue distribution of sinomenine in rats and its protein binding ability in vitro. Life Sci. 77 (25), 3197–3209.

Liu, Z., Chang, S., Guan, X., et al., 2016a. The metabolites of ambinine, a benzo[c]phenanthridine alkaloid, in rats identified by ultra-performance liquid chromatography-quadrupoletime-of-flight mass spectrometry (UPLC/Q-TOF-MS/MS). J. Chromatogr. B Analyt. Technol. Biomed. Life. Sci. 1033–1034, 226–233.

Liu, R., Gu, P., Wang, L., et al., 2016b. Study on the pharmacokinetic profiles of corynoline and its potential interaction in traditional Chinese medicine formula Shuanghua Baihe tablets in rats by LC-MS/MS. J. Pharm. Biomed. Anal. 117, 247–254.

Liu, Z., Hu, M., 2007. Natural polyphenol disposition via coupled metabolic pathways. Expert Opin. Drug Metab. Toxicol. 3 (3), 389–406.

Liu, D., Li, S., Qi, J.Q., et al., 2016. The inhibitory effects of nor-oleanane triterpenoid saponins from Stauntonia brachyanthera towards UDP-glucuronosyltransferases. Fitoterapia 112, 56–64.

Liu, X., Liu, Q., Wang, D., et al., 2010. Validated liquid chromatography-tandem mass spectrometry method for quantitative determination of dauricine in human plasma and its application to pharmacokinetic study. J. Chromatogr. B Analyt. Technol. Biomed. Life Sci. 878 (15–16), 1199–1203.

Liu, E.W., Wang, J.L., Han, L.F., et al., 2014. Pharmacokinetics study of asperosaponin VI and its metabolites caulosideside AHN saponin F and hederagenin. J. Nat. Med. 68 (3), 488–497.

Liu, K.P., Wang, L., Li, Y., et al., 2012. Preparation, pharmacokinetics, and tissue distribution properties of icariin-loaded stealth solid lipid nanoparticles in mice. Chin. Herb. Med. 4 (2), 170–174.

Liu, Q., Wang, J., Yang, L., et al., 2013a. A rapid and sensitive LC-MS/MS assay for the determination of berbamine in rat plasma with application to preclinical pharmacokinetic study. J. Chromatogr. B Analyt. Technol. Biomed. Life Sci. 929, 70–75.

Liu, Y.Z., Yang, X.B., Yang, X.W., et al., 2013b. Biotransformation by human intestinal flora and absorption-transportation characteristic in a model of Caco-2 cell monolayer of d-corydaline and tetrahydropalmatine. Zhongguo Zhong Yao Za Zhi 38 (1), 112–118.

Liu, J., Ye, H., Lou, Y., 2005b. Determination of rat urinary metabolites of icariin in vivo and estrogenic activities of its metabolites on MCF-7 cells. Pharmazie 60 (2), 120–125.

Liu, R., Zheng, L., Cheng, M., et al., 2016c. Simultaneous determination of corynoline and acetylcorynoline in human urine by LC-MS/MS and its application to a urinary excretion study. J. Chromatogr. B Analyt. Technol. Biomed. Life Sci. 1014, 83–89.

Liu, Z.Q., Zhou, H., Liu, L., et al., 2005c. Influence of co-administrated sinomenine on pharmacokinetic fate of paeoniflorin in unrestrained conscious rats. J. Ethnopharmacol. 99 (1), 61–67.

Liu, R.J., Zhu, H., Ding, L., et al., 2013. Study on pharmacokinetics of asperosaponin VI and its active metabolite in rats. Zhongguo Zhong Yao Za Zhi 38 (14), 2378–2383.

Long, L.H., Wu, P.F., Chen, X.L., et al., 2010. HPLC and LC-MS analysis of sinomenine and its application in pharmacokinetic studies in rats. Acta Pharmacol. Sin. 31 (11), 1508–1514.

Lu, Z., Chen, W., Viljoen, A., et al., 2010. Effect of sinomenine on the in vitro intestinal epithelial transport of selected compounds. Phytother. Res. 24 (2), 211–218.

Lv, X., Ge, G.B., Feng, L., et al., 2015. An optimized ratiometric fluorescent probe for sensing human UDP-glucuronosyltransferase 1A1 and its biological applications. Biosens. Bioelectron. 72, 261–267.

Ma, Y.S., Shang, Q.J., Bai, Y.C., et al., 2007. Study on pharmacokinetics of crebanine injection in rabbits. Zhongguo Zhong Yao Za Zhi 32 (7), 630–632.

Ma, H., Wang, Y., Guo, T., et al., 2009. Simultaneous determination of tetrahydropalmatine, protopine, and palmatine in rat plasma by LC-ESI-MS and its application to a pharmacokinetic study. J. Pharm. Biomed. Anal. 49 (2), 440–446.

Ma, L., Yang, X.W., 2008. Absorption of papaverine, laudanosine and cepharanthine across human intestine by using human Caco-2 cells monolayers model. Yao Xue Xue Bao 43 (2), 202–207.

Ma, H.Y., Yang, J.D., Hou, J., et al., 2017. Comparative metabolism of DDAO benzoate in liver microsomes from various species. Toxicol. In Vitro 44, 280–286.

Maas, A., Krämer, M., Sydow, K., et al., 2017. Urinary excretion study following consumption of various poppy seed products and investigation of the new potential street heroin marker ATM4G. Drug Test. Anal. 9 (3), 470–478.

Madgula, V.L., Ali, Z., Smillie, T., et al., 2009. Alkaloids and saponins as cytochrome P450 inhibitors from blue cohosh (Caulophyllum thalictroides) in an in vitro assay. Planta Med. 75 (4), 329–332.

Manda, V.K., Ibrahim, M.A., Dale, O.R., et al., 2016. Modulation of CYPs, P-gp, and PXR by Eschscholzia californica (California poppy) and its alkaloids. Planta Med. 82 (6), 551–558.

Mao, X., Peng, Y., Zheng, J., 2015. In vitro and in vivo characterization of reactive intermediates of corynoline. Drug Metab. Dispos. 43 (10), 1491–1498.

Mao, Z., Wang, X., Liu, Y., et al., 2017. Simultaneous determination of seven alkaloids from *Rhizoma Corydalis Decumbentis* in rabbit aqueous humor by LC-MS/MS: application to ocular pharmacokinetic studies. J. Chromatogr. B Analyt. Technol. Biomed. Life Sci. 1057, 46–53.

Marques, L.M.M., Callejon, D.R., Pinto, L.G., et al., 2016. Pharmacokinetic properties, in vitro metabolism and plasma protein binding of govaniadine an alkaloid isolated from *Corydalis govaniana* Wall. J. Pharm. Biomed. Anal. 131, 464–472.

Mazzanti, G., Di Sotto, A., Franchitto, A., et al., 2009. *Chelidonium majus* is not hepatotoxic in Wistar rats, in a 4 weeks feeding experiment. J. Ethnopharmacol. 126 (3), 518–524.

Meyer, G.M., Meyer, M.R., Wink, C.S., et al., 2013a. Studies on the *in vivo* contribution of human cytochrome P450s to the hepatic metabolism of glaucine, a new drug of abuse. Biochem. Pharmacol. 86 (10), 1497–1506.

Meyer, G.M., Meyer, M.R., Wissenbach, D.K., et al., 2013b. Studies on the metabolism and toxicological detection of glaucine, an isoquinoline alkaloid from *Glaucium flavum* (Papaveraceae), in rat urine using GC-MSLC-MS(n) and LC-high-resolution MS(n). J. Mass Spectrom. 48 (1), 24–41.

Miao, X.Y., 2011. Annotation of Shen Nong Ben Cao Jing. China Medical Science Press, Beijing, (Ming Dynasty).

Montani, C., Penza, M., Jeremic, M., et al., 2008. Genistein is an efficient estrogen in the whole-body throughout mouse development. Toxicol. Sci. 103 (1), 57–67.

Natesan, S., Reckless, G.E., Barlow, K.B., et al., 2008. The antipsychotic potential of l-stepholidine—a naturally occurring dopamine receptor D1 agonist and D2 antagonist. Psychopharmacology 199 (2), 275–289.

Patil, D., Gautam, M., Gairola, S., et al., 2014. Effect of botanical immunomodulators on human CYP3A4 inhibition: implications for concurrent use as adjuvants in cancer therapy. Integr. Cancer Ther. 13 (2), 167–175.

Paul, L.D., Maurer, H.H., 2003. Studies on the metabolism and toxicological detection of the *Eschscholtzia californica* alkaloids californine and protopine in urine using gas chromatography-mass spectrometry. J. Chromatogr. B Analyt. Technol. Biomed. Life Sci. 789 (1), 43–57.

Paul, L.D., Springer, D., Staack, R.F., et al., 2004. Cytochrome P450 isoenzymes involved in rat liver microsomal metabolism of californine and protopine. Eur. J. Pharmacol. 485 (1–3), 69–79.

Qian, Q., Li, S.L., Sun, E., et al., 2012a. Metabolite profiles of icariin in rat plasma by ultra-fast liquid chromatography coupled to triple-quadrupole/time-of-flight mass spectrometry. J. Pharm. Biomed. Anal. 66, 392–398.

Qian, Q., Sun, E., Fan, H., et al., 2012b. Effect of suet oil on in vivo pharmacokinetic characteristics of icariside I in extract from processed Epimedii Herba in rats. Chin. Trad. Herb. Drug 43 (10), 1981–1985.

Schep, L.J., Slaughter, R.J., Beasley, D.M., 2009. Nicotinic plant poisoning. Clin. Toxicol. 47 (8), 771–781.

Seaber, E., On, N., Dixon, R.M., et al., 1997. The absolute bioavailability and metabolic disposition of the novel anti-migraine compound zolmitriptan (311C90). Br. J. Clin. Pharmacol. 43 (6), 579–587.

Sedo, A., Vlasicová, K., Barták, P., et al., 2002. Quaternary benzo[c]phenanthridine alkaloids as inhibitors of aminopeptidase N and dipeptidyl peptidase IV. Phytother. Res. 16 (1), 84–87.

Shakya, S., Zhu, H., Ding, L., et al., 2012. Determination of asperosaponin VI in dog plasma by high-performance liquid chromatography-tandem mass spectrometry and its application to a pilot pharmacokinetic study. Biomed. Chromatogr. 26 (1), 109–114.

Shan, Y.Q., Zhu, Y.P., Pang, J., et al., 2013. Tetrandrine potentiates the hypoglycemic efficacy of berberine by inhibiting P-glycoprotein function. Biol. Pharm. Bull. 36 (10), 1562–1569.

Shen, Q., Wang, L., Zhou, H., et al., 2013. Stereoselective binding of chiral drugs to plasma proteins. Acta Pharmacol. Sin. 34 (8), 998–1006.

Shen, P., Wong, S.P., Yong, E.L., 2007. Sensitive and rapid method to quantify icaritin and desmethylicaritin in human serum using gas chromatography-mass spectrometry. J. Chromatogr. B Analyt. Technol. Biomed. Life Sci. 857 (1), 47–52.

Shi, S.J., Chen, H., Gu, S.F., et al., 2003. Pharmacokinetic-pharmacodynamic modeling of daurisoline and dauricine in beagle dogs. Acta Pharmacol. Sin. 24 (10), 1011–1015.

Shi, Q., Liang, M., Zhang, W., et al., 2007. Quantitative LC/MS/MS method and pharmacokinetic studies of columbin, an anti-inflammation furanoditerpen isolated from *Radix Tinosporae*. Biomed. Chromatogr. 21 (6), 642–648.

Silberberg, M., Morand, C., Mathevon, T., et al., 2006. The bioavailability of polyphenols is highly governed by the capacity of the intestine and of the liver to secrete conjugated metabolites. Eur. J. Nutr. 45 (2), 88–96.

Smith, M.L., Nichols, D.C., Underwood, P., et al., 2014. Morphine and codeine concentrations in human urine following controlled poppy seeds administration of known opiate content. Forensic Sci. Int. 241, 87–90.

Su, M.X., Song, M., Sun, D.Z., et al., 2012. Determination of sinomenine sustained-release capsules in healthy Chinese volunteers by liquid chromatography-tandem mass spectrometry. J. Chromatogr. B Analyt. Technol. Biomed. Life. Sci. 889-890, 39–43.

Sun, L., Chen, W., Qu, L., et al., 2013a. Icaritin reverses multidrug resistance of HepG2/ADR human hepatoma cells via downregulation of MDR1 and P-glycoprotein expression. Mol. Med. Rep. 8 (6), 1883–1887.

Sun, Y.H., He, X., Yang, X.L., et al., 2014a. Absorption characteristics of the total alkaloids from *Mahonia bealei* in an *in situ* single-pass intestinal perfusion assay. Chin. J. Nat. Med. 12 (7), 554–560.

Sun, D.L., Huang, S.D., Wu, P.S., et al., 2010. Stereoselective protein binding of tetrahydropalmatine enantiomers in human plasma, HSA, and AGP, but not in rat plasma. Chirality 22 (6), 618–623.

Sun, S.Y., Wang, Y.Q., Li, L.P., et al., 2013b. Stereoselective interaction between tetrahydropalmatine enantiomers and CYP enzymes in human liver microsomes. Chirality 25 (1), 43–47.

Sun, E., Wei, Y.J., Zhang, Z.H., et al., 2014b. Processing mechanism of Epimedium fried with suet oil based on absorption and metabolism of flavonoids. Zhongguo Zhong Yao Za Zhi 39 (3), 383–390.

Sun, Y.F., Wink, M., 2014. Tetrandrine and fangchinoline, bisbenzylisoquinoline alkaloids from *Stephania tetrandra* can reverse multidrug resistance by inhibiting P-glycoprotein activity in multidrug resistant human cancer cells. Phytomedicine 21 (8–9), 1110–1119.

Sun, E., Xu, F.J., Zhang, Z.H., et al., 2014c. Discussion about research progress and ideas on processing mechanism of traditional Chinese medicine. Zhongguo Zhong Yao Za Zhi 39 (3), 363–369.

Sun, E., Xu, F.J., Zhang, Z.H., et al., 2014d. Construction of research system for processing mechanism of traditional Chinese medicine based on chemical composition transformation combined with intestinal absorption barrier. Zhongguo Zhong Yao Za Zhi 39 (3), 370–377.

Sun, E., Zhang, Z.H., Cui, L., et al., 2014e. Discussion on research ideas of synergistic mechanism of Epimedium fried with suet oil based on self-assembled micelles formation in vivo. Zhongguo Zhong Yao Za Zhi 39 (3), 378–382.

Swanson, H.I., 2015. Drug metabolism by the host and gut microbiota: a partnership or rivalry? Drug Metab. Dispos. 43 (10), 1499–1504.

Tan, X., Jia, X.B., Chen, Y., et al., 2009. Thought and probe on basic research for Chinese materiamedica based on intestinal absorption barrier. Chin. Trad. Herb. Drug 40 (10), 1520–1524.

Tsai, P.L., Tsai, T.H., 2004. Hepatobiliary excretion of berberine. Drug Metab. Dispos. 32 (4), 405–412.

Tsai, T.H., Wu, J.W., 2003. Regulation of hepatobiliary excretion of sinomenine by P-glycoprotein in Sprague-Dawley rats. Life Sci. 72 (21), 2413–2426.

van Breemen, R.B., Li, Y., 2005. Caco-2 cell permeability assays to measure drug absorption. Expert Opin. Drug Metab. Toxicol. 1 (2), 175–185.

Vrba, J., Vrublova, E., Modriansky, M., et al., 2011. Protopine and allocryptopine increase mRNA levels of cytochromes P450 1A in human hepatocytes and HepG2 cells independently of AhR. Toxicol. Lett. 203 (2), 135–141.

Wang, K., Chai, L., Feng, X., et al., 2017a. Metabolites identification of berberine in rats using ultra-high performance liquid chromatography/quadrupole time-of-flight mass spectrometry. J. Pharm. Biomed. Anal. 139, 73–86.

Wang, W., Dolan, L.C., von Alvensleben, S., et al., 2017b. Safety of standardized *Macleaya cordata* extract in an eighty-four-day dietary study in dairy cows. J. Anim. Physiol. Anim. Nutr. doi: 10.1111/jpn.12702.

Wang, C., Li, S., Tang, Y., et al., 2012. Microdialysis combined with liquid chromatography-tandem mass spectrometry for the determination of levo-tetrahydropalmatine in the rat striatum. J. Pharm. Biomed. Anal. 64–65, 1–7.

Wang, X., Liu, X., Cai, H., et al., 2015a. Ultra high performance liquid chromatography with tandem mass spectrometry method for the determination of tetrandrine and fangchinoline in rat plasma after oral administration of Fangji Huangqi Tang and *Stephania tetrandra* S. Moore extracts. J. Sep. Sci. 38 (8), 1286–1293.

Wang, C., Wang, S., Fan, G., et al., 2010. Screening of antinociceptive components in *Corydalis yanhusuo* WT. Wang by comprehensive two-dimensional liquid chromatography/tandem mass spectrometry. Anal. Bioanal. Chem. 396 (5), 1731–1740.

Wang, X., Wang, S., Ma, J., et al., 2015b. Pharmacokinetics in rats and tissue distribution in mouse of berberrubine by UPLC-MS/MS. J. Pharm. Biomed. Anal. 115, 368–374.

Wang, C., Wu, C., Zhang, J., et al., 2015c. Systematic considerations for a multicomponent pharmacokinetic study of *Epimedii wushanensis* herba: from method establishment to pharmacokinetic marker selection. Phytomedicine 22 (4), 487–497.

Wang, Q.H., Yang, X.L., Xiao, W., et al., 2015d. Microcrystalline preparation of akebiasaponin D for its bioavailability enhancement in rats. Am. J. Chin. Med. 43 (3), 513–528.

Wei, Y., Li, P., Fan, H., et al., 2012. Metabolite profiling of four major flavonoids of *Herba Epimedii* in zebrafish. Molecules 17 (1), 420–432.

Wei, Y.J., Xue, X.L., Liu, W., et al., 2013. Metabolism study of asperosaponin VI by using zebrafish. Yao Xue Xue Bao 48 (2), 281–285.

Wen, C., Cai, J., Lin, C., et al., 2014. Gradient elution liquid chromatography mass spectrometry determination of acetylcorynoline in rat plasma and its application to a pharmacokinetic study. Xenobiotica 44 (8), 743–748.

Wong, S.P., Shen, P., Lee, L., et al., 2009. Pharmacokinetics of prenylflavonoids and correlations with the dynamics of estrogen action in sera following ingestion of a standardized Epimedium extract. J. Pharm. Biomed. Anal. 50 (2), 216–223.

Wu, J.J., Ge, G.B., He, Y.Q., et al., 2016a. Gomisin A is a novel isoform-specific probe for the selective sensing of human cytochrome p450 3A4 in liver microsomes and living cells. AAPS J. 18 (1), 134–145.

Wu, P., Huang, S., YE, Y., 2007. Difference absorption of l-tetrahydropalmatine and dl-tetrahydropalmatine in intestine of rats. Acta Pharm. Sin. 42 (5), 534–537.

Wu, H., Kim, M., Han, J., 2016b. Icariin metabolism by human intestinal microflora. Molecules 21 (9), E1158. doi: 10.3390/21091158.

Wu, C., Yan, R., Zhang, R., et al., 2013. Comparative pharmacokinetics and bioavailability of four alkaloids in different formulations from *Corydalis decumbens*. J. Ethnopharmacol. 149 (1), 55–61.

Xu, S.F., Jin, T., Lu, Y.F., et al., 2014. Effect of icariin on UDP-glucuronosyltransferases in mouse liver. Planta Med. 80 (5), 387–392.

Xu, X., Xu, M.S., Zhu, J.H., et al., 2013. Pathological changes in rats with acute *Dysosma versipellis* poisoning. Fa Yi Xue Za Zhi 29 (5), 333–336.

Xu, X.L., Yang, L.J., Jiang, J.G., 2016a. Renal toxic ingredients and their toxicology from traditional Chinese medicine. Expert Opin. Drug Metab. Toxicol. 12 (2), 149–159.

Xu, S.J., Zhu, Y.L., Yu, J.J., et al., 2016b. Pharmacokinetics of epimedin A, B, C and icariin of Chuankezhi injection in rat. Zhongguo Zhong Yao Za Zhi 41 (1), 129–133.

Yan, J., He, X., Feng, S., et al., 2014. Up-regulation on cytochromes P450 in rat mediated by total alkaloid extract from *Corydalis yanhusuo*. BMC Complement. Altern. Med. 14, 306.

Yan, X.H., Li, H.D., Peng, W.X., et al., 1997. Determination of sinomenine HCl in serum and urine by HPLC and its pharmacokinetics in normal volunteers. Yao Xue Xue Bao 32 (8), 620–624.

Yan, H., Yan, M., Li, H.D., et al., 2015. Pharmacokinetics and penetration into synovial fluid of systemical and electroporation administered sinomenine to rabbits. Biomed. Chromatogr. 29 (6), 883–889.

Yang, L., Liu, H.T., Ma, H., et al., 2007. Application of systems biology to absorption, distribution, metabolism and excretion in traditional Chinese medicine. World Sci. Tech-Modern Trad. Chin. Med. 9 (1), 98–104.

Yang, C., Xiao, Y., Wang, Z., et al., 2014a. UHPLC-ESI-MS/MS determination and pharmacokinetic study of two alkaloid components in rat plasma after oral administration of the extract of *Corydalis bungeana* Turcz. J. Chromatogr. B Analyt. Technol. Biomed. Life Sci. 960, 59–66.

Yang, X.B., Yang, X.W., Liu, J.X., 2014b. Study on material base of corydalis rhizome. Zhongguo Zhong Yao Za Zhi 39 (1), 20–27.

Yao, Y.M., Cao, W., Cao, Y.J., et al., 2007. Effect of sinomenine on human cytochrome P450 activity. Clin. Chim. Acta. 379 (1–2), 113–118.

Yap, S.P., Shen, P., Li, J., et al., 2007. Molecular and pharmacodynamic properties of estrogenic extracts from the traditional Chinese medicinal herb, Epimedium. J. Ethnopharmacol. 113 (2), 218–224.

Yates, R.A., Tateno, M., Nairn, K., et al., 2002. The pharmacokinetics of the antimigraine compound zolmitriptan in Japanese and Caucasian subjects. Eur. J. Clin. Pharmacol. 58 (4), 247–252.

Yi, S.J., Cho, J.Y., Lim, K.S., et al., 2009. Effects of *Angelicae tenuissima radix*, *Angelicae dahuricae radix* and *Scutellariae radix* extracts on cytochrome P450 activities in healthy volunteers. Basic Clin. Pharmacol. Toxicol. l105, 249–256.

Zhai, Y.K., Guo, X., Pan, Y.L., et al., 2013. A systematic review of the efficacy and pharmacological profile of *Herba Epimedii* in osteoporosis therapy. Pharmazie 68 (9), 713–722.

Zhang, Y., Dong, X., Le, J., et al., 2014a. A practical strategy for characterization of the metabolic profile of chiral drugs using combinatory liquid chromatography-mass spectrometric techniques: application to tetrahydropalmatine enantiomers and their metabolites in rat urine. J. Pharm. Biomed. Anal. 94, 152–162.

Zhang, X., Guan, J., Zhu, H., et al., 2014b. Simultaneous determination of scutellarin and tetrahydropalmatine of Deng-yan granule in rat plasma by UFLC-MS/MS and its application to a pharmacokinetic study. J. Chromatogr. B Analyt. Technol. Biomed. Life Sci. 971, 126–132.

Zhang, M., Le, J., Wen, J., et al., 2011a. Simultaneous determination of tetrahydropalmatine and tetrahydroberberine in rat urine using dispersive liquid-liquid microextraction coupled with high-performance liquid chromatography. J. Sep. Sci. 34 (22), 3279–3286.

Zhang, Z.J., Tan, Q.R., Tong, Y., et al., 2011b. An epidemiological study of concomitant use of Chinese medicine and antipsychotics in schizophrenic patients: implication for herb-drug interaction. PLoS One 6 (2), e17239.

Zhang, Y., Wang, Q.S., Cui, Y.L., et al., 2012. Changes in the intestinal absorption mechanism of icariin in the nanocavities of cyclodextrins. Int. J. Nanomed. 7, 4239–4249.

Zhang, Y.F., Yu, Y., Zhou, L.L., 2010. The pharmacokinetics study on sinomenine transdermal patch on anaesthetized Beagle dogs. Zhong Yao Cai 33 (6), 944–947.

Zhang, M.F., Zhao, Y., Jiang, K.Y., et al., 2014c. Comparative pharmacokinetics study of sinomenine in rats after oral administration of sinomenine monomer and *Sinomenium acutum* extract. Molecules 19 (8), 12065–12077.

Zhang, Z.D., Zhou, C.M., Jin, G.Z., et al., 1990. Pharmacokinetics and autoradiography of ^{3}H or ^{14}C stepholidine. Acta Pharm. Sin. 11 (4), 289–292.

Zhao, Y., Hellum, B.H., Liang, A., et al., 2012a. The in vitro inhibition of human CYP1A2, CYP2D6 and CYP3A4 by tetrahydropalmatine, neferine and berberine. Phytother. Res. 26 (2), 277–283.

Zhao, Y., Hellum, B.H., Liang, A., et al., 2015. Inhibitory mechanisms of human CYPs by three alkaloids isolated from traditional Chinese herbs. Phytother. Res. 29 (6), 825–834.

Zhong, Q., Shi, Z., Zhang, L., et al., 2017. The potential of *Epimedium koreanum* Nakai for herb-drug interaction. J. Pharm. Pharmacol. 69 (10), 1398–1408.

Zhou, J., Chen, Y., Wang, Y., et al., 2013a. A comparative study on the metabolism of *Epimedium koreanum* Nakai-prenylated flavonoids in rats by an intestinal enzyme (lactase phlorizin hydrolase) and intestinal flora. Molecules 19 (1), 177–203.

Zhou, Q., Liu, Y., Wang, X., et al., 2013b. A sensitive and selective liquid chromatography-tandem mass spectrometry method for simultaneous determination of five isoquinoline alkaloids from *Chelidonium majus* L. in rat plasma and its application to a pharmacokinetic study. J. Mass Spectrom. 48 (1), 111–118.

Zhou, J., Ma, Y.H., Zhou, Z., et al., 2015. Intestinalabsorption and metabolism of Epimedium flavonoids in osteoporosis rats. Drug Metab. Dispos. 43 (10), 1590–1600.

Zhou, X., Rong, Q., Xu, M., et al., 2017. Safety pharmacology and subchronic toxicity of jinqing granules in rats. BMC Vet. Res. 13 (1), 179.

Zou, L.W., Li, Y.G., Wang, P., et al., 2016. Design, synthesis, and structure activity relationship study of glycyrrhetinic acid derivatives as potent and selective inhibitors against human carboxylesterase 2. Eur. J. Med. Chem. 112, 280–288.

Further Reading

Chen, Y., Qu, D., Zhou, J., et al., 2015a. A microemulsion-based delivery system for enhanced absorption of epimedium flavonoids by real-time enzymolysis and mucoadhesive strategy. J. Control. Release 213, e124–e125.

Chen, Y., Xiao, X., Wang, C., et al., 2015b. Beneficial effect of tetrandrine on refractory epilepsy via suppressing P-glycoprotein. Int. J. Neurosci. 125 (9), 703–710.

Grace, O.M., Buerki, S., Symonds, M.R., et al., 2015. Evolutionary history and leaf succulence as explanations for medicinal use in aloes and the global popularity of Aloe vera. BMC Evol. Biol. 15 (29).

Zhao, M., Li, L.P., Sun, D.L., et al., 2012b. Stereoselective metabolism of tetrahydropalmatine enantiomers in rat liver microsomes. Chirality 24 (5), 368–373.

CHAPTER

6

Anticancer Chemodiversity of Ranunculaceae Medicinal Plants

ABBREVIATIONS

BBR	Berberine
CDK	Cyclin dependent kinase
DSB	Double-strand break
eEF	Eukaryotic elongation factor
EGFR	Epidermal growth factor receptor
EMT	Epithelial to mesenchymal transition

Ranunculales Medicinal Plants. http://dx.doi.org/10.1016/B978-0-12-814232-5.00006-X

ER   Estrogen receptor
ERK   Extracellular signal-regulated kinase
HCC   Hepatocellular carcinoma
HGF   Hepatocyte growth factor
HIF-1α Hypoxia-inducible factor 1α
HUVEC   Human umbilical vein endothelial cell
JNK   c-Jun NH_2-terminal kinase
LPS   Lipopolysaccharide
MAPK   Mitogen-activated protein kinase
mTOR   Mammalian target of rapamycin
PARP   Poly-ADP ribose polymerase
PCPS   *Pulsatilla chinensis* polysaccharides
PKE   *Pulsatilla koreana* extract
PSA   *Pulsatilla* saponin A
SND   Sini decoction
TCM   Traditional Chinese medicine
TQ   Thymoquinone
UPLC   Ultra-high performance liquid chromatography
VEGF   Vascular endothelial growth factor

6.1 INTRODUCTION

Bioactive natural products (plant secondary metabolites) are well known to possess therapeutic value for the prevention and treatment of various types and stages of cancer (Hao et al., 2015a,b). Ranunculaceae phytometabolites exhibit promising effects against cancer, many of which modulate signaling pathways that are key to cancer initiation and progression and enhance the anticancer potential of clinical drugs while reducing their toxic side effects. Although some Ranunculaceae compounds were isolated decades ago, this review focuses on pharmacological properties and the latest advances in molecular mechanisms and functions. We discuss our current state of knowledge of Ranunculaceae phytometabolites' adjuvant potential and in vitro/in vivo anticancer activity and highlight their abilities to modulate the hallmarks of cancer.

The Ranunculaceae family (eudicot Ranunculales) has at least 62 genera and 2200 species, and 42 genera and about 720 species are distributed throughout China, most of which are found in the southwest mountainous region (Hao et al., 2015b; Xiao, 1980). In traditional Chinese medicine (TCM), at least 13 Ranunculaceae genera are used in heat-clearing and detoxification (Qing Re Jie Du in TCM), 13 genera used in ulcer disease and sores (Yong Ju Chuang Du in TCM), and 7 genera used in swell-reducing and detoxification (Xiao Zhong Jie Du in TCM) (Hao et al., 2015b; Xiao et al., 1986). These genera may contain useful compounds that can be used to combat against cancer. Extracts/isolated compounds of at least 17 genera have shown anticancer/cytotoxic activity (Hao et al., 2013a,b,c, 2015a,b). The distribution of anticancer compounds within Ranunculaceae is not random but phylogeny related (Hao et al., 2014). For instance, *Ranunculus, Clematis, Pulsatilla, Anemone,* and *Nigella* are rich in pentacyclic triterpene saponins (Fig. 6.1); *Actaea* and *Cimicifuga* are rich in tetracyclic triterpene saponins, which are also found in *Thalictrum*; diterpenoid alkaloids are abundant in *Delphinium, Consolida,* and *Aconitum* but are also found in *Nigella* and *Thalictrum*; and isoquinoline alkaloids are abundant in *Asteropyrum, Caltha, Nigella,* Delphinieae, Adonideae, Thalictroideae, and Coptidoideae. The same class of phytometabolite, for example, saponins,

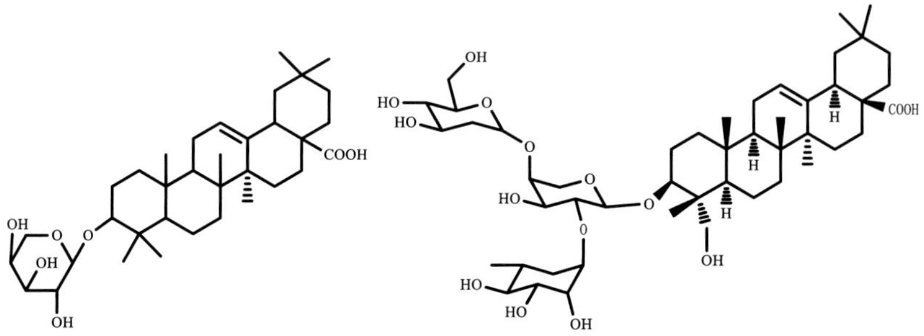

D rhamnose β-hederin (DRβ-H)*Pulsatilla* saponin D

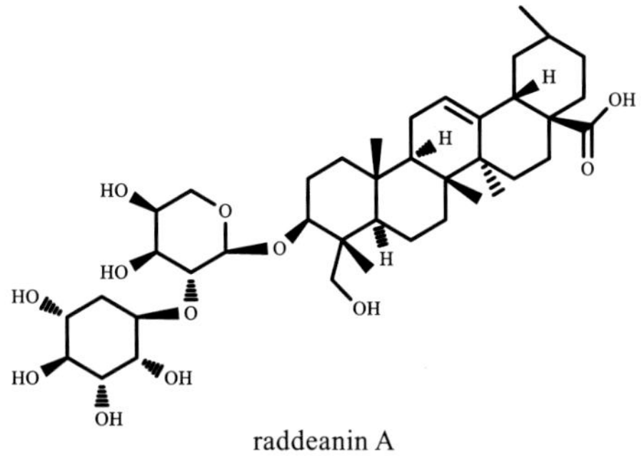

raddeanin A

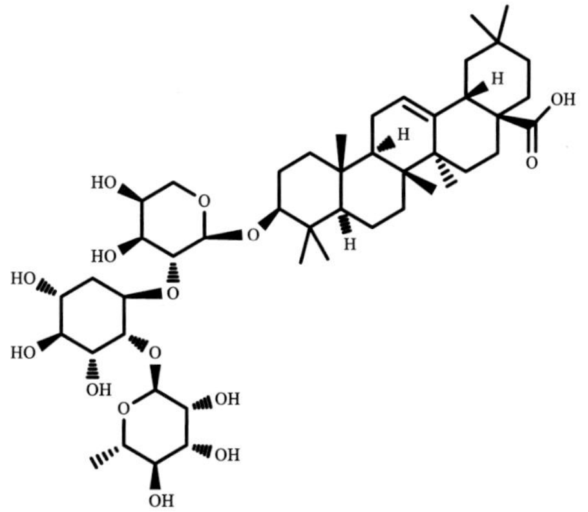

Pulsatilla saponin A

FIGURE 6.1   Representative anticancer compounds found in the family Ranunculaceae.

could exert anticancer activity via multiple pathways (Fig. 6.2) (Gaube et al., 2007), while the same signaling pathway could be the target of distinct compounds, for example, TQ (Kundu et al., 2014a,b; Schneider-Stock et al., 2014) and BBR (Ortiz et al., 2014).

Exhaustive literature search in PubMed, Google, and CNKI (http://cnki.net/) has been performed to outline the progress of molecular mechanism studies of Ranunculaceae anticancer compounds during the last decade. Search terms "anticancer," "antitumor," "apoptosis," "cell cycle," "DNA damage," "saponin," "alkaloid," "terpene," "polysaccharide," "phenolic," etc., were used, combined with "Ranunculaceae" and the names of genera.

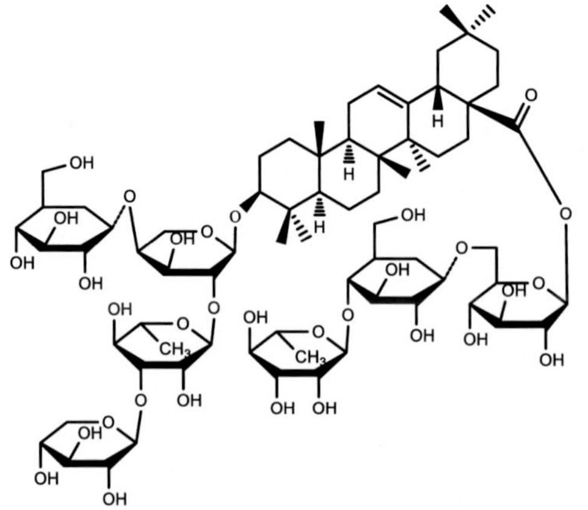

saponin B of *Anemone taipaiensis*

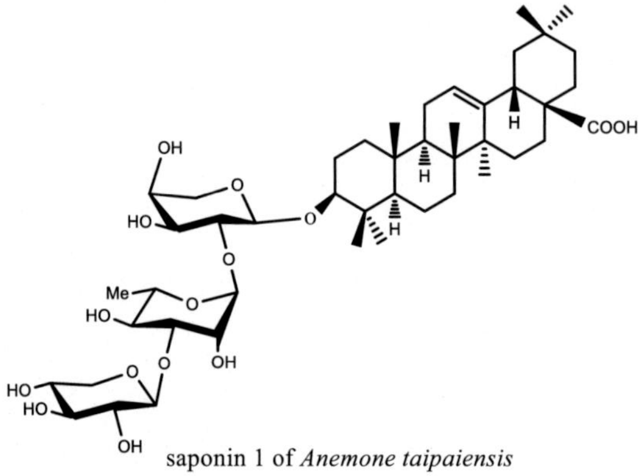

saponin 1 of *Anemone taipaiensis*

2 tetracyclic triterpene saponin:

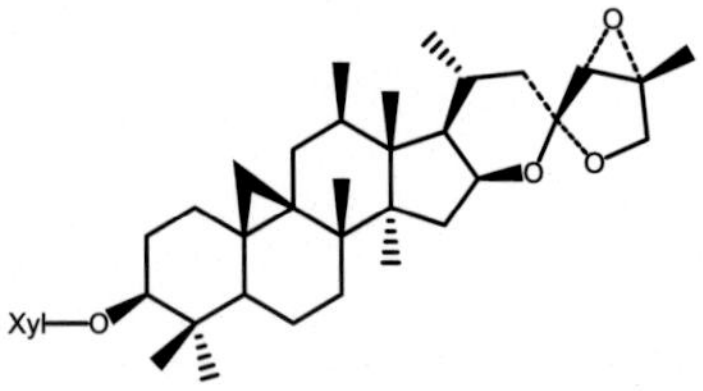

actein

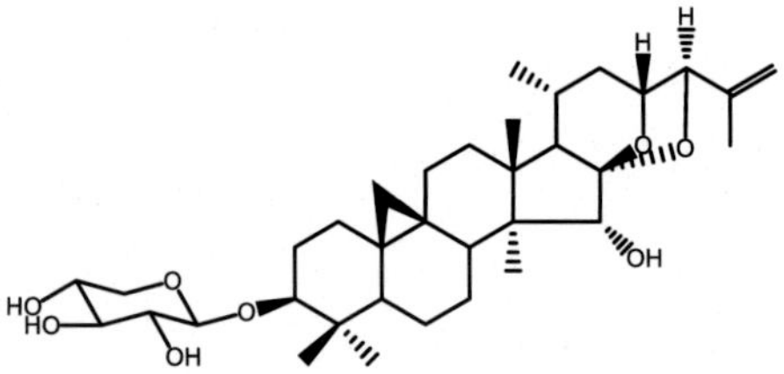

cimiside E

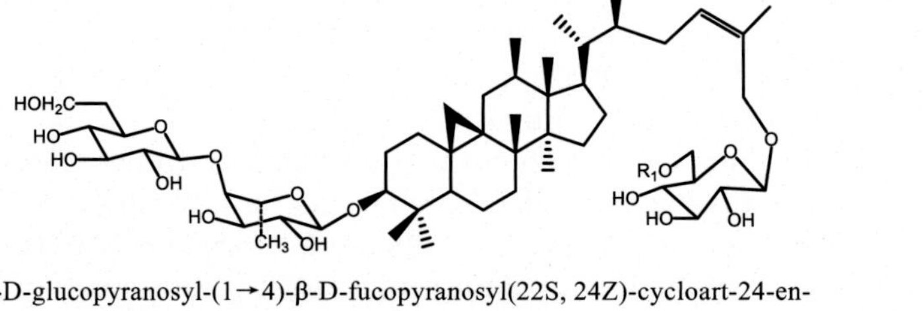

3-O-β-D-glucopyranosyl-(1→4)-β-D-fucopyranosyl(22S, 24Z)-cycloart-24-en-
3β, 22, 26-triol 26-O-β-D-glucopyranoside, R1=H

FIGURE 6.1 (*Cont.*)

3 diterpenoid:

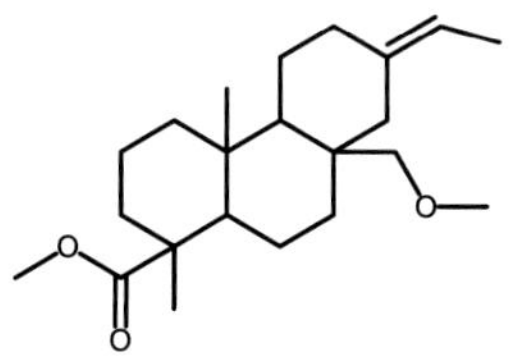

dihydroxy-isosteviol methyl ester (DIME)

4 triterpenoid:

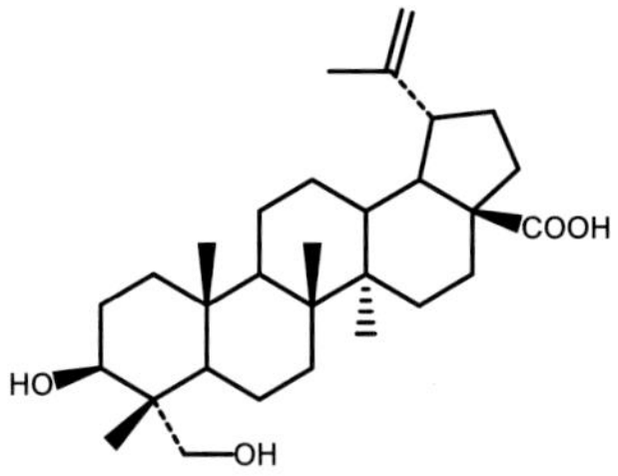

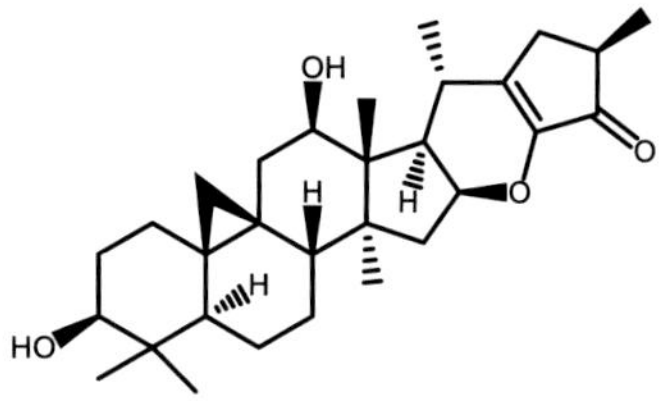

23-hydroxybetulinic acid (HBA)   cimyunnin A

5 monoterpene:

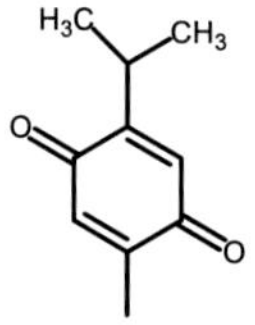

thymoquinone (TQ)

6 isoquinoline alkaloid:

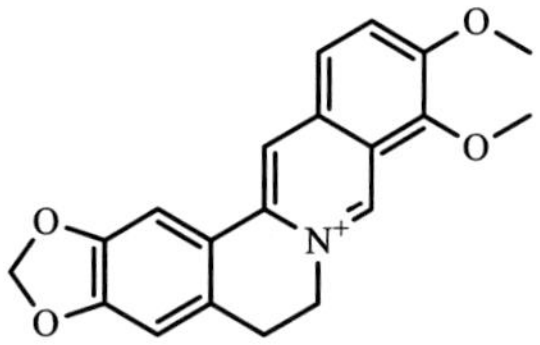

berberine (BBR)

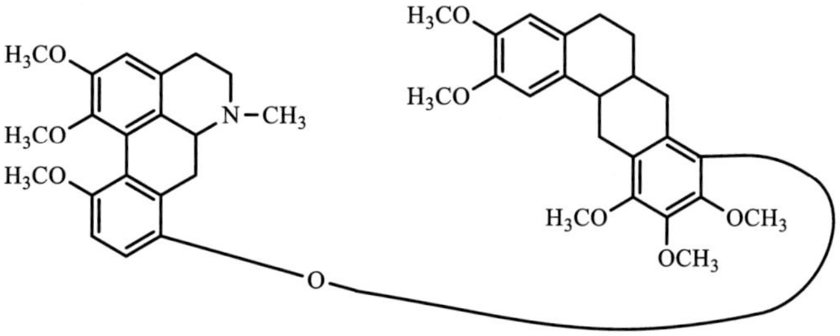

acutiaporberine

FIGURE 6.1 (*Cont.*)

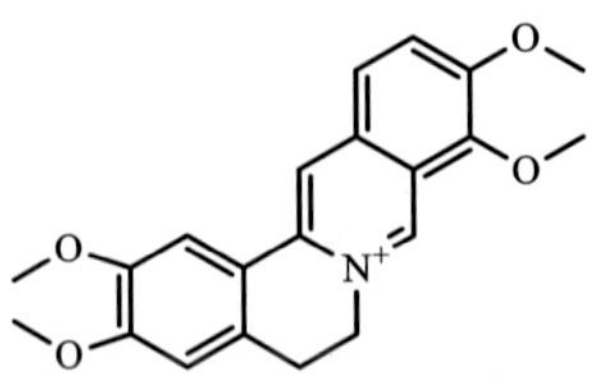

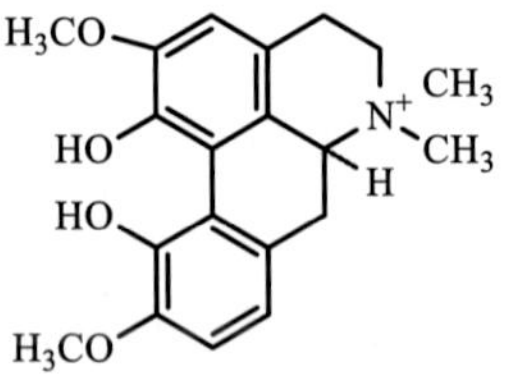

palmatinemagnoflorine

7 diterpenoid alkaloid:

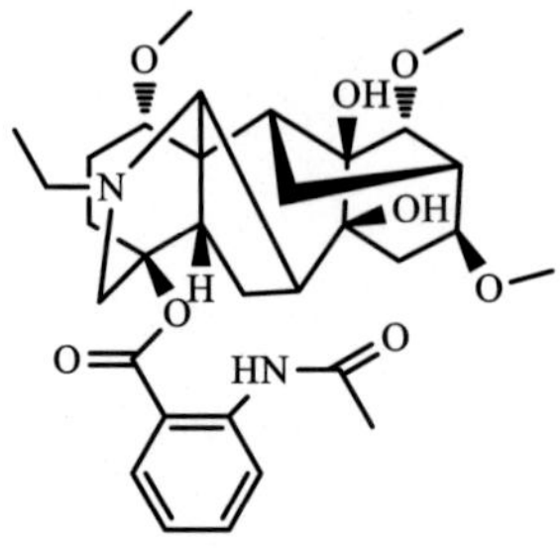

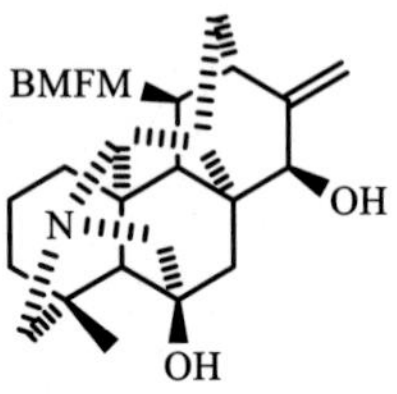

lappaconitine   pseudokobusine 11-3'-trifluoromethylbenzoate

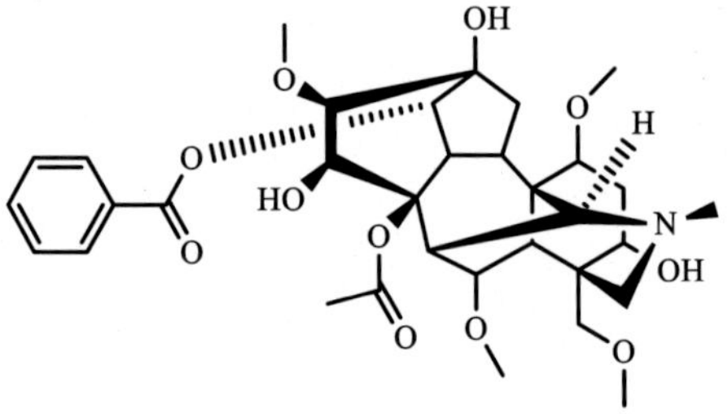

mesaconitine

8 cardioactive steroid:

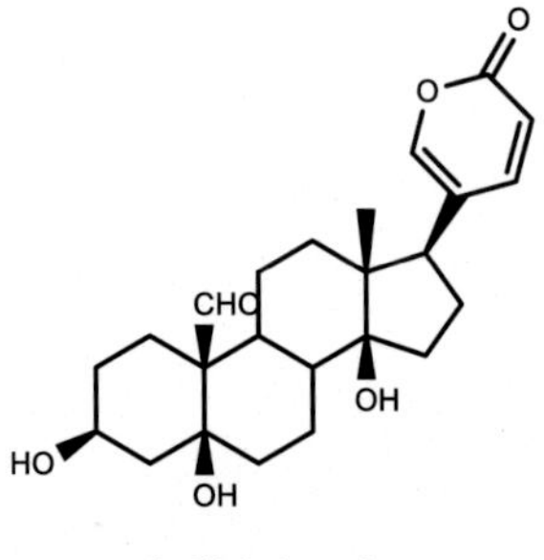

hellebrigenin

FIGURE 6.1 (*Cont.*)

9 phenolic acid:

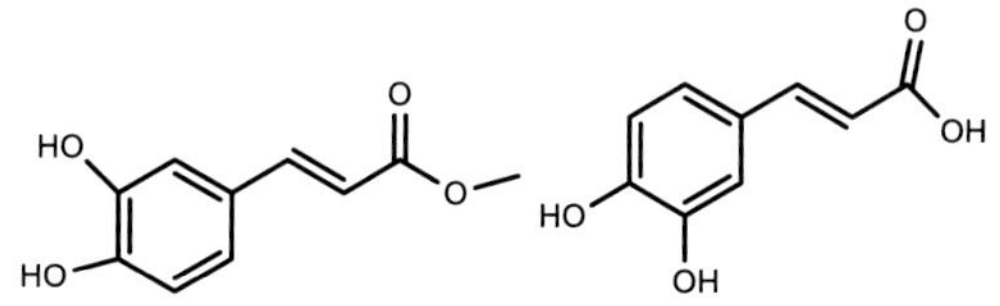

methyl caffeatecaffeic acid

10 lactone:

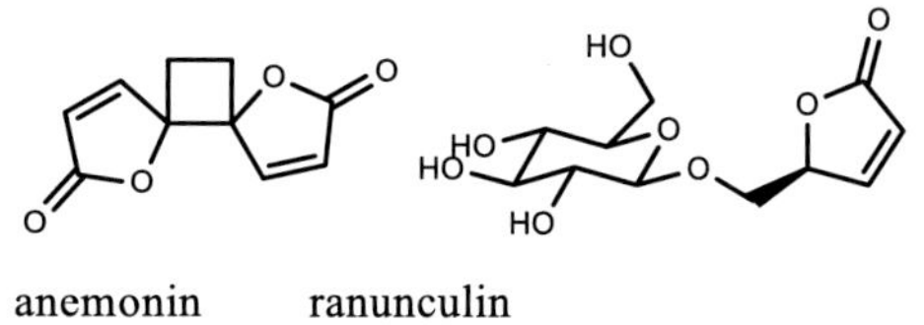

anemonin   ranunculin

FIGURE 6.1 (*Cont.*)

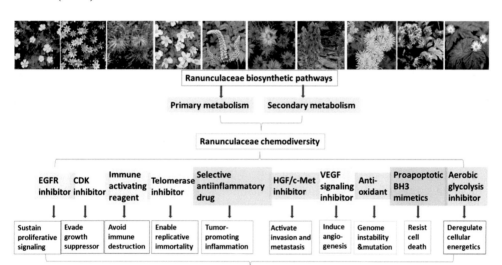

FIGURE 6.2   **The cancer hallmarks modulated by Ranunculaceae phytometabolites.** Ranunculaceae phytometabolites regulate all ten cancer hallmarks defined by Hanahan and Weinberg. The putative PARP inhibitor has not been identified from Ranunculaceae plants. Instead, various antioxidants found in Ranunculaceae might play important roles in fighting against genome instability and mutation of cancer cells.

6.2 CELL DEATH PATHWAYS

As cancer cells have evolved multiple mechanisms to resist the induction of programmed cell death (apoptosis), the modulation of apoptosis signaling pathways by Ranunculaceae compounds has been shown to constitute a key event in anticancer activities. In addition, cell cycle arrest, autophagy modulation, cell senescence, and other pathways are also involved in anticancer mechanisms of various Ranunculaceae phytometabolites.

6.2.1 Saponin

6.2.1.1 Clematis

Saponins are abundant in Ranunculaceae, especially in *Clematis, Pulsatilla, Anemone*, and Cimicifugeae (Hao et al., 2013b,c, 2015a,b), which usually exert anticancer activity via cell cycle arrest and apoptosis induction. The aglycones of *Clematis* pentacyclic triterpene saponins mainly belong to oleanolic type A; olean-3β, 28-diol type B; hederagenin type C; or hederagenin-11,13-dien type D, where types A and C are predominant (Hao et al., 2013c, 2015a). Many *Clematis* saponins have cytotoxic activity against human glioblastoma (Zhao et al., 2014a), hepatoma (Zhao et al., 2014b), cervical cancer (Li et al., 2014a), leukemia (Tian et al., 2013; Zhao et al., 2014b), gastric cancer (Tian et al., 2013; Zhao et al., 2014b), colon cancer(He et al., 2011), prostate cancer(Gong et al., 2013), etc. However, the mechanistic study is scarce. For instance, D Rhamnose β-hederin (DRβ-H), an oleanane-type triterpenoid saponin from TCM plant *Clematis ganpiniana*, inhibits phosphatidylinositol 3-kinase (PI3K)/Akt and activates extracellular signal-regulated kinase (ERK) signaling pathways in breast cancer cells (Cheng et al., 2014a). PI3K inhibitor LY294002 synergistically enhances DRβ-H-induced apoptosis while MAPK/extracellular signal-regulated kinase (MEK) inhibitor U0126 reduces the apoptosis rate. DRβ-H regulates the ratio of proapoptotic and antiapoptotic Bcl-2 family proteins. DRβ-H induces depolarization of mitochondrial membrane potential to release Apaf-1 and cytochrome C from the intermembrane space into the cytosol, where they promoted caspase-9 and -3 activation.

6.2.1.2 Pulsatilla

Pulsatilla, Anemone, and *Clematis* belong to the tribe Anemoneae, and *Pulsatilla* is evolutionarily closer to *Anemone* than to *Clematis* (Hao et al., 2015b). Saponins exhibit cytostatic and cytotoxic activity against various cancer cells, but the mechanism is not fully understood. *Pulsatilla koreana* saponin D (SB365) strongly suppresses the growth of hepatocellular carcinoma (HCC) cells in a dose-dependent manner and induces apoptosis by increasing the proportion of sub-G1 apoptotic cells from 8% to 21% through induction of expression of Bax and cleaved caspase-3 (Hong et al., 2012a). SB365 effectively suppresses the phosphorylation of PI3K downstream factors, for example, Akt, mammalian target of rapamycin (mTOR), and p70S6K, both in vitro and in vivo. SB365 suppresses the proliferation of human colon cancer and pancreatic cancer cells (Son et al., 2013a,b) and induces apoptosis by modulating the Akt/mTOR signaling pathway (Son et al., 2013a).

The K_a (absorption rate constant) and P_{app} (apparent permeability coefficient) values of *Pulsatilla* saponin D are highest in the colon (Rao et al., 2013), followed by ileum, jejunum, and duodenum. The *Pulsatilla* saponins are not transported in a concentration-dependent manner, and the transporter protein might be involved in their transport. The *Pulsatilla* saponins exhibit rapid absorption, quick elimination, and significant double-peak on the plasma concentration-time curve (Liu et al., 2013a). However, the oral bioavailability is quite low (0.55%–2.5%), due to the unfavorable molecular size (≥733.5 Da) of pentacyclic triterpene saponins, poor gut absorption, and/or extensive metabolism after absorption (Ouyang et al., 2014). Nanoformulations have to be developed to maximize the anticancer effects of *Pulsatilla* saponins.

Of the 42 different phosphoreceptor tyrosine kinases (RTKs), SB365 docks at an allosteric site on c-mesenchymal-epithelial transition factor (c-Met) and strongly inhibits its expression in gastric cancer cells (Hong et al., 2013). The activation of the c-Met signal cascade components, including Akt and mTOR, is inhibited by SB365 in a dose-dependent manner. SB365 inhibits the phosphorylation of Met proto-oncogene and the downstream signaling pathway required for growth and survival in the Met-amplified HCC827GR non-small cell lung cancer (NSCLC) cells (Jang et al., 2014). SB365 suppresses the anchorage-independent growth, migration, and invasion and induces apoptosis in HCC827GR cells. SB365 was an inducer of autophagosome formation (Zhang et al., 2015) but an inhibitor of autophagic flux. It synergistically enhances the anticancer activity of chemotherapeutic agents against cervical cancer HeLa cells. It increases the phosphorylation of ERK and inhibits the phosphorylation of mTOR and p70S6K, suggesting their roles in the effects of SB365 on autophagy. SB365 could be a promising adjuvant anticancer agent.

Pulsatilla saponin A (PSA) of *Pulsatilla chinensis* may exert its antitumor effect by inducing DNA damage and causing G2 arrest and apoptosis in multiple cancer cells (Liu et al., 2014a). PSA upregulates p53 and cyclin B protein levels and lowers Bcl-2 protein levels. PSA induces differentiation in acute myeloid leukemia in vitro, probably through the MEK/ERK signaling pathway (Wang et al., 2016b).

6.2.1.3 *Anemone*

The genus *Anemone*, evolutionarily closely related to *Pulsatilla*, is also rich in therapeutic saponins (Hao et al., 2015b). Raddeanin A, a pentacyclic triterpene saponin from *Anemone raddeana* (Liang Tou Jian in TCM), inhibits proliferation and induces apoptosis of multiple cancer cells (Guan et al., 2015; Xue et al., 2013). Raddeanin A increases Bax expression; reduces Bcl-2, Bcl-xL, and survivin expressions; and significantly activates caspase-3, -8, -9, and poly-ADP ribose polymerase (PARP) (Xue et al., 2013). Saponins B, 1, and 6 of *Anemone taipaiensis* exhibit significant anticancer activity against human leukemia, glioblastoma multiforme (GBM), and HCC (Ji et al., 2015; Li et al., 2013a; Wang et al., 2013). Saponin 1 causes characteristic apoptotic morphological changes in GBM cells (Li et al., 2013a), which was confirmed by DNA ladder electrophoresis and flow cytometry. Saponon 1 also causes a time-dependent decrease in the expression and nuclear location of NF-κB. The expression of inhibitors of apoptosis (IAP) family members, for example, survivin and XIAP, is significantly decreased by saponin 1. Moreover, saponin 1 causes a decrease in the Bcl-2/Bax ratio and initiates apoptosis by activating caspase-9 and -3 in the GBM cell lines. Thus, saponin 1 inhibits cell growth of GBM cells at least partially by inducing apoptosis and inhibiting survival signaling mediated by NF-κB.

Saponin B blocks the cell cycle at the S phase (Wang et al., 2013), induces chromatin condensation of U87MG GBM cells, and led to the formation of apoptotic bodies. Annexin V/PI assay suggests that phosphatidylserine externalization is apparent at higher drug concentrations. Saponin B activates the receptor-mediated pathway of apoptosis via the activation of Fas-l. Saponin B increases the Bax and caspase-3 ratio and decreases the protein expression of Bcl-2.

Triterpenoid saponins of *Anemone flaccida* induce apoptosis in human BEL-7402, HepG2 hepatoma cell lines, and lipopolysaccharide-stimulated HeLa cells via COX-2/PGE2 pathway (Han et al., 2013).

6.2.1.4 Cimicifugeae

Cycloartane triterpenoids and their saponins are mainly distributed in *Astragalus* (Leguminosae), Cimicifugeae, and *Thalictrum* (Ranunculaceae) and possess various bioactivities (Tian et al., 2006a). *Actaea* and *Cimicifuga* are closely related genera of the tribe Cimicifugeae (Hao et al., 2013b, 2015a,b), which are closer to *Souliea* than to *Eranthis*. *Beesia* is closer to *Anemonopsis* than to the other four genera.

9,19-Cycloartane triterpene glycosides of *Actaea asiatica* have notable cytotoxicity against HepG2 and MCF-7 cancer cells (Gao et al., 2006). The tetracyclic triterpene saponins and their aglycones display the cytotoxic activity against HCC, lung adenocarcinoma (Lu et al., 2012), gastric cancer (Guo et al., 2009), leukemia, colon cancer, breast cancer (Bao et al., 2014), etc. Cycloartane glycosides from the roots of *Cimicifuga foetida* (Sheng Ma in TCM) show significant Wnt signaling pathway inhibitory activity (Zhu et al., 2015), with IC_{50} of 3.33 and 13.34 μM, respectively. Our research group found that 23-O-acetylcimigenol-3-O-β-D-xylopyranoside can inhibit the proliferation of HepG2 cells with IC_{50} 16 μM and can induce apoptosis and G2/M cell cycle arrest (Tian et al., 2006b). This tetracyclic triterpene saponin is able to cleave PARP, regulate protein expression of Bcl-2 family, and decrease the expression of cdc 2 and cyclin B. 25-Acetyl-7,8-didehydrocimigenol 3-O-β-D-xylopyranoside is more potent than the parent compound 7,8-didehydrocimigenol 3-O-β-D-xylopyranoside in inhibiting ER-/Her2 (epidermal growth factor receptor [EGFR]2) overexpressing human breast cancer cell line MDA-MB-453 (Einbond et al., 2008). Cell cycle arrestin gastric cancer cells are induced by cimiside E of *Clematis heracleifolia* (a source plant of Sheng Ma) in S phase at a lower concentration (30 μM) and G2/M phase at higher concentrations (60 and 90 μM) (Guo et al., 2009). Cimiside E mediated apoptosis through the induction of the caspase cascade for both the extrinsic and intrinsic pathways. Our research group found that 25-anhydrocimigenol-3-O-β-D-xylopyranoside of *Souliea vaginata* (Huang San Qi in TCM) causes apoptosis and G0/G1 cell cycle arrest in HepG2 cells (Tian et al., 2006c) and exhibits a dose-dependent inhibition of tumor growth on mice implanted with H22 in vivo.

Actein, a triterpene glycoside from black cohosh (*Actaea racemosa, Cimicifuga racemosa*), strongly inhibits the growth of human breast cancer cells and induces a dose-dependent release of calcium into the cytoplasm (Einbond et al., 2013). The ER IP3 receptor antagonist heparin blocks this release. Heparin partially blocks the growth inhibitory effect, while the MEK inhibitor U0126 enhances it. Actein synergizes with the ER mobilizer thapsigargin and preferentially inhibits the growth of 293T (NF-κB) cells. Nanoparticle liposomes increase the growth inhibitory activity of actein. Actein alters the activity of the ER IP3 receptor and Na,K-ATPase, induces calcium release, modulates the NF-κB and MEK pathways, and causes cytochrome C release from mitochondria, which may partially explain its anticancer effect. Actein enhances TRAIL (tumor necrosis factor-related apoptosis-inducing ligand) effects on suppressing gastric cancer progression by activating p53/caspase-3 signaling (Yang and Ma, 2016). Actein ameliorates hepatobiliary cancer through stemness and p53 signaling regulation (Xi and Wang, 2017). KHF16 (24-acetylisodahurinol-3-O-β-D-xylopyranoside) of *C. foetida* suppresses breast cancer partially by inhibiting the NF-κB signaling pathway(Kong et al., 2016).

MCF7 (ER+/Her2 low) cells transfected for Her2 are more sensitive than the parental MCF7 cells to the growth inhibitory effects of actein (Einbond et al., 2008), indicating that Her2 plays a role in the action of actein. Actein alters the distribution of actin filaments and induces apoptosis in these cells. Actein appears to have both cytostatic and cytotoxic activity.

25-O-acetyl-7,8-didehydrocimigenol-3-O-β-d-(2-acetyl)xylopyranoside, a cycloartane triterpenoid from *C. foetida*, induces mitochondrial apoptosis via inhibiting Raf/MEK/ERK pathway and Akt phosphorylation in human breast carcinoma MCF-7 cells (Sun et al., 2016).

In human colon cancer cells (HCT-8) and in vivo in hepatoma cell (H22)-bearing mice, total glycosides of *Cimicifuga dahurica* and cisplatin show synergistic antitumor activity (Zhang et al., 2016b).

6.2.1.5 *Thalictrum*

3-O-β-D-glucopyranosyl-(1→4)-β-D-fucopyranosyl(22S,24Z)-cycloart-24-en-3β,22,26-triol 26-O-β-D-glucopyranoside(compound 1), isolated from *Thalictrum fortunei*, cause apoptosis and mitochondrial membrane potential loss in human hepatoma Bel-7402 cells (Zhang et al., 2011). The intracellular reactive oxygen species (ROS) are markedly provoked by compound 1. Compound 1 significantly increases the expression levels of cleaved caspase-3, P53, and Bax protein and decreases the expression of Bcl-2 protein.

6.2.2 Terpenoid

6.2.2.1 *Diterpene*

Dihydroxy-isosteviol methyl ester (DIME), a diterpene isolated from *Pulsatilla nigricans*, causes a significant decrease in cell viability and induces nuclear condensation and internucleosomal DNA fragmentation (Das et al., 2013). DIME interacts with calf thymus DNA, bringing apparent changes in structure and conformation.

6.2.2.2 *Triterpene*

23-Hydroxybetulinic acid (HBA) of *P. chinensis* can promote cell cycle arrest at S phase and induce apoptosis of human chronic myelogenous leukemia K562 cells via intrinsic pathways (Liu et al., 2015a). HBA disrupts mitochondrial membrane potential significantly and selectively downregulates the levels of Bcl-2 and survivin and upregulates Bax, cytochrome C, and cleaved caspase-9 and -3.

KY17 (cimigenol), a novel cycloartane triterpenoid from *Cimicifuga*, concentration-dependently induces growth inhibition and apoptosis in human colon cancer HT-29 cells (Dai et al., 2017). Cells in G2/M phase are increased, and protein levels of cleaved-caspase-8 and -3, and cleavage of PARP, are increased following KY17 treatment. Autophagy manifests as the accumulation of acridine orange, the appearance of green fluorescent protein-light-chain 3 (LC3) punctate structures, and increases levels of LC3-II protein expression. The autophagy inhibitor bafilomycin A1 enhances the induction of apoptosis by KY17.

6.2.2.3 *Monoterpene*

Thymoquinone(2-methyl-5-isopropyl-1,4-benzoquinone; TQ), a monoterpene abundant in *Nigella sativa*, downregulates E2F-1 protein and androgen receptors and causes cancer cell apoptosis (Kaseb et al., 2007). TQ downregulates expression of Bcl2 and other antiapoptotic genes (Koka et al., 2010) and upregulates apoptotic gene (caspase-3, 8, 9, and Bax) expression (El-Mahdy et al., 2005). TQ induces apoptosis in human colon cancer HCT116 cells through inactivation of STAT3 by blocking JAK2- and Src-mediated phosphorylation of EGF receptor tyrosine kinase (Kundu et al., 2014a). PAK1 is a novel kinase target of TQ (El-Baba et al., 2014).

TQ-induced conformational changes of PAK1 interrupt prosurvival MEK-ERK signaling in colorectal cancer and enhance apoptosis. TQ targets various components of intracellular signaling pathways, particularly a variety of upstream kinases and transcription factors, which are aberrantly activated during the course of carcinogenesis (Kundu et al., 2014b). The above mechanisms exemplify how TQ causes cancer cell apoptosis and death.

TQ also displays antiproliferative and cytostatic effects. For instance, TQ inhibits NF-κB activation induced by various carcinogens (Sethi et al., 2008) and controls oncogenic expression. In N-nitrosodiethylamine-induced rat HCC, TQ strongly induces G1/S arrest in cell cycle transition (Raghunandhakumar et al., 2013). TQ suppresses the activation of Akt and ERK (Yi et al., 2008) and activates c-Jun NH$_2$-terminal kinase (JNK) and p38 mitogen-activated protein kinase (MAPK) pathways (Torres et al., 2010). TQ pretreatment overcomes the insensitivity and potentiates the anticancer effect of gemcitabine through abrogation of Notch1, PI3K/Akt/mTOR-regulated signaling pathways in pancreatic cancer (Mu et al., 2015). TQ inhibits topoisomerase (Topo) IIα activity when incubated with the enzyme prior to the addition of DNA (Ashley and Osheroff, 2014) but enhances enzyme-mediated DNA cleavage ~5-fold, which is similar to the increase seen with the anticancer drug etoposide. TQ inhibits thymidine incorporation during DNA synthesis and inhibits cancer cell growth (Badary and El-Din, 2001; El-Mahdy et al., 2005).

The polo-box domain (PBD), a unique functional domain of polo-like kinase (Plk), is being targeted to develop Plk1-specific inhibitors (Shin et al., 2015). Poloxin (a synthetic derivative of TQ) and TQ, the PBD inhibitors, increase cell population in S phase and in G2/M in a p53-independent manner. Poloxin and TQ do not increase the population of cells staining positively for p-Histone H3 and MPM2, the mitotic index, but cause an increase in p21^{WAF1} and S arrest, indicating that PBD inhibitors affect interphase before mitotic entry.

Cell death may occur in several mechanisms, including apoptosis, necrosis (see subsection, Antiinflammatory Activity), and autophagy. Autophagy is a catabolic process that maintains cellular homeostasis in response to various cellular stress factors. The TQ-exposed head and neck squamous cell carcinoma cells show increased levels of autophagic vacuoles and specific autophagy markers LC3-II proteins (Chu et al., 2014). TQ treatment also increases the accumulation of autophagosomes. An in vivo BALB/c nude mouse xenograft model showed that TQ administered by oral gavage reduces tumor growth via induced autophagy and apoptosis. Bafilomycin-A1, an autophagy inhibitor, enhances TQ cytotoxicity but does not promote apoptosis. Cell viability is eradicated in autophagy-defective cells. These results imply that inhibition of autophagy is an emerging strategy in cancer therapy. In CPT-11-R LoVo colon cancer cells, TQ causes mitochondrial outer membrane permeability and activated autophagic cell death (Chen et al., 2015a). JNK and p38 inhibitors (SP600125 and SB203580, respectively) reverse TQ autophagy. TQ activates apoptosis before autophagy, and the direction of cell death is switched toward autophagy at initiation of autophagosome formation.

Glioblastoma cells may also be dependent on the autophagic pathway for survival. TQ inhibits autophagy and induces cathepsin-mediated, caspase-independent cell death in glioblastoma cells (Racoma et al., 2013). TQ induces lysosome membrane permeabilization, which results in a leakage of cathepsin B into the cytosol and mediates caspase-independent cell death that can be prevented by pretreatment with a cathepsin B inhibitor. TQ-induced apoptosis appears to be caspase independent due to failure of the caspase inhibitor z-VAD-FMK to prevent cell death and absence of the typical apoptosis signature DNA fragmentation.

6.2.3 Alkaloid

6.2.3.1 Isoquinoline Alkaloids

Thalictrum, Coptis, and *Hydrastis* are rich in isoquinoline alkaloids (Hao et al., 2015b). The anticancer effects of *Coptis Chinensis* (Huang Lian in TCM) can be ascribed to its TCM trait of removing damp-heat, fire, and toxicity (Tang et al., 2009). BBR, a major isoquinoline alkaloid of Huang Lian and other related species, significantly accumulates inside prostate cancer cells of G1 phase and enhances apoptosis (Huang et al., 2015). BBR inhibits the expression of prostate-specific antigens and the activation of EGFR and attenuates EGFR activation following EGF treatment in vitro. BBR induces apoptosis via the mitochondrial pathway in liver cancer cells (Yip et al., 2013) and increases the expression of Bax, followed by the activation of the caspase cascade. Procaspase-9 and its effectors, procaspase-3 and -7, are activated by BBR. BBR activates the transcription factor Egr-1 and consequently induces the expression of nonsteroidal antiinflammatory drug (NSAID)-activated genes (NAG-1) (Auyeung and Ko, 2009), which mediate the drug-induced proapoptotic action in HepG2 cells. Other protoberberine-type alkaloids in Huang Lian might give synergistic results for anticancer effects.

BBR significantly upregulates the mRNA expression of FoxO1 and O3a, two forkhead-box family members (Shukla et al., 2014). Their phosphorylation-mediated cytoplasmic sequestration followed by degradation is prevented by BBR-induced down-modulation of the PI3K/Akt/mTOR pathway, which promotes FoxO nuclear retention. PTEN, a tumor suppressor gene and negative regulator of the PI3K/Akt axis, is upregulated while phosphorylation of its Ser380 residue (possible mechanism of PTEN degradation) is significantly decreased in BBR-treated HepG2 cells. BBR induces a significant increase in transcriptional activity of FoxO, which effectively enhances BH3-only protein Bim expression, followed by the direct activation of proapoptotic protein Bax, as well as increased Bax/Bcl-2 ratio, mitochondrial dysfunction, caspase activation, and DNA fragmentation. Bim-silencing partially restores HepG2 cell viability during BBR exposure, implying the pivotal role of Bim in BBR-mediated cytotoxicity. The anticancer activity of *Thalictrum* alkaloids is little known. Acutiaporberine, a bisalkaloid isolated from *Thalictrum acutifolium*, may be a natural potential apoptosis-inducing agent for highly metastatic lung cancer (Chen et al., 2002).

Palmatine, not BBR, is a cell cycle blocker in G1 of A549 adenocarcinoma cells (Ji et al., 2012). BBR induces G2/M arrest in human promyelocytic leukemia HL-60 cells and murine myelomonocytic leukemia WEHI-3 cells (Lin et al., 2006), which is accompanied by increased levels of Wee1 and 14-3-3σ and decreased levels of Cdc25c, CDK1, and cyclin B1. CDK2 expression is not affected by BBR.

In melanoma cell lines, BBR at low doses (12.5–50 μM) is concentrated in mitochondria and promotes G1 arrest (Serafim et al., 2008). Higher doses (>50 μM) result in cytoplasmic and nuclear BBR accumulation and G2 arrest. DNA synthesis is not markedly affected by low doses of BBR, but 100 μM is strongly inhibitory. Notably, 100 μM of BBR inhibits cell growth with relatively little induction of apoptosis. BBR displays multiphasic effects in malignant cell lines, which are correlated with its concentration and intracellular distribution. These results help explain some of the conflicting information regarding the effects of BBR, which may be more as a cytostatic agent than a cytotoxic compound.

BBR may inhibit protein synthesis, histone deacetylase (HDAC), or Akt/mTOR pathways (Lee et al., 2014). BBR induces endoplasmic reticulum (ER) stress and autophagy, which is associated with activation of AMP-activated protein kinase (AMPK). However, BBR does not alter mTOR or HDAC activities in MDA-MB-231 human breast cancer cells. BBR induces the acetylation of α-tubulin, a substrate of HDAC6. In addition, the combination of BBR and SAHA, a pan-HDAC inhibitor, synergistically inhibits cell proliferation and induces cell cycle arrest. BBR induces both autophagy and apoptosis in HCC cells (Hou, 2011), and CD147 expression is downregulated by BBR.

BBR primarily exerts its anticancer effect by inducing cell-cycle arrest, apoptosis, and autophagy (Ortiz et al., 2014). However, BBR inhibits glioblastoma cells through induction of cellular senescence (Liu et al., 2015b). BBR drastically reduces the level of EGFR, and the downstream RAF-MEK-ERK signaling pathway is remarkably inhibited, whereas Akt phosphorylation is not altered. Pharmacologic inhibition or RNA interference of EGFR similarly induces cellular senescence of glioblastoma cells, which is rescued by the introduction of a constitutive active MAPK kinase (MKK). BBR potently inhibits the growth of tumor xenografts, which is accompanied by downregulation of EGFR and induction of senescence.

6.2.3.2 Diterpenoid Alkaloids

Diterpenoid alkaloids are abundant in *Aconitum* and *Delphinium* and are known to have anticancer activity (Hao et al., 2013a, 2015a,b). For instance, lappaconitine causes G0/G1 cell cycle arrest, apoptosis, and downregulation of cyclin E1 gene expression of NSCLC (Sheng et al., 2014). Taipeinine A, a C19-diterpenoid alkaloid from the roots of *Aconitum taipeicum*, upregulates the protein expression of Bax and caspase-3 and downregulates the expression of Bcl-2 and CCND1 (Zhang et al., 2014a). The cytotoxic activities of the *Delphinium* diterpenoid alkaloids were evaluated using the MTT method (Lin et al., 2014), and the IC_{50} values against A549 cancer cells ranged from 12.03 to 52.79 μM. Their anticancer mechanisms await further studies.

3-Isopropyl-tetrahydropyrrolo[1, 2-a]pyrimidine-2, 4(1H, 3H)-dione (ITPD), isolated from *A. taipeicum*, induces apoptosis and cell cycle arrest in S phase (Zhang et al., 2018). ITPD can mediate the mitochondrial pathway through activating caspase-3/9 and increasing the ratio of Bax/Bcl-2. Aconitine induces apoptosis in human pancreatic cancer via NF-κB signaling pathway (Ji et al., 2016).

Aconitum szechenyianum alkaloids (ASA) upregulate the expression of p38 and phosphorylated p38 MAPK (Fan et al., 2016), suggesting that ASA-induced apoptosis is associated with the p38 MAPK-mediated pathway. ASA upregulates TNF-R1 and DR5 via activation of p38 MAPK, thereby activating caspase 8, revealing that the death receptor pathway is involved in apoptosis. ASA leads to a loss of the mitochondrial out membrane potential, which upregulates p53, phosphorylated p53, and Bax, downregulates Bcl-2, causes release of cytochrome c from the mitochondria, and activates caspase-9 and -3 in A549 cell. This suggests that ASA could also induce apoptosis through the mitochondria pathway.

6.2.3.3 Others

Aurantiamide acetate, isolated from aerial parts of *Clematis terniflora*, suppresses the growth of malignant gliomas in vitro and in vivo by inhibiting autophagic flux (Yang et al., 2015).

6.2.4 Steroid

Hellebrigenin of *Helleborus*, one of bufadienolides belonging to cardioactive steroids, potently reduces the viability and colony formation of human HCC cells (Deng et al., 2014).

Hellebrigenin triggers DNA damage through DNA double-strand breaks (DSBs) and subsequently induces cell cycle G2/M arrest associated with upregulation of p-ATM, p-Chk2, p-CDK1, and cyclin B1 and downregulation of p-CDC25C. Hellebrigenin induces mitochondrial apoptosis, characterized by Bax translocation to mitochondria, disruption of mitochondrial membrane potential, release of cytochrome C into cytosol, and sequential activation of caspases and PARP. Akt expression and phosphorylation are inhibited by hellebrigenin, whereas Akt silencing with siRNA abolishes cell cycle arrest but enhances apoptosis induced by hellebrigenin. Activation of Akt by human insulin-like growth factor I could obviously attenuate hellebrigenin-induced cell death.

Furostanol derivative (25S)-22α,25-epoxyfurost-5-ene-3β,11β,26-triol 26-O-β-d-glucop yranoside (5); 20-hydroxyecdysone (6); and 3β,5β,14β-trihydroxy-19-oxo-bufa-20,22-dienolide 3-O-α-l-rhamnopyranoside, known as deglucohellebrin (7), isolated from *Helleborus caucasicus*, have a strong cytotoxic effect on Calu-1 lung cancer cells, HepG2, and Caco-2 cells (Martucciello et al., 2017). Compound 6 reduces the S-phase entry, and compound 7 induces apoptosis associated with activation of caspase-3. Compounds 6 and 7 significantly decrease protein expression of GRP78, a general ER-stress marker, suggesting proapoptotic functions.

6.2.5 Polysaccharide and Protein

An *Aconitum coreanum* polysaccharide (ACP1) fraction induces apoptosis of HCC cells via pituitary tumor transforming gene 1-mediated suppression of the P13K/Akt and activation of p38 MAPK signaling pathway and displays antitumor activity in vivo (Liang et al., 2015).

The lectin, a carbohydrate-binding protein found in the tubers of *Eranthis hyemalis* (Cimicifugeae), is an N-acetyl-D-galactosamine-specific type II ribosome inactivating protein; type II RIPs have shown anticancer properties (McConnell et al., 2015). *E. hyemalis* lectin is cytotoxic and has a cell-specific activity against the amphid neurons of *Caenorhabditis elegans*.

6.2.6 Flavonoid and Lignin

Apoptosis of breast cancer MCF-7 cells is triggered by *Trollius chinensis* flavonoids (Wang et al., 2016a). The bcl-2 and NF-κB expression is reduced, along with the increased expression of caspase-9 and -3, indicating that the inhibition of cellular proliferation occurs through activation of a mitochondrial pathway. In esophageal cancer EC-109 cells, the anticancer effects of orientin and vitexin may be associated with the regulation of the apoptosis-related gene expression of p53 and bcl-2 (An et al., 2015).

Boehmenan, a lignan of *Clematis armandii*, inhibits A431 epidermoid carcinoma cell growth via blocking p70S6/S6 kinase pathway (Pan et al., 2017). It induces apoptosis in lung cancer cells through modulation of EGF-dependent pathways (Pan et al., 2016).

6.2.7 Plant Extract

6.2.7.1 *Ranunculeae*

The plant extracts are traditionally obtained from the whole plant, root, rhizome, stem, leaf, flower, and other uncultured/cultured tissues/cells, which may contain multiple therapeutic compounds. For instance, *P. chinensis* (Bai Tou Weng in TCM), rich in pentacyclic triterpene saponins, is among the top 10 potent antimitotics (independent of toxicity) in a recent study that screens 897 aqueous extracts of commonly used natural products (0.00015–0.5 mg/mL) relative to paclitaxel for antimitotic effects on human breast cancer MDA-MB-231 cells (Mazzio et al., 2014). *P. koreana* extract (PKE) strongly suppresses the growth of HCC and anaplastic thyroid cancer cells in a dose-dependent manner (Hong et al., 2012b; Park et al., 2013). Apoptosis by PKE is observed by DAPI and TUNEL staining, and the cleaved PARP and caspase-3 are increased in Huh-7 cells (Hong et al., 2012b).

6.2.7.2 Cimicifugeae

In MDA-MB-453 human breast cancer cells, a methanolic extract of *C. racemosa* causes a significant increase in expression of ER stress (e.g., GRP78), apoptotic (GDF15), lipid biosynthetic (INSIG1 and HSD17B7) and phase I (CYP1A1) genes, and a decrease in expression of cell cycle (HELLS and PLK4) genes (Einbond et al., 2007). A lipophilic *C. racemosa* rhizome extract significantly regulates 431 genes in MCF-7 cells (Gaube et al., 2007), many of which are involved in antiproliferation and proapoptosis. The expression pattern differs from those induced by 17β-estradiol or the estrogen receptor antagonist tamoxifen. Our group found that the ethyl acetate fraction (EAF) of *C. foetida* aerial part induces G0/G1 cell cycle arrest at lower concentrations (25 μg/mL) in hepatoma cells and triggers G2/M arrest and apoptosis at higher concentrations (50 and 100 μg/mL) (Tian et al., 2007a). An increase in the ratios of Bax/Bcl-2, activation of downstream effector caspase 3, and cleavage of PARP is implicated in EAF-induced apoptosis. EAF inhibits the growth of the implanted mouse H22 tumor in a dose-dependent manner with the growth inhibitory rate of 63.32% at 200 mg/kg. Total triterpenoid glycosides of *C. dahurica* (a source plant of Sheng Ma) aerial part shows similar effects as EAF (Tian et al., 2007b). An isopropanolic extract of *C. racemosa* induces apoptosis of human prostate androgen-dependent and androgen-independent carcinoma cells, which is accompanied by the increased cytokeratin 18 (a caspase substrate) degradation (Hostanska et al., 2005).

Black cohosh extract inhibits 17β-estradiol-induced cell proliferation of endometrial adenocarcinoma cells (Park et al., 2016).

6.2.7.3 Nigella

N. sativa extract has selective cytotoxicity and apoptotic effects on U937 lymphoma cells but not ECV304 control cells (Arslan et al., 2017). In contrast, TQ has no significant cytotoxicity on both cells. *N. sativa* extract significantly increases caspase-3, BAD, and p53 gene expression in U937 cells. Both ethanol reflux extract and ultrasound-assisted ethanol extract of *Radix Semiaquilegiae* (Thalictroideae) have certain anticancer activities against human hepatoma HepG-2 and SMMC-7721 cells (Duan et al., 2013), and the antiproliferative activity increases with the increase of extract concentration.

6.2.7.4 Delphinieae

Tubeimu (*Rhizoma Bolbostemmatis*) and Fuzi (*Radix Aconiti Lateralis Preparata*) extracts have synergic effects on MDA-MB-231 and SKBR3 breast cancer cells (Chen et al., 2016).

6.2.7.5 Coptis

C. chinensis has a wide effect on cell signaling (Wang et al., 2011b), including cell cycle regulation (CDK6, CDK4, cyclin B1, cyclin E, cyclin D1, p27), cell adhesion (E-cadherin, osteopontin), differentiation, apoptosis (p-Stat3, p53, BRCA1), cytoskeleton (p-PKC α/β II, Vimentin, p-PKCα), MAPK signaling (raf-1, ERK1/2, p-p38, p-ERK), and the PI3K signaling pathway (p-Akt, Akt, p-PTEN). *C. chinensis* is a novel therapeutic drug for squamous cell carcinoma.

6.3 MicroRNAs, DNA DAMAGE, AND EPIGENETIC REGULATION

6.3.1 MicroRNA

MicroRNAs (miRs) are small 18–24-nucleotide noncoding RNAs that repress target gene expression largely by modulating translation and mRNA stability (Ayati et al., 2017). MicroRNA expression is deregulated in many types of cancer. Drug-induced microRNAs have emerged as key regulators in guiding their pharmacological effects. For instance, PEG4000-TQ-nanoparticles can expressively increase the expression of miR-34a through p53 (Bhattacharya et al., 2015). MiR-34a upregulation directly downregulates RAC1 expression followed by actin depolymerization and disrupts the actin cytoskeleton, which leads to significant reduction in the lamellipodia and filopodia formation on cell surfaces. PEG4000-TQ-nanoparticles circumvent TQ's poor aqueous solubility, thermal and light volatility, and consequently minimal systemic bioavailability, and thus might be more powerful in retarding cancer cell migration.

MiR-21-3p, a BBR-induced miRNA, directly downregulates human methionine adenosyl-transferases 2A and 2B and inhibits hepatoma cell growth (Lo et al., 2013). An overexpression of miR-21-3p increases intracellular S-adenosylmethionine content, which is disadvantageous for hepatoma cell growth. In NVP-AUY922 (second-generation Hsp90 inhibitor)-insensitive colorectal cancer cells, the combination of NVP-AUY922 and BBR causes cell growth arrest through inhibiting CDK4 expression and induction of miR-296-5p-mediated suppression of Pin1-β-catenin-cyclin D1 signaling pathway (Su et al., 2015). The miR-93 levels in cisplatin-resistant cell lines are higher than that in cisplatin-sensitive cell lines. BBR can inhibit miR-93 expression and increase its target tumor suppressor PTEN (Chen et al., 2015b). BBR modulates the sensitivity of cisplatin through miR-93/PTEN/Akt signaling pathway in ovarian cancer cells.

In multiple myeloma cells, BBR significantly downregulates miRNA clusters miR-99a ~ 125b, miR-17 ~ 92, and miR-106 ~ 25 (Feng et al., 2015). RAC1, NFκB1, MYC, JUN, and CCND1, the top five differentially regulated genes, might play key roles in the progression of multiple myeloma. Three common signaling pathways (TP53, Erb, and MAPK) link the three miRNA clusters and the five key mRNAs. Therapeutic effects of BBR on HCC, colorectal cancer, gastric cancer, ovarian cancer, and glioblastoma are also related with its regulation of target miRs (Ayati et al., 2017).

6.3.2 DNA Damage and Epigenetic Regulation

6.3.2.1 *Thymoquinone*

Human telomere DNA regulates gene transcriptions and folds up into G-quadruplex structures that inhibit telomerase overexpression in cancer cells. TQ interacts with G-quadruplex on two binding sites adjacent to the TTA loop (Salem et al., 2015). TQ is preferentially binding to G-quadruplex over duplex, which is explained by an intercalation binding mode based on π–π stacking. TQ might act as a G-quadruplex DNA stabilizer and subsequently inhibit telomerase and cancer proliferation.

TQ induces DNA damage, cell cycle arrest, and apoptosis in glioblastoma cells (Gurung et al., 2010). TQ facilitates telomere attrition by inhibiting the activity of telomerase. DNA-dependent protein kinase is a nuclear, serine/threonine protein kinase consisting of a 470-kDa catalytic subunit (DNA-PKcs) and a heterodimeric regulatory complex ku70/80. This enzyme is essential for the repair of DNA DSBs and mediates repair via phosphorylation of downstream DNA binding proteins such as p53. Telomeres in glioblastoma cells with DNA-PKcs are more sensitive to TQ-mediated effects as compared with cells deficient in DNA-PKcs.

Plant-derived antioxidants can switch to prooxidants even at low concentrations in the presence of transition metal ions such as copper. It is noted that tissue, cellular, and serum copper levels are considerably elevated in various malignancies (Zubair et al., 2013). TQ is able to cause oxidative cellular DNA breakage, which can be inhibited by copper-chelating agents and scavengers of ROS. TQ targets cellular copper in prostate cancer cells and causes a prooxidant cell death. Such a prooxidant cytotoxic mechanism could explain anticancer activity of plant-derived antioxidants.

In silico target identification suggests HDAC2 proteins as the potential targets of TQ (Attoub et al., 2013). In cancer cells, TQ treatment results in a significant HDAC2 inhibition and histone hyperacetylation (Chehl et al., 2009), providing preliminary evidence of TQ as the epigenetic regulator.

6.3.2.2 BBR and Other Alkaloids

DNA repair-deficient chicken B lymphocytes (DT40) deficient in Rev3 (Rev3−/−), a translesion DNA synthesis gene, are hypersensitive to BBR (Hu et al., 2014). Upon BBR treatment, G2/M arrest is increased in Rev3−/− cells. BBR also induced a significant increase in DSBs in Rev3−/− cells, as revealed by chromosomal aberration analysis. BBR is able to induce DNA damage, and the Rev3-associated DNA repair pathway participates in the repair processes.

When γH2AX focus formation is used to detect DNA damage in MG-63 human osteosarcoma cells, BBR is found to induce significant concentration- and time-dependent increases in DSBs (Zhu et al., 2014). BBR, followed by palmatine, appears to be the most potent DNA damage inducer in HepG2 cells (Chen et al., 2013). BBR and palmatine suppresses the activities of both Topo I and II. In BBR-treated cells, DNA damage is shown to be directly associated with the inhibitory effect of Topo II but not Topo I. DNA damage is also observed in cells treated with *Hydrastis* (goldenseal) extracts, and the extent of DNA damage is positively correlated to the BBR content. The Topo II inhibitory effect may contribute to BBR- and goldenseal-induced genotoxicity.

BBR increases the radio-sensitivity of breast cancer MCF-7 and MDA-MB-468 cells (Wang et al., 2012). The radiation-induced G2/M cell cycle delay is reduced in MCF-7 cells pretreated with BBR, as BBR causes G1/S arrest. BBR pretreatment prolongs the persistence of DSBs in the MCF-7 cell line. The protein levels of RAD51 are decreased by BBR, and in the cells pretreated with 15 μM BBR for 24 h, RAD51 protein decreases significantly at 0, 2, 6, and 24 h after X-ray exposure. BBR sensitizes human breast cancer cells to ionizing radiation by inducing cell cycle arrest and the downregulation of the homologous recombination repair protein, RAD51. BBR may be a promising radio-sensitizer for the treatment of breast cancer.

Palmatine and BBR can induce the formation of G-quadruplex as well as increase its stability (Ji et al., 2012). The unsaturated ring C, N^+ positively charges centers, and conjugated aromatic rings are key factors to increase the stabilization ability of palmatine and BBR. Interactions between BBR and pBR322 suggest BBR's biologically significant DNA-binding abilities (Saha and Khuda-Bukhsh, 2014). The possible binding sites of BBR on histone proteins are determined by molecular docking. In cervical cancer HeLa cells, BBR might modulate p53 and viral oncoproteins HPV-18 E6-E7 via epigenetic modifications. BBR can repress the expression of DNA methyltransferase (DNMT)1 and DNMT3B, which triggers hypomethylation of TP53 by changing the DNA methylation level in human multiple melanoma cell U266 (Qing et al., 2014).

BBR induces more apoptosis and cell cycle arrest but less ROS production in constitutive and rostane receptor (CAR) overexpressed mCAR-HepG2 cells (Zhang et al., 2016a). BBR inhibits expressions of CAR and its target genes CYP2B6 and 3A4. BBR enhances DNA methylation in whole genomes but reduces it in promoter region CpG sites of CYP2B6 and 3A4 genes under the presence of CAR condition. The antiproliferation of BBR might be mediated by the unique epigenetic modifying mechanism of CAR metabolic pathway, suggesting that BBR is a promising candidate in anticancer adjuvant chemotherapy.

6.3.2.3 *Phenolic Acid*

Among *C. racemosa* phenylpropanoids tested for protecting against menadione-induced DNA damage, methyl caffeate is the most potent, followed by caffeic acid, ferulic acid, cimiracemate A, cimiracemate B, and fukinolic acid (Burdette et al., 2002), all of which have antioxidant activity. The methanol extracts of *C. racemosa* show dose-dependent decreases in DNA single-strand breaks and oxidized bases induced by the quinone menadione.

6.4 OXIDATIVE PROCESS AND METABOLISM

6.4.1 Antioxidant Versus Prooxidant

Plant-derived dietary antioxidants have attracted considerable interest in recent decades for their chemopreventive and cancer therapeutic abilities. *P. chinensis* polysaccharides (PCPS) enhance superoxide dismutase (SOD) and catalase (CAT) enzyme activities and lower MDA levels in the plasma of glioma-bearing mice (Zhou et al., 2012). PCPS could relieve the liver and kidney damage of glioma-bearing mice (Zhou et al., 2012) and decrease plasma levels of aspartate aminotransferase, alanine aminotransferase, and urea.

α-Hederin, a pentacyclic triterpene saponin abundant in *N. sativa* and *Clematis*, increases the production of ROS in cancer cells (Shafiq et al., 2014) and then disturbs mitochondrial functions and causes apoptosis (Cheng et al., 2014b). Inducers of phase II enzyme play an important role in cancer chemoprevention (Hao et al., 2010). Quinone reductase (QR), a typical phase II enzyme, can convert toxic quinones to hydroquinones and reduce oxidative cycling. The oleanane saponins of *P. chinensis* exhibit more potent QR-inducing activities than the lupine saponins (Wang et al., 2011a), and the CD value (concentration required to double the QR-inducing activity of the control) of the compound with the most potent QR-inducing activity is 1.1 μM.

TQ inhibits carcinogen-metabolizing enzyme activity and oxidative damage of cellular macromolecules and attenuates inflammation (Kundu et al., 2014b). The antiproliferative and proapoptotic effects of TQ in breast cancer are mediated through p38 phosphorylation via ROS generation (Woo et al., 2013). *N. sativa* increases the activities of antioxidant enzymes, for example, SOD, CAT, and glutathione peroxidase (GPx), and protects cells against cancer (Ebru et al., 2008; Ismail et al., 2010).

BBR induces ROS production for up to 6 h of incubation with human gastric cancer SNU-5 cells (Lin et al., 2008). In PANC-1 and MIA-PaCa2 pancreatic cancer cells, apoptosis is induced by BBR via the production of ROS rather than caspase 3/7 activation (Park et al., 2015). In HepG2 cells, BBR treatment led to a pronounced increase in JNK phosphorylation and enhanced ROS generation, lipid peroxidation, decreased activities of SOD and CAT, and diminished glutathione levels (Shukla et al., 2014). BBR and d-limonene (a monoterpene) at a ratio of 1:4 exhibits a synergistic anticancer effect on gastric cancer MGC803 cells (Zhang et al., 2014b). They distinctly induce intracellular ROS generation, reduce the mitochondrial transmembrane potential ($\Delta\Psi_m$), enhance the expression of caspase-3, and decrease the expression of Bcl-2. As TQ has strong anticancer activity, whether the combination of BBR and d-limonene exerts synergistic anticancer effects merits study.

6.4.2 Metabolism

Rabbit carcinogen-metabolizing enzymes CYP1A2 and 3A4, but not CYP2E1, are significantly diminished by dietary doses of TQ (Elbarbry et al., 2012). TQ displays a significant inhibition of induced phase I CYP1A1 enzyme and increases the content of glutathione and activity of phase II enzyme glutathione-S-transferase in HepG2 cells (ElKhoely et al., 2015), which provide support for the beneficial use of TQ as a therapeutic and chemopreventive agent against liver cancer. The phase II enzyme GPx is also significantly induced by the high TQ dose in rabbits (Elbarbry et al., 2012), while the total glutathione levels are unaffected. Glutathione reductase is significantly induced, which may explain the benign effect of *N. sativa* seeds in inhibiting the generation of bioactive metabolites known to promote carcinogenesis and oxidative cell damage.

In human P-glycoprotein (MDR1, ABCB1) gene-transfected the name of a cell line (KB)/ MDR1 cells, TQ has no effect on the accumulation of daunorubicin or rhodamine 123, two fluorescent substrates of MDR1 (Nabekura et al., 2015). Acerinol, a cyclolanstane triterpenoid from *Cimicifuga acerina*, can increase the chemosensitivity of MDR1-overexpressing HepG2/ADM and MCF-7/ADR cells to chemotherapeutic drugs doxorubicin, vincristine, and paclitaxel (Liu et al., 2014b). It can also increase the retention of MDR1 substrates doxorubicin and rhodamine 123 in the above cells. Acerinol significantly stimulates the activity of MDR1 ATPase without affecting the expression of MDR1 on mRNA or protein level. Acerinol reverses the resistance of MCF-7/ADR cells to vincristine. Acerinol may be a competitive inhibitor of MDR1, and docking analysis indicates that acerinol would bind to the sites on MDR1 that partly overlapped with that of verapamil.

The ethanolic *C. racemosa* extract BNO-1055, rich in saponin, dose-dependently attenuated cellular uptake and incorporation of thymidine and BrdU and significantly inhibited cell growth after long-time exposure (Dueregger et al., 2013). These inhibitory effects of BNO-1055 can be mimicked using pharmacological inhibitors and isoform-specific siRNAs targeting the equilibrative nucleoside transporters ENT1 and 2. BNO-1055 also attenuates the uptake of clinically relevant nucleoside analogs, for example, the anticancer drugs gemcitabine and fludarabine. By inhibiting the salvage nucleoside uptake pathway, BNO-1055 potentiates the cytotoxicity of the de novo nucleotide synthesis inhibitor 5-FU without significantly altering its uptake. In MCF-7 cells, both *C. racemosa* extract and purified cycloartane saponins upregulate the expression of CYP1A1 and 1B1, two carcinogen-metabolizing enzymes (Gaube et al., 2007).

Aconite root can improve the energy metabolism in rats by influencing the metabolic process of sugar, lipid, and amino acids (Yu et al., 2011), which may be the main molecular mechanism of warming yang and dispelling cold for the treatment of the cold syndrome, common in cancer patients, according to TCM theory. *Aconitum napellus* administered around the clock induces hyperthermia overall and in a time-dependent manner (de la Peña et al., 2011), with greatest the effects during the resting span. Thus, time of day may significantly affect the outcome of *A. napellus* and other homeopathic treatments and should be considered in determining optimal dosing and treatment time in order to increase the desired outcome and decrease undesired effects. Fu Zi (lateral roots of *Aconitum carmichaelii*) can increase and maintain a dogs' body heat for at least 6 h (Chen et al., 2009). The body weight gain in aconitine-administered mice is less than that of the control group until day 22 (Wada et al., 2006). Transient rectal hypothermia occurs within 30 min after the last administration of aconitine. Then the rectal temperature gradually increases to normal. Drug metabolism of aconitine increases and the toxicity of aconitine decreases due to long-term administration of aconitine.

Fingerprints of Fu Zi, Yan Fu Zi (salted prepared aconite root), Heishunpian (processed Fu Zi), and Paofupian are obtained by ultra-high-performance liquid chromatography, and effects of Fu Zi and its processed products on rat liver mitochondrial metabolism are studied by microcalorimetry (Zheng et al., 2014). Due to inherent differences in chemical compositions, the energy metabolism of mitochondria was differentially influenced by Fu Zi and its processed products. The bioactivity sequence of the tested products was Fu Zi > Heishunpian > Paofupian > Yan Fu Zi. The compounds mesaconitine, benzoylaconitine, and benzoylhypaconitine might be the principal active components that modulate energy metabolism.

Reprogramming energy metabolism is an emerging cancer hallmark (Hanahan and Weinberg, 2011). Cancer cells rely on aerobic glycolysis, instead of mitochondrial oxidative phosphorylation, to yield ATP. Enhanced glycolysis allows the diversion of glycolytic intermediates into various biosynthetic pathways, including those generating nucleosides and amino acids, which facilitate the biosynthesis of the macromolecules and organelles required for assembling new cells. Can *Aconitum* reverse the deregulated cellular energetics of cancer cells? The TCM formula Sini decoction (SND), consisting of Fu Zi, *Glycyrrhizae Preparata Radix* and *Zingiberis Rhizoma*, significantly inhibits LDH and glycolysis in the myocardial ischemia reperfusion injury model of rat H9c2 cardiomyocyte (Li et al., 2014b). SND also modulates lipid metabolism, tricarboxylic acid cycle, and nitrogen metabolism. It is intriguing to study whether *Aconitum* and diterpenoid alkaloids therefore inhibit the aerobic glycolysis of cancer cells. To date there is also no direct evidence that *Aconitum* enhances mitochondrial oxidative phosphorylation in human cancer cells.

6.5 ANTIANGIOGENIC AND ANTIMETASTATIC EFFECTS

6.5.1 Saponin

6.5.1.1 *Pulsatilla and Anemone*

SB365 exhibits potent antiangiogenic activity and decreases the expression of hypoxia-inducible factor-1α (HIF-1α) and vascular endothelial growth factor (VEGF), a key molecule for angiogenesis (Hong et al., 2012a). SB365 suppresses tube formation and migration of human umbilical vein endothelial cells (HUVECs), as well as in vivo neovascularization in a mouse Matrigel plug assay. SB365 treatment decreases the expressions of VEGF and CD34 in the tumor tissue of an HCC xenograft model. Angiogenesis of human colon cancer and pancreatic cancer cells is also suppressed by SB365 (Son et al., 2013a,b). SB365 inhibits tube formation in hepatocyte growth factor-induced HUVECs and suppresses microvessel sprouting from the rat aortic ring, ex vivo, and blood vessel formation in the Matrigel plug assay in mice (Hong et al., 2013).

Raddeanin A significantly inhibits HUVEC proliferation, motility, migration, and tube formation (Guan et al., 2015). Raddeanin A dramatically reduces angiogenesis in chick embryo chorioallantoic membrane (CAM), restrains the trunk angiogenesis in zebrafish, and suppresses angiogenesis and growth of human HCT-15 colorectal cancer xenografts in mice. Raddeanin A suppresses VEGF-induced phosphorylation of VEGFR2 and its downstream protein kinases, including PLCγ1, JAK2, FAK, Src, and Akt. In molecular docking simulation, raddeanin A formed hydrogen bonds and hydrophobic interactions within the ATP-binding pocket of VEGFR2 kinase domain.

Raddeanin A significantly inhibits the invasion, migration, and adhesion of BGC-823 human gastric cancer cells (Xue et al., 2013). Raddeanin A can upregulate the expression of reversion-inducing, cysteine-rich protein with Kazal motifs (RECK) and E-cadherin and downregulate the expression of matrix metalloproteinase-2 (MMP-2), MMP-9, MMP-14, and RhoC.

In breast cancer cells, raddeanoside R13, extracted from *P. chinensis*, had strong antiproliferative and antimetastasis ability (Liang et al., 2016), accompanied by cell cycle arrest, apoptosis, autophagy, and reversion of epithelial-mesenchymal transition (EMT).

6.5.1.2 *Clematis*

A noncytotoxic concentration of DRβ-H, isolated from *C. ganpiniana*, markedly suppresses wound healing migration, migration through the chamber, and invasion through the matrigel (Cheng et al., 2017). DRβ-H regulates expression of RNPC1 and E-cadherin proteins of MDA-MB-231 breast cancer cells. The up-regulation of RNPC1 by DRβ-H is essential for its antimetastatic activities.

6.5.1.3 *Cimicifuga*

Actein (10, 30 mg/kg) and 26-deoxyactein (10, 30 mg/kg) significantly reduces the microvessel density in the sarcoma S180 xenograft tumor (Wu et al., 2016). Actein significantly inhibits the proliferation, reduces the migration and motility of endothelial cells (Yue et al., 2016), and suppresses the protein expressions of VEGFR1, pJNK, and pERK. The oral administration of actein at 10 mg/kg for 7 days inhibits blood vessel formation in the growth factor-containing matrigel plugs. Oral actein treatments (10–15 mg/kg) for 28 days results in decreasing mouse 4T1 breast tumor sizes and metastasis to lungs and livers. The apparent reduced angiogenic protein (CD34 and Factor VIII) expressions and downregulated metastasis-related VEGFR1 and CXCR4 gene expressions are observed in breast tumors.

6.5.2 Terpenoid

6.5.2.1 *Monoterpene*

TQ inhibits cancer angiogenesis, cell invasion, and metastasis (Jafri et al., 2010; Kundu et al., 2014b). TQ treatment decreases the transcriptional activity of the TWIST1 promoter and the mRNA expression of TWIST1, an EMT-promoting transcription factor (Khan et al., 2015).

TQ also decreases the expression of TWIST1-upregulated genes such as N-Cadherin and increases the expression of TWIST1-repressed genes such as E-cadherin, thus reducing cell migration and invasion. TQ inhibits the growth and metastasis of cancer cell-derived xenograft tumors in mice and partially attenuates the migration and invasion of TWIST1-overexpressed cell lines. Furthermore, TQ enhances the promoter DNA methylation of the TWIST1 gene in BT 549 breast cancer cells.

TQ inhibits the migration of mouse neuroblastoma (Neuro-2a) cells by downregulating MMP-2 and -9 (Arumugam et al., 2016).

6.5.2.2 Triterpene

Cimyunnin A, a triterpene isolated from the fruit of *Cimicifuga yunnanensis* and characterized by an unusual fused cyclopentenone ring G, exhibits comparable antiangiogenic activities to those of sunitinib, a clinically used first-line angiogenesis inhibitor, in the in vitro and ex vivo studies (Nian et al., 2015). The phosphorylation of VEGFR2 is suppressed by cimyunnin A in a dose-dependent manner. Dramatic downregulations of phospho-Akt (Ser473) and phospho-ERK (Thr202/Tyr204), the known downstream targets of VEGFR2, are observed at 5.0 and 10.0 µM of cimyunnin A, while total VEGFR2, ERK, and Akt remain unchanged.

6.5.3 Alkaloid

6.5.3.1 Isoquinoline Alkaloid

Isoquinoline alkaloids, such as BBR, palmatine, jatrorrhizine, and columbamine, are abundant in *Coptis*, *Thalictrum*, and *Hydrastis* and exert anticancer activity via multiple mechanisms (Hambright et al., 2015). BBR inhibits colorectal cancer invasion and metastasis via downregulation of COX-2/PGE2-JAK2/STAT3 signaling pathway (Liu et al., 2015c). The MMP-1, -2, and -9 expressions are also decreased by BBR, while MMP-7 is not affected (Lin et al., 2008, 2015c). BBR inhibits the metastatic ability of prostate cancer cells by suppressing EMT-associated genes, for example, BMP7, NODAL, and Snail (Liu et al., 2015d), which, if highly expressed, are associated with shorter survival of prostate cancer patients. BBR inhibits the expression of Id-1 (inhibitor of differentiation/DNA binding), a key regulator of HCC development and metastasis (Tsang et al., 2015). BBR can inhibit Id-1 promotor activity. Id-1 overexpression in HCC models partially rescued antiproliferative and antiinvasive activities of BBR. Palmatine, alone or in an extract, has cancer invasion inhibitory properties (Hambright et al., 2015). Synergistic inhibition of rpS6/NF-κB/FLIP axis with palmatine may have therapeutic potential for cancer with its constitutive activation.

Anoikis, or detachment-induced apoptosis, may prevent cancer progression and metastasis by blocking signals necessary for survival of localized cancer cells. Resistance to anoikis is regarded as a precondition for metastasis. BBR promotes the growth inhibition of anoikis-resistant cells to a greater extent than doxorubicin treatment (Kim et al., 2010). BBR induces cell cycle arrest at G0/G1 in the anoikis-resistant MCF-7 and MDA-MB-231 breast cancer cells.

Jatrorrhizine induces C8161 metastatic melanoma cell cycle arrest at the G0/G1 transition (Liu et al., 2013b), which is accompanied by overexpression of the cell cycle-suppressive genes p21 and p27 at higher doses. Jatrorrhizine does not induce significant cellular apoptosis at doses up to 320 µmol/l. Moreover, jatrorrhizine reduces C8161 cell-mediated neovascularization in vitro and in vivo and downregulates the expression of cadherin, a key protein in tumor vasculogenic mimicry and angiogenesis.

Columbamine induces metastatic osteosarcoma U2OS cell cycle arrest at the G2/M transition (Bao et al., 2012), which is associated with attenuating CDK6 gene expression and diminishing STAT3 phosphorylation. Columbamine does not significantly promote cell apoptosis at any dosages tested. Columbamine inhibits U2OS cell-mediated neovascularization, which is accompanied by the downregulation of MMP-2 expression and reduction of cell migration, adhesion, and invasion.

6.5.3.2 Diterpenoid Alkaloid

Hypaconitine, a diester-diterpenoid alkaloid isolated from the *Aconitum* root, inhibits TGF-β1-induced epithelial-mesenchymal transition and suppresses adhesion, migration, and invasion of lung cancer A549 cells (Feng et al., 2017).

6.5.4 Polysaccharide

ACP1 and its sulphated derivative ACP1-s significantly impairs MDA-MB-435s breast cancer cell migration (Zhang et al., 2017), and the accumulated distance and average velocity of ACP1- and ACP1-s-treated cells are reduced markedly. ACP1 and ACP1-s treatment can affect dynamic remodeling of actin cytoskeleton and suppresses phosphorylation and activation of signaling molecules.

6.5.5 Plant Extract

PKE decreases the expression of HIF-1α and VEGF in HCC cells (Hong et al., 2012b) and inhibits tube formation and migration of HUVECs. PKE potently suppresses in vivo neovascularization in a mouse Matrigel plug assay. In a mouse xenograft model, the expressions of Ki-67, VEGF, and CD31 in the tumor tissue are decreased by PKE. Angiogenesis of anaplastic thyroid cancer is also suppressed by PKE (Park et al., 2013).

The water extract of *P. koreana*, *Panax ginseng* (Ren Shen in TCM), and *Glycyrrhiza uralensis* (Gan Cao in TCM)(WEPPG) significantly inhibits fibroblast growth factor (bFGF)-induced HUVEC proliferation, adhesion, migration, and capillary tube formation (Kim et al., 2011). Expressions of various proteins, including cyclin A, p63, and KIP2, are upregulated, while nibrin and focal adhesion kinase are downregulated. The blood vessel formation in a CAM treated with WEPPG is markedly reduced. WEPPG might exert its anticancer effects via the inhibition of angiogenesis.

Aconite (*Aconiti Kusnezoffii Radix* in TCM) can induce cell differentiation, inhibit cell proliferation and migration, increase succinic dehydrogenase (SDH) activity, and gap-junction intercellular communication (GJIC) in Lewis lung cancer cells (Zhao et al., 2014c). *Coptidis Rhizoma* increases cell adhesion and decreases SDH activity and GJIC without cell differentiation, while it also suppresses cell proliferation. *Aconiti Kusnezoffii Radix* water decoction can maintain body temperature, blood oxygen saturation, red cell ATPase and blood rheology, and decrease intratumor hypoxia and capillary permeability in tumor-bearing mice, which retards cancer growth and metastasis. *Aconitum vaginatum* can inhibit the proliferation, invasion, and metastasis of A549 cells, and MMP-2 and -9 activities decrease (Xu et al., 2010). *Coptidis Rhizoma* water decoction decreases body temperature, blood oxygen saturation, red cell ATPase, blood rheology, and GJIC, and promotes intratumor hypoxia and capillary permeability, which leads to more cancer metastasis, although it also inhibits cancer growth. The hot Chinese medicine (e.g., aconite) can induce cancer cell differentiation and prevent tumor poison invagination, which might be better for lung cancer treatment than cold Chinese medicine (e.g., *Coptidis Rhizoma*).

Coptidis rhizome aqueous extract (CRAE), containing magnoflorine 2.2%, jatrorrhizine 1.68%, palmatine 4.4%, and BBR 13.8%, exhibits significant inhibition on VEGF secretion from MHCC97L and HepG2 cells at nontoxic doses (Tan et al., 2014). CRAE intervention increases the phosphorylation of eukaryotic elongation factor 2 (eEF2) in HCC cells, which blocks eEF2 activity for proceeding nascent protein synthesis. Reduction of tumor size and neovascularization are observed in mice xenograft models.

6.6 IMMUNOMODULATORY ACTIVITY

The immune destruction of cancer cells is an essential anticancer mechanism (Hanahan and Weinberg, 2011). *Ranunculus ternatus* polysaccharides enhance the proliferation of mouse thymocytes, spleen lymphocytes, and peritoneal macrophages (Lv et al., 2010). *R. ternatus* polysaccharides enhance NK cell activity and induce apoptosis in breast cancer MCF-7 cells (Niu et al., 2013; Sun et al., 2013). However, the antiproliferative effects of *R. ternatus* polysaccharides might be less pronounced than saponins in human gastric cancer BGC823 cells.

PCPS treatment to glioma-bearing mice can elevate the thymus and spleen indices (Zhou et al., 2012). A neutral polysaccharide fraction (ARP) from the rhizome of *A. raddeana* extraordinarily promotes splenocyte proliferation, NK cell and CTL activity, as well as serum IL-2 and TNF-α production in HCC-bearing mice (Liu et al., 2012). ARP has no toxicity to body weight, liver, and kidney. Moreover, it could reverse the hematological parameters induced by 5-fluorouracil to near normal.

A polysaccharide (ACP-a1) of 3.2×10^5 Da, isolated from the roots of *A. coreanum*, is mainly composed of β-D-mannose and β-D-glucose in a molar ratio of 1.2:3.5 (Li et al., 2013b). ACP-a1 significantly inhibits the growth of hepatoma H22 transplanted in mice and prolongs the survival time of H22 tumor-bearing mice. The body weight, peripheral white blood cells, thymus index, and spleen index of H22 tumor-bearing mice improve after ACP-a1 treatment. ACP-a1 could promote the secretion of serum cytokines, such as IL-2, TNF-α, and IFN-γ.

N. sativa is highlighted as a natural immune booster (Sultan et al., 2014). Modes of its actions include boosting and functioning of the immune system and activation and suppression of immune-specialized cells, interfering in several pathways that eventually lead to improved immune responses and defense systems. For instance, TQ induces the phosphorylation of proteins involved in NK cell signaling, CD28 signaling of T_H cells, and B cell receptor signaling (El-Baba et al., 2014).

6.7 ANTIINFLAMMATORY ACTIVITY

Inflammation has been identified as a significant factor in the development of solid tumors (Hanahan and Weinberg, 2011). For instance, both hereditary and sporadic chronic pancreatitis is associated with an increased risk of developing pancreatic ductal adenocarcinoma (PDA).

TQ dose- and time-dependently reduces PDA cell synthesis of monocyte chemoattractant protein (MCP)-1, TNF-α, IL-1β, and COX-2 (Chehl et al., 2009). At 24 h, TQ almost completely abolishes the expression of these cytokines, while trichostatin A (TSA, a specific HDAC inhibitor) has a less dramatic effect. TQ, but not TSA, significantly and dose-dependently reduces the intrinsic activity of the MCP-1 promoter. TQ also inhibits the constitutive and TNF-α-mediated activation of NF-κB in PDA cells and reduces the transport of NF-κB from the cytosol to the nucleus. The antiinflammatory activity of TQ might contribute to its overall anticancer effects, as does BBR (Li et al., 2015).

Ranunculaceae tribes and genera, such as *Ranunculus*, Anemoneae, *Cimicifuga*, *Helleborus*, *Nigella*, Delphinieae, *Semiaquilegia*, *Coptis*, and *Hydrastis*, are rich in both antiinflammatory and anticancer phytometabolites (Hao et al., 2015a,b). Anemonin and ranunculin, the potent antiinflammatory and anticancer compounds, are abundant in tribes Ranunculeae and Anemoneae (Hao et al., 2015b; Lee et al., 2008). Chronic inflammation is associated with tumor formation, and tumors could be portrayed as wounds that never heal (Schafer and Werner, 2008). *Ranunculus pedatus* and *R. constantinapolitanus* are found to have wound healing and antiinflammatory properties (Akkol et al., 2012). The fatty-acid fractions of *R. constantinopolitanus* exerts the antiinflammatory effects via downregulation of IL-6 and COX-2 (Fostok et al., 2009). The triterpene saponins, phenolic acid, lignans, and flavonoids of *Clematis* display various antiinflammatory activities (Hao et al., 2013c, 2015a,b). Clematis *mandshurica* inhibits IL-1β-induced MMP gene expression (Choi et al., 2014), implicating its use in combating cancer invasion. The inflammatory cells can release chemicals, notably ROS, which are actively mutagenic for nearby cancer cells, accelerating their genetic evolution toward states of intensified malignancy (Grivennikov et al., 2010). The Ranunculaceae antiinflammatory compound remarkably inhibits ROS generation and downregulates the ROS-dependent NF-κB signaling pathway (Dilshara et al., 2015).

Cycloartane-type triterpene saponins (*Cimicifuga* and *Actaea*), diterpenoid alkaloids (*Aconitum* and *Delphinium*), flavonoid glycosides (*Ranunculus* and *Trollius*) (Zhou et al., 2014), and isoquinoline alkaloids (subfamily Thalictroideae, *Coptis*, and *Hydrastis*) also display various antiinflammatory activities (Hao et al., 2015b). The cytokines (e.g., IL-6 and IFN-γ) and chemokines (e.g., MIP-1β) decrease in a serum of MCS-18 (from *Helleborus*)-treated mice (Dietel et al., 2014). *Dichocarpum* (firstly identified as an independent genus by Xiao and Wang) (Xiao and Wang, 1964), *Adonis*, *Beesia*, *Caltha*, and *Halerpestes* are less studied in modern pharmacology, although in folk medicine and TCM they are frequently used in the inflammatory diseases (Hao et al., 2015b). Taken together, Ranunculaceae plants contribute notable chemodiversity, from which promising anticancer phytometabolites and novel anticancer mechanisms can be found.

6.8 STRUCTURE-ACTIVITY RELATIONSHIP

Saponins 1–14 of *P. chinensis* show considerable cytotoxic activity (Xu et al., 2013), whereas saponins 15–36 have no significant activity, suggesting that a free carboxylic group at C-28 of aglycon is essential for their cytotoxic activity. The oleanane-type saponins show better cytotoxic activity than lupane-type saponins, and the length and linkage of glycolic chains attached to C-3 of aglycon displays an important effect to the cytotoxic potency. Oleanolic acid 3-O-α-l-rhamnopyranosyl-(1 → 2)-[β-d-glucopyranosyl-(1 → 4)]-α-l-arabinopyranoside (saponin5) exhibits the most significant cytotoxic activity.

Saponins 5–17 of *P. koreana*, with a free acidic functional group at C-28 of aglycon, exhibit moderate to considerable cytotoxic activity (Bang et al., 2005), while saponins 1–4, esterified with a trisaccharide at C-28 of aglycon, does not exhibit cytotoxic activity ($ED_{50} > 300$ μM). Oleanolic acid 3-O-α-L-rhamnopyranosyl-(1 → 2)-[β-D-glucopyranosyl- (1 → 4)]-α-L-arabinopyranoside (saponin 10) exhibits the most potent cytotoxic activity. During in vivo testing, hederagenin 3-O-α-L-rhamnopyranosyl-(1 → 2)-[β-D-glucopyranosyl- (1 → 4)]-α-L-arabinopyranoside (saponin 6) exhibits more potent anticancer activity than paclitaxel and doxorubicin. Hedragenin 3-O-β-D-glucopyranosyl-(1 → 4)-O-β-D-glucopyranosyl-(1 → 3)-O-α-L-rhamnopyranosyl-(1 → 2)-α-L-arabinopyranoside (saponin 17) exhibit potent anticancer activity. These two saponins are similarly comprised of a hederageninaglycon and a sugar sequence O-α-L-rhamnopyranosyl-(1 → 2)-α-L-arabinopyranoside at C-3 of the hederagenin, suggesting that the two elements are essential for the anticancer activity.

C ring, C-28, or C-3 modifications of PSA lead to a better balance between hemolytic toxicity ($HD_{50} > 500$ μM) and cytotoxicity toward lung cancer cells A549 (IC_{50} 4.68 μM) (Tong et al., 2017). C ring, C-28, or C-3 modifications of *Pulsatilla* saponin D lead to the discovery of compound 14 (Chen et al., 2017), which has significant cytotoxicity toward A549 cells (IC_{50} 2.8 μM) in a dose-dependent manner without hemolytic toxicity. Compound 14 induces typical G1 cell cycle arrest and apoptosis in A549 cells, and both intrinsic and extrinsic apoptosis pathways are activated.

The natural and derivatized C_{19}-diterpenoid alkaloids of *Aconitum* tested by Hazawa et al. (2009) and Wada et al. (2015) show only a slight effect or no effect against cancer cells. Most anticancer compounds are hetisine-type C_{20}-diterpenoid alkaloids (Hazawa et al., 2011), specifically kobusine and pseudokobusine analogs with two different substitution patterns, C-11 and C-11,15. Particularly, several C_{20}-diterpenoid alkaloids are more potent against multidrug-resistant KB subline KB-VIN cells. Pseudokobusine 11-3'-trifluoromethylbenzoate is a possible promising new lead meriting additional evaluation.

6.9 GENOMICS, TRANSCRIPTOMICS, PROTEOMICS, AND METABOLOMICS

It is becoming increasingly apparent that a precise and truly useful understanding of the behavior of individual phytometabolite and Ranunculaceae extracts would only occur if we are able to integrate genomic, proteomic, and other omics data sets and then distill these data. The use of omics platforms in Ranunculaceae anticancer research is still in its infancy. For instance, a PCR array was used to quantify TQ-mediated transcriptional regulation of 84 apoptosis-related genes (Sakalar et al., 2013). At a low dose (12.5 μM), TQ induces expression of four proapoptotic genes: BIK (~22.7-fold), FASL (~2.9), BCL2L10 (~2.1), and CASP1 (~2). TQ reduces the expression of an antiapoptotic gene implicated in NF-κB signaling and cancer: RELA (~8). At high doses (100 μM), TQ mediates the expression of 21 genes implicated directly in apoptosis (6 genes), TNF signaling (10), and NF-κB signaling (3), such as BIK, BID, TNFRSF10A, TNFRSF10B, TNF, TRAF3, RELA, and RELB. TQ intervenes with TNF and NF-κB signaling during TQ-mediated induction of apoptosis in cancer cells.

DNA microarray data for 12,600 genes were used to examine the antiproliferative activity of Coptidis rhizoma and eight constituent molecules against eight human pancreatic cancer cell lines (Hara et al., 2005). Twenty-seven genes showed strong correlation with the 50% inhibitory dose (ID_{50}) of *Coptis* after 72 h exposure. The test molecules were classified into two clusters, one consisting of *Coptis* and BBR and the other consisting of the remaining seven molecules. BBR can account for the majority of the antiproliferative activity of *Coptis*, and DNA microarray analyses can be used to improve our understanding of the actions of an intact herb.

Identifying human/mammalian metabolites is an integral part of the molecular mechanism investigations of anticancer phytometabolites. Eighteen metabolites of *Pulsatilla* saponin D were identified in rat plasma, urine, and feces samples by ESI-Q-TOF-MS/MS (Ouyang et al., 2014), and eight of them (M11–M18) were novel. Deglycosylation, dehydrogenation, hydroxylation, and sulfation were the major metabolic transformations of *Pulsatilla* saponin D in vivo. The metabolic information gained from metabolomic studies is relevant to the pharmacological activity of *Pulsatilla* saponin D.

Bioinformatics and cheminformatics are the indispensable part of omics platforms that contribute prolific information of molecular mechanisms of anticancer compounds. Three different in silico reverse-screening approaches proved useful for identifying the putative molecular targets of anticancer compounds in *N. sativa* (TQ, α-hederin, dithymoquinone, and thymohydroquinone) (Sridhar et al., 2014). Novel kinase targets influenced by TQ were revealed by in silico analysis of peptide array data obtained from TQ-treated HCT116 colon cancer cells (El-Baba et al., 2014). TQ induces the phosphorylation of a multitude of proteins that are involved in one or more cancer-related biological functions: apoptosis, proliferation, and inflammation. Of the 104 proteins identified, 50 were upregulated ≥2-fold by TQ and included molecules in the Akt-MEK-ERK1/2 pathway. Oncogenic PAK1 emerged as an interesting TQ target. TQ induced an increase of pERK1/2 and triggered the early formation of an ERK1/2-PAK1 complex. Modeling confirmed that TQ binds in the vicinity of Thr212 accompanied by conformational changes in ERK2-PAK1 binding. Structural modelings suggest that TQ interferes also with the kinase domain and disturbs its interaction with pPAK (Thr423), finally inhibiting MEK-ERK1/2 signaling and disrupting its prosurvival function.

6.10 CONCLUSION AND FUTURE PERSPECTIVE

This chapter focuses on the various Ranunculaceae chemical compounds that have shown promise as anticancer agents and outlines their potential mechanism of action. Deeper insights into molecular mechanisms and functions of Ranunculaceae anticancer compounds are the premise of designing analog and nanoformulation with improved absorption, distribution, metabolism, and excretion/toxicity properties and therapeutic efficacy. Studies of the representative Ranunculaceae phytometabolites, for example, TQ and BBR, are enlightening, as many anticancer mechanisms and pathways are shared by many other compounds of the same plant family. To date, little information has been reported on the anticancer effects and the underlying mechanisms of the diterpenoid alkaloids of *Aconitum* and *Delphinium* plants. Many Ranunculaceae phytometabolites have shown very promising anticancer properties in vitro but have yet to be evaluated in humans. The epigenetic transcriptional regulation with regard to Ranunculaceae compounds is little known. Gaps are present in the extensive use of high-throughput transcriptome sequencing and proteome profiling for characterizing the interaction between Ranunculaceae compounds and cancer cells, as well as the relevant regulatory network. Further studies are warranted to determine the molecular mechanisms and efficacy of Ranunculaceae compounds in treating human cancers.

Many Ranunculaceae plants contain not only anticancer phytometabolites but also compounds with free radical scavenging, immunomodulatory, and antiinflammatory activities that are helpful against cancer insurgence. However, interaction between Western drugs and herbs/botanicals should be well investigated before safe combined use, and such information must be disseminated to the allied stakeholders.

As anticancer candidates with low toxicity, Ranunculaceae compounds such as TQ, BBR, and their altered structure, as well as source plants such as *N. sativa*, *C. chinensis*, and their closely related species, will attract researchers to pursue the potential anticancer effects and the mechanisms by using innovative technologies of genomics, proteomics, and other advanced approaches. Current and future studies would consistently suggest the feasibility of mining anticancer pharmacological diversity from Ranunculaceae chemodiversity.

References

Akkol, E.K., Süntar, I., Erdoğan, T.F., et al., 2012. Wound healing and anti-inflammatory properties of *Ranunculus pedatus* and *Ranunculus constantinapolitanus*: a comparative study. J. Ethnopharmacol. 139 (2), 478–484.

An, F., Wang, S., Tian, Q., et al., 2015. Effects of orientin and vitexin from *Trollius chinensis* on the growth and apoptosis of esophageal cancer EC-109 cells. Oncol. Lett. 10 (4), 2627–2633.

Arslan, B.A., Isik, F.B., Gur, H., et al., 2017. Apoptotic effect of *Nigella sativa* on human lymphoma U937 cells. Pharmacogn. Mag. 13 (S3), 628–632.

Arumugam, P., Subramanian, R., Priyadharsini, J.V., et al., 2016. Thymoquinone inhibits the migration of mouse neuroblastoma (Neuro-2a) cells by down-regulating MMP-2 and MMP-9. Chin. J. Nat. Med. 14 (12), 904–912.

Ashley, R.E., Osheroff, N., 2014. Natural products as topoisomerase II poisons: effects of thymoquinone on DNA cleavage mediated by human topoisomerase IIα. Chem. Res. Toxicol. 27 (5), 787–793.

Attoub, S., Sperandio, O., Raza, H., et al., 2013. Thymoquinone as an anticancer agent: evidence from inhibition of cancer cells viability and invasion in vitro and tumor growth in vivo. Fundam. Clin. Pharmacol. 27 (5), 557–569.

Auyeung, K.K., Ko, J.K., 2009. *Coptis chinensis* inhibits hepatocellular carcinoma cell growth through nonsteroidal anti-inflammatory drug-activated gene activation. Int. J. Mol. Med. 24 (4), 571–577.

Ayati, S.H., Fazeli, B., Momtazi-Borojeni, A.A., et al., 2017. Regulatory effects of berberine on microRNome in cancer and other conditions. Crit. Rev. Oncol. Hematol. 116, 147–158.

Badary, O.A., El-Din, A.M.G., 2001. Inhibitory effects of thymoquinone against 20-methylcholanthrene-induced fibrosarcoma tumorigenesis. Canc. Detect. Prev. 25, 362–368.

Bang, S.C., Lee, J.H., Song, G.Y., et al., 2005. Antitumor activity of *Pulsatilla koreana* saponins and their structure-activity relationship. Chem. Pharm. Bull. 53 (11), 1451–1454.

Bao, M., Cao, Z., Yu, D., et al., 2012. Columbamine suppresses the proliferation and neovascularization of metastatic osteosarcoma U2OS cells with low cytotoxicity. Toxicol. Lett. 215 (3), 174–180.

Bao, N.M., Nian, Y., Zhu, G.L., et al., 2014. Cytotoxic 9,19-cycloartane triterpenes from the aerial parts of *Cimicifuga yunnanensis*. Fitoterapia 99, 191–197.

Bhattacharya, S., Ahir, M., Patra, P., et al., 2015. PEGylated-thymoquinone-nanoparticle mediated retardation of breast cancer cell migration by deregulation of cytoskeletal actin polymerization through miR-34a. *Biomaterials* 51, 91–107.

Burdette, J.E., Chen, S.N., Lu, Z.Z., et al., 2002. Black cohosh (*Cimicifuga racemosa* L.) protects against menadione-induced DNA damage through scavenging of reactive oxygen species: bioassay-directed isolation and characterization of active principles. J. Agric. Food Chem. 50 (24), 7022–7028.

Chehl, N., Chipitsyna, G., Gong, Q., et al., 2009. Anti-inflammatory effects of the *Nigella sativa* seed extract, thymoquinone, in pancreatic cancer cells. HPB (Oxford) 11 (5), 373–381.

Chen, D., Cao, R., He, J., et al., 2016. Synergetic effects of aqueous extracts of Fuzi (Radix Aconiti Lateralis Preparata) and Tubeimu (Rhizoma Bolbostemmatis) on MDA-MB-231 and SKBR3 cells. J. Tradit. Chin. Med. 36 (1), 113–124.

Chen, Z., Duan, H., Tong, X., et al., 2017. Cytotoxicity, hemolytic toxicity, and mechanism of action of Pulsatilla saponin D and its synthetic derivatives. J. Nat. Prod. doi: 10.1021/acs.jnatprod.7b00578.

Chen, M.C., Lee, N.H., Hsu, H.H., et al., 2015a. Thymoquinone induces caspase-independent, autophagic cell death in CPT-11-resistant lovo colon cancer via mitochondrial dysfunction and activation of JNK and p38. J. Agric. Food Chem. 63 (5), 1540–1546.

Chen, Q., Peng, W., Qi, S., et al., 2002. Apoptosis of human highly metastatic lung cancer cell line 95-D induced by acutiaporberine, a novel bisalkaloid derived from *Thalictrum acutifolium*. Planta Med. 68 (6), 550–553.

Chen, T.T., Qi, C., Guo, H., et al., 2009. The effects of Fu Zi on changes in the body heat of dogs. J. Acupunct. Meridian Stud. 2 (1), 71–74.

Chen, Q., Qin, R., Fang, Y., et al., 2015b. Berberine sensitizes human ovarian cancer cells to cisplatin through miR-93/PTEN/Akt signaling pathway. Cell. Physiol. Biochem. 36 (3), 956–965.

Chen, S., Wan, L., Couch, L., et al., 2013. Mechanism study of goldenseal-associated DNA damage. Toxicol. Lett. 221 (1), 64–72.

Cheng, L., Xia, T.S., Shi, L., et al., 2017. D Rhamnose β-hederin inhibits migration and invasion of human breast cancer cell line MDA-MB-231. Biochem. Biophys. Res. Commun. doi: 10.1016/j.bbrc.2017.11.081.

Cheng, L., Xia, T.S., Wang, Y.F., et al., 2014a. The apoptotic effect of D Rhamnose β-hederin, a novel oleanane-type triterpenoid saponin on breast cancer cells. PLoS One 9 (6), e90848.

Cheng, L., Xia, T.S., Wang, Y.F., et al., 2014b. The anticancer effect and mechanism of α-hederin on breast cancer cells. Int. J. Oncol. 45 (2), 757–763.

Choi, C.H., Kim, T.H., Sung, Y.K., et al., 2014. SKI306X inhibition of glycosaminoglycan degradation in human cartilage involves down-regulation of cytokine-induced catabolic genes. Korean J. Intern. Med. 29 (5), 647–655.

Chu, S.C., Hsieh, Y.S., Yu, C.C., et al., 2014. Thymoquinone induces cell death in human squamous carcinoma cells via caspase activation-dependent apoptosis and LC3-II activation-dependent autophagy. PLoS One 9 (7), e101579.

Dai, X., Liu, J., Nian, Y., et al., 2017. A novel cycloartane triterpenoid from Cimicifuga induces apoptotic and autophagic cell death in human colon cancer HT-29 cells. Oncol. Rep. 37 (4), 2079–2086.

Das, S., Das, J., Samadder, A., et al., 2013. Dihydroxy-isosteviol methyl ester from *Pulsatilla nigricans* induces apoptosis in HeLa cells: its cytoxicity and interaction with calf thymus DNA. Phytother. Res. 27 (5), 664–671.

de la Peña, S.S., Sothern, R.B., López, F.S., et al., 2011. Circadian aspects of hyperthermia in mice induced by *Aconitum napellus*. Pharmacogn. Mag. 7 (27), 234–242.

Deng, L.J., Hu, L.P., Peng, Q.L., et al., 2014. Hellebrigenin induces cell cycle arrest and apoptosis in human hepatocellular carcinoma HepG2 cells through inhibition of Akt. Chem. Biol. Interact. 219, 184–194.

Dietel, B., Muench, R., Kuehn, C., et al., 2014. MCS-18, a natural product isolated from *Helleborus purpurascens*, inhibits maturation of dendritic cells in ApoE-deficient mice and prevents early atherosclerosis progression. Atherosclerosis 235 (2), 263–272.

Dilshara, M.G., Lee, K.T., Lee, C.M., et al., 2015. New compound, 5-O-isoferuloyl-2-deoxy-D-ribono-γ-lacton from *Clematis mandshurica*: anti-inflammatory effects in lipopolysaccharide-stimulated BV2 microglial cells. Int. Immunopharmacol. 24 (1), 14–23.

Duan, S.P., Jin, C.L., Hao, J., et al., 2013. A study on the inhibitory effect of Radix Semiaquilegiae extract on human hepatoma HEPG-2 and SMMC-7721 cells. Afr. J. Tradit. Complement. Altern. Med. 10 (5), 336–340.

Dueregger, A., Guggenberger, F., Barthelmes, J., et al., 2013. Attenuation of nucleoside and anti-cancer nucleoside analog drug uptake in prostate cancer cells by *Cimicifuga racemosa* extract BNO-1055. Phytomedicine 20 (14), 1306–1314.

Ebru, U., Burak, U., Yusuf, S., et al., 2008. Cardioprotective effects of *Nigella sativa* oil on cyclosporine A – induced cardiotoxicity in rats. Basic Clin. Pharmacol. Toxicol. 103, 574–580.

Einbond, L.S., Mighty, J., Redenti, S., et al., 2013. Actein induces calcium release in human breast cancer cells. Fitoterapia 91, 28–38.

Einbond, L.S., Su, T., Wu, H.A., et al., 2007. Gene expression analysis of the mechanisms whereby black cohosh inhibits human breast cancer cell growth. Anticancer Res. 27 (2), 697–712.

Einbond, L.S., Wen-Cai, Y., He, K., et al., 2008. Growth inhibitory activity of extracts and compounds from Cimicifuga species on human breast cancer cells. Phytomedicine 15 (6–7), 504–511.

El-Baba, C., Mahadevan, V., Fahlbusch, F.B., et al., 2014. Thymoquinone-induced conformational changes of PAK1 interrupt prosurvival MEK-ERK signaling in colorectal cancer. Mol. Cancer 13, 201.

Elbarbry, F., Ragheb, A., Marfleet, T., et al., 2012. Modulation of hepatic drug metabolizing enzymes by dietary doses of thymoquinone in female New Zealand White rabbits. Phytother. Res. 26 (11), 1726–1730.

ElKhoely, A., Hafez, H.F., Ashmawy, A.M., et al., 2015. Chemopreventive and therapeutic potentials of thymoquinone in HepG2 cells: mechanistic perspectives. J. Nat. Med. 69 (3), 313–323.

El-Mahdy, M.A., Zhu, Q., Wang, Q.E., et al., 2005. Thymoquinone induces apoptosis through activation of caspase-8 and mitochondrial events in p53-null myeloblastic leukemia HL-60 cells. Int. J. Cancer 117, 409–417.

Fan, Y., Jiang, Y., Liu, J., et al., 2016. The anti-tumor activity and mechanism of alkaloids from *Aconitum szechenyianum* Gay. Bioorg. Med. Chem. Lett. 26 (2), 380–387.

Feng, M., Luo, X., Gu, C., et al., 2015. Systematic analysis of berberine-induced signaling pathway between miRNA clusters and mRNAs and identification of mir-99a ~ 125b cluster function by seed-targeting inhibitors in multiple myeloma cells. RNA Biol. 12 (1), 82–91.

Feng, H.T., Zhao, W.W., Lu, J.J., et al., 2017. Hypaconitine inhibits TGF-β1-induced epithelial-mesenchymal transition and suppresses adhesion, migration, and invasion of lung cancer A549 cells. Chin. J. Nat. Med. 15 (6), 427–435.

Fostok, S.F., Ezzeddine, R.A., Homaidan, F.R., et al., 2009. Interleukin-6 and cyclooxygenase-2 downregulation by fatty-acid fractions of *Ranunculus constantinopolitanus*. BMC Complement. Altern. Med. 9, 44.

Gao, J., Huang, F., Zhang, J., et al., 2006. Cytotoxic cycloartane triterpene saponins from *Actaea asiatica*. J. Nat. Prod. 69 (10), 1500–1502.

Gaube, F., Wolfl, S., Pusch, L., et al., 2007. Gene expression profiling reveals effects of *Cimicifuga racemosa* (L.) NUTT. (black cohosh) on the estrogen receptor positive human breast cancer cell line MCF-7. BMC Pharmacol. 7, 11.

Gong, Y.X., Hua, H.M., Xu, Y.N., et al., 2013. Triterpene saponins from *Clematis mandshurica* and their antiproliferative activity. Planta Med. 79 (11), 987–994.

Grivennikov, S.I., Greten, F.R., Karin, M., 2010. Immunity, inflammation, and cancer. Cell 140, 883–899.

Guan, Y.Y., Liu, H.J., Luan, X., et al., 2015. Raddeanin A, a triterpenoidsaponin isolated from *Anemone raddeana*, suppresses the angiogenesis and growth of human colorectal tumor by inhibiting VEGFR2 signaling. Phytomedicine 22 (1), 103–110.

Guo, L.Y., Joo, E.J., Son, K.H., et al., 2009. Cimiside E arrests cell cycle and induces cell apoptosis in gastric cancer cells. Arch. Pharm. Res. 32 (10), 1385–1392.

Gurung, R.L., Lim, S.N., Khaw, A.K., et al., 2010. Thymoquinone induces telomere shortening, DNA damage and apoptosis in human glioblastoma cells. PLoS One 5 (8), e12124.

Hambright, H.G., Batth, I.S., Xie, J., et al., 2015. Palmatine inhibits growth and invasion in prostate cancer cell: potential role for rpS6/NFκB/FLIP. Mol. Carcinog. 54 (10), 1227–1234.

Han, L.T., Fang, Y., Li, M.M., et al., 2013. The antitumor effects of triterpenoid saponins from the *Anemone flaccida* and the underlying mechanism. Evid. Based Complement. Alternat. Med. 2013, 517931.

Hanahan, D., Weinberg, R.A., 2011. Hallmarks of cancer: the next generation. Cell 144 (5), 646–674.

Hao, D.C., Gu, X.J., Xiao, P.G., et al., 2013a. Recent advances in the chemical and biological studies of *Aconitum* pharmaceutical resources. J. Chin. Pharm. Sci. 22 (3), 209–221.

Hao, D.C., Gu, X.J., Xiao, P.G., et al., 2013b. Recent advances in chemical and biological studies on Cimicifugeae pharmaceutical resources. Chin. Herb. Med. 5 (2), 81–95.

Hao, D.C., Gu, X.J., Xiao, P.G., et al., 2013c. Chemical and biological research of *Clematis* medicinal resources. Chin. Sci. Bull. 58 (10), 1120–1129.

Hao, D.C., Gu, X.J., Xiao, P.G., et al., 2015a. Medicinal Plants: Chemistry, Biology and Omics, first ed. Elsevier-Woodhead, Oxford, ISBN: 9780081000854.

Hao, D.C., Xiao, P.G., Chen, S.L., 2010. Phenotype prediction of nonsynonymous single nucleotide polymorphisms in human phase II drug/xenobiotic metabolizing enzymes: perspectives on molecular evolution. Sci. China Life Sci. 53 (10), 1252–1262.

Hao, D.C., Xiao, P.G., Liu, M., et al., 2014. Pharmaphylogeny vs. pharmacophylogenomics: molecular phylogeny, evolution and drug discovery. Yao Xue Xue Bao 49 (10), 1387–1394.

Hao, D.C., Xiao, P.G., Ma, H., et al., 2015b. Mining chemodiversity from biodiversity: pharmacophylogeny of medicinal plants of the Ranunculaceae. Chin. J. Nat. Med. 13 (7), 507–520.

Hara, A., Iizuka, N., Hamamoto, Y., et al., 2005. Molecular dissection of a medicinal herb with anti-tumor activity by oligonucleotide microarray. Life Sci. 77 (9), 991–1002.

Hazawa, M., Takahashi, K., Wada, K., et al., 2011. Structure-activity relationships between the *Aconitum* C20-diterpenoid alkaloid derivatives and the growth suppressive activities of Non-Hodgkin's lymphoma Raji cells and human hematopoietic stem/progenitor cells. Invest. New Drugs 29 (1), 1–8.

Hazawa, M., Wada, K., Takahashi, K., et al., 2009. Suppressive effects of novel derivatives prepared from *Aconitum* alkaloids on tumor growth. Invest. New Drugs 27 (2), 111–119.

He, Y.X., Li, L., Zhang, K., et al., 2011. Cytotoxic triterpene saponins from *Clematis mandshurica*. J. Asian Nat. Prod. Res. 13 (12), 1104–1109.

Hong, S.W., Jung, K.H., Lee, H.S., et al., 2012a. SB365 inhibits angiogenesis and induces apoptosis of hepatocellular carcinoma through modulation of PI3K/Akt/mTOR signaling pathway. Cancer Sci. 103 (11), 1929–1937.

Hong, S.W., Jung, K.H., Lee, H.S., et al., 2012b. Apoptotic and anti-angiogenic effects of *Pulsatilla koreana* extract on hepatocellular carcinoma. Int. J. Oncol. 40 (2), 452–460.

Hong, S.W., Jung, K.H., Lee, H.S., et al., 2013. SB365, *Pulsatilla* saponin D, targets c-Met and exerts antiangiogenic and antitumor activities. Carcinogenesis 34 (9), 2156–2169.

Hostanska, K., Nisslein, T., Freudenstein, J., et al., 2005. Apoptosis of human prostate androgen-dependent and -independent carcinoma cells induced by an isopropanolic extract of black cohosh involves degradation of cyto-keratin (CK) 18. Anticancer Res. 25 (1A), 139–147.

Hou, Q., 2011. Study on the Autophagy and Apoptosis Induced by Berberine in Human Hepatoma Cells with the Molecular Mechanism Research (Ph.D. dissertation) of Fourth Military Medical University.

Hu, X., Wu, X., Huang, Y., et al., 2014. Berberine induces double-strand DNA breaks in Rev3 deficient cells. Mol. Med. Rep. 9 (5), 1883–1888.

Huang, Z.H., Zheng, H.F., Wang, W.L., et al., 2015. Berberine targets epidermal growth factor receptor signaling to suppress prostate cancer proliferation in vitro. Mol. Med. Rep. 11 (3), 2125–2128.

Ismail, M., Al-Naqeep, G., Chan, K.W., 2010. *Nigella sativa* thymoquinone-rich fraction greatly improves plasma antioxidant capacity and expression of antioxidant genes in hypercholesterolemic rats. Free Radic. Biol. Med. 48, 664–672.

Jafri, S.H., Glass, J., Shi, R., et al., 2010. Thymoquinone and cisplatin as a therapeutic combination in lung cancer: in vitro and in vivo. J. Exp. Clin. Cancer Res. 29, 87.

Jang, W.J., Park, B., Jeong, G.S., et al., 2014. SB365, *Pulsatilla* saponin D, suppresses the growth of gefitinib-resistant NSCLC cells with Met amplification. Oncol. Rep. 32 (6), 2612–2618.

Ji, C., Cheng, G., Tang, H., et al., 2015. Saponin 6 of Anemone Taipaiensis inhibits proliferation and induces apoptosis of U87 MG cells. Xi Bao Yu Fen Zi Mian Yi Xue Za Zhi 31 (4), 484–486.

Ji, X., Sun, H., Zhou, H., et al., 2012. The interaction of telomeric DNA and C-myc22 G-quadruplex with 11 natural alkaloids. Nucleic Acid Ther. 22 (2), 127–136.

Ji, B.L., Xia, L.P., Zhou, F.X., et al., 2016. Aconitine induces cell apoptosis in human pancreatic cancer via NF-κB signaling pathway. Eur. Rev. Med. Pharmacol. Sci. 20 (23), 4955–4964.

Kaseb, A.O., Chinnakannu, K., Chen, D., et al., 2007. Androgen receptor–and E2F-1–targeted thymoquinone therapy for hormone-refractory prostate cancer. Cancer Res. 67, 7782–7788.

Khan, M.A., Tania, M., Wei, C., et al., 2015. Thymoquinone inhibits cancer metastasis by downregulating TWIST1 expression to reduce epithelial to mesenchymal transition. Oncotarget 6 (23), 19580–19591.

Kim, J.M., Kim, K.S., Lee, Y.W., et al., 2011. Anti-angiogenic effects of water extract of a formula consisting of *Pulsatilla koreana*, Panax ginseng and Glycyrrhizauralensis. Zhong Xi Yi Jie He Xue Bao 9 (9), 1005–1013.

Kim, J.B., Yu, J.H., Ko, E., et al., 2010. The alkaloid Berberine inhibits the growth of Anoikis-resistant MCF-7 and MDA-MB-231 breast cancer cell lines by inducing cell cycle arrest. Phytomedicine 17 (6), 436–440.

Koka, P.S., Mondal, D., Schultz, M., et al., 2010. Studies on molecular mechanisms of growth inhibitory effects of thymoquinone against prostate cancer cells: role of reactive oxygen species. Exp. Biol. Med. 235, 751–760.

Kong, Y., Li, F., Nian, Y., et al., 2016. KHF16 is a leading structure from *Cimicifuga foetida* that suppresses breast cancer partially by inhibiting the NF-κB signaling pathway. Theranostics 6 (6), 875–886.

Kundu, J., Choi, B.Y., Jeong, C.H., et al., 2014a. Thymoquinone induces apoptosis in human colon cancer HCT116 cells through inactivation of STAT3 by blocking JAK2- and Src-mediated phosphorylation of EGF receptor tyrosine kinase. Oncol. Rep. 32 (2), 821–828.

Kundu, J., Chun, K.S., Aruoma, O.I., et al., 2014b. Mechanistic perspectives on cancer chemoprevention/chemotherapeutic effects of thymoquinone. Mutat. Res. 768, 22–34.

Lee, T.H., Huang, N.K., Lai, T.C., et al., 2008. Anemonin, from *Clematis crassifolia*, potent and selective inducible nitric oxide synthase inhibitor. J. Ethnopharmacol. 116 (3), 518–527.

Lee, K.H., Lo, H.L., Tang, W.C., et al., 2014. A gene expression signature-based approach reveals the mechanisms of action of the Chinese herbal medicine berberine. Sci. Rep. 4, 6394.

Li, Y., Fu, C.M., Ren, B., et al., 2014b. Study on attenuate and synergistic mechanism between aconiti lateralis praeparata radix and glycyrrhizae radix for toxicity reduction based on metabonomic of MI-RI mouse cardiomyocytes. Zhongguo Zhong Yao Za Zhi 39 (16), 3166–3171.

Li, W., Hua, B., Saud, S.M., et al., 2015. Berberine regulates AMP-activated protein kinase signaling pathways and inhibits colon tumorigenesis in mice. Mol. Carcinog. 54 (10), 1096–1109.

Li, S.G., Huang, X.J., Li, M.M., et al., 2014a. Triterpenoid saponins from the roots of *Clematis uncinata*. Chem. Pharm. Bull. 62 (1), 35–44.

Li, H., Sun, M., Xu, J., et al., 2013b. Immunological response in H22 transplanted mice undergoing *Aconitum coreanum* polysaccharide treatment. Int. J. Biol. Macromol. 55, 295–300.

Li, J., Tang, H., Zhang, Y., et al., 2013a. Saponin 1 induces apoptosis and suppresses NF-κB-mediated survival signaling in glioblastoma multiforme (GBM). PLoS One 8 (11), e81258.

Liang, M., Liu, J., Ji, H., et al., 2015. A *Aconitum coreanum* polysaccharide fraction induces apoptosis of hepatocellular carcinoma (HCC) cells via pituitary tumor transforming gene 1 (PTTG1)-mediated suppression of the P13K/Akt and activation of p38 MAPK signaling pathway and displays antitumor activity in vivo. Tumour Biol. 36 (9), 7085–7091.

Liang, Y., Xu, X., Yu, H., et al., 2016. Raddeanoside R13 inhibits breast cancer cell proliferation, invasion, and metastasis. Tumour Biol. 37 (7), 9837–9847.

Lin, C.C., Lin, S.Y., Chung, J.G., et al., 2006. Down-regulation of cyclin B1 and up-regulation of Wee1 by berberine promotes entry of leukemia cells into the G2/M-phase of the cell cycle. Anticancer Res. 26 (2A), 1097–1104.

Lin, J.P., Yang, J.S., Wu, C.C., et al., 2008. Berberine induced down-regulation of matrix metalloproteinase-1-2 and -9 in human gastric cancer cells (SNU-5) in vitro. In Vivo 22 (2), 223–230.

Lin, C.Z., Zhao, Z.X., Xie, S.M., et al., 2014. Diterpenoid alkaloids and flavonoids from *Delphinium trichophorum*. Phytochemistry 97, 88–95.

Liu, R., Cao, Z., Pan, Y., et al., 2013b. Jatrorrhizine hydrochloride inhibits the proliferation and neovascularization of C8161 metastatic melanoma cells. Anticancer Drugs 24 (7), 667–676.

Liu, Q., Chen, W., Jiao, Y., et al., 2014a. *Pulsatilla* saponin A, an active molecule from *Pulsatilla chinensis*, induces cancer cell death and inhibits tumor growth in mouse xenograft models. J. Surg. Res. 188 (2), 387–395.

Liu, X., Ji, Q., Ye, N., et al., 2015c. Berberine inhibits invasion and metastasis of colorectal cancer cells via COX-2/PGE2 mediated JAK2/STAT3 signaling pathway. PLoS One 10 (5), e123478.

Liu, Y., Li, Y., Yang, W., et al., 2012. Anti-hepatoma activity in mice of a polysaccharide from the rhizome of *Anemone raddeana*. Int. J. Biol. Macromol. 50 (3), 632–636.

Liu, D.L., Li, Y.J., Yao, N., et al., 2014b. Acerinol, a cyclolanstane triterpenoid from *Cimicifuga acerina*, reverses ABCB1-mediated multidrug resistance in HepG2/ADM and MCF-7/ADR cells. Eur. J. Pharmacol. 733, 34–44.

Liu, Y., Song, Y., Xu, Q., et al., 2013a. Validated rapid resolution LC-ESI-MS/MS method for simultaneous determination of five pulchinenosides from *Pulsatilla chinensis* (Bunge) Regel in rat plasma: application to pharmacokinetics and bioavailability studies. J. Chromatogr. B Analyt. Technol. Biomed. Life Sci. 942–943, 141–150.

Liu, C.H., Tang, W.C., Sia, P., et al., 2015d. Berberine inhibits the metastatic ability of prostate cancer cells by suppressing epithelial-to-mesenchymal transition (EMT)-associated genes with predictive and prognostic relevance. Int. J. Med. Sci. 12 (1), 63–71.

Liu, Q., Xu, X., Zhao, M., et al., 2015b. Berberine induces senescence of human glioblastoma cells by downregulating the EGFR-MEK-ERK signaling pathway. Mol. Cancer Ther. 14 (2), 355–363.

Liu, M., Zhao, X., Xiao, L., et al., 2015a. Cytotoxicity of the compounds isolated from *Pulsatilla chinensis* saponins and apoptosis induced by 23-hydroxybetulinic acid. Pharm. Biol. 53 (1), 1–9.

Lo, T.F., Tsai, W.C., Chen, S.T., 2013. MicroRNA-21-3p, a berberine-induced miRNA, directly down-regulates human methionine adenosyltransferases 2A and 2B and inhibits hepatoma cell growth. PLoS One 8 (9), e75628.

Lu, L., Chen, J.C., Li, Y., et al., 2012. Studies on the constituents of *Cimicifuga foetida* collected in Guizhou Province and their cytotoxic activities. Chem. Pharm. Bull. 60 (5), 571–577.

Lv, X., Wang, H., Han, H., et al., 2010. Effects of polysaccharide of radix ranunculi ternati on immunomodulation and anti-oxidation. Zhongguo Zhong Yao Za Zhi 35 (14), 1862–1865.

Martucciello, S., Paolella, G., Muzashvili, T., et al., 2017. Steroids from *Helleborus caucasicus* reduce cancer cell viability inducing apoptosis and GRP78 down-regulation. Chem. Biol. Interact. 279, 43–50.

Mazzio, E., Badisa, R., Mack, N., et al., 2014. High throughput screening of natural products for anti-mitotic effects in MDA-MB-231 human breast carcinoma cells. Phytother. Res. 28 (6), 856–867.

McConnell, M.T., Lisgarten, D.R., Byrne, L.J., et al., 2015. Winter Aconite (*Eranthis hyemalis*) lectin as a cytotoxic effector in the life cycle of *Caenorhabditis elegans*. Peer J. 3, e1206.

Mu, G.G., Zhang, L.L., Li, H.Y., et al., 2015. Thymoquinone pretreatment overcomes the insensitivity and potentiates the antitumor effect of gemcitabine through abrogation of Notch1, PI3K/Akt/mTOR regulated signaling pathways in pancreatic cancer. Dig. Dis. Sci. 60 (4), 1067–1080.

Nabekura, T., Hiroi, T., Kawasaki, T., et al., 2015. Effects of natural nuclear factor-kappa B inhibitors on anticancer drug efflux transporter human P-glycoprotein. Biomed. Pharmacother. 70, 140–145.

Nian, Y., Yang, J., Liu, T.Y., et al., 2015. New anti-angiogenic leading structure discovered in the fruit of *Cimicifuga yunnanensis*. Sci. Rep. 5, 9026.

Niu, L., Zhou, Y., Sun, B., et al., 2013. Inhibitory effect of saponins and polysaccharides from Radix ranunculi ternati on human gastric cancer BGC823 cells. Afr. J. Tradit. Complement. Altern. Med. 10 (3), 561–566.

Ortiz, L.M., Lombardi, P., Tillhon, M., et al., 2014. Berberine, an epiphany against cancer. Molecules 19 (8), 12349–12367.

Ouyang, H., Zhou, M., Guo, Y., et al., 2014. Metabolites profiling of *Pulsatilla* saponin D in rat by ultra performance liquid chromatography-quadrupole time-of-flight mass spectrometry (UPLC/Q-TOF-MS/MS). Fitoterapia 96, 152–158.

Pan, L.L., Wang, X.L., Luo, X.L., et al., 2017. Boehmenan, a lignan from the Chinese medicinal plant *Clematis armandii*, inhibits A431 cell growth via blocking p70S6/S6 kinase pathway. Integr. Cancer Ther. 16 (3), 351–359.

Pan, L.L., Wang, X.L., Zhang, Q.Y., et al., 2016. Boehmenan, a lignan from the Chinese medicinal plant *Clematis armandii*, induces apoptosis in lung cancer cells through modulation of EGF-dependent pathways. Phytomedicine 23 (5), 468–476.

Park, B.H., Jung, K.H., Son, M.K., et al., 2013. Antitumor activity of *Pulsatilla koreana* extract in anaplastic thyroid cancer via apoptosis and anti-angiogenesis. Mol. Med. Rep. 7 (1), 26–30.

Park, S.Y., Kim, H.J., Lee, S.R., et al., 2016. Black cohosh inhibits 17β-estradiol-induced cell proliferation of endometrial adenocarcinoma cells. Gynecol. Endocrinol. 32 (10), 840–843.

Park, S.H., Sung, J.H., Kim, E.J., et al., 2015. Berberine induces apoptosis via ROS generation in PANC-1 and MIA-PaCa2 pancreatic cell lines. Braz. J. Med. Biol. Res. 48 (2), 111–119.

Qing, Y., Hu, H., Liu, Y., et al., 2014. Berberine induces apoptosis in human multiple myeloma cell line U266 through hypomethylation of p53 promoter. Cell Biol. Int. 38 (5), 563–570.

Racoma, I.O., Meisen, W.H., Wang, Q.E., et al., 2013. Thymoquinone inhibits autophagy and induces cathepsin-mediated, caspase-independent cell death in glioblastoma cells. PLoS One 8 (9), e72882.

Raghunandhakumar, S., Paramasivam, A., Senthilraja, S., et al., 2013. Thymoquinone inhibits cell proliferation through regulation of G1/S phase cell cycle transition in N-nitrosodiethylamine-induced experimental rat hepatocellular carcinoma. Toxicol. Lett. 223 (1), 60–72.

Rao, X., Gong, M., Yin, S., et al., 2013. Study on in situ intestinal absorption of *Pulsatilla* saponin D in rats. Chin. Trad. Herb. Drugs 44, 3515–3520.

Saha, S.K., Khuda-Bukhsh, A.R., 2014. Berberine alters epigenetic modifications, disrupts microtubule network, and modulates HPV-18 E6-E7 oncoproteins by targeting p53 in cervical cancer cell HeLa: a mechanistic study including molecular docking. Eur. J. Pharmacol. 744, 132–146.

Sakalar, C., Yuruk, M., Kaya, T., et al., 2013. Pronounced transcriptional regulation of apoptotic and TNF-NF-kappa-B signaling genes during the course of thymoquinone mediated apoptosis in HeLa cells. Mol. Cell. Biochem. 383 (1–2), 243–251.

Salem, A.A., El Haty, I.A., Abdou, I.M., et al., 2015. Interaction of human telomeric G-quadruplex DNA with thymoquinone: a possible mechanism for thymoquinone anticancer effect. Biochim. Biophys. Acta 1850 (2), 329–342.

Schafer, M., Werner, S., 2008. Cancer as an overhealing wound:an old hypothesis revisited. Nat. Rev. Mol. Cell Biol. 9, 628–638.

Schneider-Stock, R., Fakhoury, I.H., Zaki, A.M., et al., 2014. Thymoquinone: fifty years of success in the battle against cancer models. Drug Discov. Today 19 (1), 18–30.

Serafim, T.L., Oliveira, P.J., Sardao, V.A., et al., 2008. Different concentrations of berberine result in distinct cellular localization patterns and cell cycle effects in a melanoma cell line. Cancer Chemother. Pharmacol. 61 (6), 1007–1018.

Sethi, G., Ahn, K.S., Aggarwal, B.B., 2008. Targeting nuclear factor-κB activation pathway by thymoquinone: role in suppression of antiapoptotic gene products and enhancement of apoptosis. Mol. Cancer Res. 6, 1059–1070.

Shafiq, H., Ahmad, A., Masud, T., et al., 2014. Cardio-protective and anti-cancer therapeutic potential of *Nigella sativa*. Iran J. Basic Med. Sci. 17 (12), 967–979.

Sheng, L.H., Xu, M., Xu, L.Q., et al., 2014. Cytotoxic effect of lappaconitine on non-small cell lung cancer in vitro and its molecular mechanism. Zhong Yao Cai 37 (5), 840–843.

Shin, S.B., Woo, S.U., Yim, H., 2015. Differential cellular effects of Plk1 inhibitors targeting the ATP-binding domain or Polo-box domain. J. Cell. Physiol. 230 (12), 3057–3067.

Shukla, S., Rizvi, F., Raisuddin, S., et al., 2014. FoxO proteins' nuclear retention and BH3-only protein Bim induction evoke mitochondrial dysfunction-mediated apoptosis in berberine-treated HepG2 cells. Free Radic. Biol. Med. 76, 185–199.

Son, M.K., Jung, K.H., Hong, S.W., et al., 2013a. SB365, *Pulsatilla* saponin D suppresses the proliferation of human colon cancer cells and induces apoptosis by modulating the AKT/mTOR signaling pathway. Food Chem. 136 (1), 26–33.

Son, M.K., Jung, K.H., Lee, H.S., et al., 2013b. SB365, *Pulsatilla* saponin D suppresses proliferation and induces apoptosis of pancreatic cancer cells. Oncol. Rep. 30 (2), 801–808.

Sridhar, A., Saremy, S., Bhattacharjee, B., 2014. Elucidation of molecular targets of bioactive principles of black cumin relevant to its anti-tumour functionality – an In silico target fishing approach. Bioinformation 10 (11), 684–688.

Su, Y.H., Tang, W.C., Cheng, Y.W., et al., 2015. Targeting of multiple oncogenic signaling pathways by Hsp90 inhibitor alone or in combination with berberine for treatment of colorectal cancer. Biochim. Biophys. Acta 1853 (10 Pt A), 2261–2272.

Sultan, M.T., Butt, M.S., Qayyum, M.M., et al., 2014. Immunity: plants as effective mediators. Crit. Rev. Food Sci. Nutr. 54 (10), 1298–1308.

Sun, H.Y., Liu, B.B., Hu, J.Y., et al., 2016. Novel cycloartane triterpenoid from *Cimicifuga foetida* (Sheng ma) induces mitochondrial apoptosis via inhibiting Raf/MEK/ERK pathway and Akt phosphorylation in human breast carcinoma MCF-7 cells. Chin. Med. 11, 1.

Sun, D.L., Xie, H.B., Xia, Y.Z., 2013. A study on the inhibitory effect of polysaccharides from Radix ranunculus ternati on human breast cancer MCF-7 cell lines. Afr. J. Tradit. Complement. Altern. Med. 10 (6), 439–443.

Tan, H.Y., Wang, N., Tsao, S.W., et al., 2014. Suppression of vascular endothelial growth factor via inactivation of eukaryotic elongation factor 2 by alkaloids in Coptidis rhizome in hepatocellular carcinoma. Integr. Cancer Ther. 13 (5), 425–434.

Tang, J., Feng, Y., Tsao, S., et al., 2009. Berberine and *Coptidis rhizoma* as novel antineoplastic agents: a review of traditional use and biomedical investigations. J. Ethnopharmacol. 126 (1), 5–17.

Tian, X., Feng, J., Tang, H., et al., 2013. New cytotoxic triterpenoid saponins from the whole plant of Clematis lasiandra Maxim. Fitoterapia 90, 233–239.

Tian, Z., Pan, R., Chang, Q., et al., 2007a. *Cimicifuga foetida* extract inhibits proliferation of hepatocellular cells via induction of cell cycle arrest and apoptosis. J. Ethnopharmacol. 114 (2), 227–233.

Tian, Z., Si, J.Y., Chang, Q., et al., 2007b. Antitumor activity and mechanisms of action of total glycosides from aerial part of *Cimicifuga dahurica* targeted against hepatoma. BMC Cancer 7, 237.

Tian, Z., Si, J.Y., Chen, S.B., et al., 2006b. Cytotoxicity and mechanism of 23-O-acetylcimigenol-3-O-beta-D-xylopyranoside on HepG2 cells. Zhongguo Zhong Yao Za Zhi 31 (21), 1818–1821.

Tian, Z., Xiao, P.G., Wen, J., et al., 2006a. Review of bioactivities of natural cycloartane triterpenoids. Zhongguo Zhong Yao Za Zhi 31 (8), 625–629.

Tian, Z., Zhou, L., Huang, F., et al., 2006c. Anti-cancer activity and mechanisms of 25-anhydrocimigenol-3-O-beta-D-xylopyranoside isolated from *Souliea vaginata* on hepatomas. Anticancer Drugs 17 (5), 545–551.

Tong, X., Han, L., Duan, H., et al., 2017. The derivatives of Pulsatilla saponin A, a bioactive compound from *Pulsatilla chinensis*: their synthesis, cytotoxicity, haemolytic toxicity and mechanism of action. Eur. J. Med. Chem. 129, 325–336.

Torres, M.P., Ponnusamy, M.P., Chakraborty, S., et al., 2010. Effects of thymoquinone in the expression of mucin 4 in pancreatic cancer cells: implications for the development of novel cancer therapies. Mol. Cancer Ther. 9, 1419–1431.

Tsang, C.M., Cheung, K.C., Cheung, Y.C., et al., 2015. Berberine suppresses Id-1 expression and inhibits the growth and development of lung metastases in hepatocellular carcinoma. Biochim. Biophys. Acta 1852 (3), 541–551.

Wada, K., Nihira, M., Ohno, Y., 2006. Effects of chronic administrations of aconitine on body weight and rectal temperature in mice. J. Ethnopharmacol. 105 (1–2), 89–94.

Wada, K., Ohkoshi, E., Zhao, Y., et al., 2015. Evaluation of *Aconitum* diterpenoid alkaloids as antiproliferative agents. Bioorg. Med. Chem. Lett. 25 (7), 1525–1531.

Wang, T., Gong, F., Zhang, R., 2016b. Pulsatilla saponin A induces differentiation in acute myeloid leukemia in vitro. Hematology 21 (3), 182–186.

Wang, D., Han, L., Guo, Z., 2011a. Quinone reductase inducing activity of the dichloromethane/ethanol extract of the roots of *Pulsatilla chinensis*. Nat. Prod. Commun. 6 (6), 799–802.

Wang, J., Liu, Q., Yang, Q., 2012. Radiosensitization effects of berberine on human breast cancer cells. Int. J. Mol. Med. 30 (5), 1166–1172.

Wang, Y., Tang, H., Zhang, Y., et al., 2013. Saponin B, a novel cytostatic compound purified from *Anemone taipaiensis*, induces apoptosis in a human glioblastoma cell line. Int. J. Mol. Med. 32 (5), 1077–1084.

Wang, S., Tian, Q., An, F., 2016a. Growth inhibition and apoptotic effects of total flavonoids from *Trollius chinensis* on human breast cancer MCF-7 cells. Oncol. Lett. 12 (3), 1705–1710.

Wang, H., Zhang, F., Ye, F., et al., 2011b. The effect of *Coptis chinensis* on the signaling network in the squamous carcinoma cells. Front. Biosci. 3, 326–340.

Woo, C.C., Hsu, A., Kumar, A.P., et al., 2013. Thymoquinone inhibits tumor growth and induces apoptosis in a breast cancer xenograft mouse model: the role of p38 MAPK and ROS. PLoS One 8 (10), e75356.

Wu, D., Yao, Q., Chen, Y., et al., 2016. The in vitro and in vivo antitumor activities of tetracyclic triterpenoids compounds actein and 26-deoxyactein isolated from rhizome of *Cimicifuga foetida* L. Molecules 21 (8) doi: 10.3390/molecules21081001, pii: E1001.

Xi, R., Wang, L.J., 2017. Actein ameliorates hepatobiliary cancer through stemness and p53 signaling regulation. Biomed. Pharmacother. 88, 242–251.

Xiao, P.G., 1980. A preliminary study of the correlation between phylogeny, chemical constituents and pharmaceutical aspects in the taxa of Chinese Ranunculaceae. Acta Phytotax. Sin. 18 (2), 142–153.

Xiao, P.G., Wang, W.C., 1964. A new genus of Ranunculaceae—*Dichocarpum* W. T. Wang Et Hsiao. Acta Phytotax. Sin. 9 (4), 315–334.

Xiao, P.G., Wang, L.W., Lv, S.J., et al., 1986. Statistical analysis of the ethnopharmacologic data based on Chinese medicinal plants by electronic computer I. Magnoliidae. Chin. J. Mod. Dev. Trad. Med. 6 (4), 253–256.

Xu, X.F., Chen, Y.H., Wang, J., et al., 2010. Effect and mechanism of *Aconitum vaginatum* on the proliferation, invasion and metastasis in human A549 lung carcinoma cells. Zhong Yao Cai 33 (12), 1909–1912.

Xu, K., Shu, Z., Xu, Q.M., et al., 2013. Cytotoxic activity of *Pulsatilla chinensis* saponins and their structure-activity relationship. J. Asian Nat. Prod. Res. 15 (6), 680–686.

Xue, G., Zou, X., Zhou, J.Y., et al., 2013. Raddeanin A induces human gastric cancer cells apoptosis and inhibits their invasion in vitro. Biochem. Biophys. Res. Commun. 439 (2), 196–202.

Yang, Z.C., Ma, J., 2016. Actein enhances TRAIL effects on suppressing gastric cancer progression by activating p53/Caspase-3 signaling. Biochem. Biophys. Res. Commun. doi: 10.1016/j.bbrc.2016.11.162.

Yang, Y., Zhang, L.H., Yang, B.X., et al., 2015. Aurantiamide acetate suppresses the growth of malignant gliomas in vitro and in vivo by inhibiting autophagic flux. J. Cell. Mol. Med. 19 (5), 1055–1064.

Yi, T., Cho, S.G., Yi, Z., et al., 2008. Thymoquinone inhibits tumor angiogenesis and tumor growth through suppressing AKT and extracellular signal-regulated kinase signaling pathways. Mol. Cancer Ther. 7, 1789–1796.

Yip, N.K., Ho, W.S., 2013. Berberine induces apoptosis via the mitochondrial pathway in liver cancer cells. Oncol. Rep. 30 (3), 1107–1112.

Yu, H., Ji, X., Wu, Z., et al., 2011. Effects of aconite root on energy metabolism and expression of related genes in rats. Zhongguo Zhong Yao Za Zhi 36 (18), 2535–2538.

Yue, G.G., Xie, S., Lee, J.K., et al., 2016. New potential beneficial effects of actein, a triterpene glycoside isolated from Cimicifuga species, in breast cancer treatment. Sci. Rep. 6, 35263.

Zhang, Y., Bao, J., Wang, K., et al., 2015. Pulsatilla saponin D inhibits autophagic flux and synergistically enhances the anticancer activity of chemotherapeutic agents against HeLa cells. Am. J. Chin. Med. 43 (8), 1657–1670.

Zhang, H., Guo, Z., Han, L., et al., 2014a. The antitumor effect and mechanism of taipeinine A, a new C19-diterpenoid alkaloid from *Aconitum taipeicum*, on the HepG2 human hepatocellular carcinoma cell line. J. BUON 19 (3), 705–712.

Zhang, X., Li, D., Xue, X., et al., 2018. First total synthesis of a novel amide alkaloid derived from *Aconitum taipeicum* and its anticancer activity. Nat. Prod. Res. 32 (2), 128–132.

Zhang, L., Miao, X.J., Wang, X., et al., 2016a. Antiproliferation of berberine is mediated by epigenetic modification of constitutive androstane receptor (CAR) metabolic pathway in hepatoma cells. Sci. Rep. 6, 28116.

Zhang, L.L., Si, J.Y., Zhang, L.J., et al., 2016b. Synergistic anti-tumor activity and mechanisms of total glycosides from *Cimicifuga dahurica* in combination with cisplatin. Chin. J. Integr. Med.

Zhang, X.Z., Wang, L., Liu, D.W., et al., 2014b. Synergistic inhibitory effect of berberine and d-limonene on human gastric carcinoma cell line MGC803. J. Med. Food 17 (9), 955–962.

Zhang, Y., Wu, W., Kang, L., et al., 2017. Effect of *Aconitum coreanum* polysaccharide and its sulphated derivative on the migration of human breast cancer MDA-MB-435s cell. Int. J. Biol. Macromol. 103, 477–483.

Zhang, X., Zhao, M., Chen, L., et al., 2011. A triterpenoid from *Thalictrum fortunei* induces apoptosis in BEL-7402 cells through the P53-induced apoptosis pathway. Molecules 16 (11), 9505–9519.

Zhao, B., Hou, X.D., Li, H., et al., 2014c. Comparative study of Coptidis Rhizoma and Aconiti Kusnezoffii Radix on cell differentiation in lewis lung cancer. Zhongguo Zhong Yao Za Zhi 39 (14), 2732–2738.

Zhao, M., Ma, N., Qiu, F., et al., 2014a. Triterpenoid saponins from the roots of *Clematis argentilucida*. Fitoterapia 97, 234–240.

Zhao, M., Ma, N., Qiu, F., et al., 2014b. Triterpenoid saponins from the roots of *Clematis argentilucida* and their cytotoxic activity. Planta Med. 80 (11), 942–948.

Zheng, Q., Zhao, Y., Wang, J., et al., 2014. Spectrum-effect relationships between UPLC fingerprints and bioactivities of crude secondary roots of *Aconitum carmichaelii* Debeaux (Fuzi) and its three processed products on mitochondrial growth coupled with canonical correlation analysis. J. Ethnopharmacol. 153 (3), 615–623.

Zhou, X., Gan, P., Hao, L., et al., 2014. Antiinflammatory effects of orientin-2″-O-galactopyranoside on lipopolysaccharide-stimulated microglia. Biol. Pharm. Bull. 37 (8), 1282–1294.

Zhou, F., Lv, O., Zheng, Y., et al., 2012. Inhibitory effect of *Pulsatilla chinensis* polysaccharides on glioma. Int. J. Biol. Macromol. 50 (5), 1322–1326.

Zhu, Y., Ma, N., Li, H.X., et al., 2014. Berberine induces apoptosis and DNA damage in MG-63 human osteosarcoma cells. Mol. Med. Rep. 10 (4), 1734–1738.

Zhu, D.F., Zhu, G.L., Kong, L.M., et al., 2015. Cycloartane Glycosides from the roots of *Cimicifuga foetida* with Wnt signaling pathway inhibitory activity. Nat. Prod. Bioprospect. 5 (2), 61–67.

Zubair, H., Khan, H.Y., Sohail, A., et al., 2013. Redox cycling of endogenous copper by thymoquinone leads to ROS-mediated DNA breakage and consequent cell death: putative anticancer mechanism of antioxidants. Cell Death Dis. 4, e660.

Biodiversity, Chemodiversity, and Pharmacotherapy of *Thalictrum* Medicinal Plants

7.1 INTRODUCTION

The genus *Thalictrum* belongs to the subfamily Thalictroideae of Ranunculaceae and consists of the subgenera *Thalictrum* and *Lecoyerium* (Zhu and Xiao, 1989). Morphologically the former subgenus has four sections: Sect. *Leptostigma*(I), Sect. *Tripterium*(II), Sect. *Thalictrum*(III), and Sect. *Schlagintweitella*(IV). This genus has around 200 species, which are distributed in Asia, Europe, Africa, North America, and South America. Around 67 species are recorded in the flora of China and are distributed in every province and autonomous region of China, the majority of which are found in Southwest China. At least 43 species of this genus are used medicinally in China (Chen et al., 2003), many of which are used in clearing heat, dispelling dampness, diaphoresis, dysentery, bloodshot eyes, etc. (Table 7.1). In folk medicine, 14 species

Ranunculales Medicinal Plants. http://dx.doi.org/10.1016/B978-0-12-814232-5.00007-1

TABLE 7.1   A Brief Summary of Chinese *Thalictrum*

Species	Haploid chromosome no.	Medicinal part	Therapeutic efficacy	Morphology group	Alkaloid type/other phytometabolites	Note
T. acutifolium		Whole plant	Clearing heat and promoting diuresis, subdue swelling and detoxicating, hepatitis, swollen yellow body, yellow eyes, unpenetrated measles, febrile convulsion, etc.	II	I, III, IV, IX	Root (Hua Qi Shen in Chinese) as a tonic in Guangdong
T. alpinum	7,11	Root, rhizome	Heat-clearing and damp-drying, antibacterial, dysentery	IV	I, III, IV, V, VI	
T. alpinum var. elatum		Whole plant	Heat-clearing and detoxicating, measles, malnutrition of children, infantile convulsion, infantile pneumonia	IV		Root as a tonic (Yunnan); whole plant for eye disease (Sichuan)
T. angustifolium		Root	Chest tightness with nausea			
		Root			Alkaloid	
T. aquilegifolium L. var. sibiricum	7,14	Root	Carbuncles, swollen boils, hepatitis with jaundice, diarrhea, etc.	II	I, V, IX	
T. atriplex		Root, whole plant	Infectious hepatitis, carbuncles, swollen boils, etc.	I	I, III, V, VII, VIII, kaempferol/glycoside, isoquercetin	
		Root			V	
T. baicalense	7	Root	Surrogate of *coptis*, heat-clearing and damp-drying, purging intense heat and detoxicating, clearing away heart fire and tranquillizing	II	I, II, III	
		Root, rhizome	Dysentery, fever, hot and humid, vomiting, infectious hepatitis, carbuncle, swollen boils, bloodshot eyes		I, V	Tu Huang Lian in Chinese folk medicine

T. baicalense var. megalostigma		Root, rhizome	Surrogate of *Coptis*, heat-clearing and damp-drying, purging intense heat and detoxicating, stomach heat syndrome, dysentery	II	III	Ma Wei Huang Lian in Chinese folk medicine
T. cirrhosum		Root, rhizome	Heat-clearing and intense heat-purging, dysentery, gastroenteritis, aphthous ulcer	III	III, V, VII	
T. cultratum	7,21	Root, rhizome	Surrogate of *coptis*, antipyretic, anti-inflammatory, dysentery, diarrhea, viral hepatitis, influenza, measles, carbuncles, swollen boils, bloodshot eyes	III	I, III, V, VI	Ma Wei Huang Lian/Cao Huang Lian in Chinese folk medicine
T. delavayi (Fig. 7.1)	14,21	Root, rhizome	Heat-clearing and detoxicating, toothache by wind-fire, eye pain, etc.	III	I, III, V, VII, IX	Ma Wei Huang Lian in Chinese folk medicine
		Root			III	
T. delavayi var. mucronatum		Root	Heat-clearing and damp-purging, subdue swelling and detoxicating, hepatitis, unpenetrated measles, erysipelas, poor digestion, hematemesis, etc.		I, III, V, VI, IX	
T. faberi		Root, rhizome	Surrogate of *coptis*, swelling eye pain, etc.	I		Ma Wei Huang Lian in Chinese folk medicine
T. fargesii		Whole plant, root	Heat-clearing and damp-purging, subdue swelling and detoxicating, poor digestion, hematemesis, lacquer sore, erysipelas, toothache, dermatitis, child smallpox	II		
T. fauriei					V, VI	
T. finetii	21	Root, whole plant	Surrogate of *coptis*, dysentery	III	I, III, V, VII	Root (Qian Li Ma in Chinese) for cooling and refreshing in Yunnan

TABLE 7.1 A Brief Summary of Chinese *Thalictrum* (*cont.*)

Species	Haploid chromosome no.	Medicinal part	Therapeutic efficacy	Morphology group	Alkaloid type/other phytometabolites	Note
T. flavum	14,42	Root	Insecticide	III	I, III, V, VI, VII, IX	
T. foetidum	7,21	Root, rhizome	Heat-clearing and damp-drying, purging intense heat and detoxicating, conjunctivitis, swelling eye pain, infectious hepatitis, carbuncles, swollen boils, etc.	III	I, II, III, V, VI, cycloartane triterpenoid saponin, oleanane saponin	
T. foliolosum		Root	Hepatitis, dysentery, bloodshot eyes, child fever, hard penetrated smallpox, etc.	III	I, III, IV, V, VI, IX	Ma Wei Huang Lian in Chinese folk medicine
T. fortunei		Root, whole plant	Surrogate of *coptis*, furunculosis, smallpox	I	Alkaloid	Ma Wei Huang Lian in Chinese folk medicine
		Aboveground part	Cytotoxic, anticancer		Cycloartane saponin	
		Aboveground part	Proapoptotic, anticancer		Triterpenoid	
		Aboveground part			Cycloartane saponin	
T. glandulosissimum		Hairy root, rhizome	Fever with irritability and fidet, dysentery, enteritidis, acute colitis, acute pharyngitis, bloodshot eyes, carbuncle, swollen boils, etc.	III	I, III, IV, V, VII	
T. glandulosissimum var. chaotungense		Root	Surrogate of *coptis*	III	III	Ma Wei Huang Lian in Chinese folk medicine

T. honanense		Whole plant, root	Carbuncles, swollen boils	III	III, V, VI	
T. ichangense	7	Root	Heat-clearing and detoxicating, expelling wind to relieve convulsion, child white aphthous ulcer, urticaria, toothache	II	I	
		Whole plant	Dispelling cold and purging wind damp, defogging eyes, edema, etc.	I	I	Whole plant (Yan Sao Ba in Chinese) for child convulsion in Guizhou
T. isopyroides	21	Root, rhizome	Dysentery	III	I, II, III, V, VII	
T. javanicum		Whole plant	Arthritis	I	I, III, IV, V, VI, IX	Tibetan medicine
		Root, rhizome	Surrogate of Coptis, antipyretic, damp-drying, traumatic injury, eye inflammation, etc.	III	III	Stem and leaf (Yang Bu Shi in Chinese) for tabes in Sichuan
T. longistylum		Root				
T. macrorhynchum		Whole plant	Influenza	I		
T. microgynum		Root	Cool-retreating and pyrolysis, traumatic injury, etc.	II	III, V, VII	
		Whole plant	Swelling yellow body, bloodshot eyes, traumatic injury			
T. minus	21(7–42)	Root, rhizome	Dysentery, diarrhea	III	I, II, III, V, VI, VII, VIII, essential oil, cycloartane saponin	

(Continued)

TABLE 7.1   A Brief Summary of Chinese *Thalictrum* (*cont.*)

Species	Haploid chromosome no.	Medicinal part	Therapeutic efficacy	Morphology group	Alkaloid type/other phytometabolites	Note
T. minus var. hypoleucum		Root	Heat-clearing and detoxicating, dispelling wind and relieving convulsion, toothaches, acute dermatitis, eczema, white mouth sore, etc.	III		
T. omeiense		Whole plant, root	Malaria chills and fever, damp, jaundice, bellyache, diarrhea, dysentery, etc.	I	IV, VI	Shui Huang Lian in Chinese folk medicine
T. orientale		Below-ground part				
T. petaloideum	7	Root	Hepatitis with jaundice, diarrhea, dysentery, exudative epidermitis	II	I, III, VII	Whole plant/root (Hua Qi Shen in Chinese) for wind syndrome in Sichuan
T. petaloideum var. supradecompositum				II		
T. przewalskii	7,35	Flower, fruit; Root, rhizome	Hepatitis, hepatomegaly, etc. Surrogate of *Coptis*, wind-dispelling	II	Alkaloid; III, V	Tu Huang Lian in Chinese folk medicine
T. ramosum		Whole plant	Heat-clearing and blood-cooling, bloodshot eyes, dysentery, hematemesis, jaundice, etc.	I	Berberine and others	Shui Huang Lian/ Tu Huang Lian in Chinese folk medicine
T. reniforme	14	Root	Anthrax, etc.	III	Alkaloid	

Species	No.	Part used	Medicinal use			Notes
T. reticulatum		Root	Heat-clearing and detoxicating, cough, common cold, etc.	II	I, V	Cao Huang Lian in Chinese folk medicine
T. rhynchocarpum	14	Stem bark, root	Antibacterial	Similar to *T. reniforme*, III?		
T. robustum		Leaf	Antibacterial	I		
T. rutifolium		Root	Dysentery, diarrhea	III		
		Root	Heat-clearing and intense heat-purging, damp-drying and detoxicating	III		
T. simplex	14,21,28,35	Root, rhizome	Antibacterial, dysentery	III	I, II, III, VII, IX	
		Aboveground part			VIII	
T. simplex var. brevipes		Whole plant	Jaundice, diarrhea, dysentery	III		Root and rhizome (Huang Jiao Ji in Chinese) for detumescence and dehumidification in Sichuan
		Flower, fruit	Hepatitis, hepatomegaly, etc.			
T. smithii		Whole plant	Surrogate of *T. Ramosum* and *T. Omeiense*, dizziness, bellyache, dysentery, etc.	subgenus *Lecoyerium*	I, III, V, VII	Shui Huang Lian in Chinese folk medicine
T. squamiferum		Whole plant	Fever, etc.	IV	Alkaloid	Tibetan medicine
T. squarrosum	21	Whole plant	Heat-clearing and detoxicating, invigorate stomach and relieve hyperacidity, diaphoresis	III	I, III, V, apigenin, cycloartane saponin, oleanane saponin	

(Continued)

TABLE 7.1 A Brief Summary of Chinese *Thalictrum* (*cont.*)

Species	Haploid chromosome no.	Medicinal part	Therapeutic efficacy	Morphology group	Alkaloid type/other phytometabolites	Note
		Aboveground part			Diterpene glycoside, cycloartane saponin	
T. trichopus		Root	High fever in children, convulsion, psumonia, etc.	III	Alkaloid	
T. tuberiferum		Root	Skin disease, edema, kidney stone	II	III	
T. umbricola		Root/whole plant	Antipyretic, traumatic injury	II		
T. uncinulatum		Whole plant	Early measles	I	I, III	
T. virgatum		Root, rhizome	Heat-clearing and detoxicating, blood-cooling and qi-regulating, chest and diaphragm fullness, bleeding hemorrhoid, hematemesis, urticaria	III	Alkaloid	Whole plant (Yin Yang He in Chinese) for regulating Yin and Yang in Yunnan
T. viscosum		Root	Fever with irritability and fidet, dysentery, diarrhea, bloodshot eyes, sore throat, carbuncles, swollen boils	III	III	
T. wangii		Root	Detoxicating, dysentery	II	Alkaloid	
T. wuyishanicum		Root		II		

Morphology groups: I. *Sect. Leptostigma;* II. *Sect. Tripterium;* III. *Sect. Thalictrum;* IV. *Sect. Schlagintweitella.* Alkaloid types: I, aporphine; II, oxoaporphine; III, oxyberberine; IV, oxyberberine; V, bisbenzylisoquinoline; VI, aporphine-benzylisoquinoline; VII, protopine; VIII, pavine; IX, others (phenanthrene, diterpene alkaloid, etc.).

are used as the surrogate of *Coptis* (Xiao and Wang, 1965), and the root and rhizome of *Thalictrum cultratum*, *Thalictrum foliolosum*, *Thalictrum glandulosissimum var. chaotungense*, *Thalictrum finetii*, and *Thalictrum baicalense* are rich in berberine, which warrant further studies. The root and whole plant of *Thalictrum fortunei*, bitter and cold, are used in dispersing damp heat, subduing swelling and detoxicating, furunculosis, bleb, etc., and as an insecticide. *Thalictrum petaloideum*, bitter, cold, and nontoxic, is used for dispersing heat and eliminating wind, clearing stagnant heat, treating pediatric fever colds, and hard penetration of smallpox. *Thalictrum simplex*, bitter, cold, and nontoxic, is used for dispersing damp-heat, detoxification, jaundice, dysentery, asthma, measles complicated with pneumonia, rhinelcos, fever blisters, etc. In Ganzi of Sichuan Province, the whole plant of *Thalictrum squamiferum* is used by Tibetan doctors in the treatment of fever. Besides *Aconitum* (Hao et al., 2013) and *Delphinium*, *Thalictrum* is another large Ranunculaceae genus that is rich in alkaloids, but the diterpene alkaloid is rare in it and the benzylisoquinoline alkaloids (BIAs) are abundant. This chapter analyzes the molecular phylogeny of *Thalictrum* and summarizes the associated chemotaxonomic relationship. The pharmacophylogeny study contributes to the research and development of *Thalictrum* medicinal resources (Fig. 7.1).

7.2 CHEMICAL CONSTITUENTS

7.2.1 Alkaloids

Unlike *Papaver*, the phloem of *Thalictrum* has no laticiferous tube, and alkaloids are commonly accumulated in the root and rhizome (Lee et al., 2013). Around 200 BIAs are isolated from 30 Chinese *Thalictrum* species. The most common types are isoquinoline monomer and dimer, that is, B1 and B2 types. The classification of B1 is not disputed (Fig. 7.2): (1) Simple isoquinoline, (2) benzylisoquinoline, (3) aporphinoids, (4) protoberberine, (5) protopine, (6) pavine, (7) morphine, and (8) phenanthrene. B2 classification is as follows: (1) bisbenzylisoquinoline (BBI), (2) aporphine-benzylisoquinoline (ABI), (3) aporphine-pavine, (4) aporphine-substituted benzene, (5) protoberberine-benzylisoquinoline, and (6) protoberberine-aporphine. The connection location within the B2 dimer is variable, and

FIGURE 7.1   *Thalictrum delavayi*, taken in Lijiang Alpine Botanic Garden of Chinese Academy of Sciences.

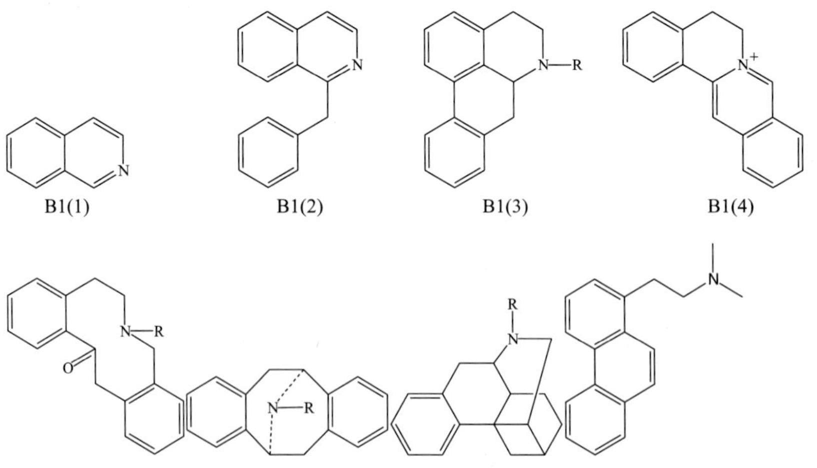

FIGURE 7.2 Skeleton of B1 alkaloids.

the regularity is not strong. The combination details of two monomers are shown in Fig. 7.3 and Table 7.2.

BBI is formed by the connection between two benzylisoquinoline units via oxygen bridges (Liu and Xiao, 1983). Two BI molecules can be connected by single, double, or triple-ether bond, but only the single- and double-ether bonds are found in *Thalictrum*. Among Chinese species, only *T. glandulosissimum* has BBI with a single-ether bond (Zhu and Xiao, 1991 a,b). The aporphine-type alkaloids are abundant in this genus, while there is not much oxoaporphine. The quaternary ammonium type is predominant in protoberberine. Oxoberberine, with the carbonyl group at C8, is found. The tetrahydroberberine derivatives are found in a few species.

The section I of subgenus *Thalictrum* has many dimer compounds, and ABI type has more than BBI type. The most abundant alkaloid of section II is of aporphine type, followed by BBI type. The morphologically primordial species of section III have less alkaloids, while the morphologically evolved species have more alkaloids. This section has more dimer compounds compared with other sections, and BBI type has significantly more than ABI type.

7.2.2 Flavonoids

Flavonoids are abundant in the leaves of *Thalictrum rhyncocarpum* (Mayeku et al., 2013) and many other species. Flavonoids are distributed extensively in *Thalictrum*, mostly in the form of flavone glycoside and flavonol glycoside (Fig. 7.3 and Table 7.1). Flavone O-glycosides such as apigenin, luteolin, and acacetin 7-O-glycosides are commonly found (Khamidullina et al., 2006). There are one or two sugar chains (bisdesmoside), and the second sugar chain binds to 4-phenolic hydroxyl group. The acetyl derivatives of flavonoid glycosides are only isolated from Asian *Thalictrum*. Flavone C-glycosides are only isolated from European *Thalictrum*. All flavonol glycosides found in this genus are 3-O-glycosides. New flavonoids are found in both alkaloid-abundant and alkaloid-absent species, indicating the nonalternate occurrence of these two compounds. Not surprisingly, phenylpropanoids and lignins are isolated from *T. baicalense*.

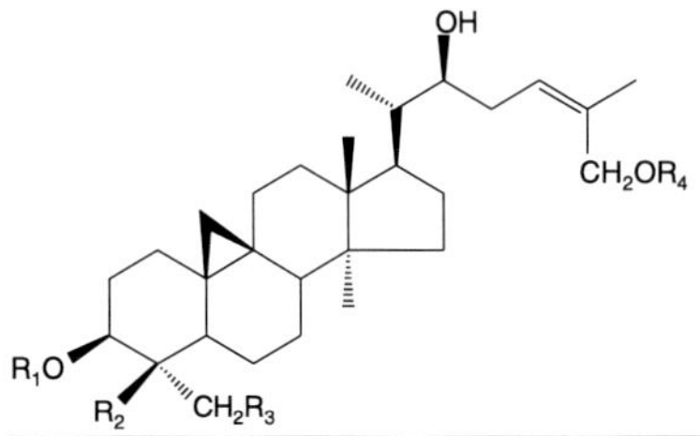

No.	R₁	R₂	R₃	R₄
1	β-D-Xyl-(1→6)-β-D-Glc-(1→4)-β-D-Fuc	CH₃	H	(6-O-acetyl)-αβ-D-Glc
2	α-L-Ara-(1→6)-β-D-Glc-(1→4)-β-D-Fuc	CH₃	H	(6-O-acetyl)-αβ-D-Glc
5	β-D-Glc-(1→4)-β-D-Fuc	CH₂OH	H	β-D-Glc
6	β-D-Glc-(1→4)-β-D-Fuc	CH₃	OH	β-D-Glc
7	β-D-Glc-(1→4)-β-D-Fuc	CH₃	H	β-D-Glc
8	β-D-Glc-(1→4)-β-D-Fuc	CH₃	H	α-L-Ara-(1→6)-β-D-Glc
9	β-D-Glc-(1→4)-β-D-Fuc	CH₃	H	β-D-Xyl-(1→6)-β-D-Glc
10	β-D-Glc-(1→4)-β-D-Fuc	CH₃	H	β-D-Qui-(1→6)-β-D-Glc
15	β-D-Glc-(1→4)-β-D-Fuc	CH₃	H	H
16	β-D-Glc-(1→6)-β-D-Glc-(1→4)-β-D-Fuc	CH₃	H	H
17	β-D-Qui-(1→6)-α-L-Rha-(1→2)-β-D-Glc-(1→4)-β-D-Fuc	CH₃	H	H
18	β-D-Glc	CH₃	H	β-D-Glc
19	β-D-Glc-(1→6)-β-D-Glc	CH₃	H	β-D-Glc
20	β-D-Glc-(1→4)-β-D-Fuc	CH₃	H	β-D-Glc
21	β-D-Glc-(1→4)-β-D-Fuc	CH₃	H	β-D-Qui-(1→6)-β-D-Glc
22	β-D-Glc-(1→4)-β-D-Fuc	CH₃	H	β-D-Qui-(1→6)-β-D-Glc
23	β-D-Glc-(1→4)-β-D-Fuc	CH₃	H	α-L-Ara-(1→6)-β-D-Glc

FIGURE 7.3 Examples of *Thalictrum* medicinal compounds (not including essential oils no. 78–89).

7.2.3 Triterpenoid Saponins

Cycloartane-type tetracyclic triterpenoids (Hao et al., 2015; Jiang et al., 2017) and oleanane-type pentacyclic triterpenoids (Fig. 7.3 and Table 7.1) are found in *Thalictrum*, which were initially isolated from *Thalictrum minus* and *Thalictrum foetidum* of east Siberia. According to the side-chain structure, cycloartane triterpenoids are divided into four groups: C-17 acyclic side chain, C-17THF (tetrahydrofuran) ring, C-20 THF ring, and C-17 cyclopentane ring. C-17

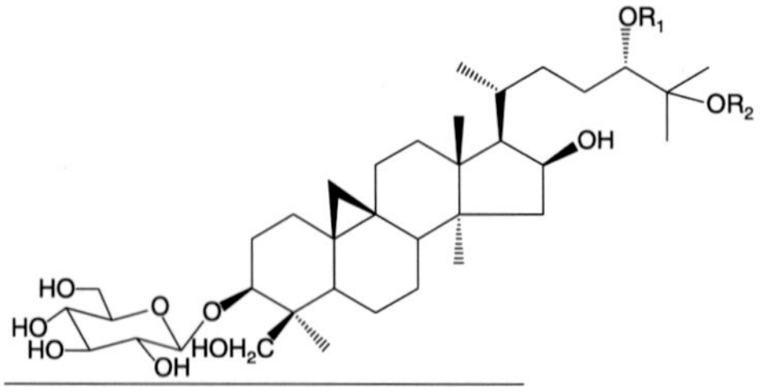

No.	R_1	R_2
3	H	β-D-Glc-(1→6)-β-D-Glc
4	H	β-D-Glc-(1→4)-β-D-Glc
11	H	β-D-Glc
12	Ac	β-D-Glc
13	Ac	β-D-Glc-(1→6)-β-D-Glc
14	Ac	β-D-Glc-(1→4)-β-D-Glc

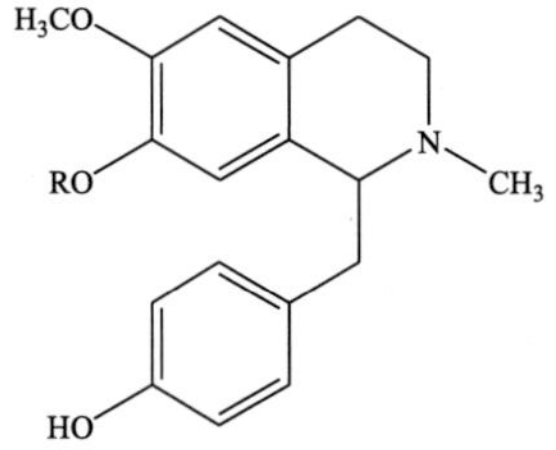

No.	R
24	CH3
65	H

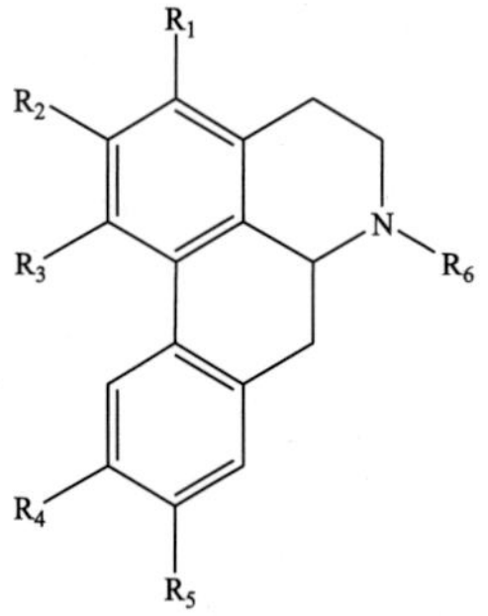

FIGURE 7.3   *(Cont.)*

No.	R_1	R_2	R_3	R_4	R_5	R_6
25	OCH_3	OCH_3	OH	OCH_3	OCH_3	CH_3
26	OCH_3	R_2+	$R_3=$	OCH_3	OCH_3	H
		OCH_2O				
62	H	OCH_3	OCH_3	OCH_3	OH	CH_3
63	H	OCH_3	OH	OCH_3	OH	CH_3
64	OCH_3	OCH_3	OCH_3	OCH_3	OH	CH_3
67	H	OCH_3	OCH_3	R_4 + R_5 =		CH_3
				OCH_2O		

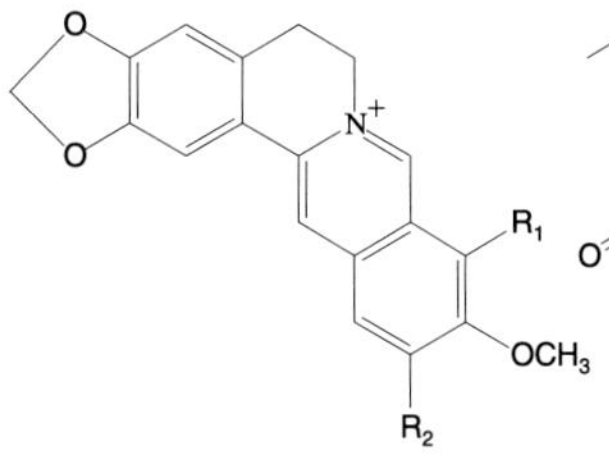

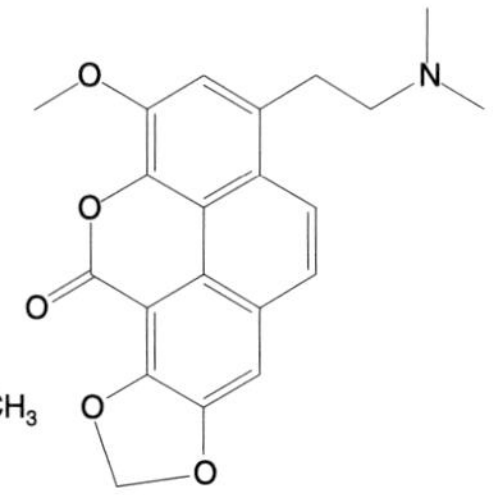

29

No.	R_1	R_2
27	OCH_3	H
28	H	OCH_3

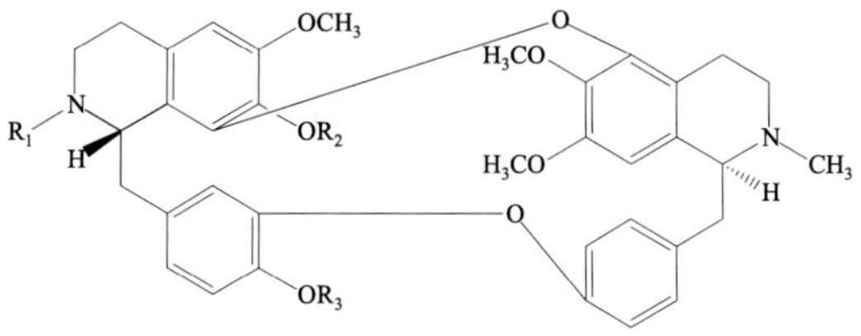

FIGURE 7.3 (*Cont.*)

No.	R₁	R₂	R₃
30	H	CH_3	CH_3
31	H	H	CH_3
32	CH_3	CH_3	H
33	H	CH_3	H
34	CH_3	H	H

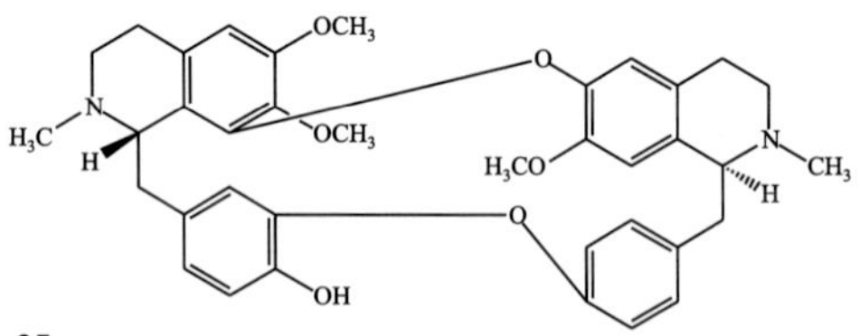

35

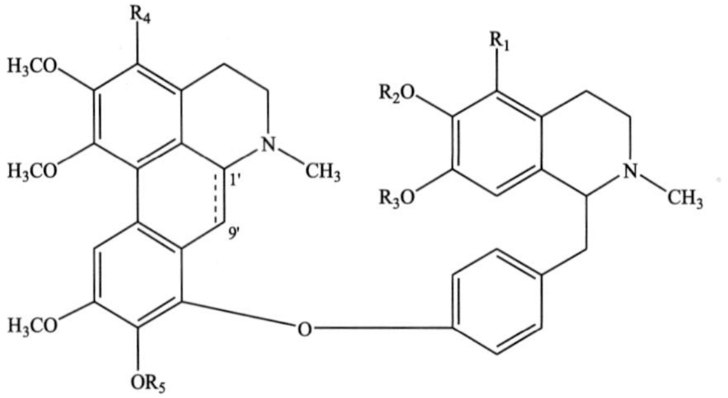

No.	R₁	R₂	R₃	R₄	R₅
36	CH_3	H	CH_3	OCH_3	$CH_3 \Delta 1'(9')$
37	H	H	H	OCH_3	CH_3
38	CH_3	H	CH_3	OCH_3	CH_3
51	H	H	CH_3	OH	CH_3
52	H	CH_3	CH_3	OH	H
53	H	H	CH_3	H	H
54	OH	CH_3	CH_3	OH	H
55	OH	CH_3	CH_3	H	H
56	H	H	CH_3	OH	H

FIGURE 7.3 *(Cont.)*

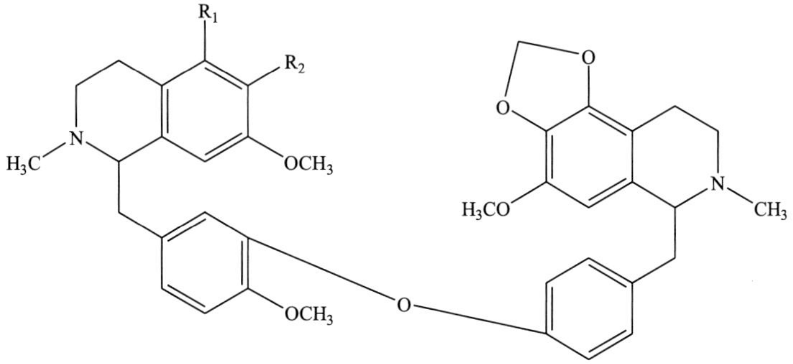

No.	R_1	R_2
39	OH	OCH_3
40	H	OCH_3
42	R_1 + R_2 =	
	OCH_2O	

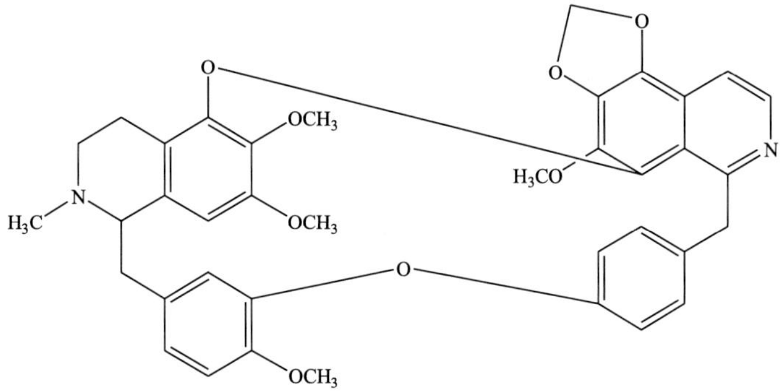

41

FIGURE 7.3 *(Cont.)*

acyclic side chain type is found in *T. minus*, *T. foetidum*, *Thalictrum smithii*, *Thalictrum foenicula-ceum*, *Thalictrum squarrosum*, etc. C-17 THF ring is found in *T. squarrosum* and *Thalictrum herba*. Triterpenoids containing the C-20 THF ring are less abundant and are found in *T. minus*. C-17 cyclopentane ring is found in *Thalictrum thunbergii*. Cycloartane triterpenoids of this genus have the C-3β orientation of substituents, and the A-ring substitution often occurs in C-29 or-30.

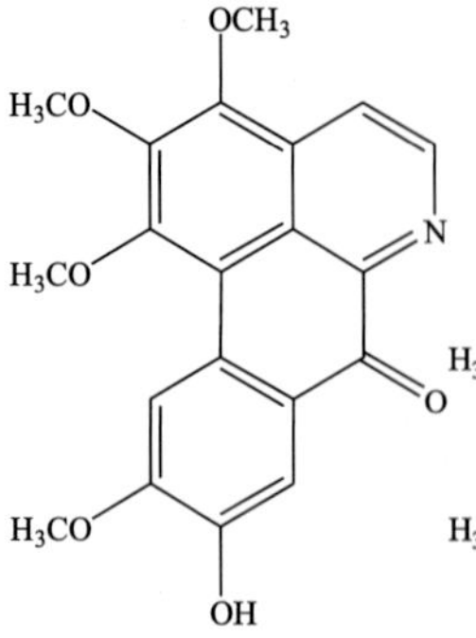

No.   R

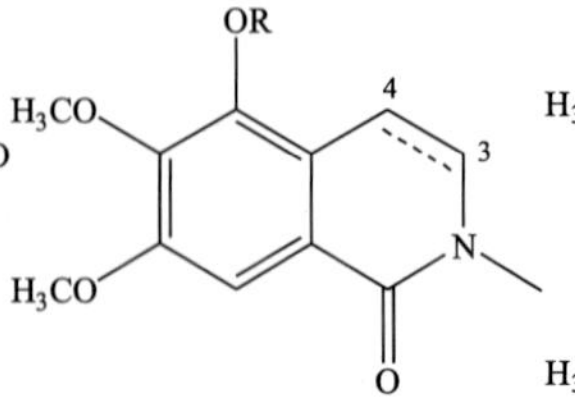

44   CH₃Δ1'(9')

45   H

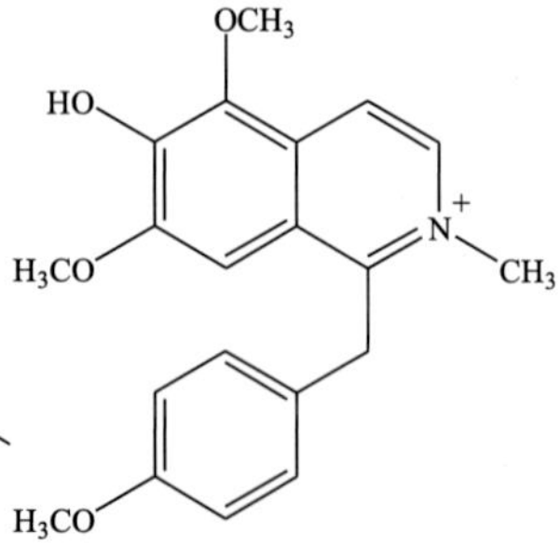

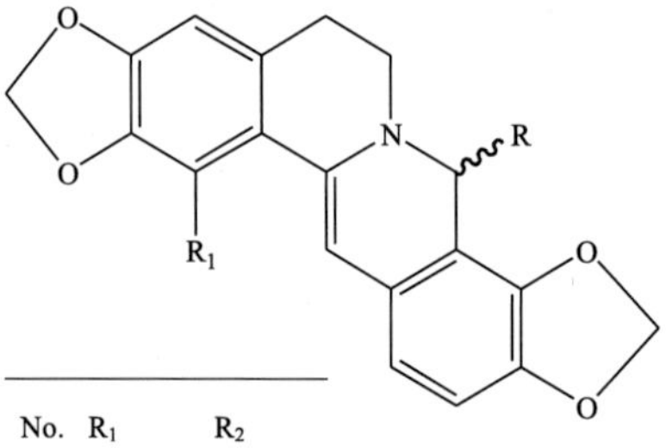

No.   R₁   R₂

47   H   =O

48   OH   OH

FIGURE 7.3   *(Cont.)*

The free pentacyclic triterpenoids, oleanolic acid and (2α,3β)-2,3-dihydroxyolean-12-en-28-oic acid are found in *Thalictrum*. The known sapogenins are oleanolic acid, 2α,3β-diacetoxy-30-hydroxyolean-24-en-28-oicacid, hederagenin, and 11α,12α-epoxyolean-28,13β-olid. The sugar chains are on C-3 and C-28. Triterpenoids are abundant in the leaves of *T. rhyncocarpum*.

7.2.4 Other Compounds

Protocatechuic acid, caffeic acid, coumaric acid, kaempferol, and β-sitosterol are isolated from the aboveground part of *Thalictrum atriplex* (Gao et al., 1999a). Dolabellane diterpene glycoside is found in *T. squarrosum* (Yoshimitsu et al., 2010). The essential oil of *T. minus*

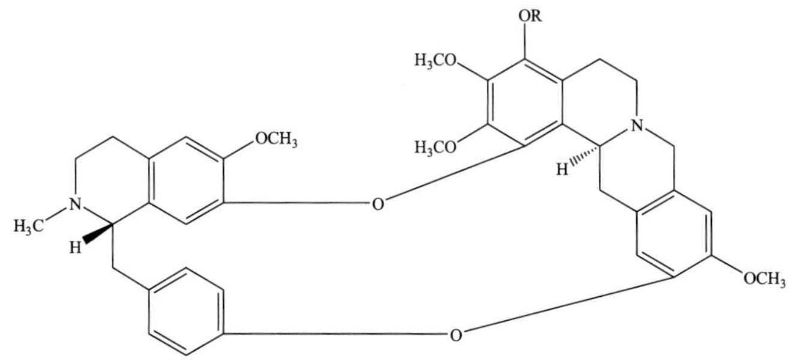

No.	R
49	H
50	CH₃

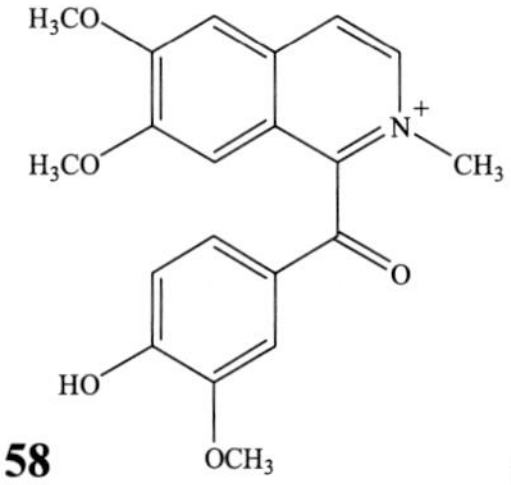

FIGURE 7.3 *(Cont.)*

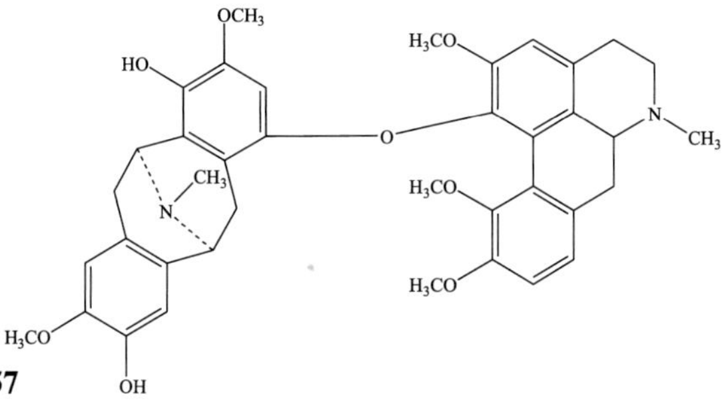

contains 12 compounds, including thymol (66.63%), para-cymene (13.05%), γ-terpinene (7.32%), carvacrol (3.70%), and 1,8-cineole (Taherpour and Maroofi, 2008). Twenty-one compounds are identified in the volatile oil of *Thalictrum ichangense* (Liu et al., 2013), for example, N-tetratetracontane, 3,5-dien-stigmastane, and octacosane. The phenolic compound thalictricoside and the cyanogenic glycoside lithospermoside are isolated from the below-ground part of *Thalictrum orientale* (Erdemgil et al., 2003).

7.3 BIOACTIVITY

7.3.1 Anticancer Activity

Dimeric alkaloids have intriguing structures and bioactivities; 11 thalifaberine-type aporphine-benzylisoquinoline alkaloids, thalicultratines A-K, a tetrahydroprotoberberine-aporphine alkaloid, thalicultratine L, and five other known ones have been isolated from the roots of *T. cultratum* (Li et al., 2017). Most alkaloids show potent cytotoxicity against human

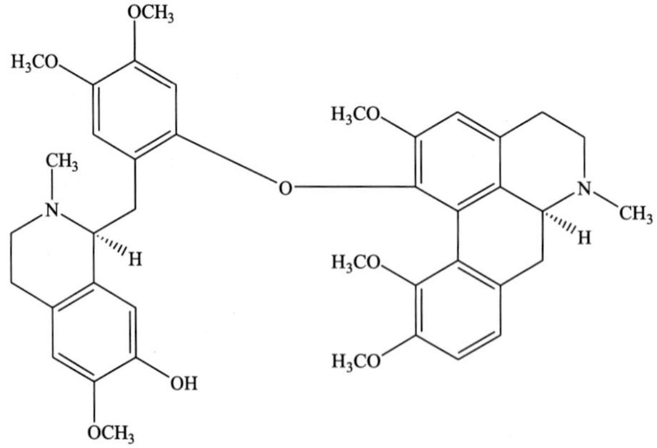

59

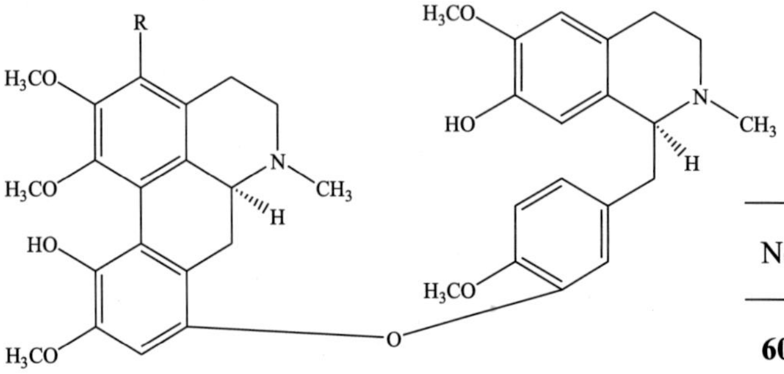

No.	R
60	H
61	OCH$_3$

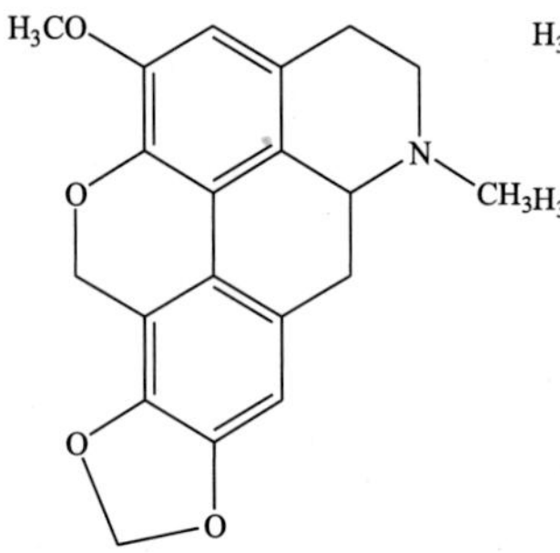

66 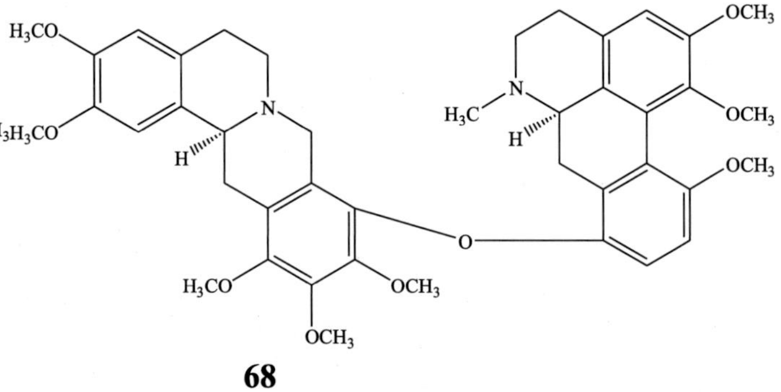   **68**

FIGURE 7.3   *(Cont.)*

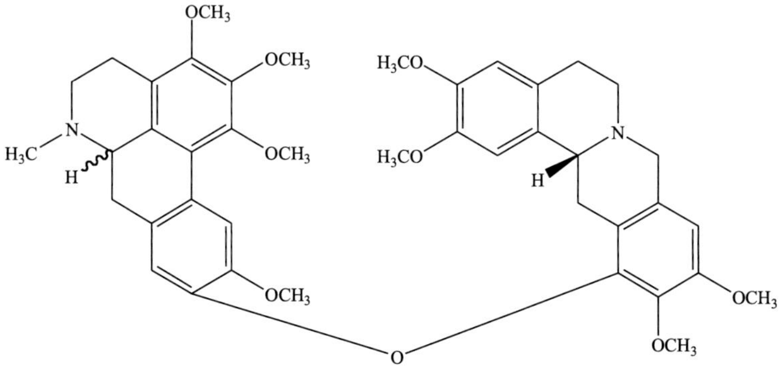

69

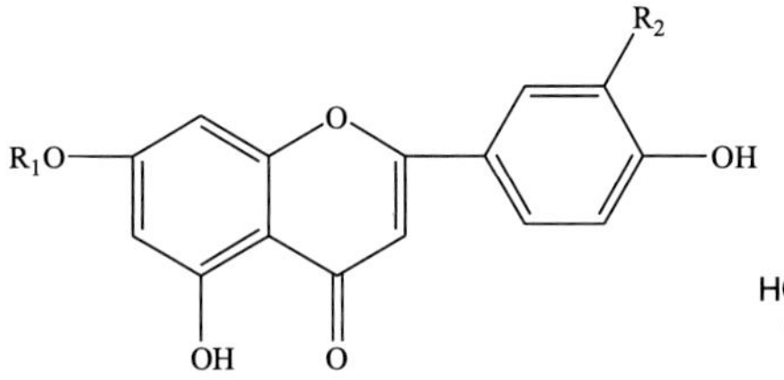

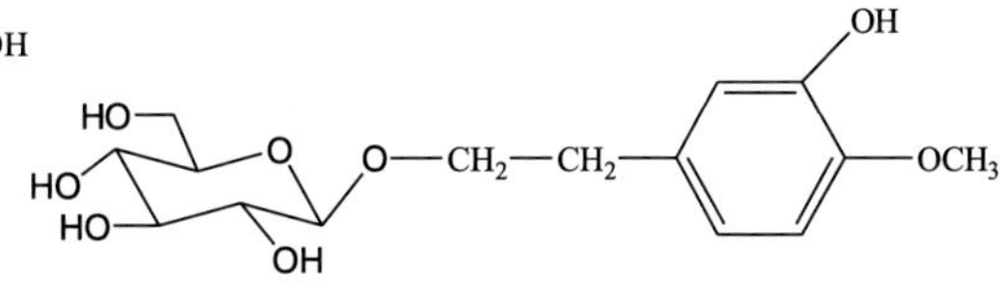

74

No.	R₁	R₂
70	β-D-Glc	H
71	α-L-Rha	H
72	β-D-Glc	OH
73	α-L-Rha	OH

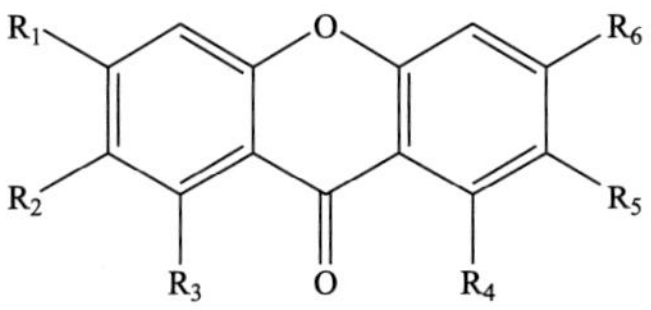

FIGURE 7.3 *(Cont.)*

No.	R_1	R_2	R_3	R_4	R_5	R_6
75	O-β-D-Glc	H	OH	H	OH	H
76	H	OH	H	OH	H	OCH$_3$

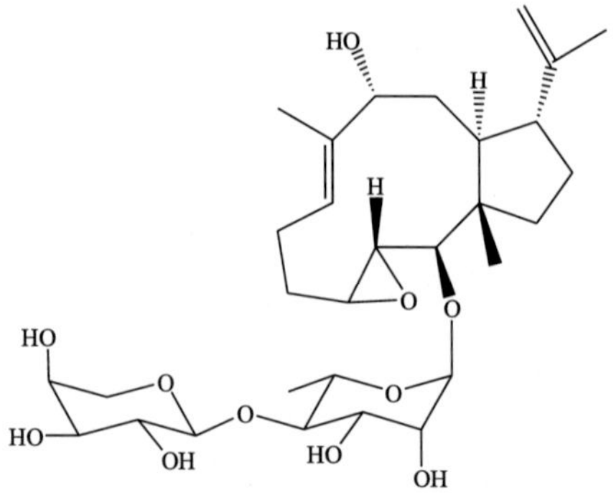

77

FIGURE 7.3   *(Cont.)*

leukemia HL-60 and prostate cancer PC-3 cells. The most potent one, compound 3, has an IC$_{50}$ of 1.06 μM against HL-60 cells. Compound 3 induces apoptosis and arrests the HL-60 cell cycle at the S phase with the loss of mitochondria membrane potential. The BIAs of *T. fortunei* exert anticancer effects via multiple mechanisms (Cheng et al., 2012). Hernandezine is an anticancer activated protein kinase (AMPK) activator (Song et al., 2017). Thaliblastine exhibits dose-dependent cytotoxic effects on HL-60, HL-60/DOX, RHE, and HD-MY-2 leukemia cells (Horvath et al., 2004), and the oligonucleosomal DNA fragmentation can be detected, suggesting apoptosis. Thaliblastine also exhibits an MDR (multidrug resistance)-phenotype reversing effect.

The BBI alkaloid acutiaporberine of *Thalictrum acutifolium* induces apoptosis of the metastatic lung cancer cells (Chen et al., 2002). The alkaloid fraction of *Thalictrum robustum* significantly inhibits HepG2 and Hela cells and induces apoptosis (Jiao et al., 2014).

Two rare chloro-containing BIAs, isolated from the whole plant of *T. foliolosum* (Li et al., 2016), exhibit moderate *in vitro* antiproliferative activity against MCF-7, PC-3, and HL-60 cells and good inhibitory effects against U937 cells. Two isoquinoline alkaloids show the strongest *in vitro* antiproliferative effects with IC$_{50}$ values of 0.93 and 1.69 μM against HL-60 cell line. Plants used for subduing and detoxicating carbuncles and boils can be a source of anticancer compounds (Table 7.1).

Cycloartane triterpenoids of *T. fortunei* had the concentration dependent cytotoxic effects in the human liver, colon and lung cancer cells (Zhang et al., 2012), and the P53-induced apoptosis pathway might be involved (Zhang et al., 2011). The volatile oil of *T. ichangense* inhibited the *in vitro* proliferation of gastric cancer cells.

TABLE 7.2 Examples of *Thalictrum* Flavonoids, Triterpenoids, Saponins, Alkaloids, and Other Compounds

No.	Species	Medicinal part	Compound	Type
1.	*T. fortunei*	Aboveground part	3-O-β-D-xylopyranosyl-(1→6)-β-D-glucopyranosyl-(1→4)-β-D-fucopyranosyl((22S,24Z)-cycloart-24-en-3β,22,26-triol 26-O-(6-O-acetyl)-β-D-glucopyranoside	A
2.		Aboveground part	3-O-α-L-arabinopyranosyl-(1→6)-β-D-glucopyranosyl-(1→4)-β-D-fucopyranosyl((22S,24Z)-cycloart-24-en-3β,22,26-triol26-O-(6-O-acetyl)-β-D-glucopyranoside	A
3.		Aboveground part	3-O-β-D-glucopyranosyl (24S)-cycloartane-3β,16β,24,25,30-pentaol 25-O-β-D-glucopyranosyl-(1→6)-β-D-glucopyranoside	A
4.		Aboveground part	3-O-β-D-glucopyranosyl (24S)-cycloartane-3β,16β,24,25,30-pentaol 25-O-β-D-glucopyranosyl-(1→4)-β-D-glucopyranoside	A
5.		Aboveground part	3-O-β-D-glucopyranosyl-(1→4)-β-D-fucopyranosyl-(22S,24Z)-cycloart-24-en-3β,22,26,30-tetraol 26-O-β-D-glucopyranoside	A
6.		Aboveground part	3-O-β-D-glucopyranosyl-(1→4)-β-D-fucopyranosyl-(22S,24Z)-cycloart-24-en-3β,22,26,29-tetraol 26-O-β-D-glucopyranoside	A
7.		Aboveground part	3-O-β-D-glucopyranosyl-(1→4)-β-D-fucopyranosyl-(22S,24Z)-cycloart-24-en-3β,22,26-triol 26-O-β-D-glucopyranoside	A
8.		Aboveground part	3-O-β-D-glucopyranosyl-(1→4)-β-D-fucopyranosyl-(22S,24Z)-cycloart-24-en-3β,22,26-triol 26-O-α-L-arabinopyranosyl-(1→6)-β-D-glucopyranoside	A
9.		Aboveground part	3-O-β-D-glucopyranosyl-(1→4)-β-D-fucopyranosyl-(22S,24Z)-cycloart-24-en-3β,22,26-triol 26-O-β-D-xylopyranosyl-(1→6)-β-D-glucopyranoside	A
10.		Aboveground part	3-O-β-D-glucopyranosyl-(1→4)-β-D-fucopyranosyl-(22S,24Z)-cycloart-24-en-3β,22,26-triol 26-O-β-D-quinovopyranosyl-(1→6)-β-D-glucopyranoside	A
11.		Aboveground part	(24S)-cycloartane-3β,16β,24,25,30-pentaol 3,25-di-O-β-glucopyranoside	A
12.		Aboveground part	24-O-acetyl-(24S)-cycloartane-3β,16β,24,25,30-pentaol3,25-di-O-β-glucopyranoside	A
13.		Aboveground part	3-O-β-D-glucopyranosyl-24-O-acetyl-(24S)-cycloartane-3β,16β,24,25,30-pentaol 25-O-β-D-glucopyranosyl(1→6)-β-D-glucopyranoside	A
14.		Aboveground part	3-O-β-D-glucopyranosyl-(24S)-cyclo-artane-3β,16β,24,25,30-pentaol 25-O-β-D-glucopyranosyl-(1→4)-β-D-glucopyranoside	A
15.	*T. squarrosum*	Aboveground part	Squarroside V	A
16.		Aboveground part	Squarroside VI	A
17.		Aboveground part	Squarroside VII	A

(Continued)

TABLE 7.2 Examples of *Thalictrum* Flavonoids, Triterpenoids, Saponins, Alkaloids, and Other Compounds (*cont.*)

No.	Species	Medicinal part	Compound	Type
18.	*T. fortunei*	Aboveground part	Thaliforoside I	A
19.		Aboveground part	Thaliforoside J	A
20.		Aboveground part	3-O-β-D-glucopyranosyl-(1→4)-β-D-fucopyranosyl-(22S,24Z)-cycloart-24-en-3β,22,26-triol 26-O-β-D-glucopyranoside	A
21.		Aboveground part	3-O-β-D-glucopyranosyl-(1→4)-β-D-fucopyranosyl-(22S,24Z)-cycloart-24-en-3β,22,26-triol 26-O-β-D-quinovopyranosyl-(1→6)-β-D-glucopyranoside	A
22.		Aboveground part	3-O-β-D-glucopyranosyl-(1→4)-β-D-fucopyranosyl-(22S,24Z)-cycloart-24-en-3β,22,26-triol 26-O-β-D-xylopyranosyl-(1→6)-β-D-glucopyranoside	A
23.		Aboveground part	3-O-β-D-glucopyranosyl-(1→4)-β-D-fucopyranosyl-(22S,24Z)-cycloart-24-en-3β,22,26-triol 26-O-α-L-arabinopyranosyl-(1→6)-β-D-glucopyranoside	A
24.	*T. flavum*	Root	(−)-Armepavine[a]	B1(2)
25.		Root	(+)-Preocoteine[a]	B1(3)
26.		Root	(+)-O-methylcassythine[a]	B1(3)
27.	*T. flavum/T. elegans/T. delavayi*	Root/whole plant/whole plant	Berberine[a]	B1(4)
28.		Root	Pseudoberberine[a]	B1(4)
29.		Root	Thaliglucinone[a]	B1(8)
30.		Root	(−)-Northalidasine[a]	B2(1)
31.		Root	(−)-Northalrugosidine[a]	B2(1)
32.		Root	(−)-Thalfoetidine[a]	B2(1)
33.		Root	(−)-Northalfoetidine	B2(1)
34.		Root	(−)-Thaligosidine[a]	B2(1)
35.		Root	(+)-Thalicberine[a]	B2(1)
36.	*T. atriplex*	Root	Dehydrothalifaberine[a]	B2(2)
37.		Root	Thalifarentine[a]	B2(2)
38.		Root	Thalifaberine[a]	B2(2)
39.		Root	Thalistine[a]	B2(1)

40.		Root	Thaliracebine[a]	B2(1)
41.		Root	Neothalfine	B2(1)
42.		Root	Thaliatrine	B2(1)
43.	T. elegans	Whole plant	1,2,3,10-tetramethoxy-9-hydroxy-4,5,6a-dehydro-7-aporphinone	B1(3)
44.	T. elegans/T. delavayi	Whole plant	Thalactamine[a]	B1(1)
45.	T. delavayi	Whole plant	5-hydro-N-methylcorydalidine	B1(1)
46.		Whole plant	1-(4-methoxybenzyl)-2-N-methyl-6-hydroxyl-5,7-dimethoxy-isoquinoline	B1(2)
47.		Root	2,3,9,10-dimethylenedioxy-8-oxoprotoberberine	B1(4)
48.		Root	2,3,9,10-dimethylenedioxy-1,8-dihydroxylprotoberberine	B1(4)
49.	T. longistylum	Root	Longiberine	B2(5)
50.		Root	O-methyllongiberine	B2(5)
51.	T. faberi	Root	3-hydroxy-6′-desmethyl-9-O-methylthalifaboramine	B2(2)
52.		Root	3-hydroxythalifaboramine	B2(2)
53.		Root	6′-desmethylthalifaboramine	B2(2)
54.		Root	3,5′-dihydroxythalifaboramine	B2(2)
55.		Root	5′-hydroxythalifaboramine	B2(2)
56.		Root	3-hydroxy-6′-desmethylthalifaboramine	B2(2)
57.	T. przewalskii	Root	Thalprzewalskiinone	B1(2)
58.	T. fauriei	Whole plant	Fauripavine	B2(3)
59.		Whole plant	Fauridine	B2(2)
60.		Whole plant	Faurithaline	B2(2)
61.		Whole plant	3-methoxyfaurithaline	B2(2)
62.	T. atriplex	Aboveground part	N-methyllaurotetanine[a]	B1(3)
63.		Aboveground part	Isoboldine[a]	B1(3)
64.		Aboveground part	Thalisopynine[a]	B1(3)
65.		Aboveground part	N-methylcocularine[a]	B1(2)
66.		Aboveground part	Demethylthalphenine[a]	B1(3)

(Continued)

TABLE 7.2 Examples of *Thalictrum* Flavonoids, Triterpenoids, Saponins, Alkaloids, and Other Compounds (*cont.*)

No.	Species	Medicinal part	Compound	Type
67.		Aboveground part	Nantenine[a]	B1(3)
68.	*T. acutifolium*	Root	Acutiaporberine	B2(6)
69.	*T. wangii*	Root	Thalibealine	B2(6)
70.	*T. fortunei*	Whole plant	Apigenin-7-O-β-D-glucopyranoside[a]	C
71.		Whole plant	Apigenin-7-O-α-L-rhamnopyranoside[a]	C
72.		Whole plant	Luteolin-7-O-β-D-glucopyranoside[a]	C
73.		Whole plant	Luteolin-7-O-α-L-rhamnopyranoside[a]	C
74.		Whole plant	2-(3-hydroxy-4-methoxyphenyl) ethyl 1-O-β-D-glucopyranoside[a]	D
75.		Whole plant	Xanthohypericoside[a]	E
76.		Whole plant	Gentisin[a]	E
77.	*T. squarrosum*	Aboveground part	Squoside A	F
78.	*T. minus*	Aboveground part	Pinene	G
79.		Aboveground part	Camphene	G
80.		Aboveground part	Myrcene	G
81.		Aboveground part	α-Terpinene	G
82.		Aboveground part	Para-cymene	G
83.		Aboveground part	1,8-cineole	G
84.		Aboveground part	γ-Terpinene	G
85.		Aboveground part	Exo-(1,7,7-trimetyl)-bicyclo[2,2,1]heptan-2-ol	G
86.		Aboveground part	Thymol	G
87.		Aboveground part	Carvacrol	G
88.		Aboveground part	Trans-caryophyllene	G
89.		Aboveground part	Bisabolene	G

A, triterpenoid; the predominant triterpenoid glycosides are of cycloartane type and oleanane type (Khamidullina et al., 2006), and only cycloartane type is listed here.
B, alkaloid; B1, isoquinoline monomer (Fig. 7.2); B1(1), simple isoquinoline, B1(2), benzylisoquinoline; B1(3), aporphinoid; B1(4), protoberberine; B1(5), protopine; B1(6), pavine; B1(7), morphine; B1(8), phenanthrene type; B2, isoquinoline dimer; B2(1), BBI; B2(2), ABI; B2(3), aporphine-pavine; B2(4), aporphine-aporphine; B2(5), protoberberine-benzylisoquinoline; B2(6), protoberberine-aporphine. C, flavonoid; two kinds of flavonoids are found, that is, flavone and flavonol. Only flavone type is listed here. D, phenolics; E, xanthones; F, diterpene glycosides; G, essential oils.
[a] Nonnovel compound, isolated from this species for the first time.

7.3.2 Antiviral Effects

The BIAs of *T. fortunei* have antiviral effects against herpesvirus, HIV, and influenza virus. The pavine alkaloid (–)-thalimonine (Thl) and its N-oxide, isolated from the Mongolian plant *T. simplex*, inhibits markedly the reproduction of the influenza virus in cell cultures of chicken embryo fibroblasts (Serkedjieva and Velcheva, 2003). At a nontoxic concentration range of 0.1–6.4 μM the alkaloid inhibits the viral reproduction in a selective and specific way. The expression of viral glycoproteins hemagglutinin (HA), neuraminidase and nucleoprotein on the surface of infected cells, virus-induced cytopathic effect, infectious virus yields, HA production, and virus-specific protein synthesis are all reduced. The inhibition is dose-related and dependant on virus inoculum. The viral reproduction is markedly inhibited when Thl is added at 4–5 h of infection. No inactivating effect on extracellular viruses is found. Thethalictuberine N-oxides have a slightly worse effect, while the synthetic analogs of thalimonine and the alkaloids of *T. foetidum* and *Thalictrum flavum* are ineffective.

7.3.3 Antibacterial, Antiinflammatory, and Antioxidant Activities

Staphylococcus xylosus, *Staphylococcus lentus*, *Staphylococcus equorum*, *Enterococcus faecalis*, and *Pantoea agglomerans* are identified as bovine mastitis pathogens. The *T. minus* extract exhibits broad spectrum antibacterial activities that vary between the bacterial species (minimal inhibitory concentration (MIC) 250–500 μg/ml) (Mushtaq et al., 2016). 5′-Hydroxythalidasine and thalrugosaminine shows promising antibacterial activity with MIC values of 64–128 μg/ml while *Staphylococcus* species are found to be the most sensitive strains. The *T. foliolosum* extracts show moderate to high activity against *Candida albicans*, *Staphylococcus aureus*, *Escherichia coli*, and *Pseudomonas aeruginosa* (Pandey et al., 2017). Extracts with high berberine content are most effective against *C. albicans* and *S. aureus*. Berberine is highly abundant in the root and rhizome of *T. cultratum*, *T. foliolosum*, *T. chaotungense*, *T. finetii*, and *T. baicalense* and its variety (Xiao and Wang, 1965). According to folk medicine, these species are of low toxicity and can be used against intestinal infection.

The cycloartane triterpenoid thalicoside A2, isolated from the aerial part of *T. minus*, has *in vitro* antibacterial and antifungal activities (Gromova et al., 2000). The stem bark and root of *T. rhyncocarpum* contain glycosides and alkaloids and are used in Kenya for antibacterial therapy. The BIAs of *T. fortunei* have antiinflammatory activities, and the involved mechanisms include the inhibitions of ROS generation, tumor necrosis factor and inducible NO synthase, and calcium channel blockage. Flavonoids in the rhizome of *T. petaloideum* might be antioxidant compounds (Aun Bulag et al., 2010). The volatile oil of *T. ichangense* has considerable antioxidant activity.

Bisbenzyltetrahydroisoquinoline alkaloids of *T. foliolosum* show the strongest antioxidant activities in 2,2′-azino-bis(3-ethylbenzothiazoline-6-sulphonic acid) (ABTS) assay (Li et al., 2016). The *T. foliolosum* extracts show relatively significant antilipid peroxidation, and β-carotene bleaching and reducing power (Pandey et al., 2017). *T. foliolosum* extracts with higher phenol and flavonoid content show better scavenging of 2,2-diphenyl-1-picrylhydrazyl (DPPH) free radicals.

7.3.4 Antiparasitic and Insecticidal Effects

Tertiary isoquinolines of *T. flavum*, particularly BBIs, are leishmanicidal against *Leishmania major* (Ropivia et al., 2010). Thalfoetidine appears as the most potent, but its new norderivative, northalfoetidine, as well as northalidasine, are of particular interest, since their potential leishmanicidal activity is not associated with a strong cytotoxicity, which is desirable for killing the parasite within macrophages. A BBI, isolated from the stem of *T. foliolosum*, exhibits almost complete inhibition of DNA topoisomerase IB of *Leishmania donovani* at 50 μM concentration (Kumar et al., 2016). It is effective in killing both wild-type and SAG-resistant promastigotes of the parasite.

The BBI northalrugosidine of *Thalictrum alpinum* has strong *in vivo* antileishmanial activity (Naman et al., 2015). Northalrugosidine shows potent *in vitro* activity against *L. donovani* promastigotes (0.28 μM) and the highest selectivity (29.3-fold) versus its general cytotoxicity against HT-29 human colon adenocarcinoma cells. It is tested *in vivo* using a murine model of visceral leishmaniasis, resulting in a dose-dependent reduction of the parasitic burden in the liver and spleen without overt toxicity effects at 2.8, 5.6, and 11.1 mg/kg per animal when administered intravenously.

T. foliolosum is used in Northeast India against malaria. The IC_{50} concentrations of its chloroform extracts for SS (chloroquine sensitive strain) and RS (chloroquine resistance strain) were 0.5 and 1.1 μg/ml respectively (Das et al., 2016).

Oleic acid and eicosyl ester of *Thalictrum javanicum* display mosquito larvicidal properties (Gurunathan et al., 2016).

7.3.5 Other Bioactivities

The BIAs of *T. fortunei* have antiarrhythmic, antiplatelet aggregation, and antihypertensive activities. They also have immunomodulatory effects. Neothalfine (BBI) of *T. atriplex* inhibits platelet aggregation (Gao et al., 2005).

Cycloartane triterpenoids have antifertility, anticancer, antimicrobial, antiinflammatory, immunomodulatory, cardiovascular, and hepatoprotective effects (Tian et al., 2006). Plants used for eliminating pathogenic heat from the blood and relieving chest tightness and nausea could be a source of compounds with cardiovascular effects (Chen et al., 2003).

7.4 PHYLOGENY

The plastid originates from cyanobacteria. Some ancestral genes are lost during the longterm evolution or are transferred to the nucleus after endosymbiosis. The comparative genomic analyses suggest that gene transfer from the plastid to the nucleus is still ongoing. The molecular evidence for recent functional gene transfer among seed plants has been documented for accD, infA, rpl22, and rpl32 (Park et al., 2015). The plastid genome of *Thalictrum coreanum* is sequenced, and infA and rpl32 genes are found to be lost. Examining the nuclear transcriptome verified that they are transferred to the nucleus. The phylogenetic distribution of rpl32 loss was investigated in 17 Thalictroideae species. The plastid-encoded rpl32 gene is likely nonfunctional in *Aquilegia, Enemion, Isopyrum, Leptopyrum, Paraquilegia, Semiaquilegia,* and 17 *Thalictrum* species due to indels that disrupt the reading frame. The nucleus-encoded rpl32 is found in *Thalictrum* and *Aquilegia,* which is highly similar to the plastid-encoded one. The phylogenetic distribution pattern and the highly similar transit peptides suggest a single transfer of the plastid-encoded rpl32 to the nucleus in the ancestor of the subfamily Thalictroideae around 20–32 Mya. Sequencing the plastid genome of *T. coreanum* helps understand the plastid genome evolution of Ranunculaceae. In contrast, the plastid genomes of *Megaleranthis saniculifolia* and *Ranunculus macranthus* reserve the functional infA and rpl32. The other two rpl32 transfers occur in unrelated angiosperm families Rhizophoraceae and Salicaceae. The transfer of rpl32 provides additional molecular evidence for the monophyly of the subfamily Thalictroideae. Five cp markers are combined to construct the phylogenetic tree of 37 Thalictroideae taxa (Fig. 7.4).

Gene duplication is one of the main drivers of genome evolution and is one of main reasons for generating genes with novel functions and new species. Gene duplication is the prerequisite of the generation of novel gene families and the functional diversification of genes and is also essential for the diversification of secondary metabolites in *Thalictrum*. Genome polyploidization frequently occurs during the evolution of *Thalictrum* (Soza et al., 2013), peaking during the recent 10.6–5.8 my and corresponding to the diversification of specific clades. The polyploid *Thalictrum* species produces more types of isoquinoline alkaloids (Kuzmanov, 1986). Due to their insecticidal activity, these species depend on wind pollination. In

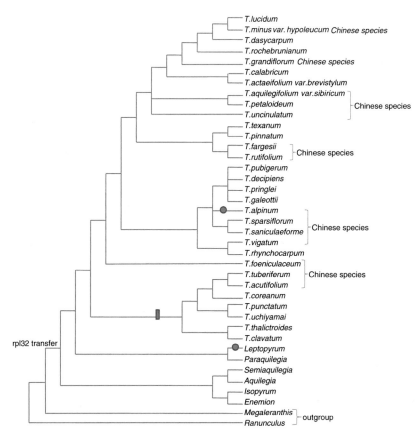

FIGURE 7.4 Phylogentic relationships of 37 Thalictroideae species, reconstructed by the ML method and five cp markers, that is, rbcL, ndhF, ndhA intron, trnL intron, and trnL-F intergenic spacer sequences. The *circle* indicates the complete loss of rpl32 from the plastid, while the rectangle indicates the shared indel event of the clade.

addition, high altitude habitats, where the polyploid species prosper, may favor wind pollinators rather than insect pollinators.

The Chinese *Sect. Thalictrum*(III, Fig. 7.5 and Table 7.1) species *T. atriplex, T. foetidum,T. flavum,* and *T. squarrosum,* which are polyploids and phylogenetically close to the evolutionarily younger polyploid species of North America, generate diversified alkaloid types (Zhu and Xiao, 1991a,b) and are traditionally used for medicine in China and adjacent countries (Xiao and Wang, 1965; Tables 7.1 and 7.3). They are distributed widely and are promising pharmaceutical resources. Morphologically, *T. smithii* of China belongs to the subgenus *Lecoyerium,* but it is closer to the Eurasia species than any New World taxa on the molecular phylogenetic tree (Soza et al., 2012, 2013) (Fig. 7.5). Whether it should be in *Lecoyerium* is worthy of further study. The traditional morphology grouping is not congruent with the molecular phylogeny, since it is obvious that the basal clade (clade I) of phylogenetic trees (Soza et al., 2012, 2013) contains both *Thalictrum rubescens* (morphology group I, Table 7.1) and *Thalictrum filamentosum, T. ichangense,* and *Thalictrum urbainii* (morphology group II), the latter three are known to more primitive in morphology. *T. alpinum* and *T. squamiferum* of morphology group IV are far apart on the molecular tree (Fig. 7.5), implying convergent evolution in morphology. The nuclear internal transcribed spacer (ITS) sequences of *Thalictrum,* submitted by various research groups, are retrieved from NCBI GenBank to construct the phylogenetic tree. It is found that in many cases multiple ITS sequences of the same species fail to cluster together (figure not shown), and in some cases they are far apart, for example, *T. cultratum, T. smithii, T. ichangense, T. rubescens,* etc. Taxa with various clonal ITS sequences are not rare (Soza et al., 2012, 2013), possibly due to the ITS multicopy, pseudogene, and concerted evolution in *Thalictrum.* The phylogenetic tree based on the four cp markers (Fig. 7.5) has good resolution, and rpl32-trnL, ndhA intron, rbcL, and trnL-trnF could be candidate DNA barcodes for the authentication of *Thalictrum* taxa. Intriguingly, three species of morphology group II, *T. ichangense, T. tuberiferum,* and *T. acutifolium,* are basal on the tree, reminiscent of the chemotaxonomic finding that *Sect. Tripterium* contains less alkaloids (Zhu and Xiao, 1991a).

7.5 DISCUSSION AND CONCLUSION

BIAs mainly exist in the relatively primitive Magnoliidae (Zhu and Xiao, 1991b). Aporphinoid, protoberberine, and BBI are relatively primitive types and are ubiquitously found in *Thalictrum.* BIA, as one of four chemotaxonomic markers, shows its utility in the phylogenetic analysis of Ranunculales (Wang et al., 2009). The ploidy levels of many species, for example,*T. simplex, T. foetidum,* and *Thalictrum delavayi,* vary. The same species consist of the normal diploids and the polyploids of variable ploidy levels. The correlation between ploidy levels and the intraspecies variation of chemodiversity warrants further study. *Thalictrum cirrhosum* and *T. cultratum* belong to morphology group III and are close to each other on the phylogenetic tree inferred from rpl32-trnL+ndhA intron+trnL-trnF (figure not shown). However, alkaloids in the former are much less than those in the latter. What are the intrinsic regulatory mechanisms? Does epigenetic transgenerational inheritance play a role in interspecies variation? Recently there has been a lot of research on the bioactivity and chemistry of *T. fortunei,* but its phylogenetic position is elusive due to the lack of molecular biology data. Many *Thalictrum* species lack DNA sequence data, and the genomic and other omics studies are in their infancy

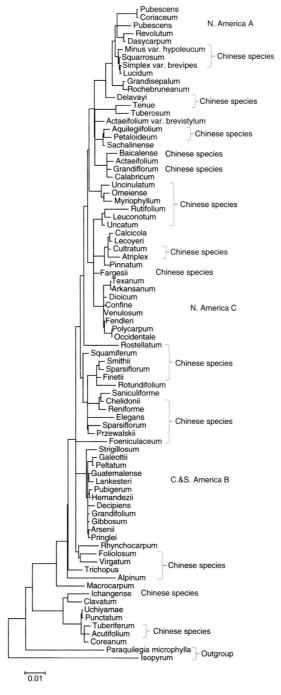

FIGURE 7.5 Phylogenetic relationships of *Thalictrum* taxa. The ML tree is inferred from cp markers rpl32-trnL+ndhAintron+rbcL+trnL-trnF. The scale bar represents the nucleotide substitution number per site.

TABLE 7.3 *Thalictrum* Plants of the Himalayas

	Species	Flower and fruit	Niche	Concomitancy species	Distribution	Ethnopharmacology
1.	*T. alpinum, T. alpinum var. microphyllum*	May–August	Alpine meadows, damp valleys, swamps, altitude 3000–5300 m	*Sedum, Polygonum, Aconitum, Rhodiola, Oxyria, Taraxacum officinalae, Rosularia, Potentilla, Tanacetum, Anaphalis* (Fig. 7.8), *Geranium* (Fig. 7.9), *Heracleum, Arisaema, Trollius*	China, India Himalayas, Bhutan, Nepal, Pakistan, Kazakhistan, Myanmar, Mongolia, Russia, Vietnam, Europe, N. America	Root decoction used as a tonic, purgative, diuretic, stimulant, dyspepsia, recuperation
2.	*T.chelidonii, T. chelidonii var. cysticarpum*	July–September	Mountain forests, altitude 2300–4500 m	*Juniper, Rhododendron, Pedicularis, Potentilla, Gaultierria, Pillea*	China, India Himalayas, Bhutan, Nepal	
3.	*T. cultratum (T. chelidonii var. cultratum)*	July–September	Grassy slopes, thickets, furrows, sometimes in forest, altitude 1700–3800 m	*Corydalis, Bergenia, Arisaema, Polygonum, Heracleum, Geranium, Potentilla, Juniperus, Moringa, Pedicularis, Nepeta, Salvia*	China, India Himalayas, Bhutan, Nepal, Pakistan	Roots used in dysentry, diarrhea, virus hepatitis, influenza, measles, carbuncles and boils, congestion of eyes
4.	*T. elegans (T. samariferum)*	July–October	Hillside, grassy slopes, altitude 2700–4000 m	*Anaphalis, Aconitum, Juniper, Polygonum, Corydalis, Berberis, Galium, Phlomis vulgaris, Rhododendron, Geranium, Potentilla, Astragalus*	China (Tibet), India Himalayas, Bhutan, Nepal, Pakistan	Plant paste used in headache
5.	*T. falconeri*	July–September	Moist and shady places, altitude 1300–3000 m	*Viola, Mazus, Geranium, Rubus*	India Himalayas	
6.	*T. foetidum (T. minus var. foetidum)*	June–August	Mountain grassy slopes, mountainous rocky boulders, altitude 350–4500 m	*Bergenia, Potentilla, Astragalus, Corydalis, Bistorta, Anaphalis, Juniperus, Polemonium*	China, India Himalayas, Afghanistan, Japan, Pakistan, Nepal, N. Asia, Europe	Tubers used for wounds, swellings, uterine tumor, paralysis, joint pain, nervous disorders, congestion of eyes, virus hepatitis, carbuncles and boils

No.	Species	Flowering time	Habitat, altitude	Associated species	Distribution	Uses
7.	*T. foliolosum*	June–October	Mountain forests or grassy slopes, altitude 1300–3200 m	*Sedum, Mazus, Potentilla, Geranium, Rubus*	China, India Himalayas, Myanmar, Nepal, Pakistan	Improve eyesight, toothache, diarrhea and piles, fever, root juice for peptic ulcer, root is diuretic, ophthalmic, purgative, for stomachache, dyspepsia, skin diseases, snake bite, leaf juice for boils and pimples
8.	*T. javanicum*	June–October	Damp place in a mountain forest, ditch, cliff, altitude 1500–3400 m	*Arisaema, Corydalis, Potentilla, Mazus, Deutzia, Ficus, Sedum, Anaphalis*	China, India Himalayas, Indonesia (Java, Sumatra), Sri Lanka	Eye disease; root for wound and trauma
9.	*T. minus var. hypoleucum (T. minus var. elatum)*	June–September	Mountain grass slope, field edge, scrub or forest, altitude 1400–2700 m	*Delphinium, Potentilla, Pedicularis, Corydalis, Trollius*	China, India Himalayas, Korea, Japan	Root for dysentery, diarrhea
10.	*T. pedunculatum*	April–July	Open mountain slopes, altitude 1800–2400 m	*Potentilla, Campanula, Coronopsis*	India Himalayas, Afghanistan, Pakistan	
11.	*T. platycarpum (T. cultratum subsp. platycarpum)*	April–July	Mountain slopes, altitude 1800–3000 m	*Anaphalis, Potentilla, Phlomis vulgaris, Pedicularis*	China (Tibet), India Himalayas, Afghanistan, Nepal, Pakistan	
12.	*T. punduanum var. glandulosum*	August–September	Moist and shady places, altitude 1800–2300 m	*Bistorta, Viola, Leucas*	India Himalayas	
13.	*T. punduanum var. punduanum*	July–August	Lithophytes on dry or moist shady places of *Quercus* forest	*Viburnum, Urtica, Geranium, Clematis, Rhodiola, Satyrium*	India Himalayas	

(Continued)

TABLE 7.3   *Thalictrum* Plants of the Himalayas (*cont.*)

	Species	Flower and fruit	Niche	Concomitancy species	Distribution	Ethnopharmacology
14.	*T. reniforme* (*T. chelidonii var. reniforme*)	June–October	Mountain grass slopes, thickets or fir forests, altitude 3100–3700 m	*Rhododendron*, mosses, liverworts	India Himalayas, Bhutan, Nepal	Root for jaundice, chancre
15.	*T. rostellatum*	June–August	Mountain forest or ditch, altitude 2500–3200 m	*Quercus, Taxus, Prunella, Ranunculus, Arisaema, Viola, Fragaria, Galium, Rhododendron*	China, India Himalayas, Bhutan, Nepal	
16.	*T. rutifolium*	June–August	Open and dry places, altitude 3000–4000 m	*Astragalus, Oxytropis* (Fig. 7.10), *Corydalis*	China (Tibet), India Himalayas, Bhutan	Clear away and purge pathogenic eat-fire, deprive the evil wetness, and detoxicate
17.	*T. saniculiforme*	July–October	Mountain grass slopes or forests, altitude 1500–2500 m	*Quercus, Ficus, Viburnum, Rhodiola, Orchis, Rubus*	China, India Himalayas, Nepal, Bhutan	
18.	*T. secundum* (*T. pauciflorum*)	June–September	Moist rock along the streams, water fall in hilly slopes, altitude 2500–3500 m	*Fragaria, Parnesia, Aquilegia, Boenninghunia, Polygonum, Hedychium, Corydalis, Potentilla, Phlomis vulgaris, Salvia, Meconopsis, Anemone*	India Himalayas, Nepal, Pakistan	Root for stomach disorders
19.	*T. squamiferum* (*T. glareosum*)	June–August	Rocky mountain slopes, rocky gravel or forest, altitude 3600–5000 m	*Corydalis, Potentilla, Poa*	China, Sikkim	Whole plant for fever
20.	*T. virgatum*	May–September	Mountain forest, rocky ledge, altitude 2300–3600 m	*Saxifraga, Rhodiola, Geranium, Sedum*	China, India Himalayas, Bhutan, Nepal	Root for stomach disease

(Hagel et al., 2015). The striking diversity of *Thalictrum* is witnessed in the Pan-Himalayas (Table 7.3). Morphologically new Chinese species are reported (Wang 2017), but it is clearly not enough to only study the Chinese taxa. To date, limited molecular phylogeny studies have contributed some meaningful information, and future molecular biology and omics studies will provide strong support for the development of *Thalictrum* drug resources (Figs. 7.6–7.10).

FIGURE 7.6   *T. petaloideum*, **taken in Lijiang Alpine Botanic Garden of Chinese Academy of Sciences (refer to Fig. 7.5).**

FIGURE 7.7   *T. finetii*, **taken in Shudu Lake of Potatso National Park, Shangri-La, Yunnan, China (refer to Fig. 7.5).**

FIGURE 7.8   *Anaphalis nepalensis*, **taken in Shudu Lake of Potatso National Park.**

FIGURE 7.9   *Geranium refractum*, taken in Bitahai Lake of Potatso National Park.

FIGURE 7.10   *Oxytropis*, taken in Tingri County, Tibet, China.

References

Aun Bulag, Tai, L.H., Hurduhu, 2010. Extraction of antioxidants from the rhizome of *Thalictrum petaloideum*. Lishizhen Med. Mater. Med. Res. 214, 872–874.

Chen, Q., Peng, W., Qi, S., et al., 2002. Apoptosis of human highly metastatic lung cancer cell line 95-D induced by acutiaporberine, a novel bisalkaloid derived from *Thalictrum acutifolium*. Planta Med. 68 (6), 550–553.

Chen, S.B., Chen, S.L., Xiao, P.G., 2003. Ethnopharmacological investigations on *Thalictrum* plants in China. J. Asian Nat. Prod. Res. 5 (4), 263–271.

Cheng, H.X., Zeng, Y.C., Jia, B.Y., et al., 2012. Screening and identification of various components in *Thalictrum fortunei* using a combination of liquid chromatography/time-of-flight tandem mass spectrometry. Pharmazie 67 (2), 106–110.

Das, N.G., Rabha, B., Talukdar, P.K., et al., 2016. Preliminary in vitro antiplasmodial activity of *Aristolochia griffithii* and *Thalictrum foliolosum* DC extracts against malaria parasite *Plasmodium falciparum*. BMC Res. Notes 9 (1), 51.

Erdemgil, F.Z., Baser, K.H., Akbay, P., et al., 2003. Thalictricoside, a new phenolic compound from *Thalictrum orientale*. Z. Naturforsch. C 58 (9–10), 632–636.

Gao, G.Y., Chen, S.B., Xiao, P.G., 1999a. Chemical constituents of *Thalictrum atriplex*. China J. Chin. Mater. Med. 24 (3), 160–161.

Gao, G.Y., Chen, S.B., Xiao, P.G., et al., 2005. Novel dimeric alkaloids from the roots of *Thalictrum atriplex*. J. Asian Nat. Prod. Res. 7 (6), 805–809.

Gao, G.Y., Wang, L.W., Xiao, P.G., et al., 1999b. Chemical constituents of *Thalictrum atriplex*. Chin. Pharm. J. 34 (3), 157–158.

Gromova, A.S., Lutsky, V.I., Li, D., et al., 2000. Thalicosides A1-A3, minor cycloartane bisdesmosides from *Thalictrum minus*. J. Nat. Prod. 63 (7), 911–914.

Gurunathan, A., Senguttuvan, J., Paulsamy, S., 2016. Evaluation of mosquito repellent activity of isolated oleic acid, eicosyl ester from *Thalictrum javanicum*. Indian J. Pharm. Sci. 78 (1), 103–110.

Hagel, J.M., Morris, J.S., Lee, E.J., et al., 2015. Transcriptome analysis of 20 taxonomically related benzylisoquinoline alkaloid-producing plants. BMC Plant Biol. 15, 227.

Hao, D.C., Gu, X.J., Xiao, P.G., et al., 2013. Recent advances in the chemical and biological studies of Aconitum pharmaceutical resources. J. Chin. Pharm. Sci. 22 (3), 209–221.

Hao, D.C., Gu, X.J., Xiao, P.G., 2015. Thalictrum medicinal resources: chemistry, biology and phylogeny. Lishizhen Med. Mater. Med. Res. 26 (7), 1731–1733.

Horvath, L.E., Konstantinov, S.M., Ilarionova, M.V., et al., 2004. Effects of the plant alkaloid thaliblastine on non-cross-resistant and sensitive human leukemia cells in relation with reversal of acquired anthracycline resistance. Fitoterapia 75 (7–8), 712–717.

Jiang, S.Q., Zhang, Y.B., Xiao, M., et al., 2017. Cycloartane triterpenoid saponins from the herbs of *Thalictrum fortunei*. Carbohydr. Res. 445, 1–6.

Jiao, K., Zhang, P., Pi, H.F., et al., 2014. The anti-tumor effect of extracts from *Thalictrum robustum* Maxim.in vitro. Chin. Hosp. Pharm. J. 34 (10), 811–815.

Khamidullina, E.A., Gromova, A.S., Lutsky, V.I., et al., 2006. Natural products from medicinal plants: non-alkaloidal natural constituents of the Thalictrum species. Nat. Prod. Rep. 23 (1), 117–129.

Kumar, A., Chowdhury, S.R., Sarkar, T., et al., 2016. A new bisbenzylisoquinoline alkaloid isolated from *Thalictrum foliolosum*, as a potent inhibitor of DNA topoisomerase IB of *Leishmania donovani*. Fitoterapia 109, 25–30.

Kuzmanov, B.A., 1986. Polyploidy and evolutionary pattern in genus *Thalictrum* L. Fitologija 31, 14–16.

Lee, E.J., Hagel, J.M., Facchini, P.J., 2013. Role of the phloem in the biochemistry and ecophysiology of benzylisoquinoline alkaloid metabolism. Front. Plant Sci. 4, 182.

Li, D.H., Guo, J., Bin, W., et al., 2016. Two new benzylisoquinoline alkaloids from *Thalictrum foliolosum* and their antioxidant and in vitro antiproliferative properties. Arch. Pharm. Res. 39 (7), 871–877.

Li, D.H., Li, J.Y., Xue, C.M., et al., 2017. Antiproliferative dimeric aporphinoid alkaloids from the roots of *Thalictrum cultratum*. J. Nat. Prod. 80 (11), 2893–2904.

Liu, C.X., Xiao, P.G., 1983. Biological activity and domestic resources of bis-benzyl quinoline-like alkaloids. Yao Xue Tong Bao 18 (5), 31–33.

Liu, M.J., Zhao, Q., Liang, N., et al., 2013. GC-MS analysis and biological activity of essential oil from leaves of *Thalictrum ichangense*. Chin. J. Exp. Trad Med. Formulae 24, 135–138.

Mayeku, P.W., Hassanali, A., Kiremire, B.T., et al., 2013. Anti-bacterial activities and phytochemical screening of extracts of different parts of *Thalictrum rhynchocarpum*. Afr. J. Tradit Complement. Altern. Med. 10 (5), 341–344.

Mushtaq, S., Rather, M.A., Qazi, P.H., et al., 2016. Isolation and characterization of three benzylisoquinoline alkaloids from *Thalictrum minus* L. and their antibacterial activity against bovine mastitis. J. Ethnopharmacol. 193, 221–226.

Naman, C.B., Gupta, G., Varikuti, S., et al., 2015. Northalrugosidine is a bisbenzyltetrahydroisoquinoline alkaloid from *Thalictrum alpinum* with in vivo antileishmanial activity. J. Nat. Prod. 78 (3), 552–556.

Pandey, G., Khatoon, S., Pandey, M.M., et al., 2017. Altitudinal variation of berberine, total phenolics and flavonoid content in *Thalictrum foliolosum* and their correlation with antimicrobial and antioxidant activities. J. Ayurveda Integr. Med., 010. doi: 10.1016/j.jaim.2017.02.

Park, S., Jansen, R.K., Park, S., 2015. Complete plastome sequence of *Thalictrum coreanum* (Ranunculaceae) and transfer of the rpl32 gene to the nucleus in the ancestor of the subfamily Thalictroideae. BMC Plant Biol. 15, 40.

Ropivia, J., Derbré, S., Rouger, C., et al., 2010. Isoquinolines from the roots of *Thalictrum flavum* L. and their evaluation as antiparasitic compounds. Molecules 15 (9), 6476–6484.

Serkedjieva, J., Velcheva, M., 2003. In vitro anti-influenza virus activity of isoquinoline alkaloids from thalictrum species. Planta Med. 69 (2), 153–154.

Song, Y., Wang, Z., Zhang, B., et al., 2017. Determination of a novel anticancer AMPK activator hernandezine in rat plasma and tissues with a validated UHPLC-MS/MS method: application to pharmacokinetics and tissue distribution study. J. Pharm. Biomed. Anal. 141, 132–139.

Soza, V.L., Brunet, J., Liston, A., et al., 2012. Phylogenetic insights into the correlates of dioecy in meadow-rues (*Thalictrum* Ranunculaceae). Mol. Phylogenet. Evol. 63 (1), 180–192.

Soza, V.L., Haworth, K.L., Di Stilio, V.S., 2013. Timing and consequences of recurrent polyploidy in meadow-rues (thalictrum, ranunculaceae). Mol. Biol. Evol. 30 (8), 1940–1954.

Taherpour, A., Maroofi, H., 2008. Chemical composition of the essential oil of *Thalectrum minus* L. of Iran. Nat. Prod. Res. 22 (2), 97–100.

Tian, Z., Xiao, P.G., Wen, J., et al., 2006. Review of bioactivities of natural cycloartane triterpenoids. China J. Chin. Mater. Med. 31 (8), 625–629.

Wang, W., Lu, A.M., Ren, Y., et al., 2009. Phylogeny and classification of Ranunculales evidence from four molecular loci and morphological data. Perspect. Plant Ecol. Evol. Syst. 11 (2), 81–110.

Wang, W.C., 2017. Five new species of Thalictrum (Ranunculaceae) from China. Guihaia 37 (4), 407–416.

Xiao, P.G., Wang, W.C., 1965. A study of the Ranunculaceous medicinal plants in China III. The medicinal plants of genus Thalictrum Linn. Acta Pharm. Sin. 12 (11), 745–749.

Yoshimitsu, H., Miyashita, H., Nishida, M., et al., 2010. Dolabellane diterpene and three cycloartane glycosides from *Thalictrum squarrosum*. Chem. Pharm. Bull. 58 (8), 1043–1046.

Zhang, X., Zhao, M., Chen, L., et al., 2011. A triterpenoid from *Thalictrum fortunei* induces apoptosis in BEL-7402 cells through the P53-induced apoptosis pathway. Molecules 16 (11), 9505–9519.

Zhang, X.T., Ma, S.W., Jiao, H.Y., et al., 2012. Two new saponins from *Thalictrum fortunei*. J. Asian Nat. Prod. Res. 14 (4), 327–332.

Zhu, M., Xiao, P.G., 1989. Study on resource utilization of Thalictrum plants. Chin. Trad Herb Drug 20 (11), 29–31.

Zhu, M., Xiao, P.G., 1991a. Chemosystematic studies on Thalictrum L. in China. Acta Phytotaxon Sin. 29 (4), 358–369.

Zhu, M., Xiao, P.G., 1991b. Distribution of benzylisoquinolines in Magnoliidae and other taxa. Acta Phytotaxon Sin. 29 (2), 142–155.

Further Reading

Al-Howiriny, T.A., Zemaitis, M.A., Gao, C.Y., et al., 2001. Thalibealine, a novel tetrahydroprotoberberine--aporphine dimeric alkaloid from *Thalictrum wangii*. J. Nat. Prod. 64 (6), 819–822.

Al-Rehaily, A.J., Sharaf, M.H.M., Zemaitis, M.A., et al., 1999. Thalprzewalskiinone, a new oxobenzylisoquinoline alkaloid from thalictrum przewalskii. J. Nat. Prod. 62 (1), 146–148.

Chen, S.B., Gao, G.Y., Xiao, P.G., et al., 2005. Bisbenzylisoquinoline alkaloids from roots of *Thalictrum atriplex*. Chin. Trad Herb Drug 36 (4), 487–489.

Gao, G.Y., Chen, S.B., Xiao, P.G., et al., 2000. Chemical constituents of *Thalictrum atriplex* (III). Chin. Trad Herb Drug 31 (5), 324–326.

Lee, S.S., Doskotch, R.W., 1999. Four dimeric aporphine-containing alkaloids from *Thalictrum fauriei*. J. Nat. Prod. 62 (6), 803–810.

Lee, S.S., Wu, W.N., Wilton, J.H., et al., 1999. Longiberine and O-methyllongiberine, dimeric protoberberine-benzyl tetrahydroisoquinoline alkaloids from *Thalictrum longistylum* (1). J. Nat. Prod. 62 (10), 1410–1414.

Li, M., Chen, X., Tang, Q.M., et al., 2001. Isoquinoline alkaloids from *Thalictrum delavayi*. Planta Med. 67 (2), 189–190.

Liang, Z.Y., Wang, Y., Yang, X.S., et al., 2004. Study on chemical constituents of *Thalictrum elegans*. Chin. Trad Herb Drug 35 (3), 243–244.

Lin, L.Z., Hu, S.F., Chu, M., et al., 1999. Phenolic aporphine-benzylisoquinoline alkaloids from *Thalictrum faberi*. Phytochemistry 50 (5), 829–834.

Wang, Y., Yang, X.S., Luo, B., et al., 2003. Chemical constituents of *Thalictrum delavayi*. Acta Bot. Sin. 45 (4), 500–503.

Zhang, X.T., Li, Y., Zhang, L.H., et al., 2009. Two new cycloartane glycosides from *Thalictrum fortunei*. Chin. Trad Herb Drug 40 (8), 1189–1191.

Zhang, X.T., Wang, H., Yin, Z.Q., et al., 2007. Chemical constituents from *Thalictrum fortunei*. J. China Pharm. Univ. 38 (1), 21–23.

Zhang, X.T., Wang, L., Ma, S.W., et al., 2013. New cycloartane glycosides from the aerial part of *Thalictrum fortunei*. J. Nat. Med. 67 (2), 375–380.

Zhang, X.T., Zhang, L.H., Ye, W.C., et al., 2006. Four new cycloartane glycosides from Thalictrum fortunei. Chem. Pharm. Bull. 54 (1), 107–110.

Biodiversity, Chemodiversity, and Pharmacotherapy of *Anemone* Medicinal Plants

ABBREVIATIONS

CDK   Cyclin-dependent kinase
COX-2   Cyclooxygenase-2
ERK   Extracellular signal-regulated kinase
HDI   Herb-drug interaction
HO   Hemeoxygenase
HUVEC Human umbilical vein endothelial cell
IC$_{50}$ Half maximal inhibitory concentration
JNK   c-jun N-terminal kinase
LPS   Lipopolysaccharide
MAPK   Mitogen-activated protein kinase

Ranunculales Medicinal Plants. http://dx.doi.org/10.1016/B978-0-12-814232-5.00008-3

MMP	Matrix metalloproteinase
PARP	Poly ADP-ribose polymerase
PI3K	Phosphoinositide 3-kinase
ROS	Reactive oxygen species
RTK	Receptor tyrosine kinase

8.1 INTRODUCTION

Anemone is a genus of more than 150 species of flowering plants in the family Ranunculaceae, native to the temperate zones of both the Northern and Southern Hemispheres. It is closely related to *Pulsatilla, Clematis*, and *Hepatica* morphologically and phytochemically (Hao et al., 2015a). More than 50 species have ethnopharmacological uses, which provide clues for modern drug discovery. Some traditional claims of *Anemone* species have been validated scientifically by preclinical and clinical studies (Liu et al., 2012; Han et al., 2009). The state-of-the-art methods have been adopted to isolate bioactive chemical constituents from *Anemone* species following bioactivity-directed fractionation (Liu et al., 2015a,b; Saito et al., 2002). Some modes of action of bioactive extracts or compounds of *Anemone* species have been established (Kong et al., 2015a,b; Guan et al., 2015). However, a comprehensive review of the *Anemone* medicinal resources is lacking. In this chapter, we summarize the ethnomedicine knowledge and the recent progress of chemodiversity and pharmacological diversity of *Anemone* medicinal plants, as well as the emerging molecular machineries and functions of medicinal compounds. Gaps are also pointed out and further work is suggested.

Exhaustive literature searches in NCBI PubMed, Google, Bing, and CNKI (http://cnki. net/) have been performed to outline the progress of *Anemone* research during the last three decades. Search terms "anti-cancer," "anti-inflammatory," "antioxidant," "saponin," "triterpene," "polysaccharide," etc., were used, combined with "*Anemone*" and the names of species.

8.2 ETHNOPHARMACOLOGY

More than 50 *Anemone* species are used in various traditional medical systems (Table 8.1, Fig. 8.1). 53 species, 9 subspecies, and 36 varieties are found in China and are distributed in most provinces except Guangdong and Hainan. According to our field survey, at least 38 species/varieties have ethnopharmacological uses. In traditional Chinese medicine (TCM) and folk medicine, *Anemone* is used for heat-clearing and detoxification [traditional remedy index (TRI) 424, distribution density of ethnopharmacological use β 30], wind-dispersing and damp-eliminating (TRI 476, β 35), warming and orifice-opening (TRI 700, β 15), pesticides (TRI 400, β 30), dysentery (TRI 1 051, β 46), malaria (TRI 356, β 30), tinea (TRI 445, β 46), ulcers and sores (TRI 1 932, β 84), arthritis (TRI 896, β 76), traumatic injury (TRI 930, β 53), pharyngolaryngitis (TRI 327, β 7), parasitic disease (TRI 424, β 30), and hepatitis (TRI 445, β 7) (Xiao et al., 1986). Correspondingly, a broad spectrum of pharmacological activities, including antitumor, antimicrobial, antiinflammatory, sedative and analgesic activities, and anticonvulsant and antihistamine effects have been observed (Sun et al., 2011). For instance, *Anemone raddeana*, distributed in the Far East, is commonly used in Northeast China for rheumatism, arthritis, skin infection, etc. Liang Tou Jian (Zhu Jie Xiang Fu in

(Continued)

TABLE 8.1 Ethnopharmacological Uses of *Anemone* Species

Species	Medicinal part	Therapeutic efficacy	Distribution	Note
A. altaica	Rhizome	Tranquilizing, orifice-opening, wind-expelling, damp-eliminating, detoxifying, pain-relieving; high fever, delirium, epilepsy, deafness with qi stagnation, dreaminess forgetfulness, chest tightness, abdominal distension, anorexia, rheumatism pain, ulcer, scabies	Europe, North Asia, China: Hubei, Henan, Shanxi, Shaanxi, Chongqing	
Amurensis	Whole plant, rhizome	Diaphoresis, liver/kidney tonifying (whole plant); Korean medicine: paralysis, menoxenia, stomachache, pertussis (rhizome)	Russia Far East, North Korea, China: Liaoning, Jilin, Heilongjiang	
A. anhuiensis	Rhizome	Traumatic injury, rheumatic arthritis	Anhui, China	
Baicalensis	Leaf	Detoxifying, vermifuge	Siberia, Korea, China: Sichuan, Gansu, Shaanxi, Qinling Mountains, Liaoning, Jilin, Heilongjiang	
A. begoniifolia	Whole plant	Wind-expelling, damp-eliminating, detoxification, pain-relieving; rheumatism, urticaria, carbuncle sore	China: Yunnan, Guangxi, Guizhou, Sichuan, Chongqing	
Biflora	Bulb	Styptic, antiphlogistic; boils, burns, cuts, and wounds	Kashmir Himalaya	
Canadensis	Root, leaf	Anthelmintic, antiaphonic, antiseptic, astringent, ophthalmic, styptic; pain in the lumbar region, crossed eyes, twitches and eye poisoning, wounds, nosebleed, sore, headache and dizziness, clear the throat	Eastern and Central N. America	
Cathayensis	Rhizome	Cancer, inflammation, analgesic, convulsion	Korea, China: Shanxi, Hebei	
A. chosenicola var. schantungensis	Root	Styptic, damp-eliminating, heat-clearing, detoxification	Shandong, China	
Cylindrical	Root, leaf, stem, fruit	Antiseptic; sore eyes (stem, fruit); headache, dizziness, wounds (root); burns (leaf)	Western N. America	

TABLE 8.1 Ethnopharmacological Uses of *Anemone* Species (*cont.*)

Species	Medicinal part	Therapeutic efficacy	Distribution	Note
A. davidii	Rhizome	Blood-activating, pain-relieving, subduing swelling, detoxicating; traumatic injury, arthritis pain, lumbar muscle strain; Tujia medicine: arthritis pain, intercostal neuralgia, traumatic injury, hematemesis, hemafecia	China: Chongqing, Tibet, Yunnan, Sichuan, Guizhou, Hunan, Hubei	
Delavayi	Rhizome	Blood-activating, stasis-scattering, tonifying kidney	China: Yunnan, Sichuan	
A. demissa	Root, fruit, whole plant	Whole plant: rheumatism, dysentery, help digestion, dyspepsia, gonorrhea, wind-cold-dampness arthralgia, joint yellow water; fruit: damp-clearing, mass-scattering, detoxifying, all kinds of cold, lump boil, snake bite	Himalayas, QTP east margin	Tibetan medicine: 素嘎益保
Demissa var. major	Root, fruit, whole plant	Whole plant: rheumatism, dysentery, help digestion, dyspepsia, gonorrhea, wind-cold-dampness arthralgia, joint yellow water; fruit: damp-clearing, mass-scattering, detoxifying, all kinds of cold, lump boil, snake bite	QTP east margin	Tibetan medicine: 素嘎益保
Demissa var. villosissima	Root, fruit, whole plant	Whole plant: rheumatism, dysentery, help digestion, dyspepsia, gonorrhea, wind-cold-dampness arthralgia, joint yellow water; fruit: damp-clearing, mass-scattering, detoxifying, all kinds of cold, lump boil, snake bite	Himalayas, QTP east margin	Tibetan medicine: 素嘎益保
Dichotoma	Rhizome	Muscle-relaxing, blood-activating, heat-clearing, detoxification, traumatic injury, dysentery, rheumatoid joint pain; skin ulcer; sore throat, cough with copious phlegm, lymphnoditis	North Asia, Europe, China: Jilin, Heilongjiang	
Drummondii	Root, seed	Abrasions, tooth ache, rheumatism; antibacterial; sex related difficulties; melancholy (root); headache (seed)	Western N. America	

A. flaccid	Rhizome	Wind-expelling, dampness-eliminating, muscle-relaxing, blood-activating; traumatic injury, arthritis pain, lumbar muscle strain	Japan, Russia Far East, China: Yunnan, Sichuan, Guizhou, Hubei, Chongqing, Hunan, Jiangxi, Zhejiang, Jiangsu, Shaanxi, Gansu	
Flaccida var. hofengensis	Rhizome	Wind-expelling, dampness-eliminating, muscle-relaxing, blood-activating; traumatic injury, arthritis pain, lumbar muscle strain	Chongqing	
Fulingensis	Rhizome	Wind-expelling, dampness-eliminating, muscle-relaxing, blood-activating; traumatic injury, arthritis pain, lumbar muscle strain	Chongqing	
A. griffthii	Rhizome, seed	Blood-activating, pain-relieving, subduing swelling, detoxicating; traumatic injury, arthritis pain, lumbar muscle strain; Tibet medicine: stomach worms, sharp pain, snake bite, cold tumor, gonorrhea, joint yellow water (seed)	Tibet, Sichuan, Chongqing (China), Sikkim, Bhutan, Nepal	Tibetan medicine: 素嘎
A. hupehensis	Rhizome, root, stem, leaf, whole plant	Heat-clearing, diuresis, detoxification, vermifuge, stasis-scattering, detumescence; dysentery, malnutrition and indigestion of children, malaria, acute jaundice hepatitis, ascariasis, furuncle carbuncle, scrofula, traumatic injury	Chongqing, Southern Shaanxi, Gansu, Zhejiang, Jiangxi, Western Hubei, Northern Guangdong, Northern Guangxi, Sichuan, Guizhou, Eastern Yunnan	
hupehensis f. alba	Rhizome	Heat-clearing, diuresis, detoxification, vermifuge, stasis-scattering, detumescence; dysentery, malnutrition and indigestion of children, malaria, acute jaundice hepatitis, ascariasis, furuncle carbuncle, scrofula, traumatic injury	Chongqing	
Hupehensis var. japonica	Rhizome	Heat-clearing, diuresis, detoxification, vermifuge, stasis-scattering, detumescence; dysentery, malnutrition and indigestion of children, malaria, acute jaundice hepatitis, ascariasis, furuncle carbuncle, scrofula, traumatic injury	Chongqing	

(Continued)

TABLE 8.1 Ethnopharmacological Uses of *Anemone* Species (*cont.*)

Species	Medicinal part	Therapeutic efficacy	Distribution	Note
A. *imbricate*	Root, fruit, whole plant, flower, stem, leaf, seed	Whole plant: expelling wind-damp, dysentery, help digestion, gonorrhea, wind-cold-dampness arthralgia, joint yellow water; fruit: damp-expelling, mass-eliminating, detoxifying, all kinds of cold, lump boil, snake bite; stomach worms, sharp pain, snake bite, cold tumor, gonorrhoea, joint yellow water (seed); gonorrhoea, joint yellow water, hypothermia, emetic (leaf); antiinflammatory, burn (stem, leaf, flower)	QTP east margin	Tibetan medicine: 素嘎盅保/素嘎
Multifida	Root, seed	Abrasions, toothache, rheumatism; antibacterial; sex-related difficulties; melancholy (root); headache (seed)	Central and Western N. America	
Narcissiflora	Leaf, root, seed	Abrasions, toothache, rheumatism; antibacterial; sex-related difficulties; melancholy (root); headache (seed)	Europe, Asia, N. America	
Nemorosa	Various parts	Headaches, tertian agues and rheumatic gout(various parts), leprosy (leaf), bring away watery and phlegmatic humors(root), lethargy, eye inflammation, malignant and corroding ulcers (root)	UK, Europe, West Asia	
Nikoensis	Leaf	Edible use	Japan	
A. *obtusiloba*	Seed, aboveground part, root, fruit	Diuresis detumescence, enriching blood, warming body, wound healing, pus drainage; antirheumatic; emetic (seed); ophthalmic; rubefacient; contusion (root); ill health, hypothermia, sore throat, chronic bronchitis, tonsillitis, hepatitis, gastric disease, dysentery, gonorrhea, arthritis pain, peripheral nerve paralysis, snake bite, stubborn dermatitis, impetigo, joint yellow water	Himalayas, QTP east margin	Tibetan medicine: 素嘎

Obtusiloba ssp. *ovalifolia*	Whole plant, aboveground part, root, fruit	Diuresis detumescence, enriching blood, warming body, wound-healing, pus drainage; styptic (whole plant), ill health, hypothermia, sore throat, chronic bronchitis, tonsillitis, hepatitis, gastric disease, dysentery, gonorrhea, arthritis pain, peripheral nerve paralysis, snake bite, stubborn dermatitis, impetigo, joint yellow water	Taibai mountain, QTP east margin	Tibetan medicine: 素嘎
Parviflora	Root, seed	Abrasions, toothache, rheumatism; antibacterial; sex related difficulties; melancholy (root); headache (seed)	Northern N. America	
Patens (*A. pulsatilla*)	Whole plant	Diaphoretic, diuretic, nervine, rubefacient; eye ailments, earache, stress, anxiety, tension, skin eruptions, rheumatism, leukorrhea, obstructed menses, bronchitis, coughs, asthma	UK, Europe	
Quinquefolia	Root	Rubefacient; rheumatism, gout, fever; vesicant, corns	Eastern N. America	
Raddeana	Rhizome	Mongolian medicine: rheumatism, low back and leg pain, phlebitis; subduing inflammation; arthralgia, chill cold, cough with copious phlegm, joint pain	Northeastern Shandong, Liaoning, Jilin, Heilongjiang, Korea, Russia Far East	Chinese Pharmacopoeia
A. reflexa	Rhizome	Open the orifices with aroma, windexpelling, dampness-eliminating, appetite-stimulating; high fever, delirium, epilepsy, deafness with qi stagnation, dreaminess forgetfulness, chest tightness, abdominal distension, anorexia, rheumatism pain, ulcer, scabies	North Korea, Siberia, Eastern Europe, Shaanxi, Eastern Jilin, Taibai mountain	

(Continued)

TABLE 8.1 Ethnopharmacological Uses of *Anemone* Species (*cont.*)

Species	Medicinal part	Therapeutic efficacy	Distribution	Note
A. rivularis	Rhizome, leaf, seed; aboveground part, root, fruit	Heat-clearing, detoxification, blood-activating, muscle-relaxing, swell-dispersing, pain-relieving; diuresis detumescence, enriching blood, warming body, wound healing, pus drainage; mumps, scrofula, carbuncle, malaria, cough, jaundice, arthritis pain, traumatic injury, stomachache, toothache; ill health, hypothermia, sore throat, chronic bronchitis, tonsillitis, hepatitis, gastric disease, dysentery, gonorrhea, arthritis pain, peripheral nerve paralysis, snakebite, stubborn dermatitis, impetigo, joint yellow water	Chongqing, Tibet, Himalayas, Sri Lanka; QTP east margin	Tibetan medicine: 素嘎
Rivularis var. *flore-minore*	Rhizome	Heat-clearing, detoxification, blood-activating, muscle-relaxing, swell-dispersing, pain-relieving; sore throat, mumps, scrofula, carbuncle, malaria, cough, jaundice, arthritis pain, traumatic injury, stomachache, toothache	Chongqing	
Rockii var. *pilocarpa*	Rhizome	Wind-expelling, dampness-removing, muscle-relaxing, blood-activating; arthritis pain, traumatic injury	Chongqing	
Rupicola	Seed	Stomach worms, sharp pain, snake bite, cold tumor, gonorrhoea, joint yellow water	Northwestern Yunnan, Western Sichuan, Southeastern and Southern Tibet, Bhutan, Nepal, Northern India	Tibetan medicine: 素嘎
Silvestris	Rhizome	Relieving oppression and masses, pus drainage, rot-eliminating, insecticide	Europe, Asia, Liaoning, Hebei, Heilongjiang, Jilin, Xinjiang, Inner Mongolia	Mongolian medicine "Xiriwusu"
Stolonifera	Leaf, stem	Edible use	China, Japan	
Taipaiensis	Rhizome	Cancer	Taibai mountain	
Tetrasepala	Seed	Stomach worms, sharp pain, snake bite, cold tumor, gonorrhoea, joint yellow water	Southern Tibet, Kashmir, Afghanistan	Tibetan medicine: 素嘎

Tibetica	Seed	Stomach worms, sharp pain, snakebite, cold tumor, gonorrhoea, joint yellow water	Tibet	Tibetan medicine: 布尔青
A. tomentosa	Rhizome	Dissipating phlegm stasis, relieving dyspepsia, detoxification, vermifuge; eparsalgia cough, traumatic injury, malnutrition and indigestion of children, malaria, dysentery, sore furuncle carbuncle, stubborn dermatitis	Chongqing, Sichuan, Gansu, Henan, Shanxi	
A. trullifolia	Root, flower	Muscle-relaxing, blood-activating, antitussive; chronic bronchitis, peripheral nerve paralysis, neuralgia, tendon complex pain	QTP east margin, Southern Tibet, Sikkim, Bhutan	Tibetan medicine: 布尔青
Trullifolia var. *linearis*	Root, flower	Muscle-relaxing, blood-activating, antitussive; chronic bronchitis, peripheral nerve paralysis, neuralgia, tendon complex pain	QTP east margin	Tibetan medicine: 布尔青
Tuberose		Anxiolytic	American Southwest	
Virginiana	Root, seed	Astringent; emetic; expectorant; TB, whooping cough, diarrhea; boils	Central and Eastern N. America	
Vitifolia	Root, leaf, rhizome	Traumatic injury, rheumatic arthralgia, enteritis, dysentery, ascariasis (rhizome); antirheumatic and vermifuge, dysentery, relieve tooth pain and headache, scabies (root), head lice (leaf)	Europe, Himalayas	

FIGURE 8.1   (A) Habitat of *Anemone*; (B) the whole plant and flower of *A. rivularis*, taken in the Alpine Botanical Garden of Shangri-La, Yunnan, China; and (C) fruit of *Anemone*, taken by the side of the Ni Yang river, Tibet, China.

Chinese), the rhizome of *A. raddeana*, is recorded in the *Pharmacopoeia of the People's Republic of China*, 2015 edition. *Anemone altaica*, extensively distributed in Europe and North Asia, is used in Northwest China for epilepsy, neurosis, and rheumatic arthralgia. The rhizome of *A. altaica* is sometimes called "Jiu Jie Chang Pu," but actually "Jiu Jie Chang Pu" originally referred to the rhizome of *Acorus tatarinowii* (Araceae), which has distinct therapeutic components (Wang et al., 1995). *Anemone vitifolia* (wild cotton) of the pan-Himalaya region is used in Southwest China for traumatic injury, rheumatic arthralgia, enteritis, dysentery, ascariasis, etc.

In Southwest China, *Anemone* has been undergoing rapid diversification following the uplift of Qinghai Tibet Plateau in the Quaternary Period and the emergence of "sky islands" (He and Jiang, 2014). Many alpine *Anemone* species are used in Tibetan medicine (Table 8.1) (Gong, 2011a; Li et al., 2015). For instance, Suga (素嘎, 锁嘎哇)is the ripe seed of *Anemone rivularis*, *A. rivularis* var. *floreminore*, *Anemone obtusiloba*, *A. obtusiloba* ssp. *ovalifolia*, and *Anemone demissa* and is used for diuresis, detumescence, enriching blood, warming body, wound-healing, and pus drainage (Li et al., 2015; Timar, 2012). Burchin (布尔青), the root and flower of *Anemone trullifolia* and *A. trullifolia* var. *linearis*, is used for muscle-relaxing, blood-activating, and antitussive. Suga Angbo (素嘎盎保), the root, fruit, and whole plant of *A. demissa*, *A. demissa* var. *major*, *A. demissa* var. *villosissima*, and *Anemone imbricata*, are efficient in antirheumatism, helping digestion, dysentery (whole plant), eliminating coldness and dampness, dissolving mass, and detoxification (fruit).

Various *Anemone* species are also used in ethnomedicine of India, Korea, Mongolia, America, Europe, etc. (Table 8.1). For instance, the *Anemone biflora* bulb is styptic and is applied on boils, burns, cuts, and wounds as antiphlogistic (Kumar et al., 2015). The root and leaves of *Anemone canadensis* is one of the most highly esteemed medicines of the Omaha and Ponca Indians (http://plants.for9.net/edible-and-medicinal-plants). A decoction of the root is used as an anthelmintic and to treat pain in the lumbar region. An infusion of the root is used as an eye wash to treat crossed eyes, twitches, and eye poisoning. In Korean medicine, the rhizome of *Anemone amurensis* is useful in paralysis, menoxenia, stomachaches, and pertussis (Lv et al., 2015). *Anemone stolonifera* and *Anemone nikoensis* are for edible use in East Asia.

8.3 PHYTOCHEMICAL COMPONENTS

Identified *Anemone* compounds include triterpenoids, saponins, steroids, lactones, fats and oils, saccharides, alkaloids, etc. (Sun et al., 2011; Cao et al., 2004; Zou et al., 2004). Oleanolic acid triterpene saponin is abundant in *Anemone* species (Table 8.2). *Anemone* contains ranunculin, anemonin, and protoanemonin, which are characteristic constituents of *Pulsatilla* and illustrate the close relationship between these two genera. *Anemone* also contains coumarins and flavonoids.

8.3.1 Saponin

Saponins are abundant in Ranunculaceae, especially in *Clematis*, *Pulsatilla*, *Anemone*, and Cimicifugeae (Hao et al., 2015a,b, 2013a,b), which usually exert anticancer activity via cell cycle arrest and apoptosis induction. The aglycones of *Clematis* pentacyclic triterpene saponins mainly belong to oleanolic type (A), olean-3β, 28-diol type (B), hederagenin type (C), or hederagenin-11,13-dien type (D), where types A and C are predominant (Hao et al., 2013a). In *Anemone*, A type (Table 8.2, Fig. 8.2) is predominant, and ursane-type triterpenoids (B type), lupane-type triterpenoids (C type), and cycloartane-type tetracyclic triterpenoids(D type) are also present (Table 8.3, Fig. 8.3).

8.3.2 Essential Oil, Volatile Compounds, and Others

Nineteen essential oil compounds are identified from the roots of *A. rivularis*, representing 96.1% of the total oil (Shi et al., 2012) (Table 8.4, Fig. 8.4). The major constituents were acetophenone (55.9%); 3-ethyl-2-methyl-hexane (16.2%); 5,6-dimethyl-decane (5.9%); and 4,5-diethyl-octane (4.4%).

The dominant benzenoid compounds in *Anemone sylvestris* anther are 2-phenylethanol and phenylacetaldehyde (Jurgens and Dotterl, 2004). Other abundant compound classes in *A. sylvestris* are fatty acid derivatives (41.8%), especially pentadecane and nonanal, and sesquiterpenoids(8.0%), for example, (E,E)-α-farnesene. Relatively low amounts of the repellent protoanemonin are found in *A. sylvestris*.

Coumarins, flavonoids, lactones, lignans, steroids, phenolic compounds, and other compounds are also detected (Table 8.4, Fig. 8.4).

TABLE 8.2 The Oleanane-Type Triterpenoids Isolated From *Anemone* Species (A Type)

No.	Type	Compound name	R_1	R_2	Taxon	Ref.
1.	A_1	Oleanolic acid	H-	H-	3,8,12,15,19,21	Wang (2012) and Tang (2013)
2.	A_1	3-Acetyloleanolic acid	H_3COC-	H-	15	Sun et al. (2011) and Cao et al. (2004)
3.	A_1	3-O-β-d-Glucopyranosyl oleanolic acid	Glc-	H-	12	Ding (2011)
4.	A_1	3-O-α-l-Rhamnopyranosyl oleanolic acid	Rha-	H-	12	Zou (2006)
5.	A_1	Raddeanoside R_0 (Fatsiaside A_1)	Ara-	H-	4,15,16,19,21	Wang et al. (2013b) and Hu et al. (2011)
6.	A_1	3-O-β-d-Xylopyranosyl oleanolic acid	Xyl-	H-	20	Wang (2012)
7.	A_1	3-O-β-d-Glucuronopyranosyl oleanolic acid	GlcA-	H-	8	Zhang (2007)
8.	A_1	3-O-β-d-Glucuronopyranosyl methyl ester oleanolic acid	GlcA methyl ester-	H-	8	Zhang (2007)
9.	A_1	3-O-α-l-Rhamnopyranosyl-(1→4)-β-d-glucopyranosyl oleanolic acid	Rha(1→4)Glc-	H-	12	Ding (2011)
10.	A_1	3-O-α-l-Rhamnopyranosyl-(1→2)-β-d-glucopyranosyl oleanolic acid	Rha(1→2)Glc-	H-	8	Ding (2011)
11.	A_1	Raddeanoside R_2	Glc(1→2)Ara-	H-	4,8,15–17	Ding (2011)
12.	A_1	Flaccidin B (Flaccidoside I)	Glc(1→2)Xyl-	H-	3,8	Zou et al. (2004)
13.	A_1	Eleatheroside K (Eleutheroside K)	Rha(1→2)Ara-	H-	7–9,15,16,21	Hou (2012)
14.	A_1	Raddeanoside R_1 (Raddeanin B)	Rha(1→4)Ara-	H-	15	Ding (2011)
15.	A_1	3-O-β-d-Glucopyranosyl-(1→4)-α-l-arabinopyranosyl oleanolic acid	Glc(1→4)Ara-	H-	15	Ding (2011)
16.	A_1	Obtusilobin	Ara(f)(1→2)Rha-	H-	13	Ding (2011)
17.	A_1	Giganteaside D	Rha(1→2)Xyl-	H-	3,8	Ding (2011)

18.	A$_1$	Narcissiflorine	Ara(f)(1→2)GlcA-	H-	12	Cao et al. (2004)
19.	A$_1$	3-O-β-l-Galactopyranosyl-(1→3)-β-d-glucopyranosyl oleanolic acid	Gal(1→3)Glc-	H-	21	Hu et al. (2011)
20.	A$_1$	3-O-α-l-Rhamnopyranosyl-(1→4)-β-d-glucuronopyranosyl oleanolic acid	Rha(1→4)GlcA-	H-	12	Ding (2011)
21.	A$_1$	3-O-α-l-Arabinopyranosyl-(1→4)-β-d-glucuronopyranosyl oleanolic acid	Ara(1→4)GlcA-	H-	12	Ding (2011)
22.	A$_1$	3-O-β-d-Glucopyranosyl-(1→3)-β-d-galactopyranosyl oleanolic acid	Glc(1→2)Gal-	H-	11	Ding (2011)
23.	A$_1$	3-O-α-l-Arabinopyranosyl-(1→3)-β-d-glucopyranosyl oleanolic acid	Ara(1→4)Glc-	H-	12	Wang (2012)
24.	A$_1$	3-O-β-d-Glucopyranosyl-(1→4)-[α-l-rhamnopyranosyl-(1→2)-α-l-arabinopyranosyl oleanolic acid	Glc(1→4)[Rha(1→2)]Ara-	H-	7	Ding (2011)
25.	A$_1$	Narcissifloridine	Ara(f)(1→2)Rha(1→4)Glc-	H-	12	Wang (2012)
26.	A$_1$	Obtusilobinin	Ara(f)(1→2)[Rha(1→4)]Glc-	H-	13	Wang (2012)
27.	A$_1$	Hupehensis saponin A	Rib(1→3)Rha(1→2)Glc-	H-	10	Wang (2012)
28.	A$_1$	Prosapogenin CP4	Rib(1→3)Rha(1→2)Ara-	H-	10,16,17,21	Wang (2012)
29.	A$_1$	Raddeanoside R$_3$ (Raddeanin A)	Rha(1→2)Glc(1→2)Ara-	H-	15	Wang (2012)
30.	A$_1$	Raddeanoside R$_6$ (Raddeanin E)	Glc(1→2)Rha(1→4)Ara-	H-	15	Wang (2012)
31.	A$_1$	Raddeanoside R$_{22}$	Glc(1→2)[Glc(1→4)]Ara-	H-	15	Wang (2012)
32.	A$_1$	3-O-α-l-Rhamnopyranosyl-(1→2)-β-d-glucopyranosyl-(1→4)-α-l-arabinopyranosyl oleanolic acid	Rha(1→2)Glc(1→4)Ara-	H-	7	Wang (2012)
33.	A$_1$	3-O-β-d-Xylopyranosyl-(1→3)-α-l-rhamnopyranosyl-(1→2)-α-l-arabinopyranosyl oleanolic acid	Xyl(1→3)Rha(1→2)Ara-	H-	7	Wang (2012)
34.	A$_1$	Huzhangoside A	Rib(1→3)Rha(1→2)Xyl-	H-	16	Wang (2012)
35.	A$_1$	Narcissiflorinine	Ara(f)(1→2)Rha(1→4)GlcA-	H-	12	Wang (2012)
36.	A$_1$	Raddeanoside R$_{13}$	Rha(1→2)[Glc(1→4)]Ara-	H-	15	Wang (2012)

(Continued)

TABLE 8.2   The Oleanane-Type Triterpenoids Isolated From *Anemone* Species (A Type) (*cont.*)

No.	Type	Compound name	R_1	R_2	Taxon	Ref.
37.	A_1	3-O-α-l-Rhamnopyranosyl-(1→2)-[α-l-arabinopyranosyl]-β-d-glucopyranosyl oleanolic acid	Rha(1→4)[Ara(1→2)]Glc-	H-	13	Wang (2012)
38.	A_1	3-O-α-l-Arabinopyranosyl-(1→2)-α-l-rhamnopyranosyl-(1→4)-[α-l-arabinopyranosyl-(1→2)]-β-d-glucopyranosyl oleanolic acid	Ara(1→2)Rha(1→4)[Ara(1→2)]Glc-	H-	12	Wang (2012)
39.	A_1	Rivularinin	Ara(f)(1→2)Rha(1→4)Glc(1→4)Glc-	H-	16	Wang (2012)
40.	A_1	Obtusilobicinin	Ara(f)(1→2)[Ara(f)(1→2)Rha(1→4)]Glc-	H-	13	Wang (2012)
41.	A_1	3-O-β-d-Xylopyranosyl-(1→3)- α-l-rhamnopyranosyl-(1→2)- β-d-glucopyranosyl-(1→4)-α-l-arabinopyranosyl oleanolic acid	Xyl(1→3)Rha(1→2)Glc(1→4)Ara-	H-	7	Wang (2012)
42.	A_1	3-O-β-d-Xylopyranosyl-(1→3)- α-l-rhamnopyranosyl-(1→2)- [β-d-glucopyranosyl-(1→4)]-α-l-arabinopyranosyl oleanolic acid	Xyl(1→3)Rha(1→2)[Glc(1→4)]Ara-	H-	7,17,19	Wang et al. (2013b)
43.	A_1	Raddeanoside R$_{23}$	Ara(1→3)Rha(1→2)[Glc(1→4)]Ara-	H-	15	Wang (2012)
44.	A_1	Hupehensis saponin B	Glc(1→3)Rib(1→3)Rha(1→2)Ara-	H-	10	Wang (2012)
45.	A_1	3-O-β-d-Xylopyranosyl-(1→2)-α-l-rhamnopyranosyl-(1→4)-[β-d-glucopyranosyl-(1→4)]-α-l-arabinopyranosyl oleanolic acid	Xyl(1→2)Rha(1→4)[Glc(1→4)]Ara-	H-	7	Wang (2012)
46.	A_1	3-O-β-d-Xylopyranosyl-(1→3)-α-l-rhamnopyranosyl-(1→2)-[β-d-glucopyranosyl-(1→4)-β-d-glucopyranosyl-(1→4)]-α-l-arabinopyranosyl oleanolic acid	Xyl(1→3)Rha(1→2)[Glc(1→4)Glc(1→4)]Ara-	H-	19	Wang (2012)
47.	A_1	Lucyoside H	Glc-	Glc-	8	Tang (2013)

No.		Name				Reference
48.	A_1	3-O-α-l-Arabinopyranosyl oleanolic acid 28-O-β-d-glucopyranoside	Ara-	Glc-	17	Wang (2014a,b)
49.	A_1	3-O-β-d-Glucopyranosyl-(1→2)-α-l-arabinopyranosyl oleanolic acid 28-O-β-d-glucopyranoside	Glc(1→2)Ara-	Glc-	17	Wang (2014a,b)
50.	A_1	Giganteasides G	Rha(1→2)Xyl-	Glc-	8	Tang (2013)
51.	A_1	Mateglycoside D	Rha(1→2)Ara-	Glc-	8	Tang (2013)
52.	A_1	Clematichinenoside A	Rib(1→3)Rha(1→2)Xyl-	Glc-	21	Wang (2012)
53.	A_1	3-O-β-d-Xylopyranosyl-(1→3)-α-l-rhamnopyranosyl-(1→2)-[β-d-glucopyranosyl-(1→4)-β-d-glucopyranosyl-(1→4)]-α-l-arabinopyranosyl oleanolic acid 28-O-β-d-glucopyranoside	Xyl(1→3)Rha(1→2)[Glc(1→4)Glc(1→4)]Ara-	Glc-	17	Wang (2014a,b)
54.	A_1	Asteryunnanoside F	H	Glc(1→6)Glc-	8	Tang (2013)
55.	A_1	Flaccidoside VII	Xyl	Glc(1→6)Glc-	8	Tang (2013)
56.	A_1	Hemsgiganoside B	GlcA-	Glc(1→6)Glc-	8	Tang (2013)
57.	A_1	Oleanolic acid 3-O-β-d-glucuronopyranosyl 28-O-β-d-glucopyranosyl-(1→6)-β-d-glucopyranoside	GlcA-	Glc(1→6)Glc-	8	Su (2007)
58.	A_1	Raddeanoside R_7 (Raddeanin F)	Glc(1→2)Ara-	Rha(1→4)Glc-	3,15	Wang (2012)
59.	A_1	Raddeanoside R_4 (Raddeanin C)	Rha(1→4)Ara-	Rha(1→2)Glc-	15	Wang (2012)
60.	A_1	Anhuienside C	Rha(1→2)Xyl-	Glc(1→6)Glc-	3,8	Wang (2012)

(Continued)

TABLE 8.2 The Oleanane-Type Triterpenoids Isolated From *Anemone* Species (A Type) (*cont.*)

No.	Type	Compound name	R_1	R_2	Taxon	Ref.
61.	A_1	3-O-α-l-Rhamnopyranosyl-(1→2)-β-d-glucopyranosyl olean-12-en-28-oic acid 28-O-β-d-glucopyranosyl-(1→6)-β-d-glucopyranosyl ester	Rha(1→2)Glc-	Glc(1→6) Glc-	8	Tang (2013)
62.	A_1	3-O-β-d-Glucopyranosyl-(1→2)-α-l-arabinopyranosyl oleanolic acid 28-O-[β-d-glucopyranosyl-(1→6)-β-d-glucopyranoside)	Glc(1→2)Ara-	Glc(1→6) Glc-	17	Wang (2014a,b)
63.	A_1	3-O-α-l-Rhamnopyranosyl-(1→2)-α-l-arabinofurnosyl olean-12-en-28-oic acid 28-O-β-d-glucopyranosyl-(1→6)-β-d-glucopyranosyl ester	Rha(1→2)Ara-	Glc(1→6) Glc-	8	Tang (2013)
64.	A_1	Tomentoside D	Ara(1→2)Xyl-	Glc(1→4) Glc-	21	Wang (2012)
65.	A_1	Raddeanoside R_5 (Raddeanin D)	Rha(1→2)Glc(1→2)Ara-	Rha(1→4) Glc-	15	Wang (2012)
66.	A_1	Raddeanoside R_{14}	Rha(1→2)[Glc(1→4)]Ara-	Glc(1→6) Glc-	9,15	Wang (2012)
67.	A_1	Anemoside A	Rib(1→3)Rha(1→2)Ara-	Rha(1→4) Glc-	16	Wang (2012)
68.	A_1	Cussonoside B	H-	Rha(1→4) Glc(1→6) Glc-	3,4,8,9,17,19	Wang (2012)
69.	A_1	Ciwujianoside C_3	Ara-	Rha(1→4) Glc(1→6) Glc-	4,17	Wang (2012)
70.	A_1	Anhuienside D	Xyl-	Rha(1→4) Glc(1→6) Glc-	3	Wang (2012)

No.		Compound				Reference
71.	A_1	Nipponoside B	Glc-	Rha(1→4)Glc(1→6)Glc-	8	Tang (2013)
72.	A_1	Oleanolic acid 3-O-β-d-glucuronopyranosyl 28-O-α-l-rhamnopyranosyl-(1→4)-β-d-glucopyranosyl-(1→6)-β-d-glucopyranoside	GlcA-	Rha(1→4)Glc(1→6)Glc-	8	Tang (2013)
73.	A_1	3-O-[3-O-Sulfonatol]-β-d-glucuronopyranosyl olean-12-en-28-oic acid 28-O-α-l-rhamnopyranosyl-(1→4)-β-d-glucopyranosyl-(1→6)-β-d-glucopyranosyl ester monosodium salt	(3'-NaSO₃)-GlcA-	Rha(1→4)Glc(1→6)Glc-	2	Zhang et al. (2012)
74.	A_1	Oleanolic acid 3-O-β-d-glucopyranosyl-(6'-butyryl) 28-O-α-l-rhamnopyranosyl-(1→4)-β-d-glucopyranosyl-(1→6)-β-d-glucopyranoside	(6'-butyryl)-Glc-	Rha(1→4)Glc(1→6)Glc-	8	Wang (2012)
75.	A_1	Oleanolic acid 3-O-β-d-glucuronopyranose methyl ester 28-O-α-l-rhamnopyranosyl-(1→4)-β-d-glucopyranosyl-(1→6)-β-d-glucopyranoside	GlcA methyl ester-	Rha(1→4)Glc(1→6)Glc-	8	Wang (2012)
76.	A_1	Anhuienside E	Rha(1→2)Glc-	Rha(1→4)Glc(1→6)Glc-	3,8	Wang (2012) and Bing et al. (2008)
77.	A_1	Ciwujianoside A_1	Glc(1→2)Ara-	Rha(1→4)Glc(1→6)Glc-	4,17,20	Wang (2012)
78.	A_1	Begoniifolide A	Glc(1→3)Ara-	Rha(1→4)Glc(1→6)Glc-	4	Wang (2012)
79.	A_1	Raddeanoside R_{19b}	Glc(1→4)Ara-	Rha(1→4)Glc(1→6)Glc-	15	Wang (2012)

(Continued)

TABLE 8.2 The Oleanane-Type Triterpenoids Isolated From *Anemone* Species (A Type) (*cont.*)

No.	Type	Compound name	R_1	R_2	Taxon	Ref.
80.	A_1	Hederasaponin B (Hederagenin B)	Rha(1→2)Ara-	Rha(1→4)Glc(1→6)Glc-	3,7,8,15,17,18,21	Wang et al. (2013d)
81.	A_1	FlaccidosideII	Rha(1→2)Xyl-	Rha(1→4)Glc(1→6)Glc-	3,8	Bing et al. (2008)
82.	A_1	FlaccidosideIII	Glc(1→2)Xyl-	Rha(1→4)Glc(1→6)Glc-	3,8,9	Wang (2012)
83.	A_1	Flaccidoside V	Rha(1→2)Qui-	Rha(1→4)Glc(1→6)Glc-	8	Tang (2013)
84.	A_1	Flaccidoside VI	Glc(1→2)Glc	Rha(1→4)Glc(1→6)Glc-	8	Tang (2013)
85.	A_1	3-O-α-l-Rhamnopyranosyl-(1→2)-β-d-glucuronopyranosyl olean-12-en-28-oic acid 28-O-α-l-rhamnopyranosyl-(1→4)-β-d-glucopyranosyl-(1→6)-β-d-glucopyranosyl ester	Rha(1→2)GlcA-	Rha(1→4)Glc(1→6)Glc-	8	Tang (2013)
86.	A_1	3-O-β-d-Xylopyranosyl-(1→2)-α-l-arabinopyranosyl oleanolic acid 28-O-α-l-rhamnopyranosyl-(1→4)-β-d-glucopyranosyl-(1→6)-β-d-glucopyranoside	Xyl(1→2)Ara-	Rha(1→4)Glc(1→6)Glc-	17	Wang (2014a,b)
87.	A_1	Koreanaside 1	Xyl(1→3)Ara-	Rha(1→4)Glc(1→6)Glc-	21	Wang (2012)
88.	A_1	Raddeanoside R_{18} (Begoniifolide D)	Glc(1→2)[Glc(1→4)]Ara-	Rha(1→4)Glc(1→6)Glc-	4	Wang (2012)

89.	Raddeanoside R$_8$	A$_1$	Rha(1→2)Glc(1→2)Ara-	Rha(1→4) Glc(1→6) Glc-	9,15	Wang (2012)
90.	Hederacholichiside E	A$_1$	Rha(1→2)[Glc(1→4)]Ara-	Rha(1→4) Glc(1→6) Glc-	7,15,17-19	Wang et al. (2013a)
91.	Raddeanoside R$_{15}$	A$_1$	Ara(1→3)Rha(1→2)Ara-	Rha(1→4) Glc(1→6) Glc-	15	Wang (2012)
92.	Huzhangoside B	A$_1$	Rib(1→3)Rha(1→2)Ara-	Rha(1→4) Glc(1→6) Glc-	10,16,21	Wang (2012)
93.	Sieboldianoside B	A$_1$	Xyl(1→3)Rha(1→2)Ara-	Rha(1→4) Glc(1→6) Glc-	7,17,19	Wang (2012)
94.	Anhuienside F	A$_1$	Glc(1→3)Rha(1→2)Xyl-	Rha(1→4) Glc(1→6) Glc-	3,8	Wang (2012)
95.	Huzhangoside C	A$_1$	Rib(1→3)Rha(1→2)Xyl-	Rha(1→4) Glc(1→6) Glc-	16,21	Wang (2012)
96.	Tomentoside C	A$_1$	Gal(1→3)Rha(1→2)Xyl-	Rha(1→4) Glc(1→6) Glc-	21	Wang (2012)
97.	3-O-β-d-Glucopyranosyl-(1→4)-α-l-rhamnopyranosyl-(1→2)-α-l-arabinopyranosyl oleanolic acid 28-O-α-l-rhamnopyranosyl-(1→4)-β-d-glucopyranosyl-(1→6)-β-d-glucopyranoside	A$_1$	Glc(1→4)Rha(1→2)Ara-	Rha(1→4) Glc(1→6) Glc-	15	Ding (2011)

(Continued)

TABLE 8.2 The Oleanane-Type Triterpenoids Isolated From *Anemone* Species (A Type) (*cont.*)

No.	Type	Compound name	R_1	R_2	Taxon	Ref.
98.	A_1	3-O-α-l-Rhamnopyranosyl-(1→2)-β-d-Glucopyranosyl-(1→4)-α-l-arabinopyranosyl oleanolic acid 28-O-α-l-rhamnopyranosyl-(1→4)-β-d-glucopyranosyl-(1→6)-β-d-glucopyranoside	Rha(1→2)Glc(1→4)Ara-	Rha(1→4) Glc(1→6) Glc-	7	Ding (2011)
99.	A_1	3-O-α-l-Arabinopyranosyl-(1→3)-α-l-rhamnopyranosyl-(1→2)-β-d-Glucopyranosyl-(1→4)-α-l-arabinopyranosyl oleanolic acid 28-O-α-l-rhamnopyranosyl-(1→4)-β-d-glucopyranosyl-(1→6)-β-d-glucopyranoside	Ara(1→3)Rha(1→2) Glc(1→4)Ara-	Rha(1→4) Glc(1→6) Glc-	15	Ding (2011)
100.	A_1	3-O-β-d-Xylopyranosyl-(1→3)-α-l-rhamnopyranosyl-(1→2)-[β-d-glucopyranosyl-(1→4)]-α-l-arabinopyranosyl 28-O-α-l-rhamnopyranosyl-(1→4)-β-d-glucopyranosyl-(1→6)-β-d-glucopyranosyl oleanolic acid	Xyl(1→3)Rha(1→2)[Glc(1→4)]Ara-	Rha(1→4) Glc(1→6) Glc-	19	Wang et al. (2013b)
101.	A_1	Raddeanoside R_{17}	Glc(1→4)Glc(1→3) Rha(1→2)Ara-	Rha(1→4) Glc(1→6) Glc-	15	Ding (2011)
102.	A_1	Raddeanoside R_{16}	Ara(1→3)Rha(1→2)[Glc(1→4)]Ara-	Rha(1→4) Glc(1→6) Glc-	15	Wang (2012)
103.	A_1	Hupehensis saponin D	Glc(1→3)Rib(1→3) Rha(1→2)Ara-	Rha(1→4) Glc(1→6) Glc-	10	Ding (2011)
104.	A_1	3-O-β-d-Glucopyranosyl-(1→4)-β-d-ribopyranosyl-(1→3)-α-l-rhamnopyranosyl-(1→2)-β-d-xylopyranosyl 28-O-α-l-rhamnopyranosyl-(1→4)-β-d-glucopyr anosyl-(1→6)-β-d-glucopyranosyl l d-glucopyranosyl oleanolic acid	Glc(1→4)Rib(1→3) Rha(1→2)Xyl-	Rha(1→4) Glc(1→6) Glc-	21	Wang (2012)

No.		Name				Reference
105.	A_1	3-O-[3-O-Sulfonato]-β-d-glucuronopyranosyl olean-12-en-28-oic acid 28-O-4-O-acety-α-l-rhamnopyranosyl-(1→4)-β-d-glucopyranosyl-(1→6)-β-d-glucopyranosyl ester monosodium salt	(3'-NaSO$_3$)-GlcA-	Ac(1→4) Rha(1→4) Glc(1→6) Glc-	2	Wang (2012)
106.	A_1	Oleanolic acid 3-O-β-d-glucuronopyranosyl 28-O-β-d-glucopyranosyl-(1→3)-α-l-rhamnopyranosyl-(1→4)-β-d-glucopyranosyl-(1→6)-β-d-glucopyranoside	GlcA-	Glc(1→3) Rha(1→4) Glc(1→6) Glc-	2	Wang (2012)
107.	A_1	3-O-[3-O-Sulfonato]-β-d-glucuronopyranosyl olean-12-en-28-oic acid 28-O-β-d-glucopyranosyl-(1→3)-α-l-rhamnopyranosyl-(1→4)-β-d-glucopyranosyl-(1→6)-β-d-glucopyranosyl ester monosodiu salt	(3'-NaSO$_3$)-GlcA-	Glc(1→3) Rha(1→4) Glc(1→6) Glc-	2	Wang (2012)
108.	A_1	Raddeanoside R$_{10}$	Rha(1→2)Glc(1→2)Ara-	Glc(1→3) Rha(1→4) Glc(1→6) Glc-	15	Wang (2012)
109.	A_1	Oleanolic acid 3-O-β-d-ribopyranosyl-(1→3)-α-l-rhamnopyranosyl-(1→2)-α-l-arabinopyranosyl 28-O-β-d-glucopyranosyl-(1→3)-α-l-rhamnopyranosyl-(1→4)-β-d-glucopyranosyl-(1→6)-β-d-glucopyranoside	Rib(1→3)Rha(1→2)Ara-	Glc(1→3) Rha(1→4) Glc(1→6) Glc-	10	Wang (2012)
110.	A_1	Oleanolic acid 3-O-β-d-ribopyranosyl-(1→3)-α-l-rhamnopyranosyl-(1→2)-[β-d-glucopyranosyl-(1→4)]-α-l-arabinopyranosyl 28-O-β-d-glucopyranosyl-(1→3)-α-l-rhamnopyranosyl-(1→4)-β-d-glucopyranosyl-(1→6)-β-d-glucopyranoside	Rib(1→3)Rha(1→2)[Glc(1→4)]Ara-	Glc(1→3) Rha(1→4) Glc(1→6) Glc-	10	Wang (2012)

(Continued)

TABLE 8.2 The Oleanane-Type Triterpenoids Isolated From *Anemone* Species (A Type) *(cont.)*

No.	Type	Compound name	R_1	R_2	Taxon	Ref.
111.	A_1	Raddeanoside R_{11}	Rha(1→2)Glc(1→2)Ara-	Glc(1→6) Glc(1→3) Rha(1→4) Glc(1→6) Glc-	12,15	Wang (2012)
112.	A_1	3-O-α-l-Rhamnopyranosyl-(1→2)-α-l-arabinopyranosyl oleanolic acid 28-O-α-l-rhamnopyranosyl-(1→4)-β-d-glucopyranosyl-(1→6)-β-d-glucopyranosyl-(1→4)-α-l-rhamnopyranosyl-(1→4)-β-d-glucopyranosyl-(1→6)-β-d-glucopyranoside	Rha(1→2)Ara-	Rha(1→4) Glc(1→6) Glc(1→4) Rha(1→4) Glc(1→6) Glc-	15,19	Wang (2012)
113.	A_1	3-O-α-l-Rhamnopyranosyl-(1→2)-β-d-glucopyranosyl-(1→2)-α-l-arabinopyranosyl oleanolic acid 28-O-α-l-rhamnopyranosyl-(1→4)-β-d-glucopyranosyl-(1→6)-β-d-glucopyranosyl-(1→4)-α-l-rhamnopyranosyl-(1→4)-β-d-glucopyranosyl-(1→6)-β-d-glucopyranoside	Rha(1→2)Glc(1→2)Ara-	Rha(1→4) Glc(1→6) Glc(1→4) Rha(1→4) Glc(1→6) Glc-	15	Wang (2012)
114.	A_1	3-O-β-d-Xylopyranosyl-(1→3)-α-l-rhamnopyranosyl-(1→2)-1β-d-glucopyranosyl-(1→4)]-α-l-arabinopyranosyl oleanolic acid 28-O-α-l-rhamnopyranosyl-(1→4)-β-d-glucopyranosyl-(1→6)-β-d-glucopyranosyl-(1→4)-α-l-rhamnopyranosyl-(1→4)-β-d-glucopyranosyl-(1→6)-β-d-glucopyranoside	Rha(1→2)[Glc(1→4)]Ara-	Rha(1→4) Glc(1→6) Glc(1→4) Rha(1→4) Glc(1→6) Glc-	15,19	Wang (2012)

#		Name				Reference
115.	A_1	Raddeanoside R_{19a}	Rha(1→2)Glc(1→2)Ara-	Rha(1→4) Glc(1→6) Glc(1→3) Rha(1→4) Glc(1→6) Glc(1→3) Rha(1→4) Glc(1→6) Glc-	15	Wang (2012)
116.	A_2	Hederagenin	H-	H-	17-19	Wang (2012)
117.	A_2	Hederagenin 3-O-β-d-glucopyranoside	Glc-	H-	11	Ding (2011)
118.	A_2	Hederagenin 3-O-α-l-arabinopyranoside	Ara-	H-	14,17,19,21	Hu et al. (2011)
119.	A_2	3-O-β-d-galactopyranosyl-(1→2)-β-d-glucopyranosyl hederagenin	Gal(1→2)Glc-	H-	11	Ding (2011)
120.	A_2	3-O-β-d-glucopyranosyl-(1→2)-α-l-arabinopyranosyl hederagenin	Glc(1→2)Ara-	H-	14	Ding (2011)
121.	A_2	3-O-β-d-glucopyranosyl-(1→2)-β-d-glucopyranosyl hederagenin	Glc(1→2)Glc-	H-	14	Ding (2011)
122.	A_2	3-O-β-d-Xylopyranosyl-(1→2)-α-l-arabinopyranosyl hederagenin	Xyl(1→2)Ara-	H-	17	Wang (2014a,b)
123.	A_2	Kalopanaxsaponin A	Rha(1→2)Ara-	H-	17-19	Wang et al. (2013b)
124.	A_2	Leontoside B	Glc(1→4)Ara-	H-	4,14,15,18,19	Wei et al. (2012)
125.	A_2	3-O-β-d-Glucopyranosyl-(1→2)-[β-d-glucopyranosyl-(1→6)]-β-d-galactopyranosyl hederagenin	Glc(1→2)[Glc(1→6)]Gal-	H-	11	Ding (2011)
126.	A_2	3-O-β-d-Glucopyranosyl-(1→4)-β-d-glucopyranosyl-(1→2)-α-l-arabinopyranosyl hederagenin	Glc(1→4)Glc(1→2)Ara-	H-	14	Ding (2011)
127.	A_2	Sapindoside B	Xyl(1→3)Rha(1→2)Ara-	H-	17	Wang (2014a,b)
128.	A_2	Pulsatiloside D	Rha(1→2)[Glc(1→4)]Ara-	H-	17-19	Wang et al. (2013b)
129.	A_2	Prosapogenin CP6	Rib(1→3)Rha(1→2)Ara-	H-	10,16,17	Ding (2011)

(Continued)

TABLE 8.2 The Oleanane-Type Triterpenoids Isolated From *Anemone* Species (A Type) *(cont.)*

No.	Type	Compound name	R_1	R_2	Taxon	Ref.
130.	A_2	Clemargenoside G	Rib(1→3)Rha(1→2)Xyl-	H-	21	Wang (2012)
131.	A_2	Hupehensis saponin C	Glc(1→3)Rib(1→3)Rha(1→2)Ara-	H-	10	Wang (2012)
132.	A_2	3-O-β-d-Glucopyranosyl-(1→3)-α-1-rhamnopyranosyl-(1→2)-[β-d-glucopyranosyl-(1→4)]-α-1-arabinopyranosyl hederagenin	Glc(1→3)Rha(1→2)[Glc(1→4)]Ara-	H-	10	Wang (2012)
133.	A_2	3-O-β-d-Xylopyranosyl-(1→3)-α-1-rhamnopyranosyl-(1→2)-[β-d-glucopyranosyl-(1→4)]-α-1-arabinopyranosyl hederagenin	Xyl(1→3)Rha(1→2)[Glc(1→4)]Ara-	H-	19	Wang (2012)
134.	A_2	Tomentoside Aa	Rib(1→3)Rha(1→2)[Glc(1→4)]Ara-	H-	21	Wang (2012)
135.	A_2	3-O-β-d-Xylopyranosyl-(1→3)-α-1-rhamnopyranosyl-(1→2)-[β-d-glucopyranosyl-(1→4)-β-d-glucopyranosyl-(1→4)]-α-1-arabinopyranosyl hederagenin	Xyl(1→3)Rha(1→2)[Glc(1→4)Glc(1→4)]Ara-	H-	19	Wang et al. (2013b)
136.	A_2	Hederagenin-28-O-β-d-glucopyranosyl ester	H-	Glc	17	Ding (2011)
137.	A_2	HN-saponin F	Ara-	Glc-	17,20	Wang (2012)
138.	A_2	3-O-β-d-Glucopyranosyl-(1→2)-β-d-galactopyranosyl hederagenin 28-O-β-d-glucopyranoside	Glc(1→2)Gal-	Glc-	11	Ding (2011)
139.	A_2	3-O-β-d-Glucopyranosyl-(1→2)-α-1-arabinopyranosyl hederagenin 28-O-β-d-glucopyranoside	Glc(1→2)Ara-	Glc-	17	Wang (2014a,b)
140.	A_2	3-O-β-d-Ribopyranosyl-(1→3)-α-1-rhamnopyranosyl-(1→2)-α-1-arabinopyranosyl hederagenin 28-O-β-d-glucopyranoside	Rib(1→3)Rha(1→2)Ara-	Glc-	17	Wang (2014a,b)
141.	A_2	Akebia saponinin D	Ara-	Glc(1→6)Glc-	17,20	Wang (2012)

142.	A$_2$	3-O-α-l-rhamnopyranosyl-(1→2)-[β-d-glucopyranosyl-(1→4)]-α-l-arabinopyranosyl hederagenin 28-O-β-d-glucopyranosyl-(1→6)-β-d-glucopyranoside	Rha(1→2)[Glc(1→4)]Ara-	Glc(1→6) Glc-	6	Wang (2012)
143.	A$_2$	Pulsatiloside C	H-	Rha(1→4) Glc(1→6) Glc-	17,20	Wang (2012)
144.	A$_2$	Begoniifolide C	CH$_3$OCOCH$_2$CO-	Rha(1→4) Glc(1→6) Glc-	4	Wang (2012)
145.	A$_2$	Cauloside D	Ara-	Rha(1→4) Glc(1→6) Glc-	4,16,17,19,20	Wang et al. (2013b)
146.	A$_2$	3-O-β-d-Glucopyranosyl hederagenin 28-O-α-l-rhamnopyranosyl-(1→4)-β-d-glucopyranosyl-(1→6)-β-d-glucopyranoside	Glc-	Rha(1→4) Glc(1→6) Glc-	11	Ding (2011)
147.	A$_2$	Cauloside F	Glc(1→2)Ara-	Rha(1→4) Glc(1→6) Glc-	17,20	Ding (2011)
148.	A$_2$	Leontoside D	Glc(1→4)Ara-	Rha(1→4) Glc(1→6) Glc-	15,19	Ding (2011)
149.	A$_2$	Kalopanaxsaponin B	Rha(1→2)Ara-	Rha(1→4) Glc(1→6) Glc-	7,17-19	Ding (2011)
150.	A$_2$	Hederacholichiside F	Rha(1→2)[Glc(1→4)]Ara-	Rha(1→4) Glc(1→6) Glc-	18,19	Wang (2012)
151.	A$_2$	Huzhangoside D	Rib(1→3)Rha(1→2)Ara-	Rha(1→4) Glc(1→6) Glc-	10,16,17,21	Wang (2012)
152.	A$_2$	Clematiganoside A	Rib(1→3)Rha(1→2)Xyl-	Rha(1→4) Glc(1→6) Glc-	21	Wang et al. (2013d)

(Continued)

TABLE 8.2 The Oleanane-Type Triterpenoids Isolated From *Anemone* Species (A Type) *(cont.)*

No.	Type	Compound name	R_1	R_2	Taxon	Ref.
153.	A_2	Sieboldianoside A	Xyl(1→3)Rha(1→2)Ara-	Rha(1→4) Glc(1→6) Glc-	17,19	Wang (2014a,b)
154.	A_2	3-O-β-d-Glucopyranosyl-(1→2)-[β-d-glucopyranosyl-(1→6)]-β-d-galactopyranosyl hederagenin 28-O-α-l-rhamnopyranosyl-(1→4)-β-d-glucopyranosyl-(1→6)-β-d-glucopyranoside	Glc(1→2)[Glc(1→6)]Gal-	Rha(1→4) Glc(1→6) Glc-	11	Ding (2011)
155.	A_2	3-O-β-d-Glucopyranosyl-(1→4)-β-d-glucopyranosyl-(1→2)-α-l-arabinopyranosyl hederagenin 28-O-α-l-rhamnopyranosyl-(1→4)-β-d-glucopyranosyl-(1→6)-β-d-glucopyranoside	Glc(1→4)Glc(1→2)Ara-	Rha(1→4) Glc(1→6) Glc-	14	Ding (2011)
156.	A_2	Hupehensis saponin E	Glc(1→3)Rib(1→3) Rha(1→2)Ara-	Rha(1→4) Glc(1→6) Glc-	10	Ding (2011)
157.	A_2	Tomentoside A	Rib(1→3)Rha(1→2) [Glc(1→4)]Ara-	Rha(1→4) Glc(1→6) Glc-	21	Wang et al. (2013d)
158.	A_2	Tomentoside B	Rib(1→3)Rha(1→2) [Glc(1→4)]Xyl-	Rha(1→4) Glc(1→6) Glc-	21	Wang et al. (2013d)
159.	A_2	3-O-β-d-Glucopyranosyl-(1→4)-β-d-glucopyranosyl-(1→4)-[α-l-rhamnopyranosyl-(1→2)]-α-l-arabinopyranosyl hederagenin 28-O-α-l-rhamnopyranosyl-(1→4)-β-d-glucopyranosyl-(1→6)-β-d-glucopyranoside	Glc(1→4)Glc(1→4) [Rha(1→2)]Ara-	Rha(1→4) Glc(1→6) Glc-	19	Wang (2014a,b)

160.	A$_2$	3-O-β-d-Xylopyranosyl-(1→3)-α-lrhamnopyranosyl-(1→2)-[β-d-glucopyranosyl-(1→4)]-α-l-arabinopyranosyl hederagenin 28-O-α-l-rhamnopyranosyl-(1→4)-β-d-glucopyranosyl-(1→6)-β-d-glucopyranoside	Xly(1→3)Rha(1→2)[Glc(1→4)]Ara-	Rha(1→4) Glc(1→6) Glc-	19	Wang (2014a,b)
161.	A$_2$	3-O-β-d-Xylopyranosyl-(1→3)-α-l-rhamnopyranosyl-(1→2)-[β-d-glucopyranosyl-(1→4)]-α-l-arabinopyranosyl hederagenin 28-O-α-l-rhamnopyranosyl-(1→4)-β-d-glucopyranosyl-(1→6)-β-d-glucopyranoside	Xly(1→3)Rha(1→2)[Glc(1→4)Glc(1→4)]Ara-	Rha(1→4) Glc(1→6) Glc-	19	Wang (2014a,b)
162.	A$_2$	3-O-β-d-Glucuronopyranosyl hederagenin 28-O-β-d-glucopyranosyl-(1→3)-α-l-rhamnopyranosyl-(1→4)-β-d-glucopyranosyl-(1→6)-β-d-glucopyranoside	GlcA-	Glc(1→3) Rha(1→4) Glc(1→6) Glc-	2	Tang (2013)
163.	A$_2$	3-O-β-d-Ribopyranosyl-(1→3)-α-l-rhamnopyranosyl-(1→2)-α-l-arabinopyranosyl hederagenin 28-O-β-d-glucopyranosyl-(1→3)-α-l-rhamnopyranosyl-(1→4)-β-d-glucopyranosyl-(1→6)-β-d-glucopyranoside	Rib(1→3)Rha(1→2)Ara-	Glc(1→3) Rha(1→4) Glc(1→6) Glc-	10	Tang (2013)
164.	A$_2$	Hederagenin 28-O-α-l-rhamnopyranosyl-(1→4)-β-d-glucopyranosyl-(1→6)-β-dglucopyranosyl-(1→4)-α-l-rhamnopyranosyl-(1→4)-β-d-glucopyranosyl-(1→6)-β-d-glucopyranoside	H-	Rha(1→4) Glc(1→6) Glc(1→4) Rha(1→4) Glc(1→6) Glc-	19	Wang (2014a,b)

(Continued)

TABLE 8.2 The Oleanane-Type Triterpenoids Isolated From *Anemone* Species (A Type) (*cont.*)

No.	Type	Compound name	R₁	R₂	Taxon	Ref.
165.	A₂	3-O-β-d-Xylopyranosyl-(1→3)-α-l-rhamnopyranosyl-(1→2)-α-l-arabinopyranosyl hederagenin 28-O-α-l-rhamnopyranosyl-(1→4)-β-d-glucopyranosyl-(1→6)-β-d-glucopyranosyl-(1→4)-α-l-rhamnopyranosyl-(1→4)-β-d-glucopyranosyl-(1→6)-β-d-glucopyranoside	Xyl(1→3)Rha(1→2)Ara-	Rha(1→4) Glc(1→6) Glc(1→4) Rha(1→4) Glc(1→6) Glc-	19	Wang (2014a,b)
166.	A₂	Hupehensis saponin F	Rib(1→3)Rha(1→2)Ara-	Rha(1→4) Glc(1→6) Glc(1→3) Rha(1→4) Glc(1→6) Glc-	10	Wang et al. (1997)
167.	A₂	Hupehensis saponin G	Rib(1→3)Rha(1→2)Ara-	Glc(1→6) Rha(1→4) Glc(1→6) Glc(1→3) Rha(1→4) Glc(1→6) Glc-	10	Wang et al. (1997)
168.	A₃	Raddeanoside Rₐ	Glc(1→4)Ara-	H-	15,19	Wei et al. (2012)
169.	A₃	27-Hydroxyoleanolic acid 3-O-β-d-glucopyranosyl (1→2)-α-l-arabinopyranoside	Glc(1→2)Ara-	H-	15	Wei et al. (2012)
170.	A₃	Raddeanoside R₁₂	Rha(1→2)Ara-	H-	15	Ding (2011)
171.	A₃	Raddeanoside R₂₀	Rha(1→2)[Glc(1→4)]ara-	H-	15	Lu et al. (2009)
172.	A₃	Raddeanoside R_b	Ara(1→3)Rha(1→2)Ara-	H-	15,19	Wei et al. (2012)
173.	A₃	Hydroxyoleanolic acid 3-O-α-l-rhamnopyranosyl (1→2)-β-d-glucopyranosyl-(1→2)-α-l-arabinopyranoside	Rha(1→2)Glc(1→2)Ara-	H-	15	Wang (2012)

No.	Group	Name				Reference
174.	A$_3$	Raddeanoside R$_{21}$	Rha(1→2)Ara-	Rha(1→4) Glc(1→6) Glc-	15	Lu et al. (2009)
175.	A$_3$	Hydroxyoleanolic acid 3-O-β-d-glucopyranosyl-(1→4)-α-l-arabinopyranosyl-28-O-α-l-rhamnopyranosyl(1→4)-β-d-glucopyranosyl(1→6)-β-d-glucopyranoside	Glc(1→4)Ara-	Rha(1→4) Glc(1→6) Glc-	15	Wang (2012)
176.	A$_3$	Raddeanoside R$_9$	Rha(1→2)Glc(1→2)Ara-	Rha(1→4) Glc(1→6) Glc-	15	Ding (2011)
177.	A$_3$	Hydroxyoleanolic acid 3-O-α-l-rhamnopyranosyl(1→2)-[β-d-glucopyranosyl-(1→4)]-α-l-arabinopyranosyl-28-O-α-l-rhamnopyranosyl(1→4)-β-d-glucopyranosyl(1→6)-β-d-glucopyranoside	Rha(1→2)[Glc(1→4)]Ara-	Rha(1→4) Glc(1→6) Glc-	15	Wei et al. (2012)
178.	A$_4$	Begoniifolide B	CH$_3$OOCH$_2$COC-	Rha(1→4) Glc(1→6) Glc-	10	Wang (2012)
179.	A$_4$	Tetrasepaloside	(2′,3′-monoaccetonide)-Rib(f)-	Rha(1→4) Glc(1→6) Glc-	20	Wang (2012)
180.	A$_4$	Anhuienside A	Glc(1→2)Glc-	H-	3	Wang (2012)
181.	A$_4$	Anhuienside B	Glc(1→2)Glc-	Rha(1→4) Glc(1→6) Glc-	3	Wang (2012)
182.	A$_5$	3β-[(O-β-d-glucopyranosyl-(1→4)-O-[a-l-rhamnopyranosyl-(1→2)]-α-l-arabinopyranosyl)oxy]-2β-hydroxyolean-12-en-28-oic acid O-α-l-rhamnopyranosyl-(1→4)-O-β-d-glucopyranosyl-(1→6)-β-d-glucopyranosyl ester	Rha(1→2)[Glc(1→4)]Ara-	Rha(1→4) Glc(1→6) Glc-	6	Mimaki et al. (2009)

(Continued)

TABLE 8.2 The Oleanane-Type Triterpenoids Isolated From *Anemone* Species (A Type) (*cont.*)

No.	Type	Compound name	R_1	R_2	Taxon	Ref.
183.	A_6	3β[(O-β-d-Glucopyranosyl-(1→4)-O-[α-l-rhamnopyranosyl-(1→2)]-α-l-arabinopyranosyl)oxy]-2β,23-dihydroxyolean-12-en-28-oic acid	Rha(1→2)[Glc(1→4)]Ara-	H-	6	Mimaki et al. (2009)
184.	A_6	3β[(O-β-d-Glucopyranosyl-(1→4)-O-[α-l-rhamnopyranosyl-(1→2)]-α-l-arabinopyranosyl)oxy]-2β,23-dihydroxyolean-12-en-28-oic acid O-α-l-rhamnopyranosyl-(1→4)-O-β-d-glucopyranosyl-(1→6)-β-d-glucopyranosyl ester	Rha(1→2)[Glc(1→4)]Ara-	Rha(1→4) Glc(1→6) Glc-	6	Mimaki et al. (2009)
185.	A_7	Anemonerivulariside A	Rib(1→3)Rha(1→2)Ara-	-	16	Anh Minh et al. (2012)
186.	A_8	Siaresinolic acid	H-	H-	19	Wang et al. (2013b)
187.	A_8	3-O-β-d-Xylopyranosyl-(1→3)-α-l-rhamnopyranosyl-(1→2)-[β-d-glucopyranosyl-(1→4)]-α-l-arabinopyra-nosyl siaresinolic acid	Xyl(1→3)Rha(1→2) [Glc(1→4)]Ara-	H-	19	Wang et al. (2013b)
188.	A_8	3-O-β-d-Glucopyranosyl-(1→4)-α-l-arabinopyranosyl siaresinolic acid 28-O-α-l-rhamnopyranosyl-(1→4)-β-d-glucopyranosyl-(1→6)-β-d-glucopyranosyl ester	Glc(1→4)Ara-	Rha(1→4) Glc(1→6) Glc-	19	Wang et al. (2013b)
189.	A_8	3-O-α-l-Rhamnopyranosyl-(1→2)-α-l-arabinopyranosyl siaresinolic acid 28-O-α-l-rhamnopyranosyl-(1→4)-β-d-glucopyranosyl-(1→6)-β-d-glucopyranoside	Rha(1→2)Ara-	Rha(1→4) Glc(1→6) Glc-	19	Wang (2014a,b)
190.	A_8	3-O-α-l-Rhamnopyranosyl-(1→2)-[β-d-glucopyranosyl(1→4)]-α-l-arabinopyranosyl siaresinolic acid 28-O-α-l-rhamnopyranosyl-(1→4)-β-d-glucopyranosyl-(1→6)-β-d-glucopyranoside	Rha(1→2)[Glc(1→4)]Ara-	Rha(1→4) Glc(1→6) Glc-	19	Wang (2014a,b)

191.	A$_8$	3-O-β-d-Xylopyranosyl-(1→3)-α-l-rhamnopyranosyl-(1→2)-α-l-arabinopyranosyl siaresinolic acid 28-O-α-l-rhamnopyranosyl-(1→4)-β-d-glucopyranosyl-(1→6)-β-d-glucopyranosyl ester	Xyl(1→3)Rha(1→2)Ara-	Rha(1→4) Glc(1→6) Glc-	19	Wang et al. (2013b)
192.	A$_8$	3-O-β-d-Xylopyranosyl-(1→3)-α-l-rhamnopyranosyl-(1→2)-[β-d-glucopyranosyl-(1→4)]-α-l-arabinopyranosyl siaresinolic acid 28-O-α-l-rhamnopyranosyl-(1→4)-β-d-glucopyranosyl-(1→6)-β-d-glucopyranosyl ester	Xyl(1→3)Rha(1→2) [Glc(1→4)]Ara-	Rha(1→4) Glc(1→6) Glc-	19	Wang et al. (2013b)
193.	A$_8$	3-O-β-d-Glucopyranosyl(1→4)-β-d-glucopyranosyl(1→4)-[α-l-rhamnopyranosyl-(1→2)]-α-l-arabinopyranosyl siaresinolic acid 28-O-α-l-rhamnopyranosyl-(1→4)-β-d-glucopyranosyl-(1→6)-β-d-glucopyranoside	Glc(1→4)Glc(1→4) [Rha(1→2)]Ara-	Rha(1→4) Glc(1→6) Glc-	19	Wang (2014a,b)
194.	A$_9$	3-O-α-l-Rhamnopyranosyl-(1→2)-[β-d-glucopyranosyl-(1→4)]-α-l-arabinopyranosyl 21α-hydroxy oleanolic acid 28-O-α-l-rhamnopyranosyl-(1→4)-β-d-glucopyranosyl-(1→6)-β-d-glucopyranoside	Rha(1→2)[Glc(1→4)]Ara-	Rha(1→4) Glc(1→6) Glc-	19	Wang (2014a,b)
195.	A$_9$	3-O-β-d-Glucopyranosyl-(1→4)-β-d-glucopyranosyl-(1→4)-[α-l-rhamnopyranosyl-(1→2)]-α-l-arabinopyranosyl 21α-hydroxy oleanolic acid 28-O-α-l-rhamnopyranosyl-(1→4)-β-d-glucopyranosyl-(1→6)-β-d-glucopyranoside	Glc(1→4)Glc(1→4) [Rha(1→2)]Ara-	Rha(1→4) Glc(1→6) Glc-	19	Wang (2014a,b)

(Continued)

TABLE 8.2 The Oleanane-Type Triterpenoids Isolated From *Anemone* Species (A Type) (*cont.*)

No.	Type	Compound name	R_1	R_2	Taxon	Ref.
196.	A_{10}	3-O-β-d-Xylopyranosyl-(1→3)-α-l-rhamnopyranosyl(1→2)-[β-d-glucopyranosyl-(1→4)]-α-l-arabinopyranosyl gypsogenin 28-O-β-d-glucopyranoside	Xyl(1→3)Rha(1→2)[Glc(1→4)]Ara-	Glc-	17	Wang (2014a,b)
197.	A_{10}	3-O-α-l-Arabinopyranosyl gypsogenin 28-O-α-l-rhamnopyranosyl-(1→4)-β-d-glucopyranosyl-(1→6)-β-d-glucopyranoside	Ara-	Rha(1→4)Glc(1→6)Glc-	17	Wang (2014a,b)
198.	A_{10}	3-O-β-d-Xylopyranosyl-(1→3)-α-l-rhamnopyranosyl-(1→2)-[β-d-glucopyranosyl-(1→4)]-α-l-arabinopyranosyl gypsogenin 28-O-α-l-rhamnopyranosyl-(1→4)-β-d-glucopyranosyl-(1→6)-β-d-glucopyranoside	Xyl(1→3)Rha(1→2)[Glc(1→4)]Ara-	Rha(1→4)Glc(1→6)Glc-	17	Wang (2014a,b)
199.	A_{11}	3β-[(O-β-d-Glucopyranosyl-(1→4)-O-[α-l-rhamnopyranosyl-(1→2)]-α-l-arabinopyranosyl)oxy]-23-hydroxyolean-18-en-28-oic acid O-α-l-rhamnopyranosyl-(1→4)-O-β-d-glucopyranosyl-(1→6)-β-d-glucopyranosyl ester	Rha(1→2)[Glc(1→4)]Ara-	Rha(1→4)Glc(1→6)Glc-	6	Mimaki et al. (2009)
200.	A_{11}	3-O-β-d-Glucopyranosyl-(1→4)-O-[α-l-rhamnopyranosyl-(1→2)]-α-l-arabinopyranosyl 23-hydroxyolean-18-en oleanolic acid 28-O-α-l-rhamnopyranosyl-(1→4)-O-β-d-glucopyranosyl-(1→6)-β-d-glucopyranoside	Glc(1→4)[Rha(1→2)]Ara-	Rha(1→4)Glc(1→6)Glc-	15	Tang (2013)
201.	A_{12}	anemoside B	Rib(1→3)Rha(1→2)Ara-	–	16	Liao et al. (2001)

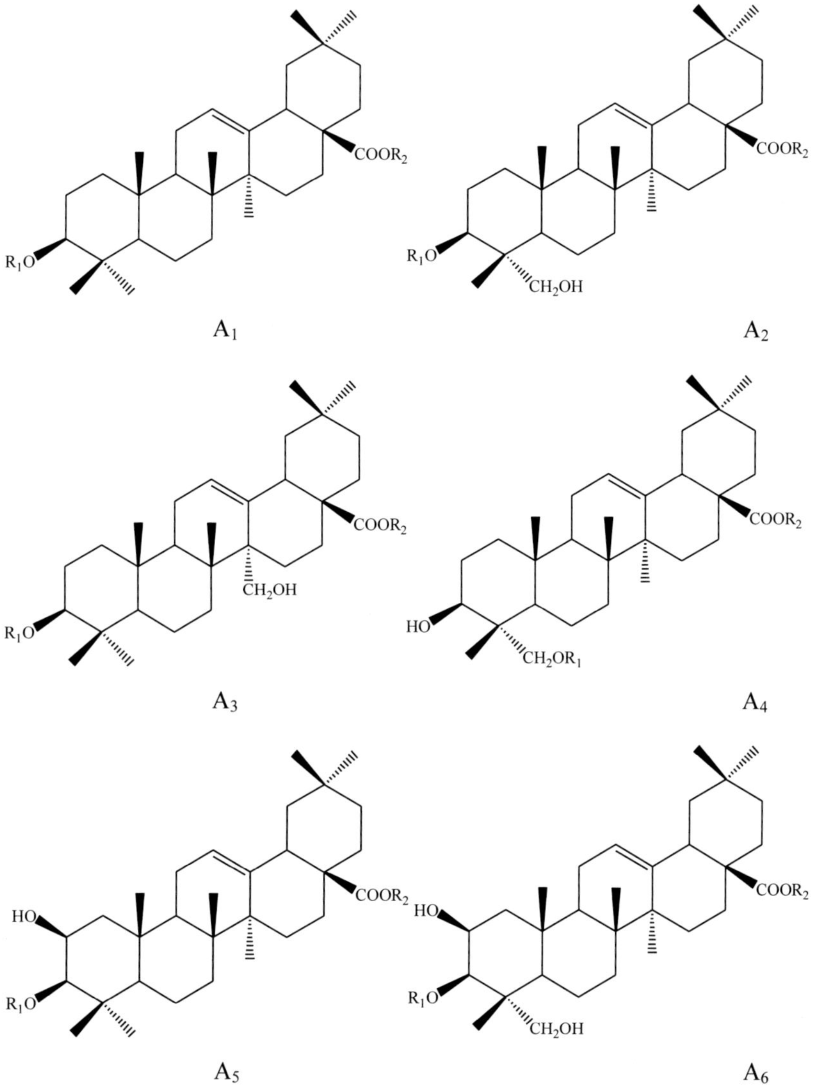

FIGURE 8.2   Basic skeletons of oleanane-type triterpenoids (A type) from *Anemone* species.

8.4 BIOACTIVITIES

8.4.1 Anticancer Activity: Cell Death Pathways, and Anticancer Targets

8.4.1.1 *Single Compound*

The genus *Anemone*, evolutionarily closely related to *Pulsatilla*, is rich in therapeutic saponins (Hao et al., 2015a). Some oleanane-type triterpenoid saponins are potently cytotoxic to the cancer cell lines (Lv et al., 2016; Bai et al., 2017). Raddeanin A (RA), a pentacyclic

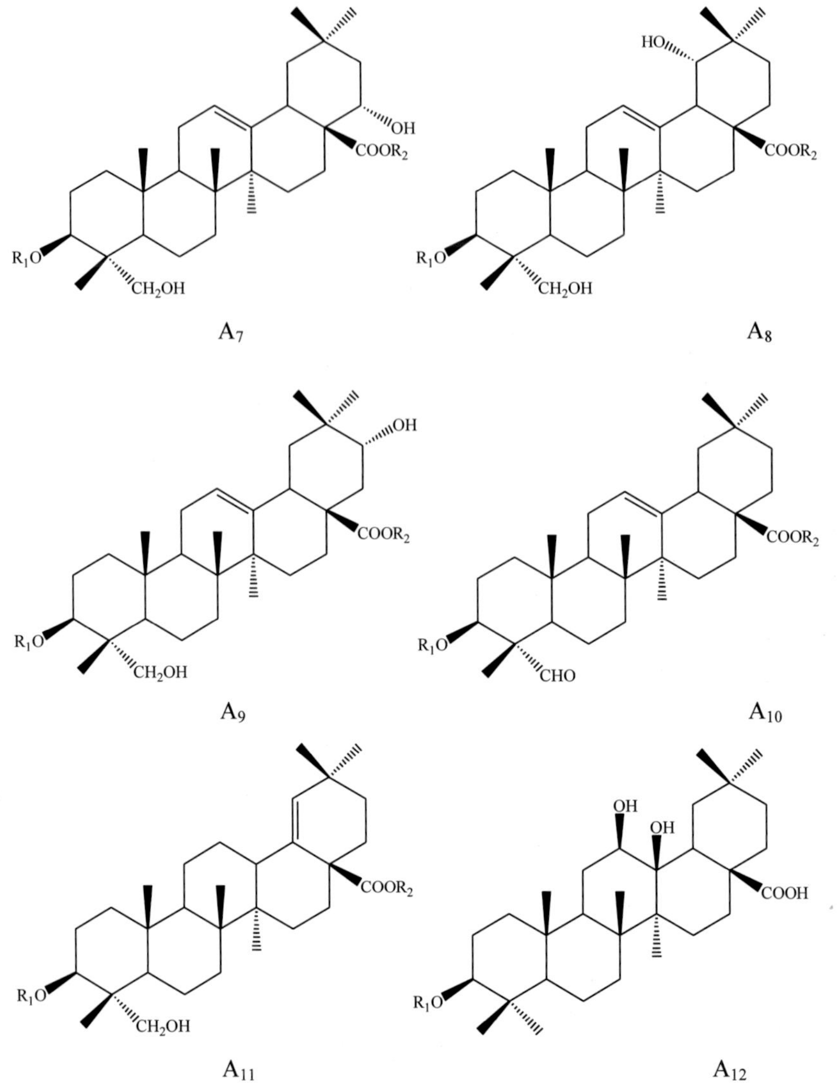

FIGURE 8.2   (*Cont.*)

triterpene saponin from *A. raddeana* (Liang Tou Jian in TCM), inhibits proliferation and induces apoptosis of multiple cancer cells (Gu et al., 2017; Guan et al., 2015; Wang et al., 2008; Xue et al., 2013). RA is detected in the heart, liver, spleen, lung, kidney, and plasma of mice following oral administration, and the abundant RA is distributed in the intestinal tract (Gu et al., 2017). RA increases Bax expression; reduces Bcl-2, Bcl-xL, and survivin expressions; and significantly activates caspase-3, caspase-8, caspase-9, and poly-ADP ribose polymerase (PARP) (Xue et al., 2013). In SGC-7901 human gastric cancer cells, RA induces apoptosis and autophagy by activating the p38 mitogen-activated protein kinase (MAPK) pathway (Teng et al., 2016), and autophagy can protect cells from apoptosis induced by RA. In QGY-7703 liver cancer cells, RA enhances cisplatin's effect by arresting

TABLE 8.3 Other Types of Triterpenoids Isolated From *Anemone* Species (B-D Types)

No.	Type	Compound name	R₁	R₂	R₃	Taxon	References
202.	B₁	Ursolic acid	H-	HOOC-	CH₃-	5	Gong (2011b)
203.	B₁	α-Amyrin acetate	H-	CH₃-	CH₃-	5	Gong (2011b)
204.	B₁	28-Hydroxy-α-Amyrin acetate	H-	HOCH₂-	CH₃-	5	Gong (2011b)
205.	B₁	2α,23-Dihydroxy-ursolic acid	OH-	HOOC-	HOCH₂-	5	Gong (2011b)
206.	B₂	2α,3β,23-Trihydroxyurs-12-en-28-oic acid	H-	H-	–	21	Hu et al. (2011)
207.	B₃	18-Hydroxyursolic acid	H-	H-	–	21	Hu et al. (2011)
208.	B₄	Flaccidoside IV	Rha(1→2)Xyl-	Glc(1→6)Glc-	–	8	Tang (2013)
209.	C	Lupeol	H-	CH₃-	–	15	Sun et al. (2011)
210.	C	Betulin	H-	HOCH₂-	–	15	Sun et al. (2011)
211.	C	Betulinic acid	H-	HOOC-	–	15,21	Hu et al. (2011)
212.	C	3β-O-α-L-Arabinopyranosyl betulinic acid 28-O-α-Lrhamnopyranosyl-(1→4)-β-D-glucopyranosyl-(1→6)-β-D-glucopyranoside	Ara-	Rha(1→4)Glc(1→6)Glc-OOC-	–	17	Wang (2014a,b)
213.	D	Cimigenol-3-O-β-D-xylopyranoside	Xyl-	–	–	1	Zou and Yang (2008c)
214.	D	Cimigenol-3-O-β-D-xylopyrano(1→3)-β-D-xylopyranoside	Xyl(1→3)Xyl-	–	–	1	Zou and Yang (2008c)

Taxon: 1, Anemone altaica; 2, A. amurensis; 3, A. anhuiensis; 4, A. begoniifolia; 5, A. cathayensis; 6, A. coronaria; 7, A. davidii; 8, A. flaccida; 9, A. hofengensis; 10, A. hupehensis; 11, A. multifida; 12, A. narcissiflora; 13, A. obtusiloba; 14, A. pulsatilla; 15, A. raddeana; 16, A. rivularis; 17, A. rivularis var. flore-minore; 18, A. rupestris; 19, A. taipaiensis; 20, A. tetrasepala; 21, A. tomentosa.
Abbreviations: Glc, β-d-glucopyranosyl; Rha, α-l-rhamnopyranosyl; Xyl, β-d-xylopyranosyl; Ara, α-l-arabinopyranosyl; Gal, β-d-galactopyranosyl; GlcA, β-d-glucuronopyranosyl; Rib, β-d-ribopyranosyl; Qui, β-d-quinovopyranosyl.

the cells in the G0/G1 phase of cell cycle and promoting their apoptosis (Li et al., 2017). The p53 and Bax, correlated with tumor cell apoptosis, are simultaneously activated, and Bcl-2 and survivin are downregulated at mRNA level. The combined use of RA increases the intracellular production of reactive oxygen species (ROS).

Saponins B, 1, and 6 of *Anemone taipaiensis* exhibit significant anticancer activity against human leukemia, glioblastoma multiforme (GBM), and HCC (Ji et al., 2015, 2016; Li et al., 2013;

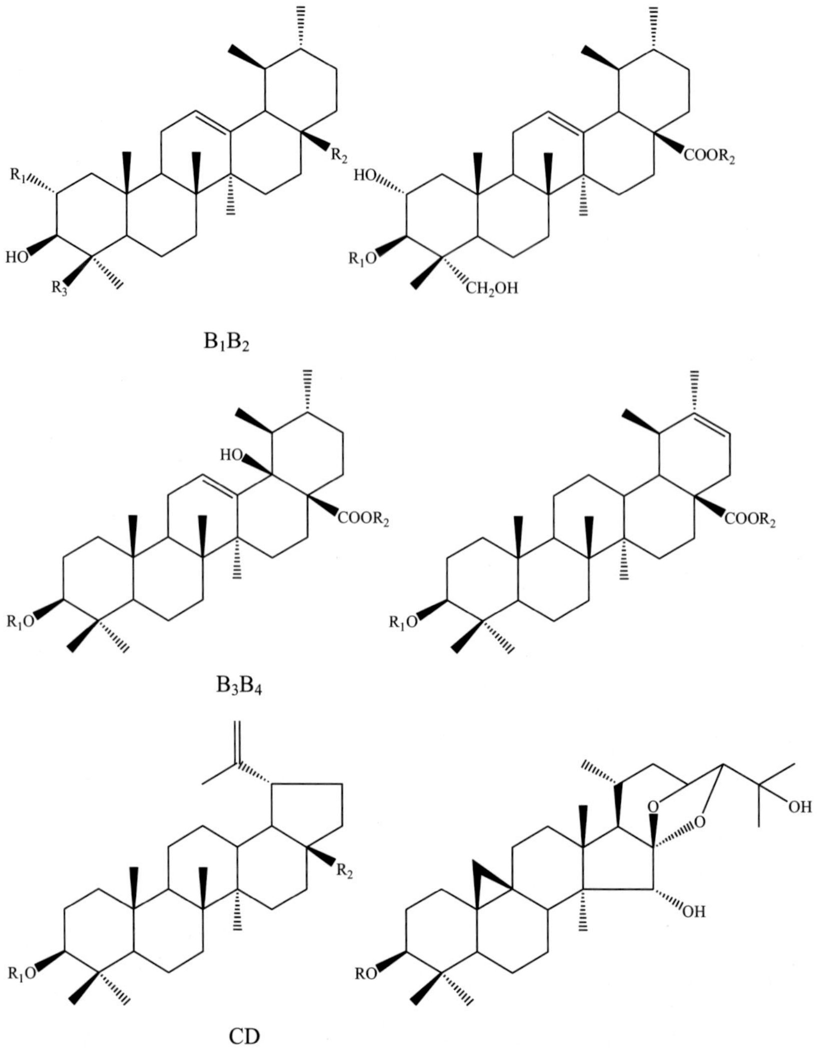

FIGURE 8.3   Basic skeletons of other type triterpenoids (B-C types) from *Anemone* species.

Wang et al., 2013a,b,c). Saponin 1 causes characteristic apoptotic morphological changes in GBM cells (Li et al., 2013), which is confirmed by DNA ladder electrophoresis and flow cytometry. Saponon 1 also causes a time-dependent decrease in the expression and nuclear location of NF-κB. The expression of inhibitors of apoptosis (IAP) family members, for example, survivin and XIAP, is significantly decreased by saponin 1. Moreover, saponin 1 causes a decrease in the Bcl-2/Bax ratio and initiates apoptosis by activating caspase-9 and caspase-3 in GBM cell lines. Thus saponin 1 inhibits cell growth of GBM cells at least partially by inducing apoptosis and inhibiting survival signaling mediated by NF-κB.

TABLE 8.4 Other Constituents Isolated From *Anemone* Species

No.	Type	Compound name	Substituent group	Taxon	Ref.
215.	I	Coumarin	$R_1 = R_2 = R_3 = R_4 = R_5 = H$	5	Gong (2011b)
216.	I	7-Ethyl-coumarin	$R_1 = R_2 = R_3 = R_5 = H$; $R_4 = OCH_2CH_3$	21	Wang (2012)
217.	I	4,7-Dimethoxyl-5-methylcoumarin	$R_1 = R_4 = OCH_3$; $R_2 = CH_3$; $R_3 = R_5 = H$	21	Ji et al. (2015) and Wang et al. (1999)
218.	I	Fraxidin	$R_1 = R_2 = H$; $R_3 = R_4 = OCH_3$; $R_5 = OH$	21	Wang (2012)
219.	I	4,5-Dimethoxyl-7-methylcoumarin	$R_1 = R_2 = OCH_3$; $R_3 = R_5 = H$; $R_4 = CH_3$	21	Ji et al. (2015) and Wang et al. (1999)
220.	I	Isofraxidin	$R_1 = R_2 = H$; $R_3 = R_5 = OCH_3$; $R_4 = OH$	21	Wang (2012)
221.	I	4-Methoxyl-5-methyl-6,7 -methylenedioxy-coumarin	$R_1 = OCH_3$; $R_2 = CH_3$; $R_3 = R_4 = O\text{-}CH_2\text{-}O$; $R_5 = H$	21	Wang et al. (1999)
222.	I	Esculetin	$R_1 = R_2 = R_5 = H$; $R_3 = R_4 = OH$	1	Wang et al. (2014)
223.	I	4, 7-Dimethoxyl-5-methyl-6 -hydroxycoumarin	$R_1 = R_4 = OCH_3$; $R_2 = CH_3$; $R_3 = OH$; $R_5 = H$	15	Ren et al. (2012)
224.	I	4,7-Dimethoxyl-5-formyl-6 -hydroxycoumarin	$R_1 = R_4 = OCH_3$; $R_2 = CHO$; $R_3 = OH$; $R_5 = H$	15	Ren et al. (2012)
225.	I	Oxypeucedanin hydrate	-	8	Zhang (2007)
226.	II	Quercetin-3-galactoside-7 -rhamnoside	$R_1 = Rha$; $R_2 = Gal$	18	Liao et al. (1999)
227.	II	Quercetin-7-rhamnoside	$R_1 = Rha$; $R_2 = OH$	18	Liao et al. (1999)
228.	II	Pelargonidin 3-[2-(xylosyl)-galactoside]	$R = H$	6	Toki et al. (2001)
229.	II	Pelargonidin 3-[2-(xylosyl)-6-(malonyl) galactoside]	$R = S_1$	6	Toki et al. (2001)
230.	II	Pelargonidin 3-[2-(xylosyl)-6-(methyl-malonyl) -galactoside]	$R = S_2$	6	Toki et al. (2001)
231.	II	(6″-O-(Pelargonidin 3-[2-(xylosyl)-galactosyl]) ((4-glucosyl) caffeoyl)-tartaryl) malonate	$R = S_3$	6	Toki et al. (2001)

(Continued)

TABLE 8.4 Other Constituents Isolated From *Anemone* Species (*cont.*)

No.	Type	Compound name	Substituent group	Taxon	Ref.
232.	II	Delphinidin 3-[2-(2-(caffeoyl) glucosyl) galactoside]-7-[6-(caffeoyl) glucoside]-3'-[glucuronide]	$R_1 = H; R_2 = GlcA; R_3 = H$	6	Saito et al. (2002)
233.	II	Delphinidin 3-[2-(2-(caffeoyl) glucosyl)-6-(malonyl) galactoside]-7-[6-(caffeoyl) glucoside]-3'-[glucuronide]	$R_1 = S_1; R_2 = GlcA; R_3 = H$	6	Saito et al. (2002)
234.	II	Delphinidin 3-[2-(2-(caffeoyl) glucosyl)-6-(2-(tartaryl) malonyl) galactoside]-7-[6-(caffeoyl) glucoside]-3'-[glucuronide]	$R_1 = S_2; R_2 = GlcA; R_3 = H$	6	Saito et al. (2002)
235.	II	Delphinidin 3-[2-(2-(caffeoyl) glucosyl)-6-(2-(tartaryl) malonyl) galactoside]-7-[6-(caffeoyl)-glucoside]	$R_1 = S_2; R_2 = H; R_3 = H$	6	Saito et al. (2002)
236.	II	Cyanidin 3-[2-(2-(caffeoyl) glucosyl)-6-(2-(tartaryl) malonyl)galactoside]-7-[6-(caffeoyl)glucoside]-3'-[glucuronide]	$R_1 = S_2; R_2 = GlcA; R_3 = OH$	6	Saito et al. (2002)
237.	III	Anemonolide	–	16	Liao et al. (2001)
238.	III	Ardicren	$R = Rha(1{\rightarrow}4)Glc(1{\rightarrow}4)[Glc(1{\rightarrow}2)]Ara-$	5	Gong (2011b)
239.	III	Protoanemonin	–	15	Sun et al. (2011)
240.	III	Anemonin	–	15	Sun et al. (2011)
241.	III	Ranunculin	–	14,15	Varitimidis et al. (2006)
242.	IV	(7S,8R,7'R,8'S)4,9,4',7'-Tetra-hydroxy-3,3'-dimethoxy-7'-O-n-butyl-7,9'-epoxylignan 9-O-β-D-glucopyranoside	–	17	Ding (2011)

(Continued)

No.	Group	Name	R-substituents	Ref.	Citation
243.	IV	(+)-Isolariciresinoil-9-O-β-D-glucopyranoside	—	1	Wang et al. (2014)
244.	IV	(+)-Pinoresinol-4-O-β-D-glucopyranoside	—	1	Wang et al. (2014)
245.	V	β-Sitosterol	R = H	1,21	Zou and Yang (2008a)
246.	V	Daucosterol	R = Glc	1,2,21	Zhang et al. (2014)
247.	V	Stigmasterol	R = H	21	Wang et al. (1999)
248.	V	3-O-β-D-Glucopyranosyl stigmasterol	R = Glc	21	Wang et al. (1999)
249.	V	Ergosterol peroxide	—	16,21	Wang et al. (1999)
250.	V	(22E,24R)-5α,8α-Epidioxy-24-methyl-cholesta-6,9(11),22-trien-3β-ol	—	16	Anh Minh et al. (2012)
251.	VI	Carboxymethyl isoferulate	—	1	Zou (2006)
252.	VI	Ferulic acid	R_1 = OH; R_2 = OCH_3; R_3 = H	1,16	Ju et al. (1986)
253.	VI	Isoferulic acie	R_1 = OCH_3; R_2 = OH; R_3 = H	1	Zou (2006)
254.	VI	Cinnamic acid	R_1 = OH; R_2 = R_3 = H	1	Zou and Yang (2008b)
255.	VI	Caffeic acid	R_1 = R_2 = OH; R_3 = H	1,16	Shao et al. (2011)
256.	VI	1-O-Caffeoyl-β-D-glucopyranoside	R_1 = R_2 = OH; R_3 = Glc	16	Anh Minh et al. (2012)
257.	VI	6-O-Feruloyl-D-glucopyranose	R_1 = OH; R_2 = OCH_3; R_3 = Glc	1	Wang et al. (2014)
258.	VI	Coumaric acid	R_1 = OH; R_2 = R_3 = H	16	Shao et al. (2011)
259.	VI	Mono-feruloyltartaric acid	—	1	Wang et al. (2014)
260.	VI	Chlorogenic acid	—	1	Wang et al. (2014)
261.	VI	Glucosyringic acid	—	1	Wang et al. (2014)
262.	VI	4-Hydroxy-benzoic acid	R_1 = H; R_2 = OH	16	Shao et al. (2011)
263.	VI	3-Hydroxy-4-methoxy-benzoic acid	R_1 = OH; R_2 = OCH_3	16	Shao et al. (2011)
264.	VI	Protocatechuic aldehyde	—	1	Zou and Yang (2008b)
265.	VI	4'-O-Methylequisetum pyrone	—	2	Zhang et al. (2014)

TABLE 8.4 Other Constituents Isolated From *Anemone* Species (*cont.*)

No.	Type	Compound name	Substituent group	Taxon	Ref.
266.	VII	γ-Terpineol	–	15	Yan et al. (1990)
267.	VII	4-Tert-butylphenetole	–	15	Yan et al. (1990)
268.	VII	2,6-Di-tert-butyl-4-methylphenol	–	15	Yan et al. (1990)
269.	VII	Benzeneacetaldehyde	R = CHO	15,16	Shi et al. (2012)
270.	VII	Phenylethyl alcohol	R = CH$_2$OH	15	Yan et al. (1990)
271.	VII	Benzaldehyde	R$_1$ = CHO; R$_2$ = R$_3$ = R$_4$ = H	16	Shi et al. (2012)
272.	VII	Benzyl alcohol	R$_1$ = CH$_2$OH; R$_2$ = R$_3$ = R$_4$ = H	16	Shi et al. (2012)
273.	VII	(R) Acetophenone	R$_1$ = COCH$_3$; R$_2$ = R$_3$ = R$_4$ = H	16	Shi et al. (2012)
274.	VII	(M) 2,4-Dihydroxy-3-methylacetophenne	R$_1$ = COCH$_3$; R$_2$ = R$_4$ = OH; R$_3$ = CH$_3$	16	Shi et al. (2012)
275.	VII	(E)-4-Phenyl-3-buten-2-one	–	16	Shi et al. (2012)
276.	VII	3-BHA	–	16	Shi et al. (2012)
277.	VII	Myristicin	–	16	Shi et al. (2012)
278.	VII	2-Methyl hexadecane	–	15	Yan et al. (1990)
279.	VII	7,9-Dimethyl hexadecane	–	15	Yan et al. (1990)
280.	VII	Nonadecanol	–	15	Yan et al. (1990)
281.	VII	Heptanoic acid	–	16	Shi et al. (2012)
282.	VII	(R) 2,3-Dimethyl heptane	–	16	Shi et al. (2012)
283.	VII	(M) 3-Ethyl-2-methyl hexane	–	16	Shi et al. (2012)
284.	VII	(M) 5-Methyl undecane	–	16	Shi et al. (2012)
285.	VII	(R) 2,4-Dimethyl undecane	–	16	Shi et al. (2012)
286.	VII	(M) 4,5-Diethyl octane	–	16	Shi et al. (2012)
287.	VII	(R) 5,6-Dimethyl decane	–	16	Shi et al. (2012)
288.	VII	2,4-Dodecadienal	–	16	Shi et al. (2012)

289.	VII	(R) Patchouli alcohol	–	16	Shi et al. (2012)
290.	VII	(M) 3-(6,6-Dimethyl-5-oxohept-2-enyl)-cyclohexanone	–	16	Shi et al. (2012)
291.	VII	Caryophyllene	–	16	Shi et al. (2012)
292.	VII	(2S,3S,4R,8E)-2-[(2′R)-2′-Hy-droxy-tetracosanoyl]-1,3,4-trihydroxy-8-oeta decene	–	1	Zou and Yang (2008b)
293.	VII	Aralia cerebroside	–	1	Zou and Yang (2008b)
294.	VII	Cirsiumaldehyde	–	1	Zou and Yang (2008a)
295.	VII	1,6,9,13-tetraoxadispiro [4,2,4,2] tetradecane-2,10-dione	–	1	Zou (2006)
296.	VII	Adenine	–	1	Wang et al. (2014)
297.	VII	Adenosine	–	1	Zou (2006)
298.	VII	Uridine	–	1	Zou (2006)
299.	VII	2,3,4,9-Tetrahydro-1H-pyridine pyrido [3,4-b] indole-3-car-boxylic acid	–	1	Wang et al. (2014)
300.	VII	Phenylalanine	–	1	Wang et al. (2014)
301.	VII	Thymidine	–	1	Wang et al. (2014)
302.	VII	13(S)-Hydroxy-7-oxo-lab-da-8,14-diene-19-oic acid-β-D-galactosyl ester	–	17	Ding (2011)
303.	VII	Tomentoside	–	21	Hu et al. (2011)
304.	VII	Linoleic acid	–	15	Sun et al. (2011)
305.	VII	Hexanoic acid	–	15	Sun et al. (2011)
306.	VII	5-hydroxy-4-oxo-pentanoic acid	R = CH$_2$OH	1	Zou (2006) and Ju et al. (1986)
307.	VII	Succinic acid	R = OH	1	Zou (2006)

(Continued)

TABLE 8.4 Other Constituents Isolated From *Anemone* Species (*cont.*)

No.	Type	Compound name	Substituent group	Taxon	Ref.
308.	VII	Palmic acid	R = OCH$_3$	1	Zou (2006)
309.	VII	Monoethyl Malonate	–	8	Zhang (2007)
310.	VII	Triacontane	–	1	Zou (2006)
311.	VII	Methyl β-D-glucopyranoside	–	1	Zou (2006)
312.	VII	β-D-Glucopyranosyl	–	2	Zhang et al. (2014)
313.	VII	α-D-Glucopyranosyl	–	2	Zhang et al. (2014)
314.	VII	Methyl-O-α-D-fructofuranoside	–	2	Zhang et al. (2014)
315.	VII	β-D-Galactitol	–	2	Zhang et al. (2014)

I, coumarins; II, flavonoids; III, lactones; IV, lignans; V, steroids; VI, phenolic compounds; VII, essential oils; VIII, other compounds.

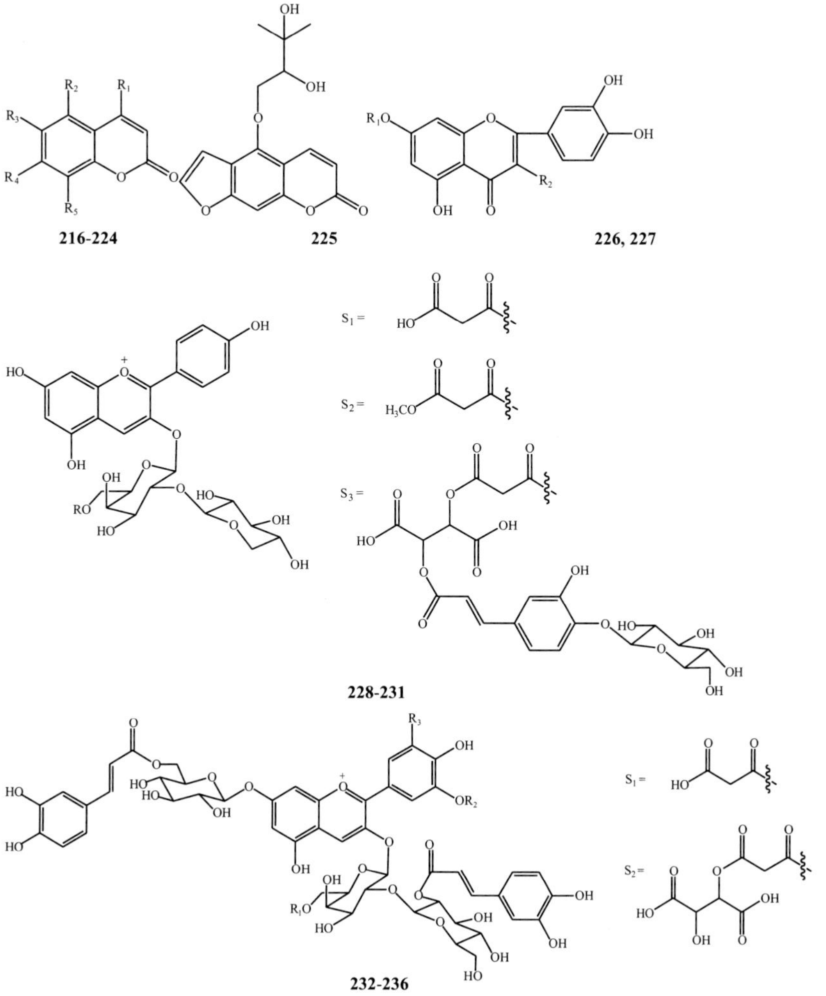

FIGURE 8.4 Other constituents isolated from *Anemone* species.

Saponin B blocks the cell cycle at the S phase (Wang et al., 2013c). Saponin B induces chromatin condensation of U87MG GBM cells and leads to the formation of apoptotic bodies. Annexin V/PI assay suggests that phosphatidylserine (PS) externalization is apparent at higher drug concentrations. Saponins B and 6 activate the receptor-mediated pathway of apoptosis via the activation of Fas-l (Ji et al., 2016). These saponins increase the Bax and caspase-3 ratio and decrease the protein expression of Bcl-2.

Triterpenoid saponins of *Anemone flaccida* induce apoptosis in human BEL-7402, HepG2 hepatoma cell lines, and lipopolysaccharide (LPS) stimulates HeLa cells via COX-2/PGE$_2$ pathway (Han et al., 2013). Flaccidoside II, one of the triterpenoid saponins of *A. flaccida*, induces apoptosis by downregulating hemeoxygenase (HO)-1 via extracellular signal-regulated kinase (ERK)-1/2 and p38 MAPK pathways (Han et al., 2016).

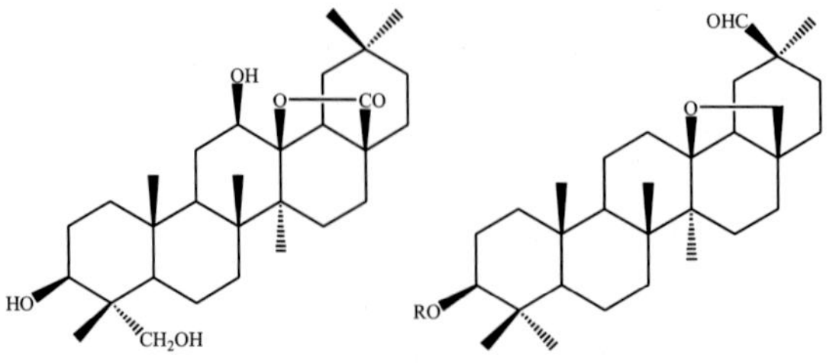

237 **238**

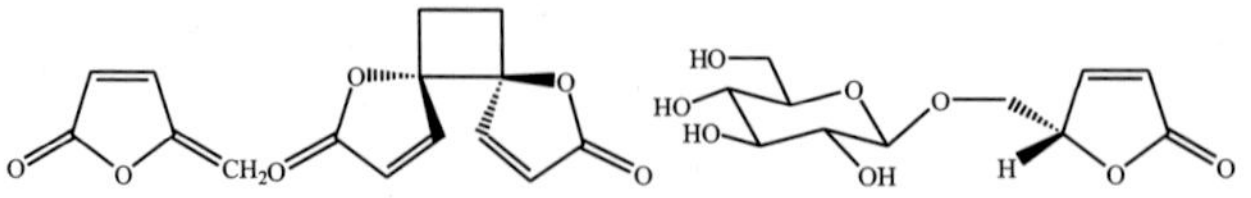

239 **240** **241**

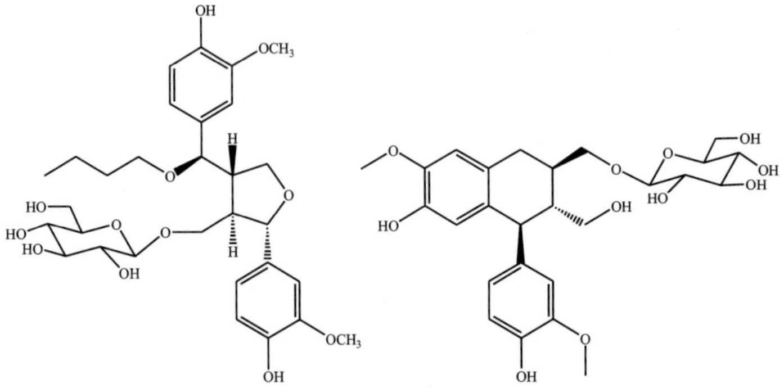

242 **243**

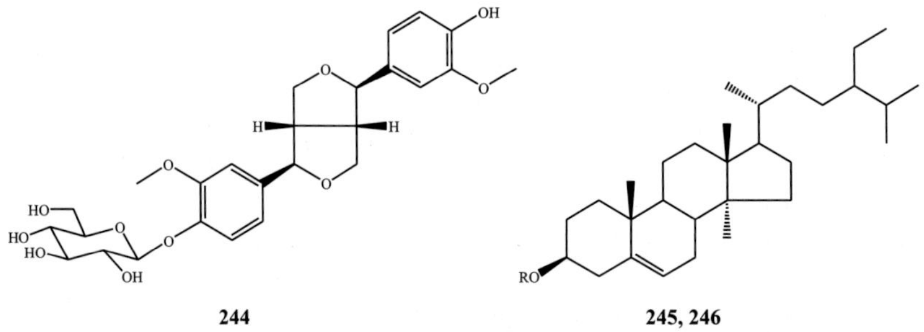

244 **245, 246**

FIGURE 8.4 (*Cont.*)

RA significantly inhibits HUVEC proliferation, motility, migration, and tube formation (Guan et al., 2015). RA dramatically reduces angiogenesis in chick embryo chorioallantoic membrane, restrains the trunk angiogenesis in zebrafish, and suppresses angiogenesis and growth of human HCT-15 colorectal cancer xenograft in mice. Raddeanin A suppresses vascular endothelial growth factor (VEGF)-induced phosphorylation of VEGF receptor 2 (VEGFR2) and its downstream protein kinases, including PLCγ1, JAK2, FAK, Src, and Akt. In molecular docking simulation, raddeanin A forms hydrogen bonds and hydrophobic interactions within the ATP-binding pocket of VEGFR2 kinase domain.

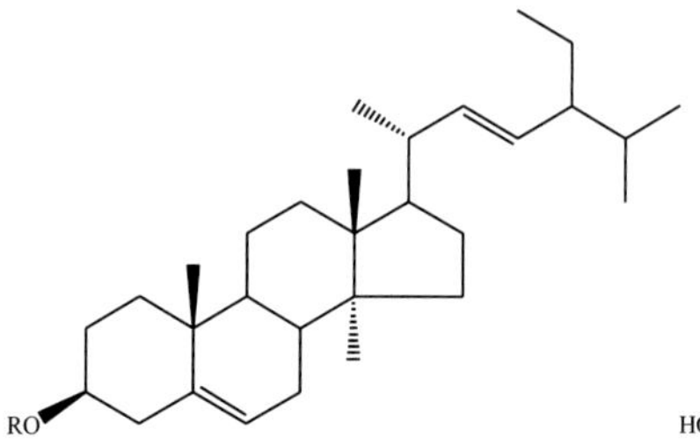

247, 248

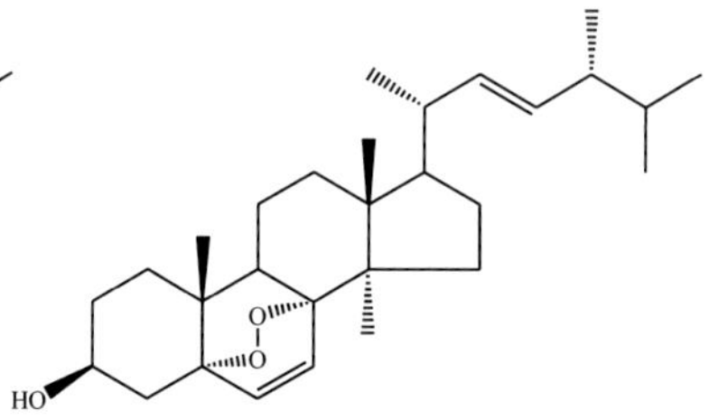

249

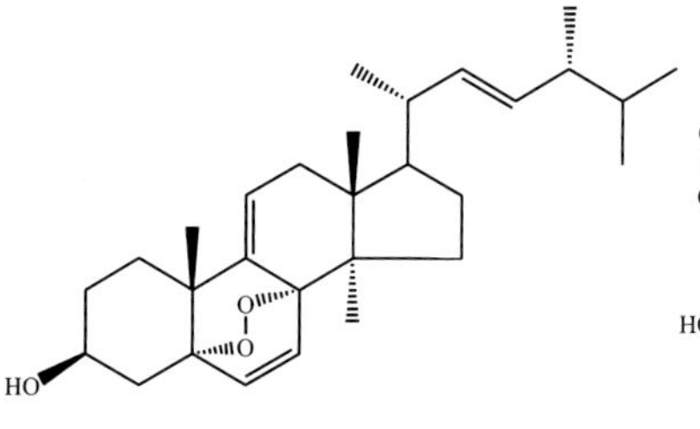

250

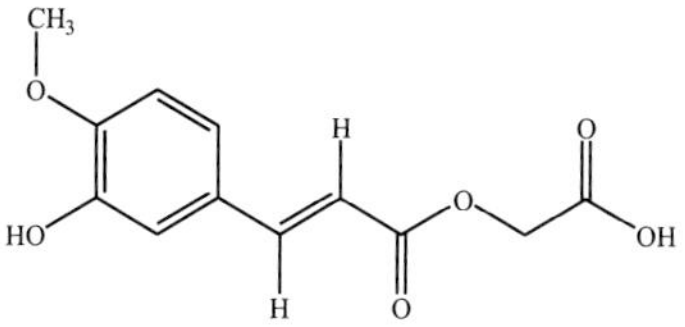

251

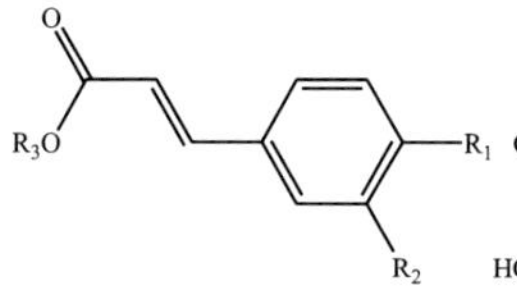

252-258

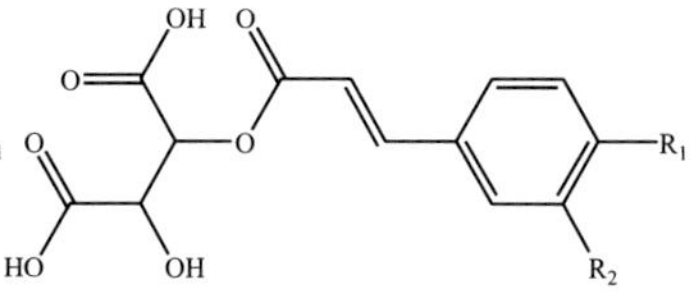

259

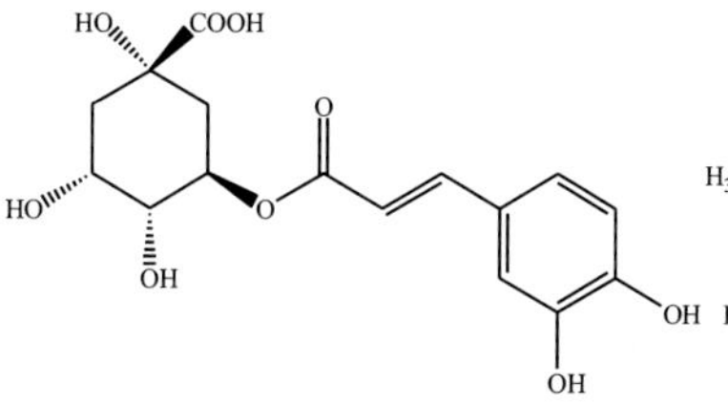

260

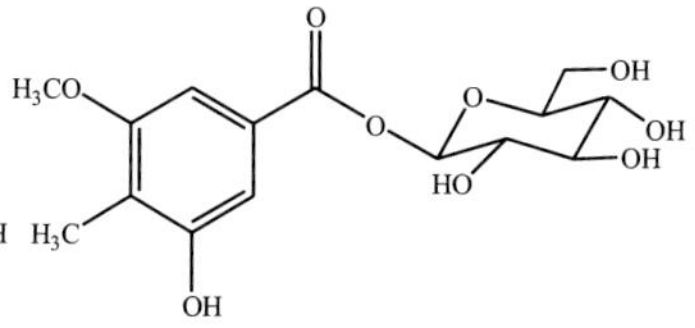

261

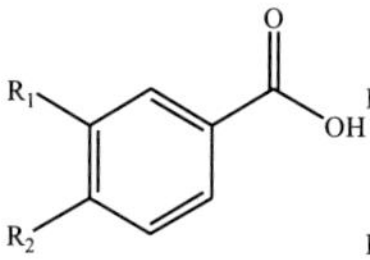

262, 263

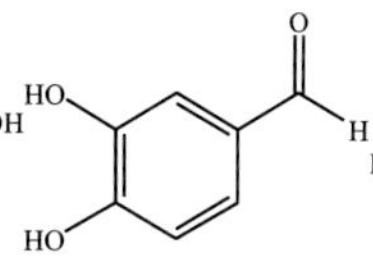

264

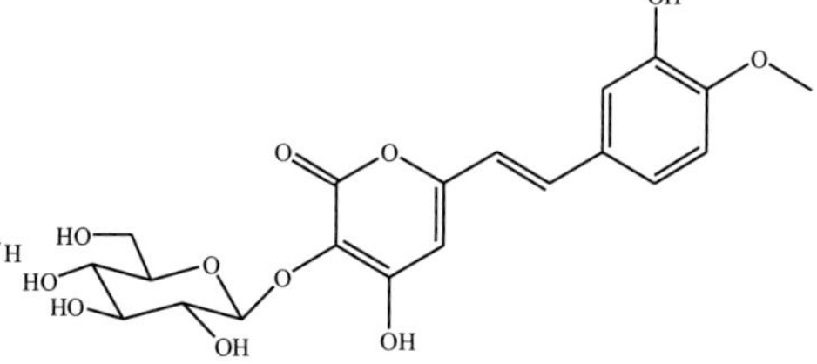

265

FIGURE 8.4   (*Cont.*)

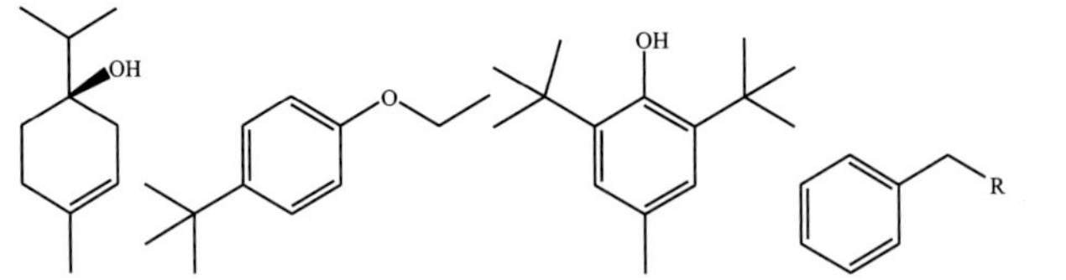

266 **267** **268** **269, 270**

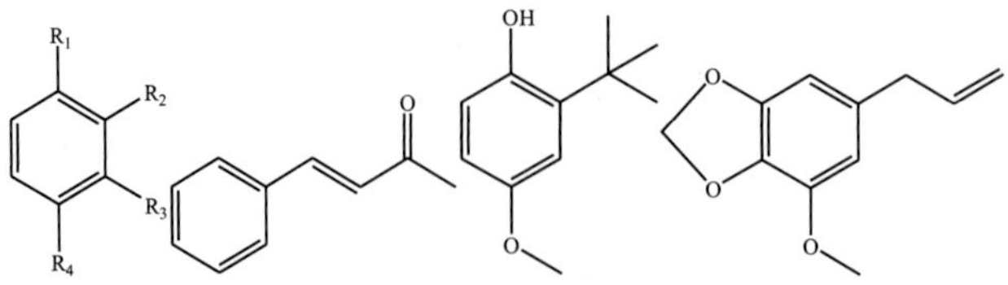

271-274 **275** **276** **277**

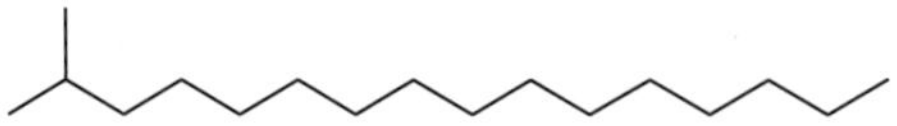

278

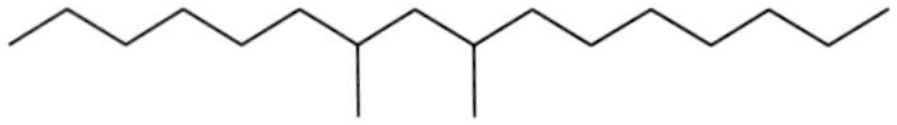

279

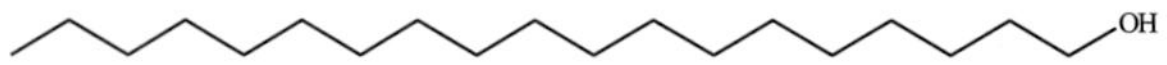

280

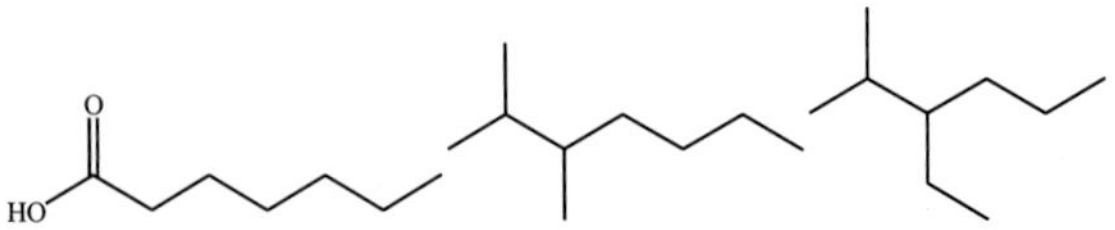

281 **282** **283**

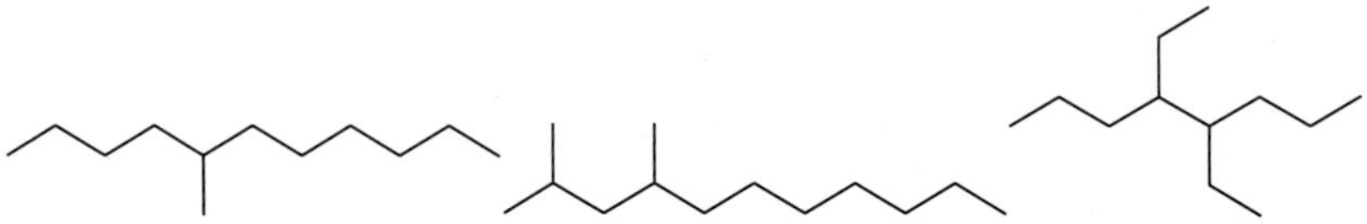

284 **285** **286**

FIGURE 8.4 (*Cont.*)

RA significantly inhibits the invasion, migration, and adhesion of the BGC-823 human gastric cancer cells (Xue et al., 2013). Raddeanin A can upregulate the expression of reversion-inducing cysteine-rich protein with Kazal motifs (RECK) and E-cadherin and downregulate the expression of matrix metalloproteinase-2 (MMP-2), MMP-9, MMP-14, and Ras homolog gene family, member C (RhoC).

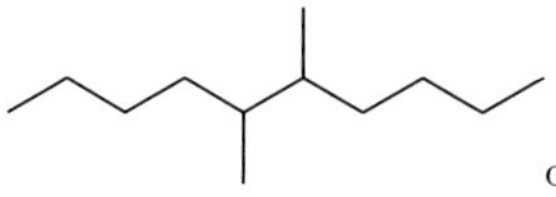

287

288

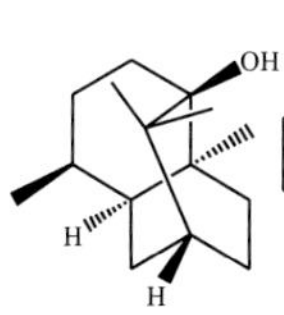

289

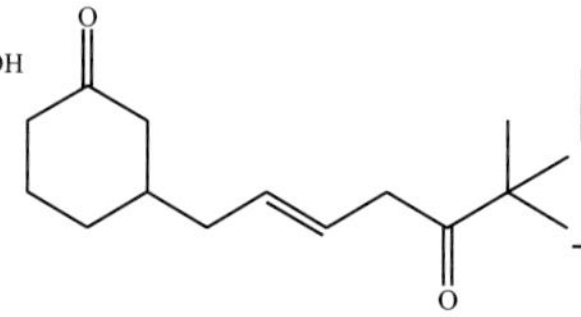

290

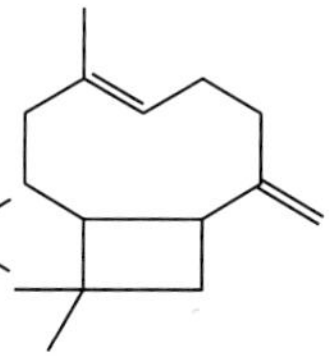

291

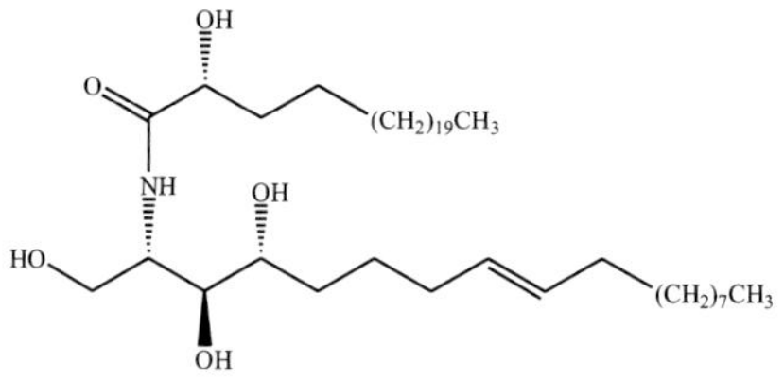

292

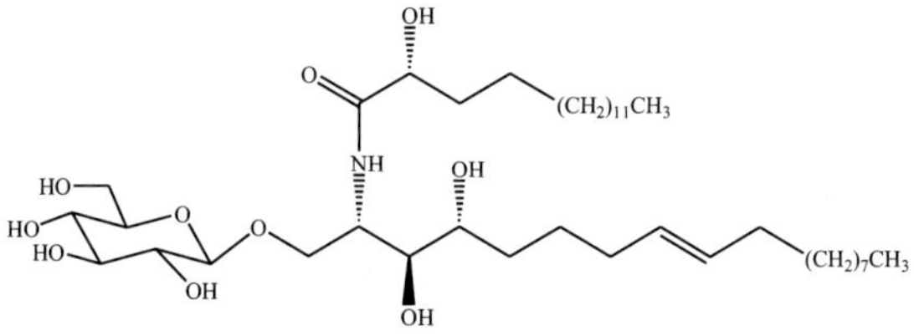

293

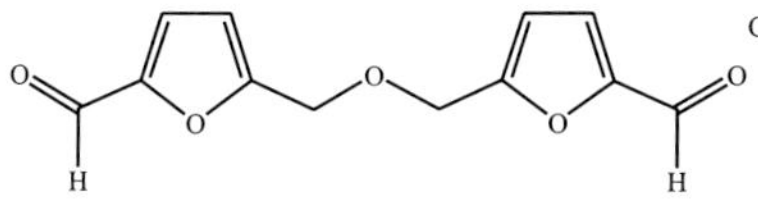

294

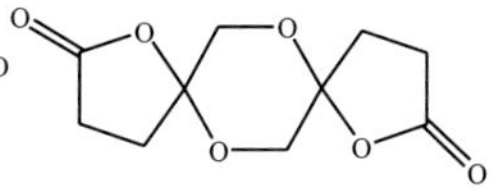

295

FIGURE 8.4 (*Cont.*)

8.4.1.2 *Plant Extract*

In a screen of 70 species of medicinal plants, the aqueous extract of *A. altaica* (AAE) has the best performance in suppressing the viability of HOS and U2OS human osteosarcoma cells in a concentration-dependent manner (Chang et al., 2015). AAE suppresses the growth of HOS and U2OS through the intrinsic apoptotic pathway, but it has no significant influence on human osteoblast hFOB cells. The high mRNA levels of apoptosis-related factors (PPP1R15A, SQSTM1, HSPA1B, and DDIT4) and cellular proliferation markers (SKA2 and BUB1B) are significantly altered by AAE treatment. AAE can upregulate the expression of a cluster of genes, especially those in the apoptosis-related factor family and caspase family.

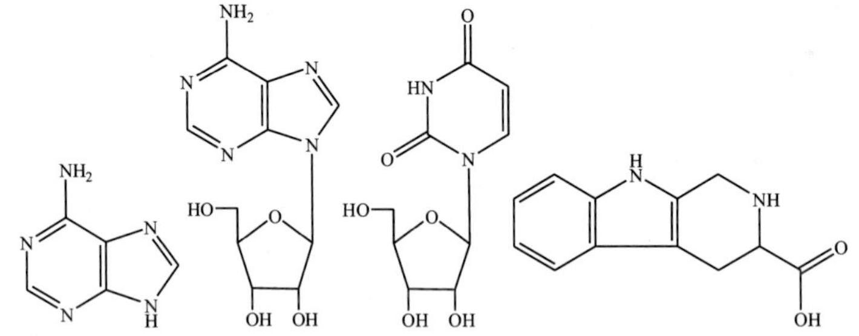

296   **297**   **298**   **299**

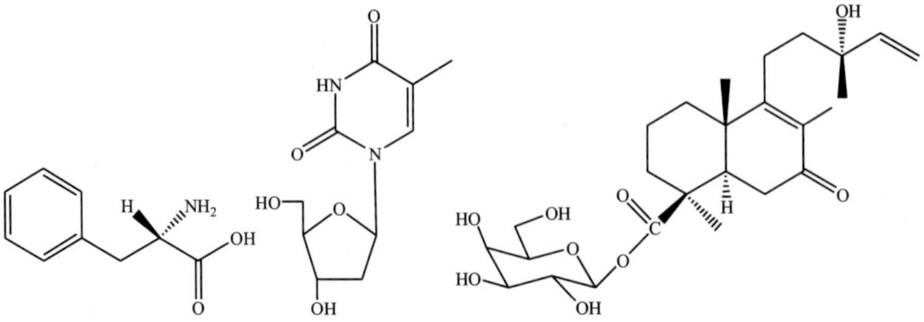

300   **301**   **302**

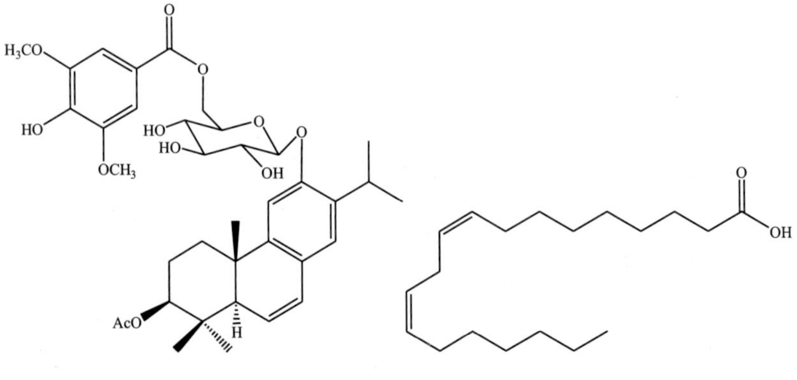

303   **304**

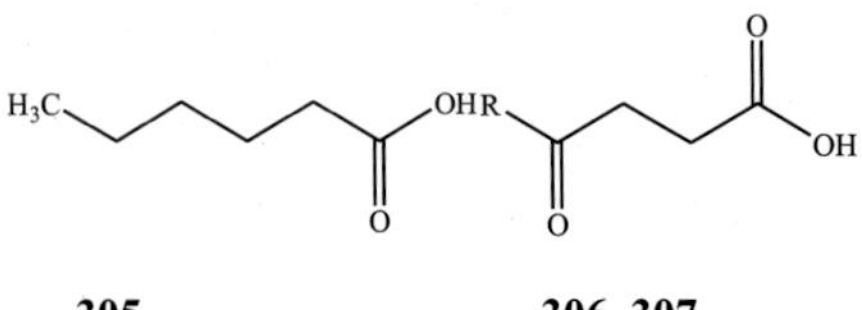

305   **306, 307**

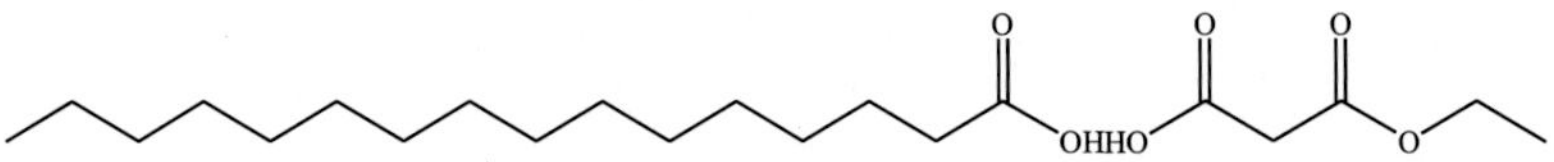

308   309

FIGURE 8.4   (*Cont.*)

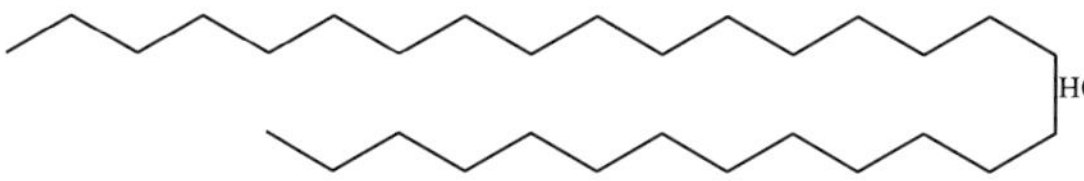

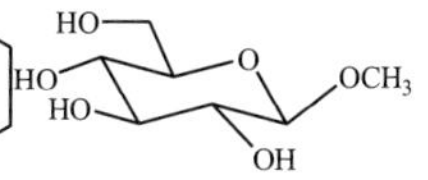

310

311

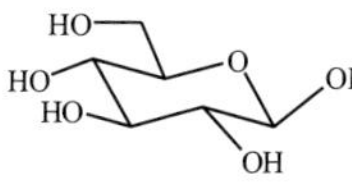

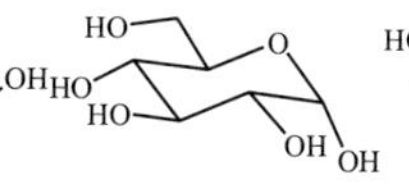

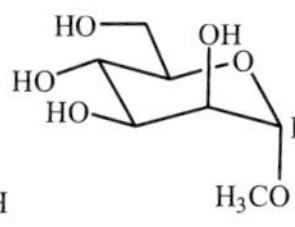

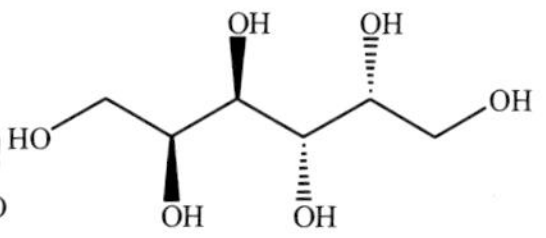

312

313

314

315

FIGURE 8.4 (*Cont.*)

The ethanol extract of the whole plant of *A. rivularis* (ARE) inhibits cancer cells, including MDA-MB321, K562, HT29, Hep3B, DLD-1, and LLC (Chung et al., 2017). ARE suppresses pyruvate dehydrogenase (PDH) kinase (PDHK) activity in vitro and inhibits aerobic glycolysis by reducing phosphorylation of PDH in human DLD-1 colon cancer and murine LLC cells. The expression of PDHK1, a major isoform of PDHKs in cancer, is not affected by ARE. ARE increases both ROS production and mitochondrial damage and suppresses the in vitro tumor growth through mitochondria-mediated apoptosis. The growth of allograft LLC cells is reduced by ARE. ARE, by inhibiting PDHK activity and tumor growth both in vitro and in vivo, is a potential candidate for developing anticancer drugs.

8.4.2 Immunomodulatory Activity

ARS, the saponins extracted from the rhizome of *A. raddeana*, shows a slight hemolytic effect and enhances significantly a specific antibody and cellular response against ovalbumin in mice (Sun et al., 2008). A neutral polysaccharide fraction (ARP) from the rhizome of *A. raddeana* extraordinarily promotes splenocyte proliferation and natural killer cell (NK) and cytotoxic T lymphocyte (CTL) activity, as well as serum interleukin 2 (IL-2) and tumor necrosis factor α (TNF-α) production in HCC-bearing mice (Liu et al., 2012). ARP has no toxicity to body weight, liver, and kidneys. Moreover, it can reverse the hematological parameters induced by 5-fluorouracil to near normal.

8.4.3 Antiinflammatory and Antioxidant Activities

Ranunculaceae tribes and genera, such as *Ranunculus*, Anemoneae, *Cimicifuga*, *Helleborus*, *Nigella*, Delphinieae, *Semiaquilegia*, *Coptis*, and *Hydrastis*, are rich in both antiinflammatory and anticancer phytometabolites (Hao et al., 2015a,b). Anemonin and ranunculin, the potent antiinflammatory and anticancer compounds, are abundant in tribes Ranunculeae and Anemoneae (Hao et al., 2015a; Lee et al., 2008).

A. flaccida (Di Wu in Chinese) crude triterpenoid saponins (AFSs) inhibit redness and swelling of the right hind paw in the type II collagen-induced arthritis (CIA) model (Liu et al., 2015a,b). The inflammatory responses are reduced by AFS treatment. The serum proinflammatory cytokines TNF-α and IL-6 are decreased in AFS-treated CIA rats at the dose of 200 and 400 mg/kg/day. AFS and its main compounds, including hederasaponin B, flaccidoside

II, and hemsgiganoside B, significantly inhibit TNF-α and IL-6 production in LPS-treated RAW264.7 cells, respectively.

Osteoclasts are bone-specialized multinucleated cells and are responsible for bone destructive diseases such as rheumatoid arthritis and osteoporosis. In RAW264.7 cells and CIA rats, the total saponin (TS) of *A. flaccida* concentration-dependently inhibits receptor activator of nuclear factor-κB ligand (RANKL)-induced osteoclast formation and bone marrow-derived macrophages (BMMs), as well as decreases the extent of actin ring formation and lacunar resorption (Kong et al., 2015a; Liu et al., 2015a,b). The RANKL-stimulated expression of osteoclast-related transcription factors are also diminished by TS, while the expression of osteoprotegerin (OPG), at both mRNA and protein levels, increases, and the ratio of RANKL to OPG in inflamed joints and sera of CIA rats decreases (Liu et al., 2015a,b). TS blocks the RANKL-triggered TRAF6 expression, as well as the phosphorylation of MAPKs and IκB-α, and inhibits NF-κB p65 DNA-binding activity. TS almost abrogates the nuclear factor of activated T-cells (NFATc1) and c-Fos expression. TS suppresses RANKL-induced osteoclast differentiation and inflammatory bone loss via the downregulation of TRAF6 level, the JNK and p38 MAPKs and NF-κB activation, and the subsequent decreased expression of c-Fos and NFATc1. Triterpenoid saponin W3 of *A. flaccida* has similar effects (Kong et al., 2015b). Therefore, TS and saponins thereof may be useful for lytic bone diseases, and further in vivo studies and clinical trials are warranted.

Two coumarins of *A. raddeana* have inhibitory effects against human leukocyte elastase (Ren et al., 2012).

3-acetyloleanolic acid (AOA), oleanolic acid (OA), raddeanoside 12 (Rd12), and Rd13, isolated from *A. raddeana*, suppresses the superoxide generation induced by N-formyl-methionyl-leucyl-phenylalanine (fMLP) in a concentration-dependent manner (Lu et al., 2001). Eleutheroside K (EK) and Rd10 significantly enhances the fMLP-induced superoxide generation in low concentrations (0.5–0.75 μM), while these compounds more efficiently suppress the superoxide generation than the other four compounds in other concentrations. Rd12 dose-dependently inhibits fMLP-induced tyrosyl phosphorylation of 123.0, 79.4, 60.3, 56.2, and 50.1 kDa proteins in human neutrophil, while Rd10 and EK enhances the tyrosyl phosphorylation of these proteins in a low-concentration range.

The superoxide generation induced by fMLP is significantly suppressed by betulin and lupeol, extracted from the roots of *A. raddeana*, depending on the concentration of the triterpenoids (Yamashita et al., 2002). The suppressive effect of betulinic acid is low. The phorbol 12-myristate 13-acetate (PMA)-induced superoxide generation is suppressed by betulin in a concentration-dependent manner but not by lupeol and betulinic acid. However, the superoxide generation induced by arachidonic acid (AA) is suppressed by lupeol, while betulin and betulinic acid weakly enhances the AA-induced superoxide generation. Lupeol and betulin suppresses tyrosyl phosphorylation of a 45.0-kDa protein in fMLP-treated human neutrophils, but betulinic acid does not. Lupeol, betulin, and betulinic acid shows no hemolytic effect, even at a concentration of 500 μM.

Five oleanolic acid triterpenoid saponins (OTSs), isolated from the rhizome of *A. raddeana*, suppresses fMLP-induced superoxide generation in a concentration-dependent manner (Wei et al., 2012). OTS-1, 2, and 4 suppresses PMA- and AA-induced superoxide generation in a concentration-dependent manner, but OTS-3 and 5 show no effect. fMLP- and PMA-induced tyrosyl or serine/threonine phosphorylation, and fMLP-, PMA-, and AA-induced translocation of p67(phox), p47(phox), and Rac to plasma membrane, are in parallel with the suppression of the stimulus-induced superoxide generation.

8.4.4 Antimicrobial Activity and Others

The antioxidant essential oil obtained from the roots of *A. rivularis* has antibacterial activity (Shi et al., 2012). The inhibition zones at 100 μg/disc and minimum inhibitory concentration (MIC) values for four bacterial strains are in the range of 11.0–20.0 mm and 125–250 μg/mL, respectively.

Saponins of *A. raddeana* rhizome are analgesic, antipyretic, anticonvulsive, antihistaminic, and sedative (Wang et al., 2017). They are used in the treatment of liver fibrosis in chronic hepatitis. However, the herb also has hemolytic effects and can be toxic.

8.5 TAXONOMY AND PHARMACOPHYLOGENY

The distribution of anticancer compounds within Ranunculaceae is not random but phylogeny related (Hao et al., 2014, 2015c; Hao and Xiao, 2015). For instance, *Ranunculus*, *Clematis*, *Pulsatilla*, *Anemone*, and *Nigella* are rich in pentacyclictriterpene saponins. *Pulsatilla*, *Anemone*, and *Clematis* belong to the tribe Anemoneae, and *Pulsatilla* is evolutionarily more close to *Anemone* than to *Clematis* (Hao et al., 2015a). *Clematis* is closer to *Naravelia* and *Anemoclema* than to *Anemone* and *Pulsatilla*. *Hepatica* is basal to all other genera of Anemoneae (Wang et al., 2009). The sister group relationship between Ranunculeae and Anemoneae is revealed by two independent groups (Wang et al., 2009; Cossard et al., 2016).

Previous phylogenies based on molecular data indicate that segregate genera from both the Northern and Southern Hemispheres (*Hepatica*, *Pulsatilla*, *Knowltonia*, *Oreithales*, and *Barneoudia*) are embedded within *Anemone* and should be subsumed within the genus. Based on a new phylogeny that substantially increases the sampling of the austral anemones (especially from Africa), (Hoot et al., 2012) analyzes combined sequence data (chloroplast *atpB-rbcL* spacer and nuclear internal transcribed spacer (ITS) regions) for 55 species of *Anemone*, using Bayesian inference, maximum likelihood (ML), and maximum parsimony. The segregate genera, *Oreithales* and *Barneoudia*, nest within *Anemone* and are included in a well-supported clade (subgenus *Anemone*, section *Pulsatilloides*) consisting largely of Southern Hemisphere species. The Mexican *Anemone mexicana* is sister to all remaining members of section *Pulsatilloides* (Fig. 8.5), which consists of two clades: a poorly supported South American and Tasmanian clade (*Anemone sellowii*, *Anemone helleborifolia*, *Anemone rigida*, *Barneoudia* and *Oreithales* species, and *Anemone crassifolia*), and a highly supported southern African clade including nine species of *Knowltonia* and eight species of *Anemone*. *Anemone antucensis* (Chile, Argentina) falls in a separate clade (subgenus and section *Anemonidium*) that is sister to *A. tenuicaulis* (New Zealand). *Anemone thomsonii* (Eastern Africa) and *Anemone somaliensis* (Somalia) are in a clade (subgenus and section *Anemone*) composed largely of Northern Hemisphere species. *A. somaliensis* is further associated with other Mediterranean tuberous anemones in subsection and series *Anemone* (*Anemone coronaria*, *Anemone hortensis*, and *Anemone pavonina*). The topology of both sections *Pulsatilloides* and *Anemonidium* suggest that anemones originated in the Northern Hemisphere and subsequently spread to the Southern Hemisphere, a pattern that is shared with other members of Ranunculaceae.

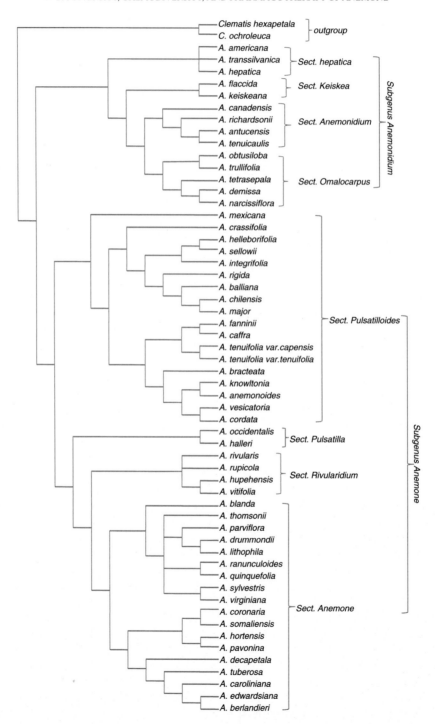

FIGURE 8.5   Phylogenetic relationship of *Anemone* inferred from ITS + *atpB-rbcL* spacer region, using the ML (maximum likelihood) method.

The taxonomic relationship of many Chinese species is elusive; we thus retrieve ITS sequences of more *Anemone* species from NCBI GenBank and reconstruct their phylogenetic relationship (Fig. 8.6). *Anemone umbrosa, Anemone reflexa, A. amurensis, A. raddeana, Anemone griffithii, A. stolonifera,* and *Anemone davidii,* belonging to the section *Anemonanthea,* cluster on the top of the phylogenetic tree, which is congruent to the morphological classification. Some non-Chinese species, for example, *Anemone virginiana, A. sylvestris, A. somaliensis,* and *Anemone tuberosa,* belonging to the section *Anemone,* are below the *Anemonanthea* species. However, some Chinese species of the section *Anemone,* for example, *Anemone hupehensis, A. begoniifolia, A. xingyiensis,* and *A. rupicola,* are closer to *A. rivularis* (Sect. *Rivularidium*), *A. orthocarpa,* and *A. hokouensis* (Sect. *Begoniifolia*). All above taxa belong to the morphological subgenus *Anemone.* The sections *Pulsatilla* and *Pulsatilloides,* belonging to the subgenus *Anemone,* have no Chinese taxa. On the other hand, five sections, belonging to the subgenus *Anemonidium,* form another major clade. The section *Himalayicae* is closer to the section *Omalocarpus* than to the section *Stolonifera.* The non-Chinese sections *Hepatica* and *Anemonidium* are basal to these sections. The sections *Himalayicae* and *Omalocarpus,* evolutionarily younger than other sections, have some important Tibetan medicinal plants, for example, *A. trullifolia, A. obtusiloba, A. demissa,* and *A. imbricata.* These two groups are still in the process of rapid radiation, corresponding to the extensive uplift of Qinghai-Tibet Plateau (QTP) during Quaternary (Hao and Xiao, 2015). The numerous morphologically and phytochemically distinct species should be investigated in detail to facilitate the sustainable development of *Anemone*-based clinical therapy.

nrITS and six plastid markers (atpB-rbcL, matK, psbA-trnQ, rpoB-trnC, rbcL, and rps16) were used to construct the phylogenetic tree of 82 taxa of tribe Anemoneae (Jiang et al., 2017). All traditional genera were resolved as monophyletic when using individual data sets, except polyphyletic *Anemone* and paraphyletic *Clematis.* The combined ITS + atpB-rbcL data set recovers monophyly of subtribes Anemoninae (i.e., *Anemone* s.l.) and Clematidinae (including *Anemoclema*). Incongruently, the concatenated plastid data set shows that one group consisting of *sect. Pulsatilla, Pulsatilloides,* and *Rivularidium* is closer to the clade *Clematis* s.l. + *Anemoclema.* There is a close relationship between *Anemoclema* and *Clematis* s.l., which includes *Archiclematis* and *Naravelia.* Nonmonophyly of *Anemone* s.l., if the inference based on the plastid data set is correct, suggests two revised genera, that is, new *Anemone* s.l. (including *Pulsatilla, Barneoudia, Oreithales,* and *Knowltonia*) and *Hepatica* (corresponding to *Anemone* subgenus *Anemonidium*).

8.6 TRANSCRIPTOMICS, PROTEOMICS, AND METABOLOMICS

Transcriptome sequencing and proteomic techniques are combined to comprehensively analyze the triterpenoid saponin biosynthetic pathway in *A. flaccid* (Zhan et al., 2016). A total of 126 putative cytochrome P450s (CYP) and 32 UDP glycosyltransferases are selected from 46,962 unigenes as the candidates of triterpenoid saponin modifiers. Four CYPs are annotated as the gene of CYP716A subfamily, the key enzyme in the oleanane-type saponin biosynthetic pathway. Based on RNA-Seq, iTRAQ proteome analysis, and quantitative RT-PCR verification, the expression level of gene and protein committed to triterpenoids biosynthesis in the leaf and the rhizome is comparable. The de novo transcriptome and proteome profiling are powerful in the discovery of candidate genes, which are related with the biosynthesis of

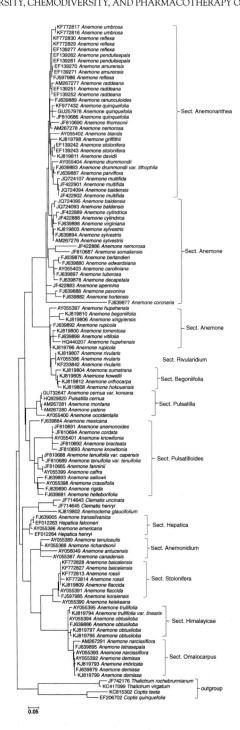

FIGURE 8.6 Phylogenetic relationship of *Anemone* ITSs inferred by the ML method. Scale bar represents 0.05 substitutions per site.

phytometabolites in nonmodel plants. The transcriptome of *A. coronaria* (Laura et al., 2015), following infection with rust, is available, allowing for the comparative transcriptomic studies. Twenty taxonomically related benzylisoquinoline alkaloid-producing plants, belonging to Ranunculaceae and closely related families, are subjected to the transcriptome sequencing and analysis (Hagel et al., 2015a). These essential data resources can be used to isolate and discover functional homologues and novel catalysts within metabolism of medicinal compounds. Orthologs can be extracted for transcriptome-based phylogeny reconstruction (Hao et al., 2012; Han et al., 2014).

Metabolomics studies provide imperative insight into the primary biochemical networks behind specialized metabolism and contribute key resources for metabolic engineering, gene discovery, and elucidation of regulatory mechanisms (Hagel et al., 2015b). Future comprehensive and thorough metabolomics investigations of *Anemone* species are warranted.

8.7 CONCLUSION AND PROSPECTS

Are triterpenoid saponins of *Anemone* plants epiphany molecules for cancer patients? Not yet. Much more in vivo evidence has to be collated in experimental animals and humans, while to date there is a lack of pharmacokinetic and pharmacodynamic data of *Anemone* anticancer compounds. The structure-activity relationship should be investigated for understanding the molecular mechanisms and the rational drug design. Some new *Anemone* taxa have been identified in Southwest China (Wang and Liu, 2008; Yuan and Yang, 2009; Wang, 2014a,b), the biodiversity center of Ranunculaceae plants. A majority of *Anemone* species have not been probed with respect to their unique biosynthetic pathways and chemodiversity. Trade-off should be attained between the conservation of endangered *Anemone* species and the utilization of *Anemone* pharmaceutical resources. The advent of the genomic era has provided important and surprising insights into the deducted genetic composition of *Anemone* species. Various innovative omics platforms would be of great help in deciphering biosynthetic pathways of *Anemone* phytometabolites, which lay a solid foundation for future synthetic biology manipulations and also help protect *Anemone* medicinal resources for sustainable utilization.

References

Anh Minh, C.T., Khoi, N.M., Thuong, P.T., et al., 2012. A new saponin and other constituents from *Anemone rivularis* Buch-Ham. Biochem. Syst. Ecol. 44, 270–274.

Bai, C., Ye, Y., Feng, X., et al., 2017. Anti-proliferative effect of triterpenoidal glycosides from the roots of *Anemone vitifolia* through a pro-apoptotic way. Molecules 22 (4), pii: E642.

Bing, F.H., Han, L.T., Zhang, G.B., et al., 2008. Chemical constituents of rhizome of *Anemone flaccida* and their immunosuppressive activities in vitro. J. China Pharm. Univ. 39 (6), 496–499.

Cao, P., Wu, F.E., Ding, L.S., 2004. Advances in the studies on the chemical constituents and biologic activities for *Anemone* species. Nat. Prod. Res. Dev. 16 (6), 581–584.

Chang, I.C., Chiang, T.I., Lo, C., et al., 2015. *Anemone altaica* induces apoptosis in human osteosarcoma cells. Am. J. Chin. Med. 43 (5), 1031–1042.

Chung, T.W., Lee, J.H., Choi, H.J., et al., 2017. *Anemone rivularis* inhibits pyruvate dehydrogenase kinase activity and tumor growth. J. Ethnopharmacol. 203, 47–54.

Cossard, G., Sannier, J., Sauquet, H., et al., 2016. Subfamilial and tribal relationships of Ranunculaceae: evidence from eight molecular markers. Plant Syst. Evol. 302, 419–431.

Ding, Y., 2011. Studies on the Chemical Constituents of *Anemone rivularis* var. *flore-minore* (Doctor's dissertation). Fourth Military Medical University.

Gong, H.D., 2011a. Investigation on traditional Tibetan medicine plant resources of *Anemone* in the eastern of Qinghai-Tibet Plateau. J. Anhui Agric. Sci. 39 (11), 6388–6391.

Gong, W.Z., 2011b. Chemical Constituents and Bioactivity of *Anemone cathayensis* (Master's thesis). Minzu University of China.

Gu, G., Qi, H., Jiang, T., et al., 2017. Investigation of the cytotoxicity, apoptosis and pharmacokinetics of Raddeanin A. Oncol. Lett. 13 (3), 1365–1369.

Guan, Y.Y., Liu, H.J., Luan, X., et al., 2015. Raddeanin A, a triterpenoid saponin isolated from *Anemone raddeana*, suppresses the angiogenesis and growth of human colorectal tumor by inhibiting VEGFR2 signaling. Phytomedicine 22 (1), 103–110.

Hagel, J.M., Mandal, R., Han, B., et al., 2015b. Metabolome analysis of 20 taxonomically related benzylisoquinoline alkaloid-producing plants. BMC Plant Biol. 15, 220.

Hagel, J.M., Morris, J.S., Lee, E.J., et al., 2015a. Transcriptome analysis of 20 taxonomically related benzylisoquinoline alkaloid-producing plants. BMC Plant Biol. 15, 227.

Han, F., Peng, Y., Xu, L., et al., 2014. Identification, characterization, and utilization of single copy genes in 29 angiosperm genomes. BMC Genomics 15, 504.

Han, L.T., Fang, Y., Cao, Y., et al., 2016. Triterpenoid saponin flaccidoside II from *Anemone flaccida* triggers apoptosis of NF1-associated malignant peripheral nerve sheath tumors via the MAPK-HO-1 pathway. Onco. Targets Ther. 9, 1969–1979.

Han, L.T., Fang, Y., Li, M.M., et al., 2013. The antitumor effects of triterpenoidsaponins from the *Anemone flaccida* and the underlying mechanism. Evid. Based Complement. Alternat. Med. 2013, 517931.

Han, L.T., Li, J., Huang, F., et al., 2009. Triterpenoidsaponins from *Anemone flaccida* induce apoptosis activity in HeLa cells. J. Asian Nat. Prod. Res. 11 (2), 122–127.

Hao, D.C., Gu, X.J., Xiao, P.G., et al., 2013a. Chemical and biological research of Clematis medicinal resources. Chin. Sci. Bull. 58 (10), 1120–1129.

Hao, D.C., Gu, X.J., Xiao, P.G., et al., 2013b. Recent advance in chemical and biological studies on Cimicifugeae pharmaceutical resources. Chin. Herb. Med. 5 (2), 81–95.

Hao, D.C., Gu, X.J., Xiao, P.G., et al., 2015b. Medicinal Plants: Chemistry, Biology and Omics. Elsevier-Woodhead, Oxford.

Hao, D.C., Ma, P., Mu, J., et al., 2012. De novo characterization of the root transcriptome of a traditional Chinese medicinal plant *Polygonum cuspidatum*. Sci. China Life Sci. 55 (5), 452–466.

Hao, D.C., Xiao, P.G., 2015. Genomics and evolution in traditional medicinal plants: road to a healthier life. Evol. Bioinform. Online 11, 197–212.

Hao, D.C., Xiao, P.G., Liu, L.W., et al., 2015c. Essentials of pharmacophylogeny: knowledge pedigree, epistemology and paradigm shift. Zhongguo Zhong Yao Za Zhi 40 (17), 3335–3342.

Hao, D.C., Xiao, P.G., Liu, M., et al., 2014. Pharmaphylogeny vs. pharmacophylogenomics: molecular phylogeny, evolution and drug discovery. Yao Xue Xue Bao 49 (10), 1387–1394.

Hao, D.C., Xiao, P.G., Ma, H., et al., 2015a. Mining chemodiversity from biodiversity: pharmacophylogeny of medicinal plants of the Ranunculaceae. Chin. J. Nat. Med. 13 (7), 507–520.

He, K., Jiang, X.L., 2014. Sky islands of southwest China I. An overview of phylogeographic patterns. Chin. Sci. Bull. 59, 585–597.

Hoot, S.B., Meyer, K.M., Manning, J.C., 2012. Phylogeny and reclassification of *Anemone* (Ranunculaceae), with an emphasis on austral species. Syst. Bot. 37 (1), 139–152.

Hou, A.W., 2012. Studies on the Fingerprint and Chemical Composition from Two Related Plants of *Anemone flaccida* (Master's thesis). Hubei University of Traditional Chinese Medicine.

Hu, H.B., Zheng, X.D., Jian, Y.F., et al., 2011. Constituents of the root of *Anemone tomentosa*. Arch. Pharm. Res. 34 (7), 1097–1105.

Ji, C., Cheng, G., Tang, H., et al., 2015. Saponin 6 of *Anemone taipaiensis* inhibits proliferation and induces apoptosis of U87 MG cells. Xi Bao Yu Fen Zi Mian Yi Xue Za Zhi 31 (4), 484–486.

Ji, C.C., Tang, H.F., Hu, Y.Y., et al., 2016. Saponin 6 derived from *Anemone taipaiensis* induces U87 human malignant glioblastoma cell apoptosis via regulation of Fas and Bcl-2 family proteins. Mol. Med. Rep. 14 (1), 380–386.

Jiang, N., Zhou, Z., Yang, J.B., et al., 2017. Phylogenetic reassessment of tribe Anemoneae (Ranunculaceae): Nonmonophyly of *Anemone* s.l. revealed by plastid datasets. PLoS One 12 (3), e0174792.

Ju, Y., Jia, Z.J., Zhu, Z.Q., 1986. Chemical constituents of *Anemone altaica*. Chin. Tradit. Herb. Drug 17 (9), 388–389.

Jurgens, A., Dotterl, S., 2004. Chemical composition of anther volatiles in Ranunculaceae: genera-specific profiles in *Anemone, Aquilegia, Caltha, Pulsatilla, Ranunculus,* and *Trollius* species. Am. J. Bot. 91 (12), 1969–1980.

Kong, X., Wu, W., Yang, Y., et al., 2015a. Total saponin from *Anemone flaccida* Fr. Schmidt abrogates osteoclast differentiation and bone resorption via the inhibition of RANKL-induced NF-κB JNK and p38 MAPKs activation. J. Transl. Med. 13, 91.

Kong, X., Yang, Y., Wu, W., et al., 2015b. Triterpenoid saponin W3 from *Anemone flaccida* suppresses osteoclast differentiation through inhibiting activation of MAPKs and NF-κB pathways. Int. J. Biol. Sci. 11 (10), 1204–1214.

Kumar, K., Sharma, Y.P., Manhas, R.K., et al., 2015. Ethnomedicinal plants of Shankaracharya Hill, Srinagar, J&K, India. J. Ethnopharmacol. 170, 255–274.

Laura, M., Borghi, C., Bobbio, V., et al., 2015. The effect on the transcriptome of *Anemone coronaria* following infection with rust (Tranzschelia discolor). PLoS One 10 (3), e0118565.

Lee, T.H., Huang, N.K., Lai, T.C., et al., 2008. Anemonin, from *Clematis crassifolia*, potent and selective inducible nitric oxide synthase inhibitor. J. Ethnopharmacol. 116 (3), 518–527.

Li, J., Tang, H., Zhang, Y., et al., 2013. Saponin 1 induces apoptosis and suppresses NF-κB-mediated survival signaling in glioblastoma multiforme (GBM). PLoS One 8 (11), e81258.

Li, M., Lei, Z., Zhong, G.Y., 2015. Analysis of varieties and standards of Ranunculaceae medicinal plants used in Tibetan medicine. Trad. Chin. Drug Res. Clin. Pharmacol. 26 (1), 133–137.

Li, J.N., Yu, Y., Zhang, Y.F., et al., 2017. Synergy of Raddeanin A and cisplatin induced therapeutic effect enhancement in human hepatocellular carcinoma. Biochem. Biophys. Res. Commun. 485, 335–341.

Liao, X., Chen, Y.Z., Ding, L.S., et al., 1999. Chemical constituents from *Anemone rupestris* ssp. *gelida*. Nat. Prod. Res. Dev. 11 (4), 1–6.

Liao, X., Li, B.G., Wang, M.K., et al., 2001. The chemical constituents from *Anemone rivularis*. Chem. J. Chin. Univ. 22 (8), 1338–1341.

Liu, C., Yang, Y., Sun, D., et al., 2015a. Total saponin from *Anemone flaccida* Fr. Schmidt prevents bone destruction in experimental rheumatoid arthritis via inhibiting osteoclastogenesis. Rejuvenation Res. 18 (6), 528–542.

Liu, Q., Zhu, X.Z., Feng, R.B., et al., 2015b. Crude triterpenoidsaponins from *Anemone flaccida* (Di Wu) exert antiarthritic effects on type II collagen-induced arthritis in rats. Chin. Med. 10, 20.

Liu, Y., Li, Y., Yang, W., et al., 2012. Anti-hepatoma activity in mice of a polysaccharide from the rhizome of *Anemone raddeana*. Int. J. Biol. Macromol. 50 (3), 632–636.

Lu, J., Sun, Q., Sugahara, K., et al., 2001. Effect of six compounds isolated from rhizome of *Anemone raddeana* on the superoxide generation in human neutrophil. Biochem. Biophys. Res. Commun. 280 (3), 918–922.

Lu, J., Xu, B., Gao, S., et al., 2009. Structure elucidation of two triterpenoid saponins from rhizome of *Anemone raddeana* Regel. Fitoterapia 80 (6), 345–348.

Lv, C.N., Fan, L., Wang, J., et al., 2015. Two new triterpenoidsaponins from rhizome of *Anemone amurensis*. J. Asian Nat. Prod. Res. 17 (2), 132–137.

Lv, C.N., Li, Y.J., Wang, J., et al., 2016. Chemical constituents from rhizome of *Anemone amurensis*. J. Asian Nat. Prod. Res. 18 (7), 648–655.

Mimaki, Y., Watanabe, K., Matsuo, Y., et al., 2009. Triterpene glycosides from the tubers of *Anemone coronaria*. Chem. Pharm. Bull. 57 (7), 724–729.

Ren, F.Z., Chen, S.H., Zheng, Z.H., et al., 2012. Coumarins of *Anemone raddeana* Regel and their biological activity. Yao Xue Xue Bao 47 (2), 206–209.

Saito, N., Toki, K., Moriyama, H., et al., 2002. Acylated anthocyanins from the blue-violet flowers of *Anemone coronaria*. Phytochemistry 60 (4), 365–373.

Shao, J.H., Zhao, C.C., Liu, M.X., 2011. Separation and structural identification on the anti-tumor activity of *Anemone rivularis*. J. Yangzhou Univ. 14 (2), 48–50.

Shi, B., Liu, W., Gao, L., et al., 2012. Chemical composition, antibacterial and antioxidant activity of the essential oil of *Anemone rivularis*. J. Med. Plant. Res. 6 (25), 4221–4224.

Su, Y.Q., 2007. Chemical Constituents and Structure Activity Relationship of *A. flaccida*, (Master's thesis). Hubei University of Traditional Chinese Medicine.

Sun, Y., Li, M., Liu, J., 2008. Haemolytic activities and adjuvant effect of *Anemone raddeana* saponins (ARS) on the immune responses to ovalbumin in mice. Int. Immuno. pharmacol. 8 (8), 1095–1102.

Sun, Y.X., Liu, J.C., Liu, D.Y., 2011. Phytochemicals and bioactivities of *Anemone raddeana* Regel: a review. Pharmazie 66 (11), 813–821.

Tang, J.Q., 2013. Studies on Triterpenoid Saponins of the Rhizomes of *Anemone flaccida* (Master's thesis). Jinan University.

Timar Tenzin Pentso, 2012. Jing Zhu Ben Cao. Shanghai: Shanghai Scientific and Technical Publishers.

Teng, Y.H., Li, J.P., Liu, S.L., et al., 2016. Autophagy protects from raddeanin A-induced apoptosis in SGC-7901 human gastric cancer cells. Evid. Based Complement. Alternat. Med. 2016, 9406758.

Toki, K., Saito, N., Shigihara, A., et al., 2001. Anthocyanins from the scarlet flowers of *Anemone coronaria*. Phytochemistry 56, 711–715.

Varitimidis, C., Petrakis, P.V., Vagias, C., et al., 2006. Secondary metabolites and insecticidal activity of *Anemone pavonina*. Z. Naturforsch. C 61 (7-8), 521–526.

Wang, J.R., Peng, S.L., Wang, M.K., et al., 1999. Chemical constituents of the *Anemone tomentosa* root. Acta Bot. Sin. 41 (1), 107–110.

Wang, M.K., Ding, L.S., Wu, F.E., 2008. Antitumor effects of raddeanin A on S180, H22 and U14 cell xenografts in mice. Ai Zheng 27 (9), 910–913.

Wang, M.K., Wu, F.E., Chen, Y.Z., 1997. Tritepenoid saponins of *Anemone hupehensis*. Phytochemistry 44 (2), 333–335.

Wang, S.L., Zhao, Z.K., Sun, J.F., et al., 2017. Review of *Anemone raddeana* rhizome and its pharmacological effects. Chin. J. Integr. Med.doi: 10.1007/s11655-017-2901-2.

Wang, W., Liu, H., Song, Y., et al., 1995. Bencaological studies on Changpu. Chin. Trad. Herb. Drug 5, 263–265.

Wang, W., Lu, A.M., Ren, Y., et al., 2009. Phylogeny and classification of Ranunculales: evidence from four molecular loci and morphological data. Perspect. Plant Ecol. Evol. Syst. 11, 81–110.

Wang, W.T., 2014a. Three new species of *Anemone* (Ranunculaceae) from Xizang. Plant Divers. Resour. 36 (4), 449–452.

Wang, W.T., Liu, B., 2008. A new section with a new species of *Anemone* (Ranunculaceae) from Mt. Xiaowutai, China. Acta Phytotaxon. Sin. 46 (5), 738–741.

Wang, X., Zhang, W., Gao, K., et al., 2013a. Oleanane-type saponins from *Anemone taipaiensis* and their cytotoxic activities. Fitoterapia 89, 224–230.

Wang, X.Y., 2014b. Studies on Bioactive Constituents from Three Medical Plants Collected on Taibai Mountain (Doctor's dissertation). Fourth Military Medical University.

Wang, X.Y., Gao, H., Zhang, W., et al., 2013b. Bioactiveoleanane-type saponins from the rhizomes of *Anemone taipaiensis*. Bioorg. Med. Chem. Lett. 23 (20), 5714–5720.

Wang, Y., Tang, H., Zhang, Y., et al., 2013c. Saponin B, a novel cytostatic compound purified from *Anemone taipaiensis*, induces apoptosis in a human glioblastoma cell line. Int. J. Mol. Med. 32 (5), 1077–1084.

Wang, Y., Kang, W., Hong, L.J., et al., 2013d. Triterpenoid saponins from the root of *Anemone tomentosa*. J. Nat. Med. 67 (1), 70–77.

Wang, Y., 2012. Studies on the Saponin Constituents of *Anemone tomentosa* (Master's thesis). Fourth Military Medical University.

Wang, Y.Z., Zeng, G., Zhang, M., et al., 2014. Chemical constituents from rhizoma of *Anemone altaica*. Chin. Trad. Herb. Drug 45 (9), 1219–1222.

Wei, S., He, W., Lu, J., et al., 2012. Effects of five oleanolic acid triterpenoid saponins from the rhizome of *Anemone raddeana* on stimulus-induced superoxide generation, phosphorylation of proteins and translocation of cytosolic compounds to cell membrane in human neutrophils. Fitoterapia 83 (2), 402–407.

Xiao, P.G., Wang, L.W., Lv, S.J., et al., 1986. Statistical analysis of the ethnopharmacologic data based on Chinese medicinal plants by electronic computer I. Magnoliidae. Chin. J. Integr. Trad. West Med. 6 (4), 253–256.

Xue, G., Zou, X., Zhou, J.Y., et al., 2013. Raddeanin A induces human gastric cancer cells apoptosis and inhibits their invasion in vitro. Biochem. Biophys. Res. Commun. 439 (2), 196–202.

Yamashita, K., Lu, H., Lu, J., et al., 2002. Effect of three triterpenoids, lupeol, betulin, and betulinic acid on the stimulus-induced superoxide generation and tyrosyl phosphorylation of proteins in human neutrophils. Clin. Chim. Acta 325 (1–2), 91–96.

Yan, Z.K., Liu, S.Y., Zhang, Y.L., et al., 1990. Preliminary study on the insecticidal activity and chemical components of *Anemone hupehensis* L. Acta Bot. Boreal. Occident. Sin. 10 (3), 141–148.

Yuan, Q., Yang, Q.E., 2009. *Anemone xingyiensis* (Ranunculaceae), a new species from Guizhou, China. Bot. Studi. 50, 493–498.

Zhan, C., Li, X., Zhao, Z., et al., 2016. Comprehensive analysis of the triterpenoid saponins biosynthetic pathway in *Anemone flaccida* by transcriptome and proteome profiling. Front. Plant Sci. 7, 1094.

Zhang, L.T., 2007. The Study on Chemical Compounds from *Anemone flaccida* and *Duchesnea indica*, (Doctor's dissertation). Tianjin University.

Zhang, Y., Huang, X.J., Wang, L., et al., 2012. New triterpenoid saponins from the rhizomes of *Anemone amurensis*. Chin. J. Chem. 30, 1249–1254.

Zhang, Y., Huang, X.J., Wang, Y., et al., 2014. Non-saponin compounds from the rhizomes of *Anemone amurensis*. Food Ind. 35 (3), 99–101.

Zou, Z.J., 2006. Studies on the Chemical Constituents of *Anemone altaica Hemisteptaly rata* and *Euphorbia latifolia*, (Doctor's dissertation). Chinese Academy of Medical Sciences.

Zou, Z.J., Liu, H., Yang, J.S., 2004. Phytochemical components and pharmacological activities of the genus *Anemone*. Chin. Pharm. J. 39 (7), 493–495.

Zou, Z.J., Yang, J.S., 2008a. Studies on chemical constituents of Rizoma *Anemones Altaicae*. Mod. Chin. Med. 10 (9), 10–11.

Zou, Z.J., Yang, J.S., 2008b. Chemical constituents from the roots of *Anemone altaica*. West China J. Pharm. Sci. 23 (3), 265–266.

Zou, Z.J., Yang, J.S., 2008c. Studies on the chemical constituents of the roots of *Anemone altaica*. J. Chin. Med. Mat. 31 (1), 49–51.

CHAPTER

9

Biodiversity, Chemodiversity, and Pharmacotherapy of *Ranunculus* Medicinal Plants

OUTLINE

9.1 INTRODUCTION

Ranunculus (Ranunculaceae) plants, around 600 species, are globally distributed (Emadzade et al., 2011; Wang, 1995). *Ranunculus* is the largest genus of Ranunculaceae and can be found on every continent, from tropical to the Arctic and Subantarctic regions. It is particularly rich in temperate and Mediterranean regions. *Ranunculus* plants survive in various environments, from low-lying wetlands to the cold alpine mountains. This genus varies in morphology and physiology, embodying a strong adaptability. According to *Flora of China*, 78 species and 9 varieties are distributed widely in China, a majority of which is in alpine regions of northwest

Ranunculales Medicinal Plants. http://dx.doi.org/10.1016/B978-0-12-814232-5.00009-5

and southwest parts. Wang Wen-Cai revises that 115 species and 4 alien naturalized species are found in China; 60 species are endemic in China, accounting for 52% of all Chinese species. *Ranunculus* is rich and complex in China, which is the second most abundant region after Europe. In China, at least eight species are used medicinally in folk medicine (Xiao, 1980); for example, the fresh whole plant is used by sticking acupoints for malaria (frequency 30/no. of species 6), asthma (11/4), rheumatic arthritis (8/3), toothaches (11/3), and jaundice (16/4); the root is orally taken for pertussis (2/1), scrofula (9/2), and traumatic injury (5/2); the whole plant is used externally for psoriasis (3/2) and as an insecticide (5/2). The representative Chinese Medicine *Ranunculus ternatus* (cat claw grass), which has a temperate nature and sweet and pungent flavor, belonging to the liver and lung meridian, with small poison, is used for reducing phlegm and resolving masses as well as for detumescence by detoxification. In other countries *Ranunculus* is also traditionally used as a drug, for instance, in Spain the whole plant of *Ranunculus parnassifolius* is used externally for antigangrene, sterilization, subduing swelling, treating tumors, etc. (Carrió et al., 2012). *Ranunculus arvensis*is is topically applied for joint pain and osteoarthritis in Turkey (Akbulut et al., 2011). The tribal communities of Chhota Bhangal, Western Himalaya, use *Ranunculus hirtellus* for curing swelling in the testes (Uniyal et al., 2006). *Ranunculus sceleratus*is is used to treat urinary disorders in Punjab–Pakistan (Umair et al., 2017), and *Ranunculus muricatus* is used to treat dental disorders in the Manoor Valley of Northern Himalaya, Pakistan (Rahman et al., 2016).

9.2 CHEMICAL CONSTITUENTS

A series of lactones, for example, protoanemonin, anemonin, ranunculin, isoranunculin, and ternatolide, are widely distributed in *Ranunculus* (Peng et al., 2006). To date, all saponins isolated from this genus are of oleanane type (Table 9.1, nos. 36–42, Fig. 9.1). Alkaloids are widely present in *Ranunculus*, most of which are structurally simple, and the major isoquinolines are of berberine type and aporphine type (Hao et al., 2015).

9.2.1 Flavonoids and Phenolics

The known flavonoids include flavonol (Table 9.1, nos. 1–8, 15, 19–28,), flavone (9–14, 17, 18), dihydroflavonol (16), and dihydroflavone (29). Flavonoids of *R. ternatus* (Mao Zhua Cao/ Xiao Mao Gen in Chinese) are its main medicinal component, and the total flavonoid glycoside content of the stem and leaf is more than 2%, which is much higher than the 0.2% in the root tuber. Sixteen flavonol glycosides and three aglycones are detected from the alcohol extract of the stem and leaf of *R. ternatus* (4, 13–27) (Chi et al., 2013). Flavonoids such asquercetin, luteolin, tricin, vitexin, orientin, isoorientin, gossypitrin, apigenin, saponaretin, etc., and their glycosides are found in *Ranunculus* (Zhong and Feng, 2011). Tricin, luteolin, 5-hydroxy-6,7-dimethoxyflavone, and 5-hydroxy-7,8-dimethoxyflavone, etc., are identified from *Ranunculus japonicus* (Mao Gen/Mao Jian Cao in Chinese) of Liupan Mountain, Ningxia (Liang et al., 2008).

Phenolic compounds are isolated from *Ranunculus chinensis* (114–116) (Zou et al., 2010).

TABLE 9.1 Examples of *Ranunculus* Flavonoids, Triterpenoids, Saponins, Alkaloids, and Other Compounds

Number	Compound	Type	Species	Medicinal part	References
1.	Kaempferol-3-O-sophoroside-7-O-β-D-glucoside	A	I	Whole plant	Wu et al. (2013)
2.	Quercetin-3-O-(2‴-E-caffeoyl)-α-L-arabinopyranosyl-(1→2)-β-D-glucoside-7-O-β-D-glucoside	A	I	Whole plant	Wu et al. (2013)
3.	Kaempferol-3,7-di-β-D-glucoside	A	I	Whole plant	Wu et al. (2013)
4.	Quercetin-7-O-β-D-glucoside	A	I, III	I Whole plant, III stem and leaf	Chi et al. (2013)
5.	Quercetin-3-O-(2‴-E-caffeoylsophoroside)-7-O-β-D-glucoside	A	I	Whole plant	Wu et al. (2013)
6.	Kaempferol-3-O-(2‴-E-caffeoylsophoroside)-7-O-β-D-glucoside	A	I	Whole plant	Wu et al. (2013)
7.	Quercetin-3-O-(2‴-E-ferulylsophoroside)-7-O-β-D-glucoside	A	I	Whole plant	Wu et al. (2013)
8.	Kaempferol-3-O-(2‴-p-coumarylsophoroside)-7-O-β-D-glucoside	A	I	Whole plant	Wu et al. (2013)
9.	Apigenin-8-C-α-L-arabinopyranosyl-6-C-β-D-glucoside	A	I	Whole plant	Wu et al. (2013)
10.	Apigenin-6-C-α-L-arabinopyranosyl-8-C-β-D-glucoside	A	I	Whole plant	Wu et al. (2013)
11.	Tricin7-O-β-D-glucopyranoside	A	I	Whole plant	Nazir et al. (2013)
12.	Orientin	A	II	Whole plant	Zhong and Feng (2011)
13.	Luteolin	A	II, III	II Whole plant, III stem and leaf	Chi et al. (2013)
14.	Apigenin	A	II, III	II Whole plant, III stem and leaf	Chi et al. (2013)
15.	Quercetin	A	II, III	II Whole plant, III Stem and leaf	Chi et al. (2013)
16.	Dihydromyricetin	A	III	Stem and leaf	Chi et al. (2013)
17.	Schaftoside	A	III	Stem and leaf	Chi et al. (2013)
18.	Isoschaftoside	A	III	Stem and leaf	Chi et al. (2013)

(Continued)

TABLE 9.1 Examples of *Ranunculus* Flavonoids, Triterpenoids, Saponins, Alkaloids, and Other Compounds (*cont.*)

Number	Compound	Type	Species	Medicinal part	References
19.	Rutin	A	III	Stem and leaf	Chi et al. (2013)
20.	Isoquercitrin	A	III	Stem and leaf	Chi et al. (2013)
21.	Quercitrin	A	III	Stem and leaf	Chi et al. (2013)
22.	Myricetin	A	III	Stem and leaf	Chi et al. (2013)
23.	Kaempferol-7-O-β-D-glucoside	A	III	Stem and leaf	Chi et al. (2013)
24.	Quercetin-4′-O-β-D-glucoside	A	III	Stem and leaf	Chi et al. (2013)
25.	Isorhamnetin-7-O-β-D-glucoside	A	III	Stem and leaf	Chi et al. (2013)
26.	Kaempferol	A	III	Stem and leaf, root	Zhong and Feng (2011)
27.	Isorhamnetin	A	III	Stem and leaf	Chi et al. (2013)
28.	Quercetin3-O-(2-t-p-coumaroyl)-β-D-glucopyranosyl-(1→2)-β-D-glucopyranoside-7-O-β-D-glucopyranoside	A	IV	Leaf	Prieto et al. (2004)
29.	Sternbin	A	III	Root	Zhang et al. (2006)
30.	3β-acetoxy-(20S,22E)-dammaran-22-en-25-ol	B	III	Root	Zhao et al. (2008)
31.	Ursolic acid	B	III	Root	Zhao et al. (2008)
32.	Oleanolic acid	B	III	Root	Zhao et al. (2008)
33.	Betulinic acid	B	III	Root	Zhao et al. (2008)
34.	3-Epiocotillol	B	III	Root	Zhao et al. (2008)
35.	Dimmarenediol II acetate	B	III	Root	Zhao et al. (2008)
36.	3-O-[β-D-glucopyranosyl(1→3)-α-L-arabinopyranosyl]-28-O-[α-L-rhamnopyranosyl(1→4)-β-D-glucopyranosyl-(1→6)-β-D-glucopyranosyl] hederagenin	B	V	Whole plant	Wegner et al. (2000)
37.	3-O-[β-D-glucopyranosyl(1→3)-β-D-glucopyranosyl] oleanolic acid [α-L-rhamnopyranosyl(1→4)-β-D-glucopyranosyl-(1→6)-β-D-glucopyranosyl] ester	B	V	Whole plant	Wegner et al. (2000)

TABLE 9.1   Examples of Ranunculus Flavonoids, Triterpenoids, Saponins, Alkaloids, and Other Compounds (*cont.*)

Number	Compound	Type	Species	Medicinal part	References
38.	3-O-(β-D-glucopyranosyl)-28-O-[α-L-rhamnopyranosyl(1→4)-β-D-glucopyranosyl-(1→6)-β-D-glucopyranosyl] hederagenin	B	V	Whole plant	Wegner et al. (2000)
39.	3-O-[β-D-glucopyranosyl-(1→2)-α-L-arabionpyranosyl]-28-O-[α-L-rhamnopyranosyl(1→4)-β-D-glucopyranosyl-(1→6)-β-D-glucopyrano-syl] hederagenin	B	V	Whole plant	Wegner et al. (2000)
40.	3-O-(α-L-arabionpyranosyl)-28-O-[α-L-rhamnopyranosyl(1→4)-β-D-glucopyranosyl-(1→6)-β-D-glucopyranosyl] hederagenin	B	V	Whole plant	Wegner et al. (2000)
41.	3-O-(β-D-glucopyranosyl) oleanolic acid [α-L-rhamnopyranosyl-(1→4)-β-D-glucopyranosyl-(1→6)-β-D-glucopyranosyl] ester	B	V	Whole plant	Wegner and Hamburger (2002)
42.	3-O-(α-arabinopyranosyl-1′)-28-O-[β-glucopyranosyl-1‴′→6‴′(α-rhamnopyranosyl-1′)2S-O-(β-glucopyranosyl-1‴)] hederagenin	B	VI	Leaf	
43.	(25R)-26-[(α-L-rhamnopyranosyl)oxy]-22α-methoxyfurost-5-en-3β-yl O-β-D-glucopyranosyl-(1→3)-O-[6-acetyl-β-D-glucopyranosyl-(1→3)]-O-β-D-glucopyranoside	C	IV	Leaf	Prieto et al. (2007)
44.	Ternatusine A	D	III	Root	Zhan et al. (2013)
45.	11-O-β-D-glucopyranosyl rutaecarptine (ternato-side C)	D	III	Root	Zhang et al. (2007)

(*Continued*)

TABLE 9.1   Examples of *Ranunculus* Flavonoids, Triterpenoids, Saponins, Alkaloids, and Other Compounds (*cont.*)

Number	Compound	Type	Species	Medicinal part	References
46.	11-O-α-L-rhamnopyranosyl-(1→6)-β-D-glucopyranosyl rutaecarptine (ternatoside D)	D	III	Root	Zhang et al. (2007)
47.	4-oxo-5-(O-β-D-glucopyranosyl)-pentanoic acid-1-O-butylester (ternatoside A)	E	III	Root	Tian et al. (2006)
48.	4-oxo-5-(O-β-D-glucopyranosyl)-pentanoic acid-methyl ester	E	III	Root	Xiong et al. (2008a)
49.	Benzyl alcoholO-B-D-glucopyranoside	E	III	Root	Xiong et al. (2008b)
50.	(R)-3-[3-hydroxy-4-(O-β-D-glucopyranosyl)phenyl]-2-hydroxypropanoic acid butyl ester(ternatoside B)	E	III	Root	Tian et al. (2006)
51.	(2S)-3-O-β-D-galactopyranosyl-1,2-di-O-[(9Z,12Z,15Z)-octadeca-9,12,15-trienoyl]-sn-glycerol	E	V	Whole plant	Wegner et al. (2000)
52.	(2S)-3-O-[α-D-galactopyranosyl-(1→6)-β-D-galactopyranosyl]-1,2-di-O-[(9Z,12Z,15Z)-octadeca-9,12,15-trienoyl]-sn-glycerol	E	V	Whole plant	Wegner et al. (2000)
53.	Protocatechuic acid	F1	I, VIII, VII	I Whole plant, VIII种子, VII whole plant	Wu et al. (2013)
54.	Vanillic acid	F1	III	Root	Deng et al. (2013)
55.	p-hydroxybenzonic acid	F1	VIII	Seed	Li et al. (2010)
56.	Ferulic acid	F1	I	Whole plant	Wu et al. (2013)
57.	P-coumaric acid	F1	I,III	I Whole plant, III root	Xiong et al. (2008a)
58.	Caffeic acid	F1	I	Whole plant	Wu et al. (2013)
59.	4-O-D-glucopyranosyl-p-coumaric acid	F1	III	Root	Tian et al. (2006)
60.	Gallic acid	F1	III,VIII	III Root, VIII seed	Deng et al. (2013)
61.	4-[formyl-5-(hydroxymethyl)-1H-pyrrol-1-yl] butanoic acid	F1	III	Root	Qi et al. (2012)
62.	Ellagic acid	F1	VIII	Seed	Li et al. (2010)

TABLE 9.1   Examples of Ranunculus Flavonoids, Triterpenoids, Saponins, Alkaloids, and Other Compounds (*cont.*)

Number	Compound	Type	Species	Medicinal part	References
63.	O-phthalic acid	F1	III	Root	Chen et al. (2006)
64.	4-O-β-D-glucopyranosylcaffeic acid	F1	III	Root	Zhang et al. (2006)
65.	Hexadecanoic acid	F2	VII,III	VII Whole plant, III root	Peng et al. (2011)
66.	4-oxo-pentanoic acid	F2	III	Root	Xiong et al. (2008b)
67.	Succinic acid	F2	III	Root	Xiong et al. (2008a)
68.	Nonanedioic acid	F2	III	Root	Xiong et al. (2008b)
69.	Myristic acid	F2	III	Root	Tian et al. (2004)
70.	Pentadecanoic acid	F2	III	Root	Chen et al. (2006)
71.	Palmitelaidic acid	F2	III	Root	Chen et al. (2006)
72.	Heptadecanoic acid	F2	III	Root	Chen et al. (2006)
73.	Octadecanoic acid	F2	III	Root	Chen et al. (2006)
74.	8-Jeceric acid	F2	III	Root	Chen et al. (2006)
75.	9-Jeceric acid	F2	III	Root	Chen et al. (2006)
76.	Linoleic acid	F2	III	Root	Chen et al. (2006)
77.	α-Linolenic acid	F2	III	Root	Chen et al. (2006)
78.	Arachidic acid	F2	III	Root	Chen et al. (2006)
79.	Behenic acid	F2	III	Root	Chen et al. (2006)
80.	Tricosanoic acid	F2	III	Root	Chen et al. (2006)
81.	Methyl 5-hydroxy-4-oxo-pentanoate	G	III	Root	Xiong et al. (2008a)
82.	Methyl hydrogen succinate	G	III	Root	Xiong et al. (2008b)
83.	Succinic acid monoethyl ester	G	III	Root	Xiong et al. (2008a)
84.	3,4-Dihydroxybenzoic acid methyl ester	G	III	Root	Xiong et al. (2008b)
85.	Methyl 4-Hydroxybenzoate	G	III	Root	Zhang et al. (2006)
86.	Palmitic acid ethyl ester	G	III	Root	Tian et al. (2004)
87.	Diethylhexyl phthalate (DEHP)	G	III	Root	Zhao et al. (2010)
88.	Henicosanoicacid methyl ester	G	III	Root	Zhao et al. (2010)
89.	Mono-butyl phthalate	G	III	Root	Chen et al. (2005)
90.	Glycerol-β-palmitate	G	III	Root	Zhao et al. (2010)
91.	Glycerol-β-steariate	G	III	Root	Zhao et al. (2010)
92.	β-stiosterol	H	I,VII, VIII, III	I, VII Whole plant VIII Seed, III root	Peng et al. (2011)

(*Continued*)

TABLE 9.1   Examples of *Ranunculus* Flavonoids, Triterpenoids, Saponins, Alkaloids, and Other Compounds (*cont.*)

Number	Compound	Type	Species	Medicinal part	References
93.	Stigmasterol 3-O-β-ᴅ-glucopyranoside	H	III	Root	Xiong et al. (2008a)
94.	Stigmasterol	H	III,VII	III Root, VII whole plant	Tian et al. (2004)
95.	Campasterol	H	III	Root	Tian et al. (2004)
96.	Daucosterol	H	III	Root	Zhao et al. (2010)
97.	Stigmast-4-en-3,6-dione	H	VII	Whole plant	Gao et al., 2005
98.	Tanshinol	I	I	Whole plant	Wu et al. (2013)
99.	Methyl 3, 4-dihydroxy-phenyl-lactate	I	I	Whole plant	Wu et al. (2013)
100.	Protocatechuyl aldehyde	I	I,VII	Whole plant	Nazir et al. (2013)
101.	Methyl(R)-3-[2-(3,4-dihydroxybenzoyl)-4,5-dihydroxyphenyl]-2-hy-droxypropanoate	I	III	Root	Deng et al. (2013)
102.	N-butyl (R)-3-[2-(3,4-dihydroxybenzoyl)-4,5-dihydroxyphenyl]-2-hydroxypropanoate	I	III	Root	Deng et al. (2013)
103.	1-Docosene	I	VII	Whole plant	Peng et al. (2011)
104.	Emodin	I	VII	Whole plant	Peng et al. (2011)
105.	Ketologanin	I	VIII	Seed	Li et al. (2010)
106.	5-Hydroxymethylfuralde-hyde	I	III	Root	Chen et al. (2005b)
107.	p-Hydroxybenzaldehyde	I	III	Root	Xiong et al. (2008b)
108.	Phillygenin	I	III	Root	Zhao et al. (2010)
109.	Methyl 3,4,5-Trihydroxy-benzoate	I	IX	Root	Khan et al. (2006)
110.	R(+)-4-methoxydalbergione	I	IX	Root	Khan et al. (2008)
111.	R(+)-dalbergiophenol	I	IX	Root	Khan et al. (2008)
112.	Scoparone	I	VII	Whole plant	Gao et al. (2005)
113.	5-Hydroxymethylfuroic acid	I	III	Root	Chen et al. (2005a)
114.	Ranunchinesin A	I	VIII	Aerial part	Zou et al. (2010)
115.	Oresbiusin A	I	VIII	Aerial part	Zou et al. (2010)
116.	Ternatoside B	I	VIII	Aerial part	Zou et al. (2010)

A, flavonoid; B, triterpenoid; C, steroid saponin; D, alkaloid; E, glycoside; F, acid (F1 organic acid; F2 fatty acid); G, ester; H, sterol; I, others.

I, *R. muricatus*; II, *R. japonicus*; III, *R. ternatus* (root tuber is "cat claw grass"); IV, *R. lanuginosus*; V, *R. fluitans* Lamk.; VI, *R. ficaria*; VII, *R. sceleratus*; VIII, *R. chinensis*; IX, *R. repens*.

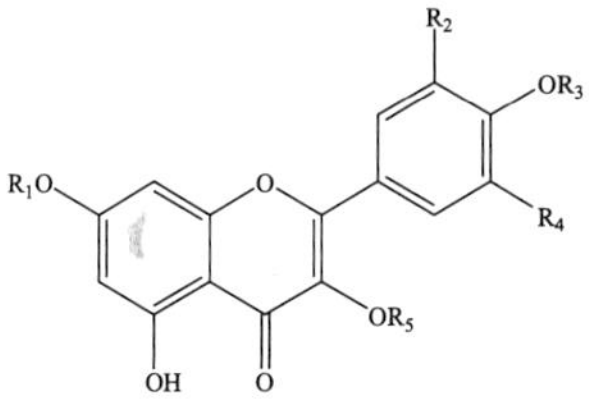

No.	R_1	R_2	R_3	R_4	R_5
1	β-D-Glc	H	H	H	槐糖
2	β-D-Glc	OH	H	H	(2‴-E-caffeoyl)-α-L-Ara-(1→2)-β-D-Glc
3	β-D-Glc	H	H	H	β-D-Glc
4	β-D-Glc	OH	H	H	H
5	β-D-Glc	OH	H	H	2‴-E-咖啡酰槐糖
6	β-D-Glc	H	H	H	2‴-E-咖啡酰槐糖
7	β-D-Glc	OH	H	H	2‴-E-阿魏酰槐糖
8	β-D-Glc	H	H	H	2‴-p-香豆酰槐糖
15	H	OH	H	H	H
19	H	H	H	H	β-D-Glc-(1→2)-α-L-Rha
20	H	H	H	H	β-D-Glc
21	H	H	H	H	α-L-Rha
22	H	H	H	OH	H
23	β-D-Glc	H	H	H	H
24	H	H	β-D-Glc	H	H
25	β-D-Glc	OCH₃	H	H	H
26	H	H	H	H	H
27	H	OCH₃	H	H	H
28	β-D-Glc	OH	H	H	(2-t-p-coumaroyl)-β-D-Glc-(1→2)-β-D-Glc

FIGURE 9.1 Flavonoids, triterpenoids, saponins, alkaloids, and other compounds (1–116) of *Ranunculus* plants.

9.2.2 Alkaloids

Multiple alkaloid glycosides, for example, indolopyridoquinazoline alkaloid glycosides (45, 46) (Zhang et al., 2007), are found in the root of *R. ternatus* (Tian et al., 2006). Ternatusine A (44) (Zhan et al., 2013), a new pyrrole derivative with a rare epoxyoxepino ring, is isolated from the root of *R. ternatus*.

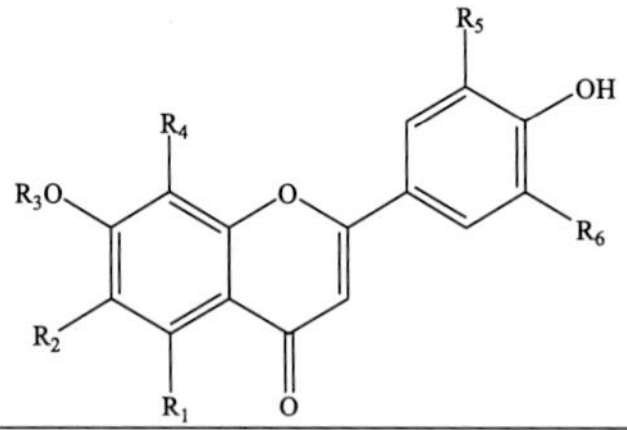

No	R_1	R_2	R_3	R_4	R_5	R_6
9	OH	β-D-Glc	H	α-L-Ara	H	H
10	OH	β-D-Glc	H	β-D-Glc	H	H
11	OH	H	β-D-Glc	H	OCH$_3$	OCH$_3$
12	OH	H	H	β-D-Glc	OH	H
13	OH	H	H	H	OH	H
14	OH	H	H	H	H	H
17	H	β-D-Glc	H	α-L-Ara	H	H
18	H	α-L-Ara	H	β-D-Glc	H	H

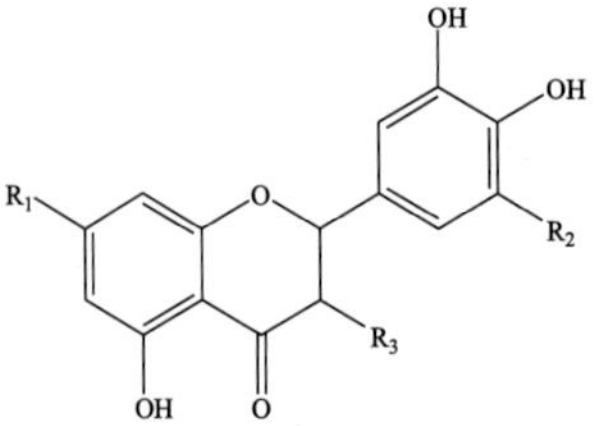

No	R_1	R_2	R_3
16	OH	OH	OH
29	OCH$_3$	H	H

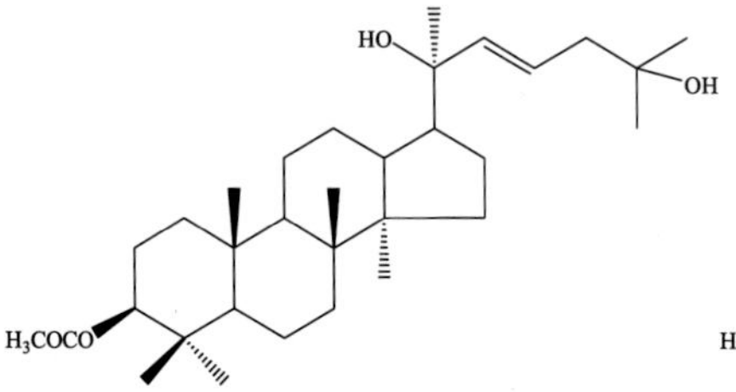

30

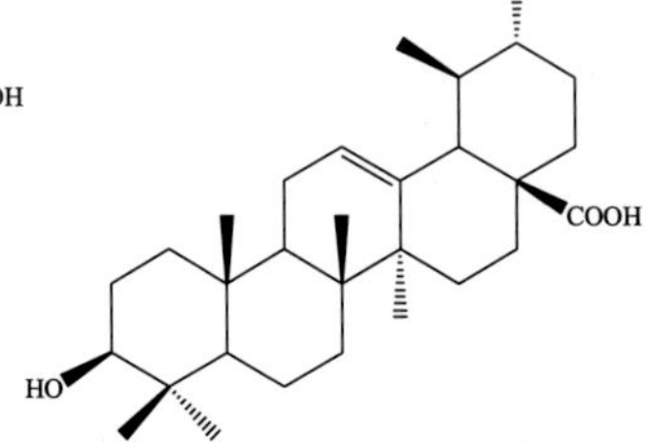

31

FIGURE 9.1   (*Cont.*)

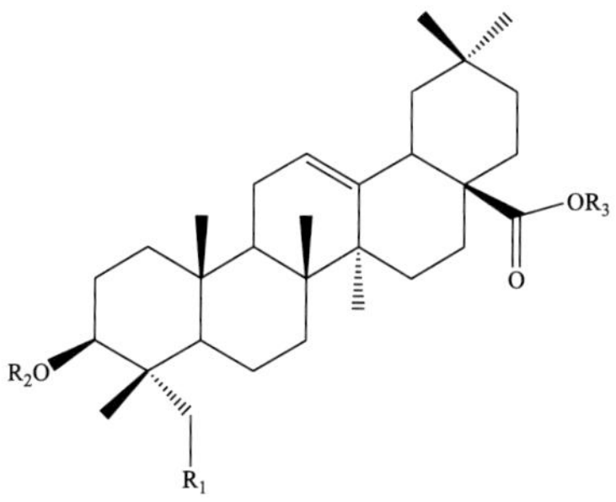

No	R$_1$	R$_2$	R$_3$
32	H	H	H
36	OH	β-D-Glc-(1→3)-α-L-Ara	α-L-Rha-(1→4)-β-D-Glc-(1→6)-β-D-Glc
37	H	β-D-Glc-(1→3)-α-L-Ara	α-L-Rha-(1→4)-β-D-Glc-(1→6)-β-D-Glc
38	OH	β-D-Glc	α-L-Rha-(1→4)-β-D-Glc-(1→6)-β-D-Glc
39	OH	β-D-Glc-(1→2)-α-L-Ara	α-L-Rha-(1→4)-β-D-Glc-(1→6)-β-D-Glc
40	H	α-L-Ara	α-L-Rha-(1→4)-β-D-Glc-(1→6)-β-D-Glc
41	H	β-D-Glc	α-L-Rha-(1→4)-β-D-Glc-(1→6)-β-D-Glc
42	OH	α-L-Ara	β-D-Glc-(1→6)-α-L-Rha-2S-β-D-Glc

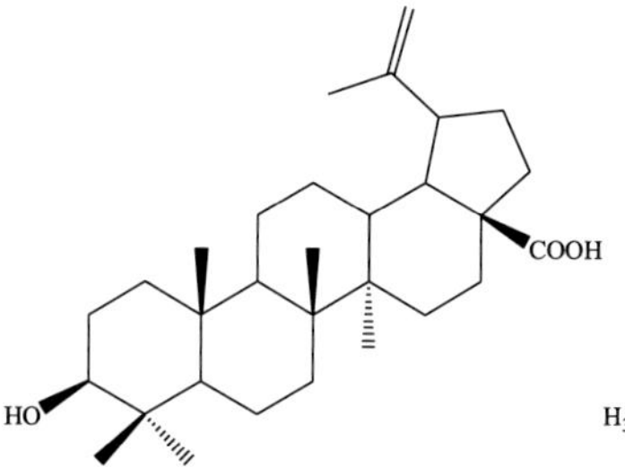

33

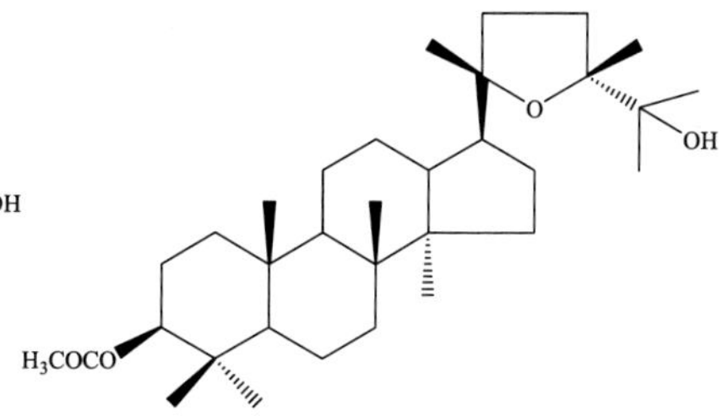

34

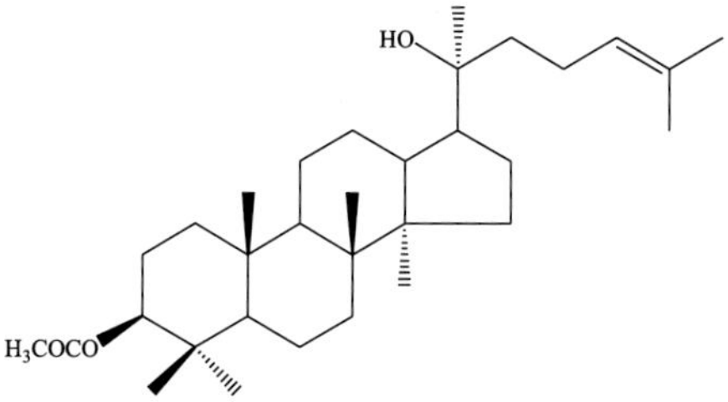

35

FIGURE 9.1 *(Cont.)*

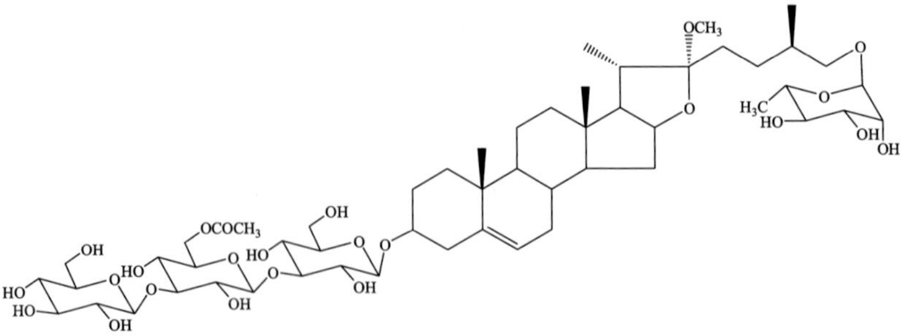

43

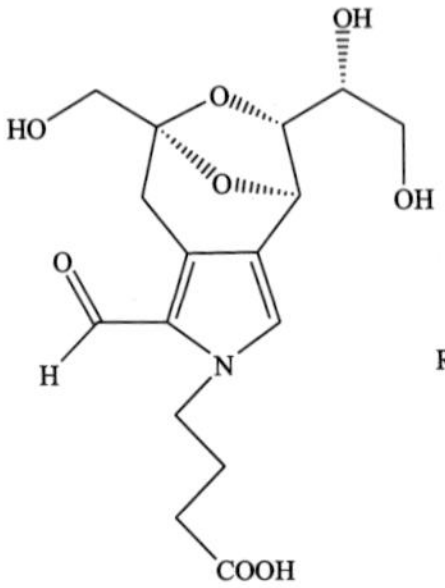

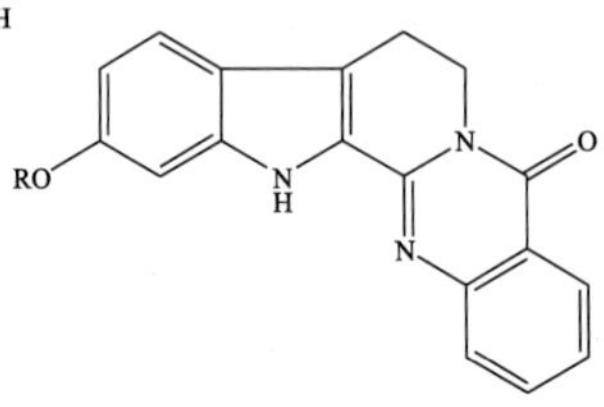

44

No	R
45	β-D-Glc
46	α-L-Rha-(1→6)-β-D-Glc

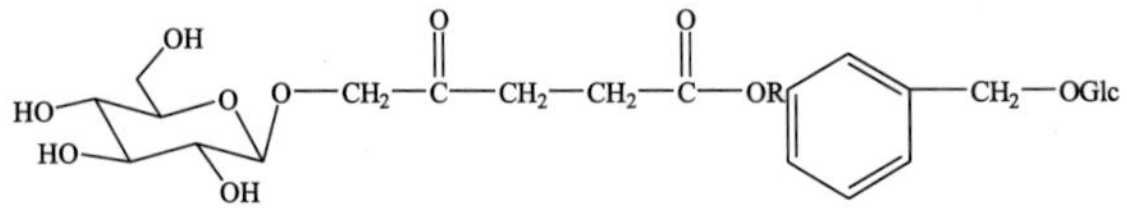

49

No	R
47	(CH$_2$)$_3$CH$_3$
48	CH$_3$

FIGURE 9.1 (*Cont.*)

9.2.3 Triterpenoids and Saponins

Ursolic acid (31), oleanolic acid (32), betulinic acid (33, lupinane type), 3-epiocotillol acetate, dimmarenediol II acetate, and 3β-acetoxy-(20S, 22E)-dammaran-22-en-25-ol (34, 35, 30, dammarane type) are isolated from *R. ternatus* (Zhao et al., 2008).

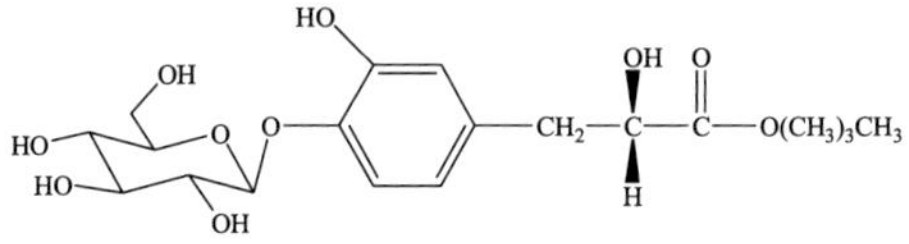

50

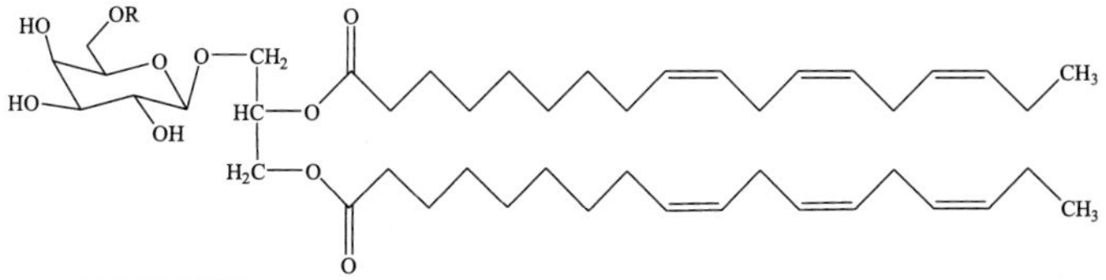

No	R
51	H
52	α-D-Gal

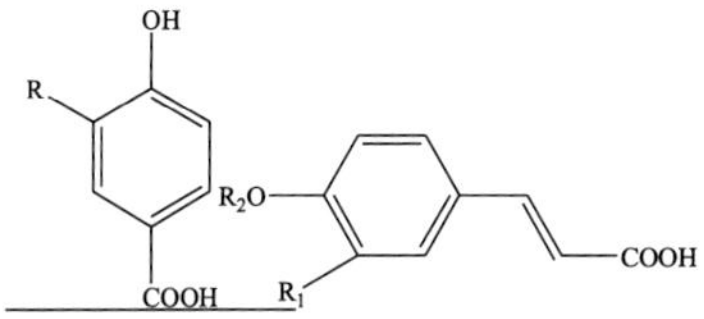

No	R
53	OH
54	OCH₃
55	H

No	R₁	R₂
56	OCH₃	H
57	H	H
58	OH	H
59	H	β-D-Glc
64	OH	β-D-Glc

| **6061** | **62** | **63** |

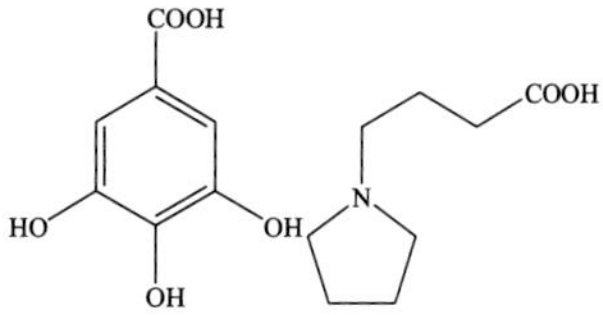

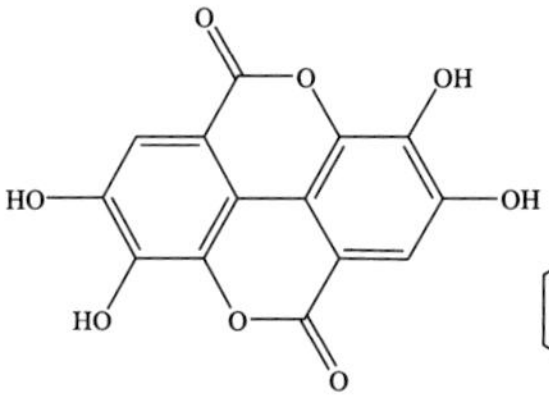

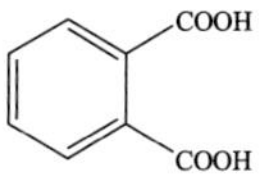

No	Formula
65	CH₃(CH₂)₁₄COOH

FIGURE 9.1 *(Cont.)*

66	CH₃COCH₂CH₂COOH
67	HOOCCH₂CH₂COOH
68	HOOC(CH₂)₇COOH
69	CH₃(CH₂)₁₂COOH
70	CH₃(CH₂)₁₃COOH
71	C₆H₁₃CH=CHC₇H₁₄COOH
72	CH₃(CH₂)₁₅COOH
73	CH₃(CH₂)₁₆COOH
74	C₉H₁₉CH=CHC₆H₁₂COOH
75	C₈H₁₇CH=CHC₇H₁₄COOH
76	C₅H₁₁CH=CHCH₂CH=CHC₇H₁₄COOH
77	C₂H₅CH=CHCH₂CH=CHCH₂CH=CHC₇H₁₄COOH
78	CH₃(CH₂)₁₈COOH
79	CH₃(CH₂)₂₀COOH
80	CH₃(CH₂)₂₁COOH

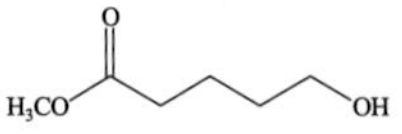

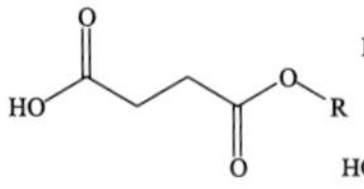

No	R
82	CH₃
83	CH₂CH₃

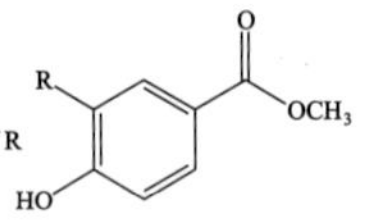

No	R
84	OH
85	H

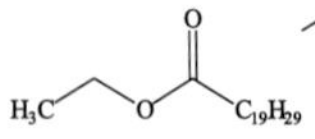

86 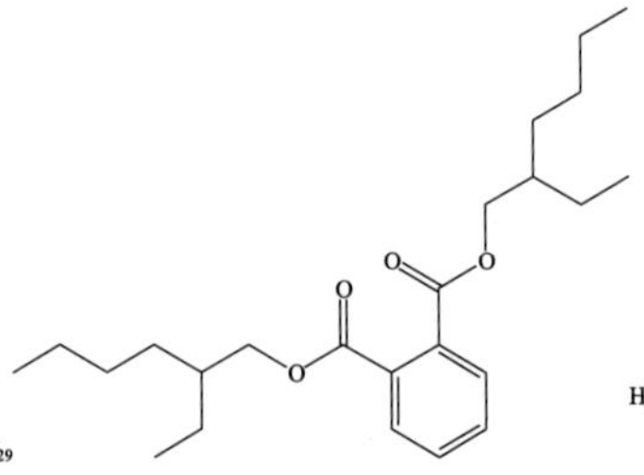   87 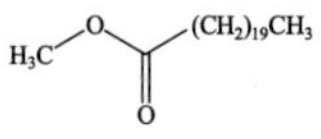   88

FIGURE 9.1   *(Cont.)*

9.2.4 Lipids and Volatile Compounds

Ranunculus is a common nectariferous plant, for example, the pollen kitt of *Ranunculus bulbosus* has various lipid compounds (Piskorski et al., 2011), including octanal, fatty acid amides, saturated and unsaturated hydrocarbons, and secondary alcohols. The volatile compounds, for example, esters, terpenoids, and protoanemonin, are found in the pollen. Sixty-one components of essential oil extracted from the root of *R. ternatus* are identified by conventional vapor distillation (Zhang et al., 2006), mainly esters, alkanes, and aromatic compounds. The volatile components in the anther of different *Ranunculus* species are similar (Jürgens and Dötterl, 2004), including protoanemonin, other fatty acid derivatives, benzenoid aromatic hydrocarbon, monoterpene, and sesquiterpene. The intraspecies variation of the chemical composition of anther volatiles is much lower than the interspecies one, rationalizing its use in chemotaxonomy.

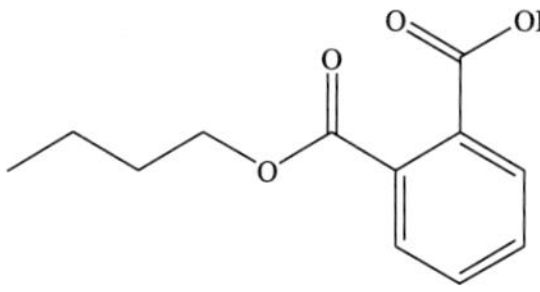

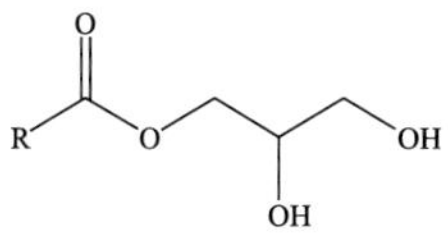

No	R
90	$(CH_2)_{14}CH_3$
91	$(CH_2)_{16}CH_3$

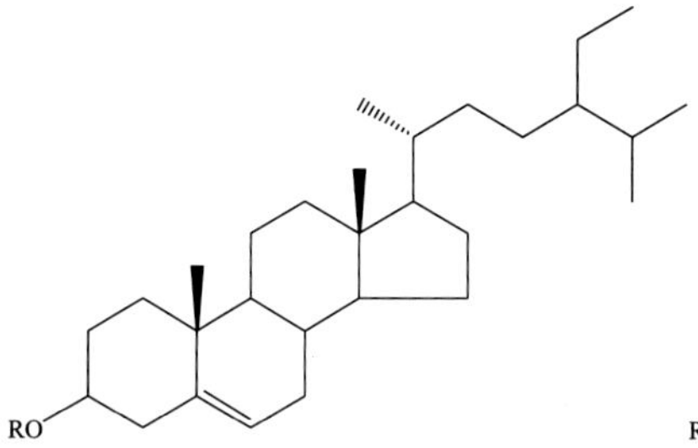

No	R
92	H
96	β-D-Glc

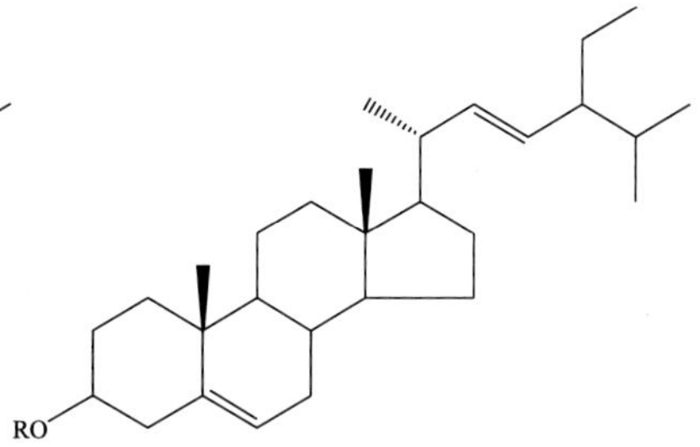

No	R
93	β-D-Glc
94	H

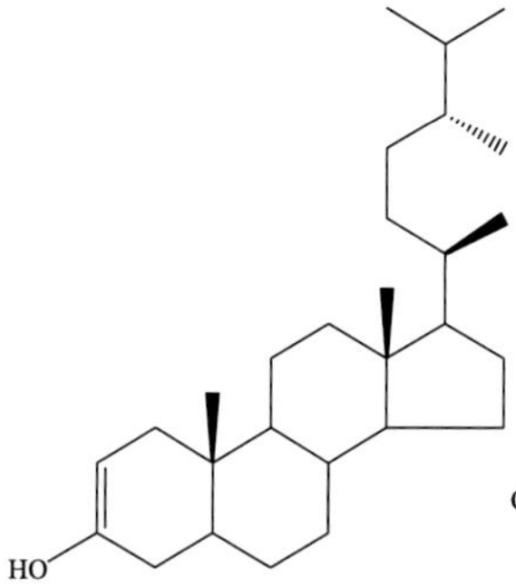

FIGURE 9.1 *(Cont.)*

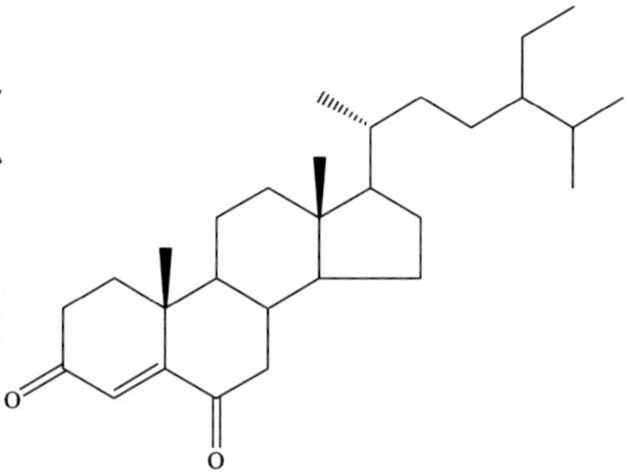

Ranunculus nipponicus var. submerses is a wild edible plant in Japan. The major essential oil components of its fresh aerial part are phytol (41.94%), heptadecane (5.92%), and geranyl propionate (5.76%), while the dry plant is rich in β-ionone (23.54%), 2-hexenal (8.75%), and dihydrobovolide (4.81%) (Nakaya et al., 2015). The fresh and dried oils have green-floral and citrus-floral odor, respectively. Phenylacetaldehyde (green, floral odor, flavor dilution factor

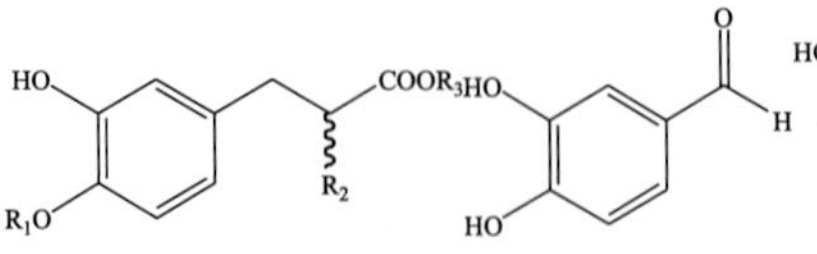

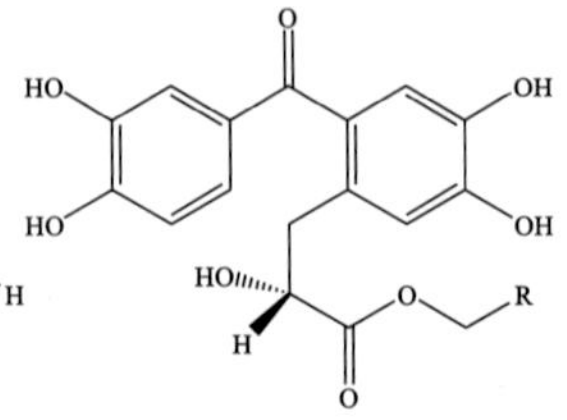

100

No	R_1	R_2	R_3
98	H	β-OH	H
99	H	β-OH	CH_3
114	H	α-OH	$CH_2CHOHCH_2OH$
115	H	α-OH	CH_3
116	β-D-Glc	α-OH	$(CH_2)_3CH_3$

No	R
101	H
102	$(CH_2)_2CH_3$

$H_2C{=}CH{-}(CH_2)_{19}CH_3$

103

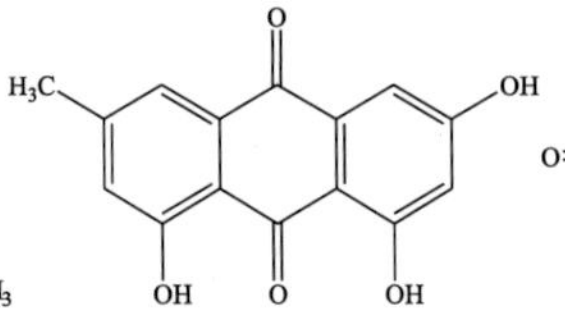

104

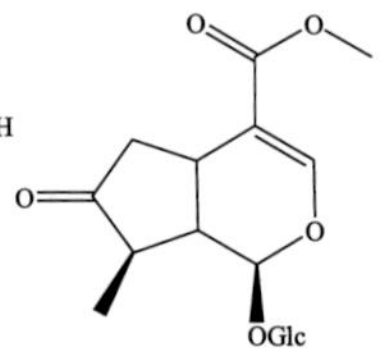

105

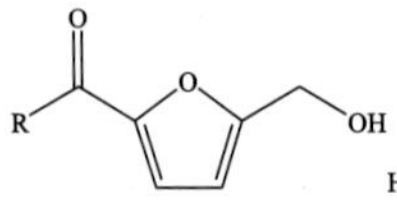

No	R
106	H
113	OH

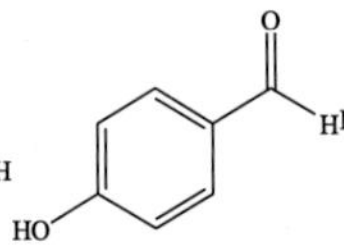

107

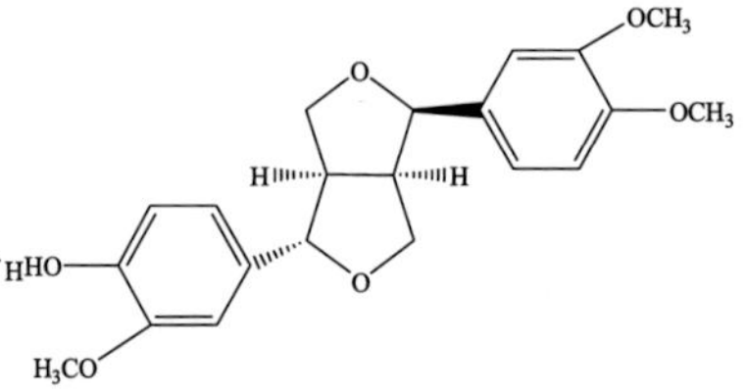

108

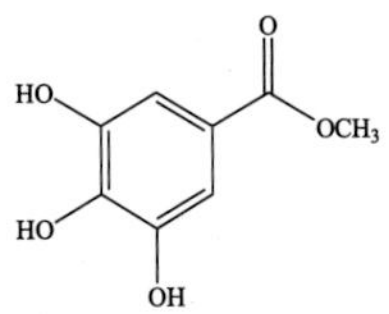

109

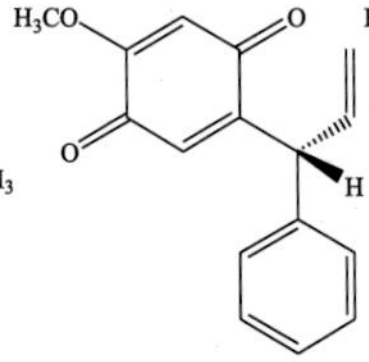

110

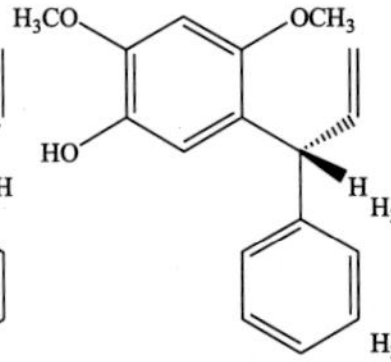

111

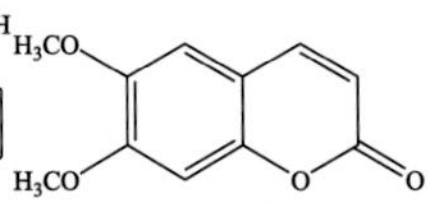

112

FIGURE 9.1 *(Cont.)*

8) and β-ionone (violet-floral odor, FD-factor 8) are the most characteristic odor compounds of the fresh oil. β-Cyclocitral (citrus odor, FD-factor 64) and β-ionone (violet-floral odor, FD-factor 64) are the most characteristic odor compounds of the dried oil. These compounds contribute to the flavor of *R. nipponicus var. submerses*.

The aerial part of *Ranunculus constantinopolitanus* is East Mediterranean (Fostok et al., 2009), and *R. ternatus* (Chen et al., 2006) contains palmitic acid, C18:2 and C18:1 isomers and stearic acid (1:5:8:1). Mono- and digalactosyldiacylglycero lipids of *Ranunculus fluitans* are tensioactive (Wegner and Hamburger, 2002). Divinyl ether fatty acid, synthesized by the leaves of *Ranunculus lingua* and *Ranunculus peltatus*, could participate in the interaction between plant and pathogens (Hamberg, 2002).

9.2.5 Other Compounds

Ent-kaurane-type diterpene glycosides and benzophenone are isolated from the aerial part of *R. muricatus* (Wu et al., 2015). *R. ternatus* has macro- and microelements, for example, K, Fe, Ca, Cr, Mg, Mn, Zn, Co, Cu, Ni, Se, and Sr (Chen et al., 2005a). Pb, Cd, and As are very little. The Zn/Cu ratio is around 3 in *R. ternatus* root tuber, which is reminiscent of the fact that the anticancer Chinese medicines have higher Zn level and lower Cu level. Examples of phenolics (98–100, 107, 109, 111, 114–116), benzophenone (101, 102), alkene (103), anthraquinone (104), iridoid (105), furan derivative (106, 113), lignin (108), benzoquinone (110), and coumarin (112) are shown in Table 9.1 and Fig. 9.1.

R. ternatus has a water-soluble polysaccharide HB-1 with the molecular weight of 23930. The molar ratio of Glc, Ara, and Gal is 16.071:2.722:1, and the proportion of the absolute configuration $-^1$Glc (A), $-^1$Glc^4- (B), (C), and $-^1$Gal^6- (D) is 16%, 62%, 14% and 8%, respectively (Huang et al., 2014). The repetitive unit likely consists of 3 As, 3 Cs, 13 Bs, and 1 D. HB-1 is supposed to have seven repetitive units.

Three rarely occurring furfural fructosides are isolated from the root of *R. ternatus* (Feng et al., 2017), along with other two heterocyclic compounds 4-(2-((2S-2,3-dihydroxypropoxy) methyl)-5-formyl-1H-pyrrol-1-yl)butanoic acid and 3S,4S-4,5,8-trihydroxy-3-(prop-1-en-2-yl) isochroman-1-one. Four hybrids of γ-amino acids and sugars are isolated from roots of *R. ternatus* (Feng et al., 2016), which have potential tail-to-tail ether-connected (6,6-ether-connected) modes in the sugar moiety.

9.3 BIOACTIVITY

Protoanemonin is strongly antibacterial and can be used in dispersing swelling and resolving toxin (traditional remedy index 326; Xiao et al., 1986), dysentery (3291), carbuncles and boils (944), pharyngolaryngitis, tuberculosis, scrofula, etc.

9.3.1 Anticancer Activity

Apigenin-4'-O-α-L-rhamnopyranoside of *Ranunculus sieboldii* inhibits BEL-7407 and A549 cancer cells, with IC$_{50}$43 and 77 μg/ml, respectively (Pan et al., 2004); scopoletin and scoparone inhibits KB and HL-60 leukemia cells, and luteolin shows cytotoxic activity toward these four cancer cell lines. The total saponin of *R. ternatus* enhances the immunoregulatory function of normal mice and significantly inhibits the proliferation of MCF-7 breast cancer cells and induces apoptosis (Yin et al., 2008). The polysaccharide of *R. ternatus* induces the apoptosis of MCF-7 cells and increases the activity of NK cells (Sun et al., 2013), thus inhibiting cancer cell growth. The polysaccharide-protein complex of *R. ternatus* RTG-III enhances the inhibition of immune cells toward HL-60 (Chen et al., 2004).

The methanol extract (ME) and water extract (WE) of *R. arvensis* displays in vitro anticancer activity, with IC_{50} 20.27 ± 1.62 and 93.01 ± 1.33 µg/disc respectively (Bhatti et al., 2015b). Brine shrimp lethality assay shows that LC_{50} values of acetone extract, methanol-acetone extract, and ME are 384.66 ± 9.42 µg/ml, 724.11 ± 8.01 µg/ml, and 978.7 ± 8.01 µg/ml respectively.

9.3.2 Antiinflammatory, Antioxidant, and Analgesic Activities

The antioxidant phenylpropanoids and flavonoids of *Ranunculus auricomus* protect the plant from light stress (Klatt et al., 2016). Protoanemonin has strong irritability; in locally external uses, it shows vascular dilation, promoting local blood circulation and increasing local permeability, etc. (Zheng et al., 2006). The total glycoside of *R. japonicus* increases the pain threshold of a mouse's body surface against heat stress and prolongs pain reaction time on the hot plate; swelling of rat feet and xylene-induced auricular swelling are alleviated (Wang et al., 2009). Ranuncoside, isolated from *R. muricatus* (Raziq et al., 2017), potently scavenges the DPPH free radicals (IC_{50} 56.7 ± 0.43 µM) and strongly inhibits the activities of lipoxygenase (IC_{50} 63.9 ± 0.17 µM) and xanthine oxidase (IC_{50} 43.3 ± 0.22 µM).

In folk medicine of Turkey, the ME of *Ranunculus pedatus* substantially promotes wound healing (Akkol et al., 2012), and the ME of *R. pedatus* and *Ranunculus constantinapolitanus*(100 mg/kg) has antiinflammatory activity. ME and WE of *R. pedatus* and ME of *R. constantinapolitanus* significantly increase the hydroxyproline content of treated tissues. The polysaccharide of *R. ternatus* dose-dependently enhances the proliferation of mice thymus cells and spleen lymphocytes, as well as phagocytosis of macrophages (Lv et al., 2010); it has a certain reduction ability and strong ability to remove •OH and superoxide anion. The antioxidant activity of *Ranunculus marginatus var. trachycarpus* and *Ranunculus sprunerianus* extracts is positively correlated with the contents of total phenolics and total flavonoids.

R. arvensis has long been used in Pakistan and Iran to treat various diseases, such as arthritis, asthma, hay fever, rheumatism, psoriasis, gut diseases, and rheumatic pain (Boroomand et al., 2017). The chloroform, chloroform:methanol, methanol, methanol:acetone, acetone, methanol:water, and water extracts of *R. arvensis* have been examined for DPPH (1,1-diphenyl-2-picrylhydrazyl) free radical scavenging assay and hydrogen peroxide scavenging assay (Bhatti et al., 2015a); ME shows the strongest antioxidant activity (IC_{50} 34.71 ± 0.02). Total flavonoids and phenolics range 0.96–6.0 mg/g of extracts calculated as rutin equivalent and 0.48–1.43 mg/g of extracts calculated as gallic acid equivalent respectively. The antioxidant rutin and caffeic acid are most abundant.

R. constantinopolitanus has various fatty acids, which downregulate IL-6 level of LPS-triggered SCp2 cells andCOX-2 expression of IL-1 treats mouse intestinal epithelia Mode-K cells. The ME of Spanish *R. peltatus* subsp. *baudotii* selectively inhibits COX-1 pathway (Prieto et al., 2008), and 200 µg/mL completely inhibits the release of inflammation mediator

12-(S)-HHTrE from cell culture, while it has no effect on 5-LOX and 12-LOX. The n-hexane, chloroform, and ethyl acetate fractions of the ME inhibits LTB4 release by rat peritoneal neutrophils, but more polar fractions are inactive and did not increase the 5-LOX activity. In vivo studies show that ME of *R. peltatus* subsp. *baudotii* alleviates the dinitrofluorobenzene (DNFB)-induced edema by 40% but fails to inhibit the edema caused by oxazolone. The results agree with the traditional viewpoint that *Ranunculus* species growing in waterside lowlands can be used as a topical antiinflammatory remedy without the prominent irritant action of nonaquatic *Ranunculus* species. Various extracts of the aboveground parts of *R. sceleratus* show antiinflammatory or neutral effects (Prieto et al., 2003); the in vitro experiments show that nonpolar extracts inhibit the generation of eicosanoid, and polar extracts, containing flavonoids and lactones, enhance the synthesis of 5(S)-HETE, LTB4 and 12(S)-HHTrE.

9.3.3 Antimicrobial, Antiparasitic, and Insecticidal Activities

The crude extract of *R. muricatus* shows the strongest inhibitions of *Micrococcus luteus*, followed by acetone extract fraction (Khan et al., 2016); the crude extract has strong antifungal activity against *Candida albicans*. Tricin 7-O-β-D-lucopyranoside, protocatechuic aldehyde, isoscopoletin, anemonin, and β-sitosterol, isolated from *R. muricatus*, has strong antimicrobial activity against various bacteria and *Aspergilus niger* (Nazir et al., 2013), and tricin 7-O-β-D-lucopyranoside shows maximal antimicrobial activity. The phenolic compound in the root of *Ranunculus repens* significantly inhibits bacterial urease (Khan et al., 2006). Protoanemonin inhibits the three kinds of virulence factors of *Pseudomonas aeruginosa* O1, which is related to dosage (Lv et al., 2013). Protoanemonin might act on the quorum sensing system of *P. aeruginosa* and inhibit the expression of the virulence factor, thus decreasing the virulence of this bacterium. Four extracts of *R. marginatus var. trachycarpus* and *R. sprunerianus* had antibacterial activity against eight bacteria. The ME of the whole plant *Ranunculus myosuroudes*, which grows in Lebanon, inhibits eight bacteria and *C. albicans* (Barbour et al., 2004). A benzophenone (101), isolated from the root of *R. ternatus*, exhibits strong activity against tuberculosis, and the activity of a 1:1 mixture of benzophenone and gallic acid is better than benzophenone alone(Deng et al., 2013). The inhibitory effects of pyrrole butyrate of *R. ternatus* on the standard strains of *Mycobacterium tuberculosis*, multidrug resistant strains, and extensively resistant strains were noteworthy (Qi et al., 2012). The WE of the aboveground part of *R. arvensis* inhibits multiple fungi (Hachelaf et al., 2013). Some traditional antiinfection plants can be used singly or in combination in South Africa, and WE of *Ranunculus multifidus* (0.02 mg/ml) has the best inhibitory activity against sexually transmitted infection pathogens *C. albicans*, *Ureaplasma urealyticum*, *Oligella ureolytica*, *Trichomonas vaginalis*, *Gardnerella vaginalis*, and *Neisseria gonorrhoeae* (Naidoo et al., 2013).

2-Deoxy-D-ribotide-1,4-lactone; isomaltol-α-D-glucoside; vanillic acid 4-O-β-D-glucoside; salicin; 3,4-dimethoxybenzoic acid; and caffeic acid isolated from *R. ternatus* have a certain resistance to drug-resistant tuberculosis (Xiong et al., 2016a). *R. ternatus* fatty acids (R)-3-hydroxy-11-methoxy-11-oxoundecanoic acid, palmitic acid, ethyl palmitate, adipic acid, and stearic acid have in vitro inhibitory activity against INH+RFP-resistant *M. tuberculosis* (Xiong et al., 2016b).

The ethanol extract (EE) of *Ranunculus tricophyllus* kills *Trypanosoma brucei* rhodesiense and *Leishmania donovani* (Orhan et al., 2006) and is nontoxic to the mammalian L6 cells. *R. sceleratus* inhibits the growth of *Trypanosoma cruzi*, which is stronger than allopurinol (Schinella et al., 2002), and it was innocuous to the rat polimorphonuclear cells. In Turkey, *Ranunculus asiaticus* is used to kill the book pest (Yavuz, 2006). The whole plant extract of *R. myosuroudes* has insecticidal activity against adults and second nymphal instars of cotton whitefly (Hammad et al., 2014). Furostanol saponins (43) of *Ranunculus lanuginosus* leaves show a significant feeding deterrent activity against a potential predator, *Myrmica rubra* ant workers (Prieto et al., 2007).

Apigenin 4'-O-alpha-rhamnopyranoside, apigenin 7-O-beta-glucopyranosyl-4'-O-alpha-rhamnopyranoside, tricin 7-O-beta-glucopyranoside, tricin, and isoscopoletin, isolated from traditionally used *R. sieboldii* and *R. sceleratus*, has inhibitory activity against HBV replication (Li et al., 2005). Protocatechuyl aldehyde inhibits HSV-1 replication.

9.3.4 Effects on the Cardiovascular System

Total glycoside of *R. japonicus* (TGRJ) of medium-to-high dosage decreases the blood pressure of renal hypertension in rats ($P < 0.05$) (Liu et al., 2012), and the serum NO concentration is significantly increased ($P < 0.01$), while the concentration of angiotensin (Ang) II is not significantly affected. Ang II significantly promotes the internal flow of extracellular calcium, and different doses of TGRJ could inhibit the increase of AngII-mediated intracellular calcium ions. TGRJ has a protective effect on myocardial ischemia induced by posterior pituitary hormone in mice. The EE of *R. japonicus* significantly inhibits the increase of free calcium in rabbit aortic smooth muscle cells induced by NE (Cai and Li, 2004), reducing intracellular calcium concentration to make vasodilation.

Isoproterenol (ISO, 10 μmol/L) or AngII (1 μmol/L) is used to induce cardiac myocyte hypertrophy (Dai et al., 2015) in newborn rats. Pretreating cells with TGRJ (0.3/L) for 30 min significantly inhibits the cell size increase after stimulation, the total protein content, and protein synthesis. TGRJ inhibits hypertrophy-associated gene expression, namely, atrial natriuretic peptide (ANP), B-type natriuretic peptide, and β-myosin heavy chain (β-MHC). TGRJ inhibits ISO or AngII-induced chronic $[Ca^{2+}]_i$ rise, and corrects the downregulation of ISO or AngII-induced sarcoplasmic/endoplasmic reticulum Ca^{2+} ATPase (SERCA) 2a expression/activity.

The heart rate, left ventricular systolic pressure, maximal increase rate of left ventricular systolic pressure, and coronary blood flow are significantly reduced in ischemia-reperfusion of isolated rat heart, and LDH and CK levels increase, with infarct area 58.78% (Gao et al., 2014). TGRJ improves these indexes and has a protective effect on myocardium.

9.3.5 Other Effects

Asiaticoside and hydroxyl asiaticoside of Asia *Ranunculus* can promote wound-healing and are used for trauma, surgical trauma, burns, scar lumps, vaginal diseases, etc. (Salas Campos et al., 2005). Phenolic compounds (109–111), isolated from the root of *R. repens*, inhibits xanthine oxidase (Khan et al., 2008).

9.3.6 Toxicity and Safety

The contact dermatitis caused by *Ranunculus* is reported at home and abroad (Ge et al., 2012; Uçmak et al., 2014), especially when whole-grass external treatment is used for rheumatism and arthritis. The causative compounds are protoanemonin and sesquiterpene lactone. In Turkeys, *R. arvensis*is is commonly used for local treatment of joint pain, muscle pain, burns, lacerations, edema, abscess drainage, hemorrhoids, and warts (Kocak et al., 2016). Protoanemonin is heat-resistant and can cause chemical burns even after boiling. The metabolomics analyses of three UK *Ranunculus* species suggest that protoanemonin could cause equine grass sickness (Michl et al., 2011), and the concurrent higher iron and heavy metal contents might be also relevant (Edwards et al., 2010). The intake of *R. japonicus* by livestock causes gastrointestinal inflammation, nephritis, hernia pain, dysentery, hematuria, and the final spasm to death (Liang et al., 2008). The intake of *R. bulbosus* led to abortion in cattle and horses (Swerczek, 2016).

9.4 PHYLOGENETIC RELATIONSHIP

Illumina transcriptome sequencing (RNA-seq) is used to study the floral transcriptomes of *R. auricomus* complex (Pellino et al., 2013); SNP and insertion-deletion (indel) polymorphisms are mined. Annotated genes with open reading frames (ORFs) are analyzed for signatures of divergent versus stabilizing selection. A comparison between five genotypes suggests that two apomictic genotypes (hexaploid) are evolved from *Ranunculus carpaticola* and *Ranunculus cassubicifolius* (tetraploid), and *R. cassubicifolius* is from the hybridization between *R. carpaticola* and *Ranunculus notabilis*. Many genes associated with reproduction, including meiosis and gametogenesis, have elevated dN (nonsynonymous substitution rate)/dS (synonymous substitution rate) ratios. It is challenging to reconstruct the reticulate evolutionary history of plants; the multicopy intergenic spacer ITS nrDNA is commonly used to analyze the hybrid, but the accuracy of its inference is affected by concerted evolution, high intraspecies polymorphism, and the alteration of reproduction mode. The relevance of the secondary structure changes remains controversial. The extent of polymorphism within and between sexual species and their hybrid derivatives in the *R. auricomus* complex is investigated to test morphology-based hypotheses of hybrid origin and parentage of taxa (Hodacˇ et al., 2014). Direct sequencing of ITS from three sexual species, their synthetic hybrids, and one sympatric natural apomicts is conducted, and ITS copies of four representative individuals are cloned. Phylogenetic network analyses indicate additivity of parental ITS variants in both synthetic and natural hybrids. The triploid synthetic hybrids are genetically much closer to their maternal progenitors, probably due to ploidy dosage effects. The natural hybrids are genetically and morphologically closer to the paternal progenitor species. Secondary structures of ITS1-5.8S-ITS2 are rather conserved in all taxa. The similarities in ITS polymorphisms suggest that the natural apomict *R. variabilis* is an ancient hybrid of the diploid sexual species *R. notabilis* and *R. cassubicifolius*. The additivity pattern shared by *R. variabilis* and the synthetic hybrids suggests that *R. variabilis* originated from ancient hybridization. Concerted evolution of ITS copies in *R. variabilis* is incomplete, probably due to a shift to asexual reproduction. Under the comprehensive inter- and intraspecific sampling, ITS polymorphisms are powerful for elucidating reticulate evolutionary histories.

Transcriptome sequencing was performed for submerged aquatic plant *Ranunculus bungei*, terrestrial relative to *Ranunculus cantoniensis* and *Ranunculus brotherusii* (Chen et al., 2015), and 126,037, 140,218 and 114,753 contigs were obtained respectively. Bidirectional best hit method and OrthoMCL method identified 11,362 and 8,174 1:1:1 orthologous genes (one ortholog is represented in each species) respectively. dN/dS analyses found 14 genes of *R. bungei* potentially involved in the adaptive transition from terrestrial to aquatic habitats. Some homologs of these genes in model plants are involved in vacuole protein formation, regulating "water transport process" and "microtubule cytoskeleton organization." The taxonomically challenging *R. cantoniensis* complex includes *R. chinensis* (2×), *Ranunculus trigonus* (2×), *Ranunculus silerifolius var. silerifolius* (2×), *R. silerifolius var. dolichanthus* (2×), *R. cantoniensis* (4×), *Ranunculus diffusus* (4×), *Ranunculus vaginatus* (5×), and *R. sieboldii* (6×, 8×) (Li, 2016). The ITS genotypes of *R. shuichengensis* and *R. repens* were unique, and they were not included in the above complex. *Ranunculus vaginatus* (5×) might originate from the hybridization between *R. diffusus* and *R. sieboldii* (6×). In Puer of Yunnan Province, *R. chinensis*, *R. trigonus*, *R. silerifolius var. silerifolius*, and *R. silerifolius var. dolichanthus* hybridize mutually to form seven hybrid genotypes. A total of 1687 orthologs were identified based on the transcriptome data sets of four taxa of *R. cantoniensis* complex. The transcriptome of *R. sceleratus* was sequenced (Zhao et al., 2016); 3455 orthologs were recovered for *R. sceleratus*, *R. bungei*, *R. cantoniensis*, and *R. brotherusii* using OrthoMCL. After removing the clusters that were not consistent with phylogeny, blast, or alignment, 884 clusters were retained. These orthologs mined from transcriptome data sets are useful in reconstructing phylogeny of *Ranunculus*.

454 pyrosequencing and other high-throughput sequencing platforms can be used to identify microsatellite markers of various *Ranunculus* species, including *R. bulbosus* (Matter et al., 2012) and *R. cantoniensis* complex (Li, 2016), which are useful in population-level analysis.

The early-diverging eudicots include five major lineages: Ranunculales, Trochodendrales, Buxales, Proteales, and Sabiaceae. The complete plastomes of eight early-diverging eudicot taxa, including *Epimedium sagittatum* (Berberidaceae), *Euptelea pleiosperma* (Eupteleaceae), *Akebia trifoliata* (Lardizabalaceae), *Stephania japonica* (Menispermaceae), and *Papaver somniferum* (Papaveraceae), are sequenced (Sun et al., 2016); they are compared with plastomes of the early-diverging eudicots, for example, *Nandina*, *Megaleranthis*, *Ranunculus*, and *Mahonia*. All newly sequenced plastomes share the same 79 protein-coding genes, 4 rRNA genes, and 30 tRNA genes, except for that of *Epimedium*, in which infA is pseudogenized and clpP is highly divergent and possibly a pseudogene. The boundaries of the plastid inverted repeat (IR) vary significantly across early-diverging eudicots. Eighteen genes are likely present in the IR region of the eudicot ancestor. Maximum likelihood phylogenetic analysis of 79 genes and 97 taxa (Fig. 9.2) largely agree with previously inferred early-diverging eudicot relationships. The plastome-scale phylogenetic analyses are important references for mining new pharmaceutical resources.

The European continent has nine *Pulsatilla* species, five of which are only in mountain regions of Southwest and South-central Europe (Szczecińska and Sawicki, 2015). The other four species prefer lowlands of North-central and Eastern Europe. Research efforts focus on the biology, ecology, and hybridization of rare and endangered species. Genomic resources, including complete plastid genomes and nuclear rRNA clusters, should be developed for conservation and sustainable utilization. Three sympatric *Pulsatilla* species are most commonly found in Central Europe. Six complete plastid genomes and nuclear rRNA clusters are sequenced for these species. Four junctions between single-copy regions and IRs and junctions between the identified locally collinear blocks are confirmed by Sanger sequencing. Among

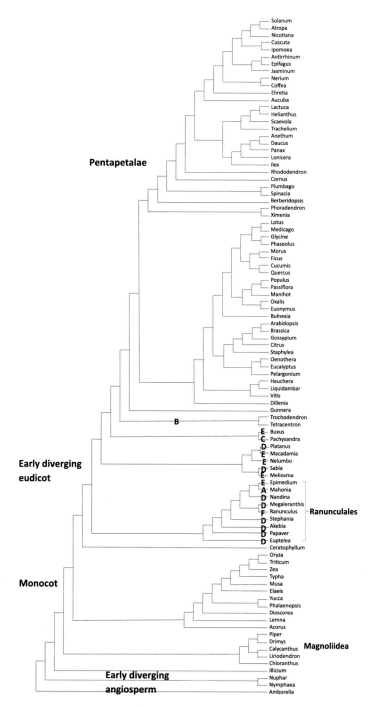

FIGURE 9.2   Seventy-nine plastid gene sequences of 97 taxa are used to construct the maximum likelihood phylogenetic tree. (A–F) six IR types; (E) typical IR type of eudicots.

120 unique genes of *Pulsatilla* genomes, 21 are duplicated in IR region. Comparative plastid genomes of *Pulsatilla, Aconitum,* and *Ranunculus* species reveal variations in the genome structure, but the gene content remains constant. Among five regions of the rRNA cluster, only ITS2 can differentiate closely related *Pulsatilla patens* and *Paxillus vernalis.* The resulting data can be used to identify regions that are particularly useful for barcoding, phylogenetic, and phylogeographic studies. The investigated taxa can be identified at each stage of development based on their species-specific SNPs. The identified nonsynonymous mutations could play an important role in adaptations to changing environments.

Anemoneae, to which *Clematis* and *Pulsatilla* belong, is closely related with Ranunculeae. The chloroplast (cp) genome of *Clematis terniflora* is 159,528 bp (Li et al., 2015). The phylogenetic analysis of 32 taxa shows its strong sister relationship with *Ranunculus macranthus.* The cp genome of more Ranunculeae and Anemoneae species should be sequenced to advance population and phylogenetic studies of these medicinal plants. Cp rbcL+matK are useful DNA-barcoding markers for the rapid identification of the Canadian Arctic flora, including *Ranunculus* (Saarela et al., 2013).

Oceanic island endemics typically exhibit very restricted distributions, which is reminiscent of the geoherb (Daodi medicinal material). In Macaronesia, only one endemic angiosperm species, *Ranunculus cortusifolius,* has a distribution spanning the surrounding archipelagos (Williams et al., 2015). Is *R. cortusifolius* a single widespread Macaronesian endemic species with a single origin? Cp (matK-trnK, psbJ-petA) and ITS sequences are used to infer the phylogeny and divergence time. The single origin of *R. cortusifolius* is supported. Populations in different regions has distinct genotypes. The large distances between archipelagos are effective barriers to dispersal, promoting allopatric diversification at the molecular level. Isolation is not accompanied by marked morphological diversification, which may be explained by the typical association of *R. cortusifolius* with stable and climatically buffered laurel forest communities. This study is enlightening to the exploration of genetic mechanisms of geoherbs.

Radix Ranunculi Ternati originates from two *Ranunculus* species; the predominant one is *R. ternatus,* but *Ranunculus polii* is used in a few regions. On the ITS-based phylogenetic tree, these two species cluster together (Fig. 9.3), indicating their close evolutionary relationship. Flora of China, according to morphological features, divided *Ranunculus* into *Sect. Auricomus* (Spach) Tamura, *Sect. Echinella* DC., *Sect. Ficariifolium* L. Liou, *Sect. Flammula* (Spach) Tamura, *Sect. Hecatonia* (Ovcz.) Tamura, *Sect. Ranunculastrum* DC. Prodr., *Sect. Ranunculus,* and *Sect. Xanthobatrachium* Ovcz. However, on the phylogenetic tree based on cp psbJ/petA+trnK/ matK and nr ITS sequences, *R. arvensis* and *R. muricatus* of *Sect. Echinella* cluster with *Sect. Ranunculus,* but they are far away from each other (Fig. 9.4), and *R. muricatus* is within *R. cantoniensis* complex; *R. sceleratus* clusters with *Sect. Xanthobatrachium* and is not an independent clade; and *Ranunculus regelianus* of *Sect. Ranunculastrum* is within *Sect. Ranunculus.* The molecular grouping is similar on the ITS tree (Figs. 9.3 and 9.5. As the morphological grouping is not congruent with the molecular marker–based grouping, chemotaxonomy data and DNA sequence data of more species should be combined for comprehensive analysis. *Sect. Auricomus,* to which *R. ternatus* belongs, contains the most species, including the 53 domestic species, which are widespread nationwide especially in the southwest mountainous area; studies of their medicinal composition should be highlighted.

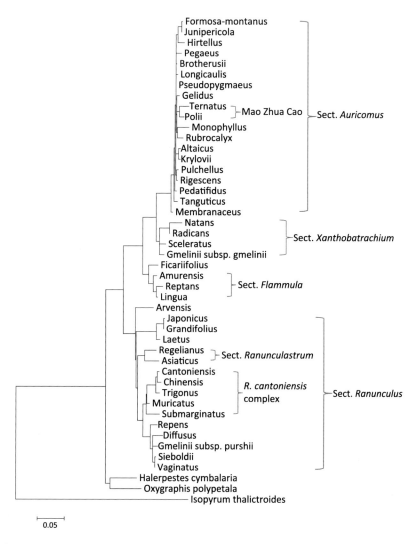

FIGURE 9.3   Phylogenetic relationship of *Ranunculus* nuclear ITS sequences inferred by the ML method.

9.5 CONCLUSION

The globalization process is speeding up the disappearance of traditional medicinal knowledge; along with the national survey of Chinese traditional medicine resources (Guo et al., 2013), the knowledge and experience of folk medicine should be actively collected and excavated. The traditional pharmaceutical knowledge accumulated by human practice over the past thousands of years is not only a cultural and biological heritage but also a valuable resource for the development of new drugs. Medicinal compounds such as protoanemonin

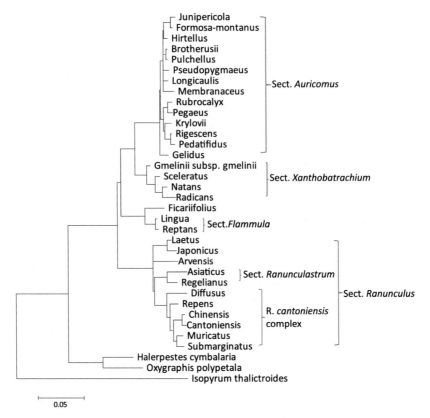

FIGURE 9.4 Phylogenetic relationship of *Ranunculus* plants inferred from cp psbJ/petA+trnK/matK and nuclear ITS sequences by the ML method.

FIGURE 9.5 *R. tanguticus*, taken in Shika snow mountain of Shangri-La, Diqing Tibetan Autonomous Prefecture, China (refer to Fig. 9.3).

and anemonin widely exist in various plants of *Ranunculus*, and this genus is widely distributed in nature, adaptable, and resource intensive; future research should not be limited to the few species that have a record of folk curative effect, and the research and development of more species should be strengthened from the world scope in order to expand the source of medicine.

References

Akbulut, S., Semur, H., Kose, O., et al., 2011. Phytocontact dermatitis due to *Ranunculus arvensis* mimicking burn injury: report of three cases and literature review. Int. J. Emerg. Med. 4, 7.

Akkol, E.K., Süntar, I., Erdoğan, T.F., et al., 2012. Wound healing and anti-inflammatory properties of *Ranunculus pedatus* and *Ranunculus* constantinapolitanus: a comparative study. J. Ethnopharmacol. 139 (2), 478–484.

Barbour, E.K., Al Sharif, M., Sagherian, V.K., et al., 2004. Screening of selected indigenous plants of Lebanon for anti-microbial activity. J. Ethnopharmacol. 93 (1), 1–7.

Bhatti, M.Z., Ali, A., Ahmad, A., et al., 2015a. Antioxidant and phytochemical analysis of *Ranunculus arvensis* L. extracts. BMC Res. Notes 8, 279.

Bhatti, M.Z., Ali, A., Saeed, A., et al., 2015b. Antimicrobial, antitumor and brine shrimp lethality assay of *Ranunculus arvensis* L. extracts. Pak. J. Pharm. Sci. 28 (3), 945–949.

Boroomand, N., Sadat-Hosseini, M., Moghbeli, M., et al., 2017. Phytochemical components, total phenol and mineral contents and antioxidant activity of six major medicinal plants from Rayen, Iran. Nat. Prod. Res.doi: 10.1080/14786419.2017.1315579.

Cai, S., Li, X.K., 2004. The effect of the extract from *Ranunculus japonicus* on $[Ca^{2+}]_i$ inside rabbit VSMC by serologic pharmacological test. J. Chin. Med. Mater. 27 (10), 741–743.

Carrió, E., Rigat, M., Garnatje, T., et al., 2012. Plant ethnoveterinary practices in two pyrenean territories of catalonia (Iberian peninsula) and in two areas of the balearic islands and comparison with ethnobotanical uses in human medicine. Evid. Based Complement. Altern. Med., 896295.

Chen, J., Yao, C., Ouyang, P.K., 2005a. Determination of constant and trace elements in *Ranunculus ternatus* Thunb by ICP-AES. Spectrosc. Spectral Anal. 25 (4), 560–562.

Chen, J., Yao, C., Xia, L.M., Ouyang, P.K., 2006. Determination of fatty acids and organic acids in *Ranunculus ternatus* Thunb using GC-MS. Spectrosc. Spectral Anal. 26 (8), 1550–1552.

Chen, L.Y., Zhao, S.Y., Wang, Q.F., et al., 2015. Transcriptome sequencing of three *Ranunculus* species (Ranunculaceae) reveals candidate genes in adaptation from terrestrial to aquatic habitats. Sci. Rep. 5, 10098.

Chen, Y., Dai, L., Shen, Y.S., 2004. Purification and properties of a polysaccharide RTG-III from *Ranunculus ternali*. Chin. Pharm. J. 40 (5), 339–341.

Chen, Z., Tian, J.K., Cheng, Y.Y., 2005b. Studies on chemical constituents of *Ranunculus ternatus* (II). Chin. Pharm. J. 40 (18), 1373–1375.

Chi, Y.M., Li, Y., Zhang, Y., et al., 2013. Analysis of flavonoids and components of stems & leaves of *Ranunculus ternatus* Thunb. Using ultra performance liquid chromatography-quadrupole tandem time of flight mass spectrometry. Chin. J. Chromatogr. 31 (9), 838–842.

Dai, H.L., Jia, G.Z., Zhao, S., 2015. Total glycosides of *Ranunculus japonicus* prevent hypertrophy in cardiomyocytes via alleviating chronic Ca^{2+} overload. Chin. Med. Sci. J. 30 (1), 37–43.

Deng, K.Z., Xiong, Y., Zhou, B., et al., 2013. Chemical constituents from the roots of *Ranunculus ternatus* and their inhibitory effects on *Mycobacterium tuberculosis*. Molecules 18 (10), 11859–11865.

Edwards, S.E., Martz, K.E., Rogge, A., et al., 2010. Edaphic and phytochemical factors as predictors of Equine Grass Sickness cases in the UK. Front. Pharmacol. 1, 122.

Emadzade, K., Gehrke, B., Linder, H.P., et al., 2011. The biogeographical history of the cosmopolitan genus *Ranunculus* L (Ranunculaceae) in the temperate to meridional zones. Mol. Phylogenet. Evol. 58 (1), 4–21.

Feng, Z.M., Zhan, Z.L., Yang, Y.N., et al., 2016. Naturally occurring hybrids derived from γ-amino acids and sugars with potential tail to tail ether-bonds. Sci. Rep. 6, 25443.

Feng, Z.M., Zhan, Z.L., Yang, Y.N., et al., 2017. New heterocyclic compounds from *Ranunculus ternatus* Thunb. Bioorg. Chem. 74, 10–14.

Fostok, S.F., Ezzeddine, R.A., Homaidan, F.R., et al., 2009. Interleukin-6 and cyclooxygenase-2 downregulation by fatty-acid fractions of *Ranunculus constantinopolitanus*. BMC Complement. Altern. Med. 9, 44.

Gao, X.W., Liu, Y., Yang, Z.C., et al., 2014. Protective effect of total glycosides of *Ranunculus japonicus* on myocardial ischemic-reperfusion injury in isolated rat hearts. Zhong Yao Cai 37 (8), 1429–1433.

Gao, X.Z., Zhou, C.X., Zhang, S.L., et al., 2005. Studies on the chemical constituents in herb of *Ranunculus sceleratus*. China J. Chin. Mater. Med. 30 (2), 124–126.

Ge, M.J., Zhao, Y.H., Deng, H.T., et al., 2012. Six cases of local skin injury caused by applying *Ranunculus*. Chin. J. Burns 28 (4), 259–260.

Guo, L.P., Lu, J.W., Zhang, X.B., et al., 2013. Formulation of technical specification for national survey of Chinese materia medica resources. China J. Chin. Mater. Med. 38 (7), 937–942.

Hachelaf, A., Zellagui, A., Touil, A., et al., 2013. Chemical composition and analysis antifungal properties of *Ranunculus arvensis* L. Pharmacophore 4 (3), 89–91.

Hamberg, M., 2002. Biosynthesis of new divinyl ether oxylipins in *Ranunculus* plants. Lipids 37 (4), 427–433.

Hammad, E.A., Zeaiter, A., Saliba, N., et al., 2014. Bioactivity of indigenous medicinal plants against the cotton whitefly, *Bemisia tabaci*. J. Insect Sci. 14, 105.

Hao, D.C., Gu, X.J., Xiao, P.G., 2015. *Ranunculus* medicinal resources: chemistry, biology and phylogeny. Lishizhen Med. Mater. Med. Res. 26 (8), 1990–1993.

Hodac˘, L., Scheben, A.P., Hojsgaard, D., et al., 2014. ITS polymorphisms shed light on hybrid evolution in apomictic plants: a case study on the *Ranunculus auricomus* complex. PLoS One 9 (7), e103003.

Huang, X., Zhao, Y., Jin, X., 2014. Structural characterisation of a polysaccharide from radix *Ranunculus ternati*. Iran J. Pharm. Res. 13 (4), 1403–1407.

Jürgens, A., Dötterl, S., 2004. Chemical composition of anther volatiles in Ranunculaceae: genera-specific profiles in *Anemone, Aquilegia, Caltha, Pulsatilla, Ranunculus,* and *Trollius* species. Am. J. Bot. 91 (12), 1969–1980.

Khan, F.A., Zahoor, M., Khan, E., 2016. Chemical and biological evaluation of *Ranunculus muricatus*. Pak. J. Pharm. Sci. 29 (2), 503–510.

Khan, W.N., Ali, I., Gul, R., et al., 2008. Xanthine oxidase inhibiting compounds. Chem. Nat. Comp. 44 (1), 74–75.

Khan, W.N., Lodhi, M.A., Ali, I., et al., 2006. New natural urease inhibitors from *Ranunculus repens*. J. Enzyme Inhib. Med. Chem. 21 (1), 17–19.

Klatt, S., Hadacek, F., Hodac˘, L., et al., 2016. Photoperiod extension enhances sexual megaspore formation and triggers metabolic reprogramming in facultative apomictic *Ranunculus auricomus*. Front. Plant Sci. 7, 278.

Kocak, A.O., Saritemur, M., Atac, K., et al., 2016. A rare chemical burn due to *Ranunculus arvensis*: three case reports. Ann. Saudi Med. 36 (1), 89–91.

Li, J.S., Li, J., Yin, H.L., et al., 2010. Study on chemical constituents of *Ranunculus chinensis* Bunge. Bull. Acad. Mil. Med. Sci. 34 (1), 68–70.

Li, H., Zhou, C., Pan, Y., et al., 2005. Evaluation of antiviral activity of compounds isolated from *Ranunculus sieboldii* and *Ranunculus sceleratus*. Planta Med. 71 (12), 1128–1133.

Li, M., Yang, B., Chen, Q., et al., 2015. The complete chloroplast genome sequence of *Clematis terniflora* DC (Ranunculaceae). Mitochondrial DNA 2015, 1–3.

Li, T.J., 2016. A hybrid Zone and Reticulate Evolution in *Ranunculus cantonensis* Polyploid Complex (Ph.D. dissertation). Southwest University.

Liang, Y.F., Chen, Z.T., Liu, L.H., 2008. Studies on chemical constituents of *Ranunculus japonicus*. China J. Chin. Mater. Med. 33 (19), 2201–2203.

Liu, S.J., Liu, Y., Zhang, X., et al., 2012. Effect of TGRJ on blood press, NO and Ang II in renal hypertensive rats. J. Chin. Med. Mater. 35 (6), 953–955.

Lv, X.H., Wang, H.M., Han, H.X., et al., 2010. Effects of polysaccharide of Radix Ranunculi Ternati on immunomodulation and anti-oxidation. China J. Chin. Mater. Med. 35 (14), 1862–1865.

Lv, Z.M., Deng, L., Liao, F., 2013. Effect of protoanemonin on expression of virulence factors regulated by quorum sensing system of *Pseudomonas aeruginosa*. Drugs Clin. 28 (2), 155–157.

Matter, P., Pluess, A.R., Ghazoul, J., et al., 2012. Eight microsatellite markers for the bulbous buttercup *Ranunculus bulbosus* (Ranunculaceae). Am. J. Bot. 99 (10), e399–401.

Michl, J., Modarai, M., Edwards, S., et al., 2011. Metabolomic analysis of *Ranunculus* spp. as potential agents involved in the etiology of equine grass sickness. J. Agric. Food Chem. 59 (18), 10388–10393.

Naidoo, D., van Vuuren, S.F., van Zyl, R.L., et al., 2013. Plants traditionally used individually and in combination to treat sexually transmitted infections in northern Maputaland, South Africa: antimicrobial activity and cytotoxicity. J. Ethnopharmacol. 149 (3), 656–667.

Nakaya, S., Usami, A., Yorimoto, T., et al., 2015. Characteristic chemical components and aroma-active compounds of the essential oils from *Ranunculus nipponicus var. submersus* used in Japanese traditional food. J. Oleo Sci. 64 (6), 595–601.

Nazir, S., Li, B.S., Tahir, K., et al., 2013. Antimicrobial activity of five constituents isolated from *Ranunculus muricatus*. J. Med. Plant Res. 7 (47), 3438–3443.

Orhan, I., Sener, B., Atici, T., et al., 2006. Turkish freshwater and marine macrophyte extracts show in vitro antiprotozoal activity and inhibit FabI, a key enzyme of *Plasmodium falciparum* fatty acid biosynthesis. Phytomedicine 13 (6), 388–393.

Pan, Y.X., Zhou, C.X., Zhang, S.L., et al., 2004. Constituents from *Ranunculus sieboldii* Miq. J. Chin. Pharm. Sci. 13 (2), 92–96.

Pellino, M., Hojsgaard, D., Schmutzer, T., et al., 2013. Asexual genome evolution in the apomictic *Ranunculus auricomus* complex: examining the effects of hybridization and mutation accumulation. Mol. Ecol. 22 (23), 5908–5921.

Peng, T., Xing, Y.J., Zhang, Q.J., et al., 2011. Chemical constituents in herb of *Ranunculus sceleratus*. Chin. J. Exp. Trad. Med. Formulae 17 (6), 66–68.

Peng, Y., Chen, S.B., Xiao, P.G., et al., 2006. Preliminary pharmaphylogenetic study on *Ranunculaceae*. China J. Chin. Mater. Med. 31 (13), 1124–1128.

Piskorski, R., Kroder, S., Dorn, S., 2011. Can pollen headspace volatiles and pollen kitt lipids serve as reliable chemical cues for bee pollinators? Chem. Biodivers. 8 (4), 577–586.

Prieto, J.M., Braca, A., Morelli, I., et al., 2004. A new acylated quercetin glycoside from *Ranunculus lanuginosus*. Fitoterapia 75 (6), 533–538.

Prieto, J.M., Recio, M.C., Giner, R.M., et al., 2008. In vitro and in vivo effects of *Ranunculus peltatus subsp. baudotii* methanol extract on models of eicosanoid production and contact dermatitis. Phytother. Res. 22 (3), 297–302.

Prieto, J.M., Recio, M.C., Giner, R.M., et al., 2003. Pharmacological approach to the pro- and anti-inflammatory effects of *Ranunculus sceleratus* L. J. Ethnopharmacol. 89 (1), 131–137.

Prieto, J.M., Schaffner, U., Barker, A., et al., 2007. Sequestration of furostanol saponins by *Monophadnus* sawfly larvae. J. Chem. Ecol. 33 (3), 513–524.

Qi, Z.J., Ying, H.M., Li, J.J., et al., 2012. Study on quality control of Ranunculi Ternati Radix. Chin. Pharm. J. 47 (2), 101–103.

Rahman, I.U., Ijaz, F., Iqbal, Z., et al., 2016. A novel survey of the ethno medicinal knowledge of dental problems in Manoor Valley (Northern Himalaya). Pak. J Ethnopharmacol. 194, 877–894.

Raziq, N., Saeed, M., Ali, M.S., et al., 2017. A new glycosidic antioxidant from *Ranunculus muricatus* L. (Ranunculaceae) exhibited lipoxygenasae and xanthine oxidase inhibition properties. Nat. Prod. Res. 31 (11), 1251–1257.

Saarela, J.M., Sokoloff, P.C., Gillespie, L.J., et al., 2013. DNA barcoding the Canadian Arctic flora: core plastid barcodes (rbcL+matK) for 490 vascular plant species. PLoS One 8 (10), e77982.

Salas Campos, L., Pastor Amorós, T., Martín Campos, R., et al., 2005. Therapeutic uses for Asian yellow-flowered crowfoot, genus *Ranunculus*: the case of Blastoestimulina. Rev. Enferm 28 (1), 59–62.

Schinella, G.R., Tournier, H.A., Prieto, J.M., et al., 2002. Inhibition of *Trypanosoma cruzi* growth by medical plant extracts. Fitoterapia 73 (7–8), 569–575.

Sun, D.L., Xie, H.B., Xia, Y.Z., 2013. A study on the inhibitory effect of polysaccharides from *Radix Ranunculus ternate* on human breast cancer MCF-7 cell lines. Afr. J. Tradit Complement. Altern. Med. 10 (6), 439–443.

Sun, Y., Moore, M.J., Zhang, S., et al., 2016. Phylogenomic and structural analyses of 18 complete plastomes across nearly all families of early-diverging eudicots, including an angiosperm-wide analysis of IR gene content evolution. Mol. Phylogenet. Evol. 96, 93–101.

Swerczek, T.W., 2016. Abortions in Thoroughbred mares associated with consumption of bulbosus buttercups (*Ranunculus bulbosus* L.). J. Am. Vet. Med. Assoc. 248 (6), 669–672.

Szczecińska, M., Sawicki, J., 2015. Genomic resources of three *Pulsatilla* species reveal evolutionary hotspots, species-specific sites and variable plastid structure in the family *Ranunculaceae*. Int. J. Mol. Sci. 16 (9), 22258–22279.

Tian, J.K., Sun, F., Cheng, Y.Y., 2006. Chemical constituents from the roots of *Ranunculus ternatus*. J. Asian Nat. Prod. Res. 8 (1–2), 35–39.

Tian, J.K., Wu, L.M., Wang, A.W., et al., 2004. Studies on chemical constituents of *Ranunculus ternatus* I. Chin. Pharm. J. 39 (9), 661–663.

Uçmak, D., Ayhan, E., Meltem Akkurt, Z., et al., 2014. Presentation of three cases with phyto contact dermatitis caused by Ranunculus and *Anthemis genera*. J. Dermatolog. Treat. 25 (6), 467–469.

Umair, M., Altaf, M., Abbasi, A.M., 2017. An ethnobotanical survey of indigenous medicinal plants in Hafizabad district, Punjab–Pakistan. PLoS One 12 (6), e0177912.

Uniyal, S.K., Singh, K.N., Jamwal, P., et al., 2006. Traditional use of medicinal plants among the tribal communities of Chhota Bhangal Western Himalaya. J. Ethnobiol. Ethnomed. 2, 14.

Wang, R.L., Tan, Y.Z., Luo, S.B., 2009. Anti-inflammatory and analgesic effect of total glycosides of *Ranunculus japonicus*. Lishizhen Med. Mater. Med. Res. 20 (2), 290–292.

Wang, W.C., 1995. A revision of the genus *Ranunculus* in China (I). Bull. Bota Res. 15 (2), 137–180.

Wegner, C., Hamburger, M., 2002. Occurrence of stable foam in the upper Rhine River caused by plant-derived surfactants. Environ. Sci. Technol. 36 (15), 3250–3256.

Wegner, C., Hamburger, M., Kunert, O., et al., 2000. Tensioactive compounds from the aquatic plant *Ranunculus fluitans* L. (Ranunculaceae). Helv. Chim. Acta 83 (7), 1454–1464.

Williams, B.R., Schaefer, H., De Sequeira, M.M., et al., 2015. Are there any widespread endemic flowering plant species in Macaronesia? Phylogeography of *Ranunculus cortusifolius*. Am. J. Bot. 102 (10), 1736–1746.

Wu, B.L., Qin, F.M., Zhou, G.X., 2013. Studies on chemical constituents of *Ranunculus muricatus* Linn. Nat. Prod. Res. Dev. 25 (6), 736–738.

Wu, B.L., Zou, H.L., Qin, F.M., et al., 2015. New ent-kaurane-type diterpene glycosides and benzophenone from *Ranunculus muricatus* Linn. Molecules 20 (12), 22445–22453.

Xiao, P.G., 1980. A preliminary study of the correlation between phylogeny, chemical constituents and pharmaceutical aspects in the taxa of Chinese Ranunculaceae. Acta Phytotaxonom Sin. 18 (2), 142–153.

Xiao, P.G., Wang, L.W., Lv, S.J., et al., 1986. Computer statistical analysis of traditional therapeutic effects of medicinal plants in China I, Magnoliidae. Chin. J. Int. Trad. West Med. 6 (4), 253–256.

Xiong, Y., Chang, M.Y., Zhang, C.H., et al., 2016b. Fatty Acids from *Ranunculus ternatus*. J. Trop. Subtrop. Bot. 24, 348–351.

Xiong, Y., Chen, H., Deng, M.Z., et al., 2016a. Constituents from *Ranunculus ternatus* and their inhibitory effects on multidrug-resistant tuberculosis. J. Chin. Med. Mater. 39, 775–777.

Xiong, Y., Deng, K.Z., Gao, W.Y., et al., 2008a. Studies on chemical constituents of *Ranunculus ternatus*. China J. Chin. Mater. Med. 33 (8), 909–911.

Xiong, Y., Deng, K.Z., Guo, Y.Q., et al., 2008b. Studies on chemical constituents of flavonoids and glycosides in *Ranunculus ternatus*. Chin. Trad. Herb Drug 39 (10), 1449–1451.

Yavuz, M., 2006–2009. Kabikaj-(see symbol test) and the buttercup: *Ranunculus asiaticus* L. Yeni Tip Tarihi Arastirmalari 12–15, 135.

Yin, C.P., Fan, L.C., Zhang, L.D., et al., 2008. The inhibiting effect of extracts in Radix Ranunculi Ternati on the growth of human breast cancer cells in vitro. Chin. J. Hosp. Pharm. 28 (2), 243–245.

Zhan, Z.L., Feng, Z., Yang, Y., et al., 2013. Ternatusine A, a new pyrrole derivative with an epoxyoxepino ring from *Ranunculus ternatus*. Org. Lett. 15 (8), 1970–1973.

Zhang, H.S., Yue, X.F., Zhang, Z.Q., 2006. Extraction of volatile oil from *Ranunculus ternatus* and GC-MS analysis of its chemical constituents. China J. Chin. Mater. Med. 31 (7), 609–611.

Zhang, L., Yang, Z., Tian, J.K., 2007. Two new indolopyridoquinazoline alkaloidal glycosides from *Ranunculus ternatus*. Chem. Pharm. Bull. 55 (8), 1267–1269.

Zhao, S.Y., Chen, L.Y., Wei, Y.L., et al., 2016. RNA-seq of *Ranunculus sceleratus* and identification of orthologous genes among four *Ranunculus* species. Front. Plant. Sci. 7, 732.

Zhao, Y., Ruan, J.L., Wang, J.H., et al., 2008. Chemical constituents of radix *Ranunculus ternate*. Nat. Prod. Res. 22 (3), 233–240.

Zhao, Y., Ruan, J.L., Wang, J.H., et al., 2010. Studies on chemical constituents of *Ranunculus ternatus*. J. Chin. Med. Mater. 33 (5), 722–724.

Zheng, W., Zhou, C.X., Zhang, S.L., et al., 2006. Studies on the chemical constituents in herb of *Ranunculus japonicus*. China J. Chin. Mater. Med. 31 (11), 892–894.

Zhong, Y.M., Feng, Y.F., 2011. Advances in studies on flavonoids and lactones in plants of *Ranunculus* Linn. Chin. Trad. Herb Drug 42 (4), 825–828.

Zou, Y.P., Tan, C.H., Wang, B.D., et al., 2010. Phenolic compounds from *Ranunculus chinensis*. Chem. Nat. Comp. 46 (1), 19.

Further Reading

Tian, J.K., Sun, F., Cheng, Y.Y., 2005. Two new glycosides from the roots of *Ranunculus ternatus*. Chin. Chem. Lett. 16 (7), 928–930.

Zhang, X.G., Tian, J.K., 2006. Studies on chemical constituents of Ranunculus ternatus (III). Chin. Pharm. J. 41 (19), 1460–1462.

Index